THE

PSYCHOBIOLOGY

OF GENE EXPRESSION

Neuroscience and Neurogenesis in Hypnosis and the Healing Arts

Dreams and the Growth of Personality (1972/1985)

Hypnotic Realities (1976, with Milton H. Erickson)

Hypnotherapy: An Exploratory Casebook (1979, with Milton H. Erickson)

Experiencing Hypnosis (1981, with Milton H. Erickson)

The Collected Papers of Milton H. Erickson on Hypnosis (1980, editor)
Volume I: The Nature of Hypnosis and Suggestion
Volume II: Hypnotic Alteration of Sensory, Perceptual, and Psychophysiological Processes
Volume III: Hypnotic Investigation of Psychodynamic Processes
Volume IV: Innovative Hypnotherapy

Healing in Hypnosis. Volume 1: The Seminars, Workshops, and Lectures of Milton H. Erickson (1984, edited with Margaret O. Ryan)

Life Reframing in Hypnosis. Volume 2: The Seminars, Workshops, and Lectures of Milton H. Erickson (1985, edited with Margaret O. Ryan)

Mind-Body Communication in Hypnosis. Volume 3: The Seminars, Workshops, and Lectures of Erickson (1986, edited with Margaret O. Ryan)

Creative Choice in Hypnosis. Volume 4: The Seminars, Workshops, and Lectures of Erickson (1991, edited with Margaret O. Ryan)

The Psychobiology of Mind-Body Healing: New Concepts of Therapeutic Hypnosis (1986)

Mind-Body Therapy: Ideodynamic Healing in Hypnosis (1988, with David Cheek)

The February Man: Evolving Consciousness and Identity in Hypnotherapy (1989, with Milton H. Erickson)

The Twenty Minute Break: The Ultradian Healing Response (1991, with David Nimmons)

Ultradian Rhythms in Life Processes: A Fundamental Inquiry into Chronobiology and Psychology (1992, edited with David Lloyd)

The Symptom Path to Enlightenment: The New Dynamics of Self-Organization in Hypnotherapy (1996)

Dreams, Consciousness, Spirit: The Quantum Experience of Self-Reflection and Co-Creation. Revised third edition of Dreams and the Growth of Personality (2000)

A NORTON PROFESSIONAL BOOK

THE

PSYCHOBIOLOGY

OF GENE EXPRESSION

Neuroscience and Neurogenesis in Hypnosis and the Healing Arts

ERNEST LAWRENCE ROSSI, PH.D.

W. W. Norton & Company
New York | London

For information about permission to reproduce selections from this book, write to
Permissions, W. W. Norton & Company, Inc., 500 Fifth Avenue, New York, NY 10110

Library of Congress Cataloging-in-Publication Data

Rossi, Ernest Lawrence.
   The psychobiology of gene expression : neuroscience and neurogenesis in hypnosis and
the healing arts / Ernest Lawrence Rossi.
      p. cm.
   "A Norton professional book."
   Includes bibliographical references and index.
   ISBN 0-393-70343-6
      1. Neuropsychology. 2. Behavior genetics. 3. Gene expression. 4. Developmental
neurobiology. 5. Biological psychiatry. 6. Hypnotism. 7. Medicine, Psychosomatic. I.
Title.

   QP360 .R665 2002
   612.8–dc21   2002016631

W. W. Norton & Company, Inc., 500 Fifth Avenue, New York, NY 10110
www.wwnorton.com

W. W. Norton & Company Ltd., Castle House, 75/76 Wells Street, London W1T 3QT

1 2 3 4 5 6 7 8 9 0

This book is dedicated to my wife, Kathryn Lane Rossi,
who traveled with me through the dark night of the soul
in the conception and creation of this volume,
until it finally released us to the champagne sunlight of freedom
kayaking along Baywood.

CONTENTS

List of Illustrations | IX

Acknowledgments | XIV

Preface | XV

Part I The Psychobiology of Gene Expression in Human Experience | 1

1 Gene Expression and Human Experience
New Models of Development, Creativity, and the Healing Arts | 6

2 Behavioral State-Related Gene Expression
How Gene Expression Modulates Human Experience | 42

3 Activity-Dependent Gene Expression
How Human Experience Modulates Gene Expression | 107

4 Dreaming, Gene Expression, and Neurogenesis
Self-Reflection and the Co-Creation of Experience | 152

5 Psychosocial Genomics and the Healing Arts
Replaying Gene Expression via Therapeutic Hypnosis and the Humanities | 188

Part II: The Psychodynamics of Gene Expression in the Healing Arts | 253

6 Positive Psychology Replaying The Four-Stage Creative Cycle
How to Enjoy Creating a Great Day and Night | 257

7 The Experiential Theater of Demonstration Therapy: Part A
The Preparation and Incubation Stages of Creative Work | 299

8 The Experiential Theater of Demonstration Therapy: Part B
The Insight and Verification Stages of Creative Work | 339

9 Implicit Processing Heuristics in the Healing Arts
The Language of Facilitating Creative Experience | 392

10 Novel Approaches to Activity-Dependent Creative Work
Exercises in Symmetry-Breaking, Self-Reflection, and Co-Creation | 425

Epilogue
The Psychosocial Genomics of Creative Experience | 481

References | 485

Name Index | 537

Subject Index | 544

LIST OF ILLUSTRATIONS

| Boxes |

2.1 Toward mathematical models of psychosocial genomics | 63

4.1 The Feigenbaum scenario as a mathematical model of psychobiological dynamics | 181

5.1 The contrast between linear and nonlinear dynamics in psychology | 204

5.2 Toward a mathematical model of the chronobiology therapeutic hypnosis | 205

5.3 The continuum of therapeutic hypnosis ranging from high-phase to low-phase | 207

9.1 Typical implicit processing heuristics to initiate stage one, preparation | 399

9.2 Key implicit processing heuristics initiating stage-two, incubation | 403

9.3 Implicit processing heuristics facilitating safe accessing and creative replay | 404

9.4 Implicit processing heuristics facilitating natural variation with therapeutic dissociations | 405

9.5 Implicit processing heuristics reframing negativity and confusion | 406

9.6 Implicit processing heuristics facilitating therapeutic dissociations, creative replay, and reframing of the negative | 407

9.7 Implicit processing heuristics facilitating permissive and creative inner work | 408

9.8 Four-step, time-binding implicit processing heuristics | 409

9.9 Four-step implicit processing heuristic utilizing the affect bridge | 410

9.10 Implicit processing heuristics facilitating private creative inner work | 411

9.11 Implicit processing heuristics facilitating surprise, insight, and conscious selection | 412

9.12 Implicit processing heuristics facilitating the novelty-numinosum-neurogenesis effect | 413

9.13 Implicit processing heuristics supporting the "aha!" moment of creative insight | 414

9.14 Implicit processing heuristics selecting and stabilizing stage three insights | 416

9.15 Implicit processing heuristics facilitating and selecting creative moments | 417

9.16 Stage-four implicit processing heuristic to validate continuing creative work | 418

9.17 Implicit processing heuristics facilitating co-creation | 419

10.1 The four-stage creative cycle facilitated by activity-dependent symmetry-breaking using the palms-facing-each-other process | 438

10.2 Activity-dependent creative facilitation: The palms-up symmetry-breaking hand process | 445

10.3 Replaying and reframing a problem into a creative resource using the palms-down process | 449

10.4 Transforming a symptom into a signal using the arm-lever process | 456

10.5 Imagery facilitating the ultradian creative cycle | 459

10.6 Visible and hidden psychodynamics in the creative cycle using arms-in-front-and-back process | 461

10.7 The Buddha beneficence and fear-not psychodynamic process | 463

10.8 A time-binding implicit processing heuristic in ideodynamic signaling | 465

10.9 Milton Erickson's utilization of minimal touch cues | 471

10.10 Full-body four-stage creative replay in individual and group activity | 475

| Tables |

1.1 The wide range of time frames in the activation, functions, and domains of gene expression | 8

2.1 Examples of complex adaptive systems of significant ultradian rhythms replaying creative human experience | 60

3.1 Ericksonian approaches to psychotherapy updated with neuroscience | 145

5.1a Age-related changes in gene expression in rat muscle | 224

5.1b Caloric restriction-induced alterations in gene expression in rat muscle | 227

6.1 The creative choice between the ultradian healing response and the ultradian stress syndrome | 292

7.1 Implicit processing heuristics that may facilitate gene expression during stages one and two of the creative cycle | 337

8.1 Implicit processing heuristics that may facilitate gene expression and neurogenesis during stages three and four of the creative cycle | 391

| FIGURES |*

1.1a   Maternal deprivation and ODC gene expression | 15

1.1b   Normalization of ODC gene expression after return to mother | 15

1.2   The tenfold way of the gaia-gene-body-mind complex adaptive system | 21

1.3   Temporal waves of gene expression in CNS development*

2.1   Psychosocial genomics as a four-stage model of mind-body communication and healing | 45

2.2   Common elements in the design of circadian oscillatory loops*

2.3   Regulation of elements in the neurospora oscillator*

2.4   Identity and regulation of elements in the mammalian oscillator and their roles in entrainment*

2.5–2.10   Pychobiological rhythms that have their source in gene expression | 54–58

2.11   Four-stage creative process in the chronobiology of consciousness, sleep, and dreams | 68

3.1   The human brain with enlarged cross section of hippocampus | 112

3.2   Time parameters of short- and long-term memory*

3.3   Proposed sequence of mechanisms involved in expression of LTP*

3.4   Reversible neurogenesis at the level of dendritic growth in the brain | 118

3.5   Ultradian time parameters of posttraumatic stress in the hippocampus | 122

3.6   The consciousness, novelty, numinosum, gene expression, neurogenesis cycle | 139

4.1   Phenomenological equation of self-reflection in a dream | 164

4.2   Self-reflection and reality-creating choice points in dreams | 168

4.3   Phenomenological equation of implicit processing in a dream | 172

4.4   Dynamics of divergence and convergence in a dream | 174

4.5a   Phenomenological equation of co-creative process of identity and personality maturation | 176

4.5b   Phenomenological equation of the psychosynthesis of new identity | 176

4.5c   The connection between the psychological concept of identity and the mathematical concept of symmetry-breaking | 177

5.1   A four-stage psychogenomic model of the psychosomatic network | 200

5.2   Cellular-genetic dynamics of the interleukin-2 pathway*

5.3   A comparison of changes in gene expression of IL-2 and IL-10 in ultradian time | 219

* An asterisk (*) denotes the figure is located in the color insert.

5.4–5.5 Typical 90–120-minute ultradian time parameters of early-activated genes in response to extracellular signaling | 219

5.6–5.7 Ultradian time parameters in the intermediate- and late-activated ranges of gene expression | 221

5.8 Downregulation of suppressed genes in ultradian time (<24 hours) | 221

5.9 PET image of gene expression in the liver, intestines, and bladder of the human body*

5.10 Venn diagram of genes accessed by BSGE, ADGE, and other psychological conditions ameliorated through hypnosis | 231

5.11 The central role of immediate-early genes (IEGs) in psychobiology | 237

6.1 The four-stage creative psychobiological cycle | 261

6.2 Testosterone levels in dominant and subordinate male baboons in response to stress | 262

6.3 The breakout heuristic in art, science, myth, and psychotherapy | 267

6.4 Typical incubation time for stage two of the creative process | 270

7.1 Stage one: An activity-dependent approach to the psychogenomic facilitation of the creative cycle with therapeutic hypnosis | 311

7.2 Stage two: Psychobiological arousal and the novelty-numinosum-neurogenesis effect | 327

8.1 Stage three: Shadow-boxing in activity-dependent experiences may facilitate gene expression | 365

8.2 Stage four: Validating self-affirmation with support of the therapeutic community | 389

9.1 The psychosocial genomics of the four-stage creative cycle | 421

10.1 Mind-brain space of the sensory-motor homunculus | 442

ACKNOWLEDGMENTS

I gratefully acknowledge the poetry, personalities, and prescience of the researchers laboring in the laboratories of mind and matter who have inspired my motivation for writing this book. I weep with gratitude for those unsung writers of children's books on fairy tales, myths, philosophy, nature, and science who first set me on my path. I deeply appreciate my mentors Galileo Galilei, Sigmund Freud, Carl Jung, Franz Alexander, and Milton H. Erickson. I owe a continuing debt of gratitude to the Milton H. Erickson Foundation and the American Society of Clinical Hypnosis and their institutes throughout the world for their continuing inspiration and support. I thank the Society for Clinical and Experimental Hypnosis for presenting me with the Ernest and Josephine Hilgard Award for Best Theoretical Paper in 2001 (Rossi, 2000c), which came at a critical time to consolidate my determination to persevere in writing this book even when my own will faltered.

I thank the National Science Foundation for the grants that enabled me to attend the Calculus In Context workshops and summer institutes on current development in teaching computer mathematics. I acknowledge the generosity of the Department of Energy of America in allowing me to attend their Human Genome Program Contractor-Grantee workshops for full access to the creative edge of genomics research. I sincerely thank the Affymetrix, Incyte, and Silicon Genetics Corporations for the privilege of attending their training workshops in functional genomics and the use of the GeneSpring data mining software by Anoop Grewal that made major contributions to the theory and research reported in this book.

A very special thanks to my percipient editor, Margaret Ryan, and to the famous Central California Coastal artist, Ginny Mancuso, who made the delightful human illustrations that so well express the new psychotherapeutic processes developed in Chapters 7, 8, and 10. Dabney Ewin, M.D., former president of The American Society of Clinical Hypnosis, was a comrade-in-arms in the final proofreading stage of this book when he often understood better than I what I was trying to say.

PREFACE

Three fundamental discoveries of current neuroscience will forever change the way we understand human nature. The first is that novelty, enriching life experiences, and physical exercise can activate neurogenesis—new growth in the brain—throughout our entire lifetime.

The second is that such experiences can turn on gene expression within minutes throughout the brain and body to guide growth, development, and healing in ways that could only be described as miraculous in the past. Like lightning and thunderclap, these two discoveries are so startling and unexpected that we hardly know what to make of them.

The third discovery follows as a natural implication of the first two. We now *really* know that "every recall is a reframe." That is, whenever we recall an important memory, nature opens up the possibility for us to reconstruct it on a molecular-genomic level within our brain. That is, we are constantly engaged in a process of creating and reconstructing the structure of our brain and body on all levels, from mind to gene. The profound implications of these three discoveries stretch our imagination. They suggest:

- how we can use our consciousness to co-create ourselves;
- new approaches to stress, psychotherapy, and the healing arts;
- a new bridge between the arts, humanities, science, and spirit.

This book is written from the perspective of a psychotherapist whose daily work involves all kinds of people coping with all kinds of problems. Psychotherapy engages our deepest conceptions of what life is all about, what our highest values and worst fears are, and how to advance to the next level of personal development. The psychotherapist is, above all, a generalist in exploring and facilitating the pathways of personal development, consciousness, and healing. The new neuroscientific discoveries about enriching life experiences, neurogenesis, and gene expression are poised to profoundly expand our understanding of the role of culture, the humanities, psychotherapy, and the healing arts in the creation of who and what we are. We are learning how our thoughts and emotions can modulate how our brain, body, and genes interact in ordinary everyday life to create our lives in ways that previous generations could not have understood. The current challenge for therapists is to integrate this new knowledge and recreate the foundations of psychotherapy with a positive perspective.

Our genes are not deeply buried in our biology, remote from our daily consciousness and concerns. On the contrary, our genes express themselves every moment of our lives in response to everything that stirs our curiosity, wonder, and fascination. Our genes are expressed in continually changing dramas that flow with meaningful life events. Our genes are switched on and off in response to our conscious efforts to cope with outer stresses as well as to our inner hopes, wishes, fantasies, and dreams. This responsivity is why the concept of "genetic determinism" is such a myth. Nature and nurture are cooperative partners that coordinate gene expression and neurogenesis to create our life experiences and continually update our memories in fresh ways, whether we are aware of it or not.

Significant life events can turn on genes that lead to the synthesis of proteins that, in turn, generate new neurons and connections between them in the brain. Our daily and hourly life experiences, thoughts, emotions, and behavior can modulate gene expression and neurogenesis in ways that change the physical structure and development of the brain. These are the profound facts of current neuroscience that we explore in this book. The meaning of these facts is still a matter of conjecture and debate, however, as the Human Genome Project shifts its gears at the present time. Up to now the Human Genome Project has been concerned with the task of discovering the structure of the individual genes that make up the totality of our genome. Now that this structural phase is coming to a successful conclusion, attention is shifting to the process of annotation: What are the roles of our individual genes, and how does the society of our genes interact with external events to create the drama of our daily life experience?

This new area of research that explores environmental-gene interactions has been called the "new horizon in health" by the Committee on Future

Directions for Behavioral and Social Research at the National Institutes of Health in the United States of America (Singer & Ryff, 2001):

> A substantial body of research reveals that specific genes can be expressed at different times in an organism's life. Whether a particular gene is expressed and the degree to which it is expressed depends strongly on the environmental conditions experienced by the organism. Such gene expression is implicated in both positive and negative health.... Neither genes nor environment dominates development; rather there is a continual interaction between genes and the environment. Phenotype [observable form, function, and behavior of all living things] emerges as a function of this constant dialogue, and any effort to ascribe percentage values to isolated variables is likely to be biologically meaningless. . . . As modern genetics identifies individual or multiple genes associated with many human diseases, such advances will only underscore the importance of understanding the environmental factors that regulate gene expression of these and other genes. . . . Scientifically, the key task is to define the pathways that lead to disease. This requires understanding first how genes and related factors might be associated with the onset of particular disease outcomes and, second, tracking relevant mediating conditions. The behavioral and social sciences are essential to advancing knowledge of these environmental conditions. Full explication of health pathways thus hinges on integrative multidisciplinary research. (pp. 63-64)

This is the point at which *The Psychobiology of Gene Expression* enters the scene. Now that we know how significant life experiences can turn on gene expression and neurogenesis to continually update the brain and body in ways that modulate our consciousness, memory, learning, and behavior in health and illness–well, what now? Are these to remain abstract facts safely sequestered in academic textbooks, or can we take these facts into the mainstream of human affairs? Can we use these facts to create a new vision of the role of culture, the humanities, and the arts in facilitating novelty, neurogenesis, and healing?

In the course of exploring these new questions, I generated hypotheses about the associations between novelty, gene expression, neurogenesis, and the *numinosum*—the experience of fascination, mystery, and tremendousness that motivates our lives. The numinosum was originally described by Rudolph

Otto (1923–1950) as a positive emotion of wonderment accompanying our sense of adventure in exploring the world, our creative endeavors in the arts and sciences, and our spiritual quest for the ultimate. The experience of the numinosum can turn negative, however, when our quest leads to misadventure and excessive stress. Chronic stress can alter gene expression, unfortunately, in ways that generate "stress proteins." Such stress proteins can interfere with optimal performance and, ultimately, produce psychosomatic illness.

When the normal checks and balances of our family, cultural, and social structures fail to channel chronic stress and wayward gene expression in a creative manner, we have a need for psychotherapy and the healing arts. This is the essence of the new theory and practice of the healing arts developed in this book. In Part I we review the profoundly significant leading-edge neuroscientific research on the psychobiology of gene expression and neurogenesis that leads us to a new vision of the role of consciousness and creativity in the humanities and the healing arts. We make a bold effort to integrate the language and insights of the humanities, psychology, and molecular biology to create a new foundation for understanding the psychobiology of culture and the healing arts. Along the way we make a few tentative efforts to explore elementary mathematical models that may facilitate these new insights for some students. Readers who are not interested in mathematics as a common language that integrates all the sciences, however, need not fear. All mathematics have been sequestered in boxes that are not required reading to comprehend and facilitate the practical processes of psychotherapy and healing developed in this book.

In Part II we begin to explore how we may creatively facilitate the psychodynamics of gene expression, neurogenesis, and healing in psychotherapy. *No claims can be made for the clinical efficacy of these innovative approaches to psychotherapy and the healing arts at the present time, however.* The clinical research that is needed to establish their therapeutic potential has not yet been done. The major mission of this book is to organize and present the facts of current neuroscience to generate new views of the role of consciousness and culture in human development and the healing arts so that they may be assessed by a new generation of therapists and researchers.

Ernest Lawrence Rossi
Baywood Park–Los Osos, California

THE PSYCHOBIOLOGY OF GENE EXPRESSION IN HUMAN EXPERIENCE

Part One

THE PSYCHOBIOLOGY OF GENE EXPRESSION IN HUMAN EXPERIENCE

Gene expression is a new concept in psychobiology that is explored here as a bridge between brain growth, behavior, and creative human experience. We review current research documenting how human relationships and social interactions can activate genes within the brain to facilitate neurogenesis that encodes new memory, learning, and behavior. Throughout Part I we explore a new idea about the *mind-body problem*—the *explanatory gap* between the sciences of mind and matter—that is resolved by an emerging science that we will call "psychosocial genomics." We will learn how the interplay between *behavioral state-related gene expression* (nature) and *experience- or activity-dependent gene expression* (nurture) generates the interconnections between biology and psychology that are now open for experimental investigation and therapeutic application in psychotherapy and the healing arts.

In Chapter 1 we develop a new way of looking at the relation between genes and human experiencing that is very different from the academic disciplines of behavioral genetics and evolutionary psychology. We outline the facts of current neuroscience and functional genomics that create the new discipline of *psychosocial genomics: how the subjective experiences of human consciousness, our perception of free will, and social dynamics can modulate gene expression, and vice versa.* We carefully explore the new terrain of the Human Genome Project for the kinds of research that can generate a new understanding of Gaia, genes, mind, and the matter of life. We orient ourselves to how we can use this data to create a new vision of the essential role of culture and the humanities as well as psychotherapy and the healing arts in facilitating the human condition.

In Chapter 2 we take our first steps in exploring the surprising and little known research on *behavioral state-related gene expression: how behavioral states, such as sleeping, dreaming, consciousness, vigilance, stress, emotional arousal, and depression, are associated with different patterns of gene expression.* We learn how a special class of genes called *immediate early genes* can respond to significant life events and psychosocial cues in an adaptive manner within minutes! We propose a new idea about the possibility of utilizing immediate early genes and behavioral state-related gene expression as a bridge between mind, brain, and body that can facilitate our understanding of the psychobiological underpinnings of psychotherapy and the healing arts. We illustrate this possibility with many mini-case histories and touch on the new kinds of clinical research that are now needed to validate these innovative approaches.

In Chapter 3 we explore another special class of genes that is responsive to psychosocial cues and significant life events. These *experience- or activity-dependent genes* generate the synthesis of proteins and neurogenesis in the brain that encodes new memory, learning, and behavior. Our daily and hourly life experiences, sensations, thoughts, images, emotions, and behavior can modulate gene expression and neurogenesis in ways that actually can change the physical structure and functioning of our brain.

Chapter 4 presents current research documenting how significant life events are reviewed and replayed in our dreams to continually update our consciousness and foster the evolution of our personality. This replay of significant life events in our dreams actually turns on an immediate early gene, called "Zif-268," that generates a neural growth factor that optimizes neurogenesis and brain growth. We explore the implications of *the dream-protein hypothesis,* in which REM sleep is recognized as a state of heightened psychobiological arousal wherein we can self-reflect and co-create ourselves. We illustrate this view with the fascinating dream series of a young woman coping with emotional transitions of her marriage that inspire a profound spiritual development in her.

In Chapter 5 we document how immediate early genes, behavioral state-related gene expression, and activity-dependent gene expression are implicated as the processes that facilitate a deep psychobiological approach to therapeutic hypnosis and holistic healing. In a series of 10 hypotheses we review the essentials of gene expression in neurogenesis and healing and how the placebo response might facilitate genuine healing at the molecular level. We propose a new worldview based on the discoveries of the Human Genome Project—discoveries that are setting the stage for a profound expansion of our understanding of life. The essential mission of the humanities and the healing arts in the new millennium is to explore how to use this new world-view for the practical facilitation of the health and well-being of ourselves and our planet.

In a seminal paper, Eric Kandel (1998), who received the Nobel Prize in Medicine in 2000, outlined "A New Intellectual Framework for Psychiatry" that well describes our essential mission in Part I of this book.

> Insofar as *psychotherapy or counseling is effective and produces long-term changes in behavior, it presumably does so through learning, by producing changes in gene expression that alter the strength of synaptic connections and structural changes that alter the anatomical pattern of interconnections between nerve cells of the brain.* As the resolution of brain imaging increases, it should eventually permit quantitative evaluation of the outcome of psychotherapy. . . . Stated simply, *the regulation of gene expression by social factors makes all bodily functions, including all functions of the brain, susceptible to social influences. These social influences will be biologically incorporated in the altered expressions of specific genes in specific nerve cells of specific regions of the brain.* These socially influenced alterations are transmitted culturally. They are not incorporated in the sperm and egg and therefore are not transmitted genetically. (p. 460, italics added)

With these words as our inspiration, let us now explore the theory and research of optimizing gene expression and neurogenesis to facilitate brain growth and healing in Part I. In Part II we will focus on the actual practice of constructing and reconstructing a better brain in everyday life as well as the positive, creative experiences of the arts and sciences. This will serve as our foundation for creating new scientific approaches to the psychosocial genomics of healing in alternative and complimentary medicine, therapeutic hypnosis, and the many schools of psychotherapy.

Chapter 1

GENE EXPRESSION
AND HUMAN EXPERIENCE

New Models of Development, Creativity, and the Healing Arts

"If the promise of the revolutionary enterprise of reading our own instruction book is going to play out for the benefit of humankind, we need your expertise and your skills. It is fair to say that we're embarked on a rather remarkable adventure. The goal of the Human Genome Project is not just to get that instruction book, but also to understand how it works, in both health and disease. So we can think of this as a milestone, of having a working draft of the human gene sequence in front of us, as the end of the beginning."

—Francis Collins, Human Genome Project Director
Address to the American Psychological Association, 2000

"You have to think ahead. Science goes where you imagine it."

—Judah Folkman
"Folkman Looks Ahead"

Utilizing the new science of functional genomics currently emerging from the Human Genome Project, we will begin by illustrating creative models of how the society of genes can be understood in new ways that optimize human development and creativity in normal daily living as well as psychotherapy and the healing arts. Our approach is to trace the circular pathways of communication that flow among environment, gene expression, body, mind, and spirit. Our mission is to learn how to use human consciousness and our perception of free will to optimize the natural dynamics of gene expression in health and well-being in ordinary life as well as creative work in the arts, sciences, and humanities.

FUNCTIONAL GENOMICS AND HUMAN EXPERIENCE

Until now, most research in the Human Genome Project has been conducted on the purely biological and biochemical levels to uncover new genes and their molecular structure. As this initial *structural phase* of the Human Genome Project approaches a conclusion with the identification of a currently estimated 30,000 human genes, however, attention is now shifting to the second or *annotation phase* that explores the functions of genes. *Functional genomics* is the new science that explores how genes express themselves and interact in health and illness—that is, how networks of genes are turned on and off in response to signals from all parts of the body as well as the outer environment. Since this book is about the *psycho*biology of gene expression, we will focus primarily on the psychological, social, and cultural signals that modulate gene expression, and vice versa.

We have learned recently, for example, that many ordinary states and aspects of everyday life, such as waking, sleeping, dreaming, work, stress, play, and especially motherhood, are all associated with uniquely individual patterns of gene expression. In the research literature the subject matter addressed in this book has been called by several names, such as "behavioral state–related gene expression," "activity-dependent gene expression," *and* "experience-dependent gene expression." These apparently different expressions all focus on varying nuances of a surprising but fundamental idea of functional genomics: *Most of our genes are not independent biological determinants of behavior but active players responding quickly, from one moment to the next, to the cues, challenges, and contingencies of our ever-changing daily experience.* Our thoughts, emotions, and behavior modulate *gene expression* in health and optimal performance as well as stress and illness. We introduce this new perspective on the relatively brief time frames of gene expression in daily experience by comparing them with the much longer time frames of Darwin's evolution and Mendel's classical genetics, as illustrated in Table 1.1.

We provide a thumbnail sketch of each level of gene expression here as an introduction to the relationships between gene expression and the phenomenology of human experiencing in the following chapters.

TABLE 1.1 | THE WIDE RANGE OF TIME FRAMES IN THE ACTIVATION, FUNCTIONS, AND DOMAINS OF GENE EXPRESSION.

Gene expression	Approximate time	Major function	Research domain
Evolution	Eons	Origins	Darwin
Inheritance	Generations	Replication	Mendel
Development	A lifetime	Growth	Embryology
Housekeeping	Daily	Metabolism	Functional genomic
Clock genes	Circadian	Synchronization	Chronobiology
Late activated	4–8 hours	Immune	Immunology
Intermediate & early active	1–2 hours	Environmental responses	Psycho–neuro–immunology
Behavioral state-related	Hours	Wake, sleep, dream, mood	Psychology
Activity-dependent	Minutes to hours	Memory, learning	Neuroscience
Immediate early	Seconds, minutes	Arousal, stress	Psychobiology

Darwin and the Evolution of Human Experience

Evolutionary studies of the origin of species begun by Darwin (1859) explore how his *principles of natural variation and selection* can account for the emergence of new forms of life. The evolution of species is usually believed to require *time frames of eons and millennia*. Modern studies document, however, how critical transitions in the environment, such as a severe drought can alter the course of evolution in a generation or two. Moreover, even Darwin theorized, "It may metaphorically be said that *natural selection is a daily and hourly scrutinizing, throughout the world*," (Weiner, 1994, p. 6, italics added). This was no mere metaphor. Current neuroscience documents how novel experiences in everyday life can change gene expression in hours and even minutes to influence our health, performance, and well-being. Darwinian variation and natural selection operate in the evolution of human consciousness whenever we have significant, creative, and exhilarating life experiences (Zeki, 2001). In Chapter 3 we will present detailed research on these newly recognized connections between psychological experience and gene expression that will profoundly deepen our understanding of the meaning of culture and our pursuit of wonder and the unknown in the arts, sciences, and humanities as well as psychotherapy and the healing arts.

Mendel and Classical Genetics: Physical Traits and Psychological Processes

Classical or Mendelian genetics is concerned with the laws of inheritance. It is concerned with how the units of heredity—the genes that make up *the genotype,* wrapped up in chromosomes within each cell of the body—are linked together and transmitted from one generation to the next as the biological basis of observable traits—*the phenotype*—such as the texture of peas in a pod, eye color in humans, and many patterns of behavior such as activity levels, aggression, and sexuality. Mendelian genetics studies the distribution of such traits over *the broad time frame of generations.* Mendel (1865/1965) required seven years, for example, to collect his data on the crossbreeding of peas for his first publication on the distribution of their physical traits over many generations. The popular but erroneous idea about genes is that they are independent biological determinants—the autocratic ruler and source of physical traits, inherited abilities, dysfunctions, etc.—a view that gives rise to *the nature versus nurture debate* (Singer & Ryff, 2001). Is human experience determined primarily by nature (genes) or nurture (life experiences)? In this book we explore the discoveries that are currently resolving this controversy in a new way. We will learn how genes interact with the environment to modulate human experiencing, and vice versa. Above all, we will explore how our *subjective states of mind, consciously motivated behavior, and our perception of free will can modulate gene expression to optimize health.*

Human Development and the Life Cycle

The central mystery of development—how a single fertilized egg can grow into a unique human being—is now understood as the unfolding of waves of interacting patterns of gene expression. Although most cells of the body contain all our genes, relatively few subsets of these genes need to be expressed to generate the proteins that define each cell of the body. The *structure, function,* and *identity* of any particular cell of the body are due to the particular subset of genes that is expressed during the cell's formation and daily activity. *The structure, function, and identity of cells are intimately associated with circular loops of communication, via gene expression, on all levels of human development.* Embryonic development, infancy, childhood, adolescence, adulthood, and death itself are now understood as outcomes of interactive patterns of communication between environment and genes over the course of a lifetime.

HOUSEKEEPING AND THE GENE EXPRESSION/PROTEIN SYNTHESIS CYCLE

A core of genes is in constant activity in most cells of the body to maintain the basic *daily housekeeping functions* of life, such as metabolism, energy production, growth, respiration, waste removal, etc. Many other genes, by contrast, are turned on or off in coordinated cycles of response to signals that ultimately come from moment-to-moment changes in the environment. Genes in complex multicellular forms of life (such as human beings) usually do not act independently, as promoted by popular but often misleading ideas about "selfish genes" (Dawkins, 1976). In the continuous process of creative adaptation, the society of genes (Ridley, 1996) expresses itself in "family" patterns of cooperative and coordinated responsiveness to signals that ultimately come from the outside world (Ridley, 1999, 2001).

Genes respond to such signals by making a template of themselves called *messenger RNA* (mRNA). Messenger RNA is a code, a pattern, for making a protein in an adaptive response to important signals. Proteins, in turn, make up the molecular machinery of life. Proteins are the building blocks of many of the basic *structures* of the cell. Proteins may take the form of enzymes that facilitate metabolism by generating the *energy* of the cell in the form of adenosine triphosphate (ATP). Proteins also make up many of the *informational messenger molecules* of life, such as hormones, growth factors, and cytokines in the immune system. Here the time frame of the basic *gene expression-protein synthesis cycle* (usually abbreviated to "gene expression") is about 1.5–2 hours. As will be discussed in Chapter 2, these relatively brief (ultradian) cycles are usually coordinated within the daily 24-hour (circadian) rhythm.

Clock Genes and the Chronobiology of Life

The psychobiology of gene expression finds a major research and experimental database in the so-called "clock genes" that coordinate the chronobiology of life. Clock genes set the time cycles of many common everyday psychobiological states, such as waking, sleeping, and dreaming. The time parameters of the gene expression cycle set the pace and rhythm of many patterns of mind-body communication, health, and healing. The fundamental life processes of metabolism, homeostasis, growth, energy, information flow, behavior, memory, and meaning—in both health and illness—all dance to the basic rhythm of gene expression and the synthesis of their "cognate" proteins over a circadian or daily time frame.

Early, Intermediate, and Late-Activated Genes in Psychoimmunology

There is a wide range of time frames for the activation and expression of many genes associated with the central nervous system, the endocrine system, and the immune system. *Early activated genes,* for example, reach a peak of expression in about an hour, while *intermediate activated genes* peak in about two hours. Early activated genes are of special interest because they mediate some relationships between mind and body that are addressed by the field of *psychoneuroimmunology.* Psychosocial stress, for example, can turn off the early activated interleukin-2 gene so that the immune system cannot communicate well, and we are left more vulnerable to all sorts of opportunistic infections. Positive psychosocial experiences, on the other hand, can turn on the interleukin-2 gene within an hour or two to facilitate molecular communication, healing, and health. Late activated genes, by contrast, require as much as 4–8 hours to reach their peak levels of expression. As is illustrated in detail in the following chapters, the turning on and off of cascades of gene expression begins within a few minutes of receiving important psychological and social signals and may continue for hours, days, weeks—or even a lifetime.

Behavioral State-Related Gene Expression and Arousal

Behavioral state-related gene expression refers to a special class of genes that is associated with changing states of arousal, such as waking, sleeping, and dreaming. These genes are often expressed in association with clock genes in a wide range of human experiences, including emotional arousal, crisis, and the sense of triumph, on the one hand, or stress, despair, and depression, on the other hand. The new approaches to creativity, psychotherapy, and the healing arts explored throughout this book attempt to engage our natural patterns of behavioral state-related gene expression to facilitate our normal circadian (about 24 hours) and ultradian (less than 20 hours) rhythms of mind-body communication and healing. Behavioral state-related gene expression is a little closer to nature in the nature–nurture equation. *Experience- or activity-dependent gene expression,* by contrast, is a little closer to the nurture side of the equation.

Activity-Dependent Gene Expression and Neurogenesis

Neuroscience research is currently exploring how the conscious experience of *novelty, environmental enrichment, and voluntary physical exercise* can modulate gene expression to encode new memory and learning. As noted, this is called "experience- or activity-dependent gene expression." Vivid conscious experience can turn on genes that code for proteins that lead to neurogenesis— the generation of new neurons and their connections in the brain. This new growth within the brain is the anatomical and molecular basis of our ever-changing memory, learning, and behavior.

Most novel, interesting, surprising, and arousing psychosocial stimuli associated with new adventures, for example, can induce the expression of activity-dependent genes within a few minutes. The concentrations of mRNA, transcribed from these genes, typically peak within 20 minutes, and their translation into fresh proteins usually requires an hour and a half or two. These time parameters of activity-dependent gene expression are set in motion by the stimulating effects of work, play, social activities, and even dreaming after a day of novel experiences. Activity-dependent gene expression is associated with motivation, particularly with our conscious *numinous experiences* of wonder, mystery, fascination, curiosity, and creativity. This association is the basis for what we will call the *novelty-numinosum-neurogenesis effect,* which proposes that a major function of novelty and numinous experiences in cultural activities (art, dance, drama, education, literature, music, myth, painting, spiritual rituals, storytelling, etc.) is to turn on activity-dependent gene expression to facilitate neurogenesis and the growth of the brain.

Consciousness and Psychosocial Genomics

The interplay between *behavioral state-related gene expression* and *activity-dependent gene expression* generates much of the psychobiology of consciousness. Consciousness can be understood as a series of continuously emerging experiential states that focus attention to evoke activity-dependent gene expression and neurogenesis in the dynamics of self-reflection and the co-creation of the self. We propose the creation of a new discipline, to be called *"psychosocial genomics,"* to explore how the psychological dramas and social encounters of everyday life can turn on activity-dependent gene expression and neurogenesis in ways that optimize performance, health, and well-being.

Immediate Early Genes, Arousal, and Stress

It is now recognized that a special class of *immediate early genes* (sometimes called "third messengers") do not require the full cycle of fresh protein synthesis for their expression. Because of this feature, they can be activated immediately—that is, within seconds to a minute or two—in response to important signals from the outside world. While about a hundred immediate early genes (IEGs) are now known, the full range of their functions are still not well understood. They are of central interest in psychosocial genomics, however, because of their major role in mediating psychological arousal and optimal performance as well as the stress response. IEGs typically turn on other "target genes" that generate protein synthesis for adaptive responses to stressful events. We will explore the hypothesis that IEGs play a central role, at the cellular-genomic level, in mediating the bridge between mind and matter in both health and illness. IEGs mediate the entire range of gene expression, from normal development to the dysfunctions that we try to heal via mind-body or psychosomatic medicine. A mother's touch, for example, is needed to activate the genes that code for proteins that generate the normal physical and psychological development of a baby. This profound interaction between human touch, gene expression, and infant development is well illustrated by the tale of "A Sister's Helping Hand."

A Sister's Helping Hand: How Touch Can Turn on Gene Expression

The heartwarming true story of "A Sister's Helping Hand" that appeared in the *Reader's Digest* (May, 1995, pp. 155–56) provides a vivid example of gene expression via human experience. The story begins with the premature birth of twins. Each of the twins was immediately placed in a separate incubator in accordance with the normal hospital rules. One of the twins, the weaker of the two, was not expected to live, however. A sympathetic nurse, following her heartfelt intuition and sense of sheer desperation, defied hospital rules by placing the two babies together in one incubator. Unexpectedly the healthier twin threw an arm over her sister in an endearing embrace. The smaller baby's heart soon stabilized, her temperature returned to normal, and she survived. The twins thrived together and now, at home, they still sleep snuggling together.

How can we account for what appears to be a heart warming miracle wrought by a newborn sister's helping hand? No one was around at the time to measure gene expression in the twins during their early life crisis. Current research in gene expression and human experience, however, is now finding an answer to how healing by touch is possible. Saul Schanberg and his colleagues (Kuhn & Schanberg, 1998; Kuhn et al., 1991) at Duke University, for example, discovered how maternal touch can activate *immediate early genes* such as *c-myc* and *max,* which in turn activate a target gene called ODC (ornithine decarboxylase). Turning on the ODC gene leads to the synthesis of proteins that contribute to physical growth and maturation at the cellular level (Bartolome et al., 1999; Russell, 1985; Wang et al., 1996). Figure 1.1a illustrates Schanberg's (1995) experimental finding that deprivation of maternal touch for 10 or 15 minutes results in a dramatic drop in ODC gene expression and the physical growth of 10-day-old rat pups. Within two hours ODC activity is down 40%—where it remains until maternal touch returns. Figure

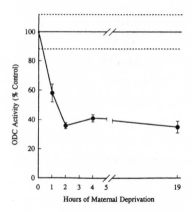

FIGURE 1.1A | Deprivation of maternal touch for 10 or 15 minutes results in a drop in ODC gene expression and the physical growth of 10-day-old rat pups.

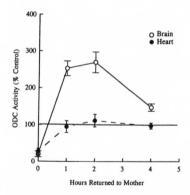

FIGURE 1.1B | Full recovery of heart rate and an over compensation to 300% of normal ODC gene expression in the brain when the touch deprived pups are returned to their mother.

1.1b graphs the full recovery of heart rate and an overcompensation to 300% of normal ODC gene expression in the brain after the touch-deprived pups were returned to their mother. A graduate student's stroking of the pups lightly with a soft, tufted artist's paintbrush for 15 minutes was enough to turn on the ODC gene and other genes and hormones associated with biological growth as well.

It has been known for some time that human infant orphans fail to thrive and grow physically when isolated in an institution without the normal amount of touch, even when all other needs for warmth, food, and care are provided. This condition has been called *psychosocial dwarfism or nonorganic failure-to-thrive* (Gardner, 1972). When pediatric nurses supplied infants suffering from this deprivation with tender loving touch, however, their growth pattern returned to normal within hours. The nonorganic failure-to-thrive diagnosis was also documented by social workers investigating homes where the environment was described as psychosocially inadequate. It was found that failure-to-thrive babies in these homes had abnormally low growth hormone levels, which are associated with low ODC gene expression activity (Powell et al., 1967, 1973). Once the babies received adequate maternal touch, however, ODC gene expression, growth hormone, and physical growth returned to normal. Simple administration of growth hormone alone, without continued maternal touch, failed to improve the growth of these failure-to-thrive babies (Rayner & Rudd, 1973).

Clinical observations of the relationship between psychosocial deprivation and altered physiological and behavioral responsiveness in human infants were, in part, the inspiration for George Engel's early and widely cited formulation of *the biosocial approach to psychosomatic medicine.* Engel and Reichsman (1956) reported on the psychological depression in an infant who experienced prolonged social deprivation when isolated in a hospital ward with a gastric fistula, a genetic condition in which stomach contents flow through a hole in the abdomen. The infant baby girl became withdrawn and noncommunicative until Engel and Reichsman began to play with her, while collecting and measuring the volume and contents of her draining gastric fluid. They reported the accidental discovery of immediate and obvious changes in the infant's stomach secretion during this play. When the infant was withdrawn or frightened by a stranger, her gastric flow would cease. When playfully engaged and relaxed, the stomach secretions would flow abundantly. Engle and Reichsman discovered that too little social stimulation was just as significant for optimal infant development as the finding that excessive stimulation induced stress (Alexander, 1950; Lynch, 2000; Sternberg, 2000).

Kuhn and Schanberg (1998) summarized the relationship between maternal deprivation, gene expression, and development as follows:

We have shown that maternal separation initiates a complex adaptive biobe-havioral response in preweaning rat pups that includes (1) a decrease in the synthesis of ornithine decarboxylase [ODC], an obligatory enzyme for normal cell growth and development, (2) a reduction of DNA synthesis, an index of cell multiplication, (3) abnormal patterns of neuroendocrinal secretion, and (4) a suppression of cell responses to growth hormone, prolactin, and insulin, three major trophic [growth-inducing] hormones. . . . Recently we have shown that in the absence of "nurturing touch" the brain initiates the suppression of ornithine decarboxylase gene transcription by interfering with the cell's ability to transduce the activating signal induced by the growth promoting hormones. Studies indicate that central endorphinergic pathways may mediate this action. This is accomplished by the down regulation of specific Immediate Early Genes (c-myc and max) that normally promote the synthesis of this critical growth-regulatory enzyme. These studies of short-term maternal separation not only demonstrated that maternal care is a critical regulator of pup physiology and biobehavioral development, but that there are marked similarities between this animal model of maternal separation and the delay in growth and development observed in children with the deprivation syndrome or in touch-deprived premature human neonates. Our identification of a specific type of nurturing touch as a neonatal growth requirement led us to test supplemental tactile stimulation in isolated, very premature, human babies. The result of our intervention with message was dramatic. Infants not only showed marked gains in weight and behavioral development, but also significant enhancement in sympatho-adrenal maturation. We suggest that animal models of maternal deprivation can be used to understand the integrative processing of appropriate sensory input, CNS function, and end-organ physiology required to maintain normal development. (p. 216)

These associations between gene expression, psychobiological development, and human behavior are examples of the *complex adaptive systems* typical of all life processes (Solé & Goodwin, 2000). Many of the apparent mysteries and miracles of life have their source in the complexity of the natural psychobiological dynamics of gene expression across so many levels, ranging from molecule to mind, that they are difficult to sort out. A new way of exploring this complexity, "data mining," utilizes computer software to uncover previously unknown relationships in the data of functional genomics.

DATA MINING, HUMAN DEVELOPMENT,
AND BEHAVIOR WITH COMPUTER SOFTWARE

The central role of ODC gene expression can be explored further with the aid of the new types of computer software that are currently being developed to organize and annotate data from the Human Genome Project. This software can now access the data banks from gene expression studies throughout the world. Using one such computer software program, GeneSpring (www.sigenetics.com), I discovered that there are at least 28 other genes that are expressed along with the ODC gene. I was very excited to learn, for example, that the brain-derived neurotropic factor (BDNF) gene is turned on in association with the ODC gene. This immediately generated a new and unexpected hypothesis about an association between maternal touch, brain development, and the actual molecular pathways and dynamics between them! It also invites a host of clinical studies about how maternal touch could be used to facilitate human brain development.

The association between ODC and BDNF gene expression also has fascinating implications for exploring the reverse of neurogenesis: the atrophy of brain cells in response to the environment and psychosocial stimulation. Duman and Neslter (2000) recently reviewed how subjective and emotional states of consciousness associated with stress, addictions, and mood disorders (such as depression, and other psychiatric conditions) are involved in these same patterns of ODC and BDNF gene expression.

The discovery of the therapeutic facilitation of gene expression via touch and psychosocial interactions may go a long way toward explaining the so-called mysteries of the indigenous and shamanistic healing practices of many cultures (Achterberg, 1985; Greenfield, 1994, 2000) as well as alternative, complementary, and holistic approaches that are becoming so popular at the present time (Achterberg et al., 1994). It is estimated that 40,000 nurses and caregivers in the United States have eagerly embraced "therapeutic touch," which is explained by an erroneous theory attributing its value to "energy fields" (Rosa et al., 1998)—an interesting parallel to Mesmer's false attribution of healing to "animal magnetism" in the early days of hypnosis about 200 years ago.

Touch and "therapeutic passes" have a long and honorable place in the history of healing in hypnosis (Edmonston, 1986; Erickson & Rossi, 1981).

The "king's touch" was eagerly sought by the masses in the Middle Ages as a method of healing. With our newly recognized scientific associations between touch and gene expression for physical growth, healing (ODC gene), and neurogenesis (BDNF), we can now appreciate how more than naiveté was involved in this seemingly gullible attitude. We gain our first intimations of how the novelty and numinosity of rare and highly prized psychosocial experiences can evoke a genuinely healing placebo response on the molecular-genomic level. We will explore these possibilities of psychosocial genomics as the deep psychobiological basis of therapeutic hypnosis and the healing arts in Chapter 5. In Part II of this book, we will make a more systematic exploration of how the dynamics of psychosocial genomics could be facilitated in developing a new generation of creative approaches to healing and problem-solving in psychotherapy.

From our current perspective we can hypothesize that the arts of healing in the traditional and spiritual rituals of many cultures, from ancient times to the present, involve touch and many deeply felt experiences that facilitate at least three classes of gene expression: *immediate early, behavioral state-related,* and *experience- or activity-dependent.* The true story of "A Sister's Helping Hand" is just one modern example of how human experience can turn on gene expression and protein synthesis in a variety of life situations involving both health and illness. Such observations from ancient times to the present are at the heart of the new psychobiological approaches to facilitating gene expression, neurogenesis, and healing that will be explored in the following chapters.

Anecdotal accounts of human touch and healing, such as the story of "A Sister's Helping Hand," are highly controversial because they are unreliable in the sense that they are difficult to replicate in an objective manner. Indeed, these anecdotal accounts are always subjective to one degree or another, even when great efforts are made to design objective research. There are many Internet sites that catalogue and discuss the possibilities of optimizing healing via touch-facilitated gene expression in infants and children (Field, 1995), teens and adults (Ironson et al., 1996), and senior citizens (Meek, 1993) in a wide variety of conditions ranging from addictions, anorexia, anxiety, arthritis, and attention deficit disorder to cancer, chronic pain, and sleeping disturbances (amritherapy.com/research.htm). Impressive as many of these studies may be, however, there is still no general agreement on how to interpret them (www.rccm.org.uk/massbasic.htm). The critical analyses of such studies focuses on the usual suspects: experimental design, statistical analysis, and inadequate controls. The problems with experimental design and controls continue to undermine the validity and reliability of research in this area of alternative and complementary mind-body medicine (www.Nichd.nih.gov/Cochrane/Viickers/Vickers.htm).

In great part these experimental difficulties may be attributed to the fact that gene expression and human experience are rarely related in simple, linear patterns of cause and effect. To the contrary, they are usually replayed as the complex adaptive systems of circular, nonlinear dynamics that typify life processes on all levels, from the cellular-genomic to the phenomenology of consciousness (Solé & Goodwin, 2000). A new approach to exploring these complex interactions between gene expression and human experiencing is the use of DNA microarrays ("gene chips") in the process of data mining and knowledge discovery, which will be discussed in the following sections (Hamadeh & Afshari, 2000).

THE COMPLEX ADAPTIVE SYSTEM
OF GENE EXPRESSION IN HUMAN EXPERIENCE

Studies relating gene expression and human experience can be interpreted from many perspectives: anthropological, cultural, social, psychological, and the biological. Our approach to coping with the highly controversial status of research is to recognize the true complexity of the interaction between human experience and gene expression as an adaptive evolutionary system in social life (Axelrod & Cohen, 1999). Psychobiological observations always involve the circular, nonlinear dynamics of complex adaptive systems (Prigogine, 1980, 1997; Prigogine and Stengers, 1984; Rossi, 1996a, 2000a). This means that the easily observed cause-and-effect determinism of classical Newtonian physics, which is modeled by billiard balls bouncing off each other in predictable linear paths, is of limited usefulness in understanding living systems. *Complex adaptive systems are holistic organizations in nature that are able to maintain their integrity and evolve over time with emergent characteristics that are more than the sum of their parts.* All living organisms display emergent characteristics of complex adaptive systems. The psychobiological processes

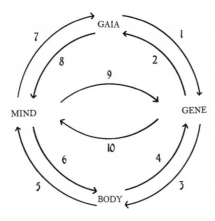

FIGURE 1.2 | The creative cycle replaying the complex adaptive system of Gaia, gene, body, and mind along the tenfold way of evolution.

explored in this book—from the levels of gene expression and neurogenesis to consciousness and the healing arts—are all examples of complex adaptive systems.

The key to developing this new complex adaptive systems model is to find our way along the major circular pathways of communication between gene expression and human experience. The complexities of the Human Genome Project and the new science of psychosocial genomics require a broad overview of the territory we are exploring. We need a map to understand the history and significance of the many lines of research that are now coming together. A map, using the simplest possible four-letter words, is what I describe as the ten-fold way of the *Gaia-gene-body-mind complex adaptive system* (Figure 1.2). This creative cycle, which generates and replays the process of evolutionary adaptation between an organism's genes and the environment, has been described at the level of informational flow within the cell by the molecular biologist Loewenstein (1999) in his wonderfully accessible book, *The Touchstone of Life*. The conceptual map of Figure 1.2 presents a perspective of the ten pathways that represent the new domain of the psychobiology of gene expression, or *psychosocial genomics*. First we will outline some of the current lines of scientific research on each path of the ten-fold way in preparation for the greater detail to come in the following chapters.

PATHWAYS ONE AND TWO:
THE GAIA-GENE CYCLE OF EVOLUTION

Gaia is the name the ancient Greeks used for the goddess, Earth. The concept of a Gaia-gene complex adaptive system to describe the evolution of life on earth is represented by the first two paths of the ten-fold way in Figure 1.2. The Gaia hypothesis originally proposed that the entire earth is one large, living organism (Lovelock, 1988). The atmosphere, the oceans, the climate, the crust and green mantle of the planet evolved to a state of dynamic equilibrium initiated, at least in part, by the behavior of living organisms. Lovelock (1988) explains:

> Specifically, the Gaia hypothesis said that the temperature, oxidation state, acidity, and certain aspects of the rocks and waters are at any time kept constant, and that this homeostasis is maintained by active feedback processes operated automatically and unconsciously by the biota. Solar energy sustains comfortable conditions for life. The conditions are only constant in the short-term and evolve in synchrony with the changing needs of the biota as it evolves. Life and its environment are so closely coupled that evolution concerns Gaia, not the organisms or the environment taken separately [p. 19]. . . . The outer boundary is the earth's atmospheric edge to space. Within the planetary boundary, entities . . . grow ever more intense as the inward progress goes from Gaia to ecosystems, to plants and animals, to cells and to DNA. The boundary of the planet then circumscribes a living organism, Gaia, a system made up of all the living things and their environment. (p. 40)

Gaia is now a rather specialized term that designates an evolutionary theory of how life and its environment modulate each other as a complex adaptive system involving the transformations of matter, energy, and information (Margulis, 1998; Margulis & Sagan, 1986). Epochal discoveries about the origin of the universe in the Big Bang 15 million years ago are made by studying patterns in the cosmic microwave background radiation that still accounts for about 1% of the noise on our radio and televisions sets. In the original hot plasma of cosmic creation, light and matter were an undifferentiated fluid. After about 300,000 years the universe cooled enough for the plasma to condense and allow enough differentiation so that photons of light could separate and shine forth for the first time. The Milky Way that we can see with the unaided eye on a clear night contains 100 billion stars arranged in a disk orbiting around a central hub. Our galaxy is but one among billions of complex adaptive systems evolving together in the cosmos.

The further evolution of the cosmos continued via cooling, which fostered progressive *symmetry-breaking* in a series of *bifurcations* that led to the differentiation and creation of the quasars—and, finally, to the fiery synthesis of the elements in furnace of the stars that our bodies are made of today. Whatever the details, it is now clear that processes of *progressive symmetry-breaking, polarization, duality, division, and synthesis* were involved in the acts of creation that led to the formation of earth about five billion years ago. In Part II, particularly in Chapter 10, we will use these same explanatory principles to describe and illustrate the creative dynamics of the psychotherapeutic process.

The self-organizing molecules of life that are the origin of DNA and genes began their evolution about four billion years ago. As a background for understanding the evolution of gene expression in human experience, we will begin with the work of Charles Darwin on the origin of species and Gregor Mendel on classical genetics. We will review the explanatory principles of natural variation and selection in the pioneering research of Darwin and Mendel because they are fundamental for understanding the creative dynamics of gene expression, neurogenesis, and healing that are developed in the new therapeutic approaches introduced later in this book.

Charles Darwin's Natural Variation and Selection in the Struggle for Existence

Considering the magnitude of the achievement and influence of Charles Darwin's *The Origin of Species by Means of Natural Selection* in 1859, it is truly remarkable that he conceived these ideas before we had any real understanding of genes as the units of heredity. Although the general perception is that Darwin described the evolution of the physical organism, a careful reading of his original words reveals that he actually covered what we now call the psychobiological domain. In the final chapter of *The Origin of Species*, Darwin recapitulates his conclusions about the evolution of physical characteristics of the body as well as instincts—that is, what we would today call behavioral or psychological processes:

As this whole volume is one long argument, it may be convenient to the reader to have the leading facts and inferences briefly recapitulated. That many and grave objections may be advanced against the theory of descent with modification through natural selection, I do not deny. I have endeavored to give them their full force. Nothing at first can appear more difficult to believe than that the more complex organs and instincts should have been perfected, not by means superior to, though analogous with, human reason, but by the accumulation of innumerable slight variations, each good for the individual possessor. Nevertheless, this difficulty, though appearing to our imagination insuperably great, cannot be considered real if we admit the following propositions, namely—that graduations in the perfection of any organ or instinct, which we may consider, either do now exist or could have existed, each good of its kind—that all organs and instincts are, in ever so slight a degree, variable,—and lastly, that there is a struggle for existence leading to the preservation of each profitable deviation of structure or instinct. The truth of these propositions cannot, I think, be disputed. (pp. 375–76)

While Darwin's concept of "the struggle for existence" in the wild is well known, it is not usually recognized that his initial observations were made on the variability of inherited traits in domesticated animals common on the farms of his day as well as ours:

Under domestication we see much variability. . . . Variability is governed by many complex laws—by correlation of growth, by use and disuse, and by the direct action of the physical conditions of life. . . . Man does not produce variability; he only unintentionally exposes organic beings to new conditions of life, and then nature acts on the organization, and causes variation. But man can and does select the variations given to him by nature, and thus accumulates them in any desired manner. . . . Why, if man can by patience select variations most useful to himself, should nature fail in selecting variations useful, under changing conditions of life, to her living products? What limit can be put to this power, acting during long ages and rigidly scrutinizing the whole constitution, structure and habits of each creature,—favoring the good and rejecting the bad? I can see no limit to this power, in slowly and beautifully adapting each form to the most complex relations of life. The theory of natural selection, even if we looked no further than this, seems to me to be in itself probable. (pp. 381–83)

In his later volume on *The Descent of Man and Emotions,* Darwin made further contributions that set the stage for modern theories of consciousness, behavior, meaning, and reinforcement that are well expressed by the psychologist Richard DeGrandpre (2000):

> The historical and anti-teleological underpinnings of reinforcement, which of course lies at the heart of Skinner's radical behaviorism, have been said to be homologous to [Darwin's] natural selection (Glenn et al., 1992; Palmer & Donahoe, 1992; Skinner, 1981). Natural selection has along with other evolutionary processes, led to species fantastically well suited for their ecological niche (Weiner, 1994). The diversity and complexity of life, in fact, is so impressive that many have trouble imagining how this could be largely attributable to the invisible hand of a few generic processes (i.e., heritable variation in traits and tendencies, selection of a subset of the traits or tendencies in the population, and retention of them over time through the preservation of their underlying DNA). . . . Mayr (1988, p. 43) wrote, "natural selection rewards past events . . . but does not plan for the future." This is precisely what gave evolution by natural selection its flexibility. With the environment changing incessantly, natural selection . . . never commits itself to a future goal (p. 726). . . .
>
> The infant who shakes her right leg and, as a consequence, jiggles an overhead mobile and quickly learns to repeat the performance. . . . Although this example is of explicit reward, it is suggestive of something much larger, which is that all motor learning, from the increasingly accurate motions of eyes, hands, and feet of a newborn to the walking and talking of a toddler, is the product of natural contingencies built into the child's world. (p. 730)

The infant who shakes a leg, jiggles an overhead mobile, and quickly learns to repeat the performance are examples of what we would now call *experience-* or *activity-dependent gene expression and neurogenesis in the encoding of new memory, learning, and behavior.* We will explore the type of research that is needed to assess this possibility in Chapter 3, which is concerned with neurogenesis and the numinous experiences of excitement, fascination, mystery, and wonder that occasionally accompany our more surprising experiences of self-stimulation and self-reflection. The homology between Darwin's concept of *natural selection* in the origin of species and the modern psychological concept of *natural contingencies* in reinforcing selective aspects of consciousness, meaning, memory, and behavior is a fundamental aspect of the new psychobiological approaches to psychotherapy and the healing arts developed in Part II.

Consciousness, Meaning, and
Free Will via Darwinian Variation and Selection

Current humanistic studies describe consciousness and free will as emergent properties of nature and nurture. But emergent from *what*, and *how*? In a series of books the Nobel laureate Gerald Edelman (1987) and his associates have extended Darwin's dynamic principles of variation and natural selection to the molecular, cellular, embryonic, and neural developmental levels to coordinate a plausible picture of how consciousness, meaning, and free will could emerge in an entirely naturalistic manner. Current research to study the early evolution of the genetic code and genetic algorithms using computer simulations (Rietman, 1994) are providing a new empirical data base establishing the reality of Darwin's twin principles of variation and natural selection in the dynamics of living systems on all levels, from the molecular-genomic to the subjective human experiences of consciousness. In Part II we will explore new approaches to facilitating consciousness, meaning, and free will by optimizing variation and natural selection on the phenomenological level by coping with stress via creative experience, psychotherapy, and the healing arts.

Gregor Mendel: Peas, Combinationslehr and Bioinformatics

To those who believe science has lost the soul of nature, let us present a prophetic poem written by the adolescent Gregor Mendel as a paean to Johann Gutenberg regarding "the highest goal of earthly ecstasy" (Henig, 2000).

Yes, his laurels shall never fade,
Though time shall suck down by its vortex
Whole generations into the abyss,
Though naught but moss-grown fragments
Shall remain of the epoch

In which the genius appeared. . . .
May the might of destiny grant me
The supreme ecstasy of earthly joy,
The highest goal of earthly ecstasy,
That of seeing, when I arise from the tomb,
My art thriving peacefully
Among those who are to come after me.

There was no pining for romantic love or visions of heroism by the sword in this profoundly prescient poem by the young Mendel. No wonder this lad eventually became *The Monk in the Garden,* "thriving peacefully" as he cultivated and counted his peas (Henig, 2000). For those of us who come after Mendel, the father of genetics, it is sobering to learn that this sensitive youth had a difficult time of it: He twice failed his qualifying examinations to become a high school science teacher. He never did qualify. However, the good Abbot Napp of the St. Thomas monastery in Brünn, the provincial capital of Moravia, did send the young Mendel to study science for two years at the Royal Imperial University of Vienna "to fit himself" better.

At the university Mendel showed a talent and enthusiasm for physics and *Combinationslehr*—the mathematics of combination theory that could be used to determine all the possible arrangements of any group of natural or symbolic objects. Little did Mendel know at the time that *Combinationslehr* would become the key that motivated his scientific approach of counting the distributions of hereditary traits between generations of *Pisum sativum,* the peas that he carefully cultivated and crossbred in his garden in the Augustinian monastery of St. Thomas. Mendel's early illustrations of peas and the original tree or fork-like bifurcating diagram used to map the distributions of hereditary traits between generations, which can still be found in basic textbooks of biology today, are the outcome of his early study of *Combinationslehr.* Combination theory is the foundation of modern *bioinformatics*—the new mathematics currently being developed to handle the data management challenges of modern functional genomics at the molecular level. In the following chapters we will generalize the principles of bioinformatics into a new field, psycho-bioinformatics, to serve as the emerging database for the clinical applications of psychosocial genomics in the healing arts.

PATHWAYS THREE AND FOUR:
THE CREATIVE REPLAY OF THE GENE-PROTEIN CYCLE
AND THE EVOLUTION OF THE BODY

Some of the most prescient speculations about the evolution of self-organizing molecules and genes were first formulated two generations ago by one of the founders of quantum physics, Erwin Schrödinger, in his famous book *What is Life?* (1944):

> A small molecule might be called "the germ of a solid." Starting from such a small solid germ, there seem to be two different ways of building up larger and larger associations. One is the comparatively dull way of repeating the same structure in three directions again and again. That is the way followed in a growing crystal. Once periodicity is established, there is no definite limit to the size of the aggregate. The other way is that of building up a more and more extended aggregate without the dull device of repetition. That is the case of more and more complicated organic molecules in which every atom, and every group of atoms, plays an individual role, not entirely equivalent to that of many others (as is the case in a periodic structure). We might quite properly call that an aperiodic crystal or solid and express our hypothesis by saying: We believe a gene—or perhaps the whole chromosome fiber—to be an aperiodic solid. (pp. 64–65)

Schrödinger's early idea of the gene, the basic unit of life and heredity, as an aperiodic solid may seem a bit obscure to modern students. Yet this concept inspired a series of discoveries culminating in the molecular structure of DNA and genes, for which Watson and Crick were awarded the Noble prize in 1958. A case can be made for the view that our modern concepts of information, self-organization, and emergence in evolutionary studies of the origin of life on a molecular level were initiated, in part, by Schrödinger's work. The essence of Schrödinger's insight was that organic processes are not a mere repetition of "the comparatively dull way of repeating the same structure in three directions again and again . . . like a growing crystal." In contrast to the dull, repetitive, and strictly regular periodicity of atoms doing the same thing over and over again in inorganic processes like the formation of a crystal, life processes involved "more and more complicated organic molecules in which every atom, and every group of atoms, plays an individual role, not entirely equivalent to that of many others (as is the case in a periodic structure)."

Schrödinger is saying, in effect, that life involves "an individual role" for "complicated organic molecules" that may now manifest behavior that is not mere, unchanging repetition of the past. This "individual role" implies there is creativity in the "aperiodic structure" of individual replays and resynthesis of complicated organic molecules such as DNA and RNA that we now recognize as the "code of life." What Schrödinger called "aperiodic" is what we now call "the nonlinear dynamics of complex adaptive systems." All living process are essentially creative replays of complex adaptive systems on all levels, from the molecular-genomic to the psychosocial. This leads us to propose that Schrödinger's *wave equation,* which describes the behavior of the basic structure and dynamics of the quantum level of reality, is a source of the psychobiology of gene expression, neurogenesis, and healing that can be facilitated by replaying cycles of creative experience in the new approaches to psychotherapy, as developed in the following chapters.

All living cells store their evolutionary history as genetic information that is encoded in the double helix of the DNA molecule. Genes literally come to life when they creatively replay and express this information in response to signals from the environment to transcribe their stored DNA information into messenger RNA in the nucleus of the cell. Messenger RNA is like a digital tape that is then shipped out into the cytoplasm of the cell, where it is translated into proteins that are the molecular work engines of life. How did this two step process of *transcription* and *translation,* called the "dogma of molecular biology" by its discoverers, Watson and Crick (1953 a, b), evolve? Freeman Dyson (1999), an emeritus professor of physics at the Institute for Advanced Study in Princeton, has recently summarized current theories of the origins of life that could well serve as a guide in our search for a synergetic synthesis of molecules and mind in the therapeutic arts.

Dyson credits the mathematician John von Neumann with the insight that any self-replicating automaton, including life itself, must consist of at least two interacting processes: hardware and software. Computers, for example, may be described as the interaction between the *software* that contains information and *hardware* that processes the information. In the living cell the software is the nucleic acid comprising the genes that code for evolutionary memory and *replication.* The hardware of life is primarily proteins that comprise most of the molecular structures of the machinery of *metabolism.*

Genes consist of a type of nucleic acid called *deoxyribonucleic* (DNA) that makes up what we now call *the code of life.* Another type of nucleic acid *ribonucleic* (RNA), plays multiple roles. Messenger RNA (mRNA) is the molecule that conveys the genetic information of the DNA in the nucleus of the cell to the protein-making component in the cytoplasm of the cell. Transfer RNA (tRNA) brings amino acids to the ribosomes to manufacture the proteins. It is well to note here that these molecular dynamics of life are all parts of the grand, complex adaptive system that includes psychology at the phenomenological level—what we call *human experience.*

RNA's highly versatile nature suggests that it has played a pivotal role in the origin of life. This theory of the origin of life is called *"RNA world."* Exploratory studies of the origin of self-replicating molecules in RNA world quickly led to the time parameters of life in gene expression and metabolism. The time parameters of the essentially sigmoid growth curves of RNA evolution in the test tube range between one and four hours, depending on experimental conditions. This is surprisingly similar to the one-and-a-half to two-hour cycle of ODC gene expression in response to maternal touch we reviewed earlier in the research of Schanberg and others. In later chapters we will find these same time parameters in the basic rest-activity cycle (Lloyd & Rossi, 1992), which characterizes the creative replays of complex adaptive systems on all levels, from gene expression to the psychobiology of human experience in waking, sleeping, dreaming, memory, learning and the healing arts.

PATHWAYS FIVE AND SIX:
MIND-BODY COMMUNICATION AND HEALING

Pathways five and six bring us to the relationships between mind and molecule that are the heart of our continuing journey (Rossi, 1968, 1986/1993, 1996a, 2000a, 2000b, 2000c, 2000d). In an effort to blow away some of the hoary vestiges of time-worn efforts to explain "mind," we will explore the concept of *state-dependent memory, learning, and behavior* (SDMLB) as an experiential analogue of mind. The state-dependent concept enables us to escape the reductionistic trap of saying that mind is nothing but the action of molecules. The phenomenon of SDMLB enables us to create a new self-reflective psychobiology that explores how mind modulates molecules, and vice versa (Pert et al., 1985, 1989; Rossi, 1986/1993, 1996a).

State-dependent memory develops whenever mind-modulating substances, such as alcohol, caffeine, barbiturates, amphetamines, or cocaine, are ingested and metabolized in the brain. For the past 50 years, one of the most routine studies conducted by pharmaceutical companies is to test whether an experimental drug has a state-dependent effect on memory, learning, and behavior. Any drug that passes through the blood-brain barrier usually modifies memory, learning, and behavior in a state-dependent manner. That is, when the drug is present in the brain, the animal responds to training with a certain kind or degree of memory, learning, and behavior. When the drug is metabolized so that it is no longer present in the brain, it is found that what was learned under the influence of the drug is partially lost; it is usually not as strong, so we say that the animal has a partial amnesia. Put the drug back into the brain, however, and the memory, learning, and behavior tend to return to their original strength. Researchers explain that the memory, learning, and behavior are now *state-dependent*—that is, their expression is dependent on the psychobiological state of the brain. Memory, learning, and behavior are now present or absent, depending on the presence or absence of a drug in the brain.

All this may be interesting, but one could still draw a blank and say, "Well, so what?" The really important finding that makes the state-dependent concept so significant is that we have learned that it is not only drugs that have this effect. We now know that there are hundreds of the body's own natural messenger molecules (stress hormones, neuropeptides, neurotransmitters, growth factors, mitogens, etc.) that can modulate memory, learning, emotions, and behavior in a state-dependent manner. The implications of the finding that stress hormones, for example, are involved in the activation and suppression of gene expression, leading to state-dependent modulations in memory, learning, and behavior, are profound (Rossi, 1996a; Rossi & Cheek, 1988).

The state-dependent concept functions as a psychobiological bridge between the creative replay of the complex adaptive dynamics of gene expression and psychological experience in the healing arts. The classical phenomena of hypnosis, multiple personality, neurosis, the posttraumatic stress disorders, psychosomatic symptoms, and mood disorders can all be understood as manifestations of state-dependent phenomena (Rossi, 1996a). Under stress, certain patterns of memory, learning, and behavior—symptoms—are learned and encoded by the release of stress hormones that modulate gene expression throughout the body. Once the outer stressor is removed, the body's stress molecules are metabolized (that is, broken down by normal biochemistry) and the person apparently recovers and seems symptom-free. Reintroduce stress to varying degrees, however, and the brain and body respond by releasing the hormones that re-evoke the state-dependent symptoms. The therapeutic applications of SDMLB, via the creative replay of *behavioral state-related gene expression* and *activity-dependent gene expression* are explored in detail in Chapters 2, 3, 7, and 8.

PATHWAYS SEVEN AND EIGHT:
GAIA, GENE, BODY, AND MIND: A COOPERATIVE ENTERPRISE

While the popular perception of Darwinian evolution focuses on competition and the survival of the fittest, life would not be possible without *cooperation* on all levels from Gaia and gene to body and mind. Lynn Margulis's many clear examples of how the dynamics of cooperation and symbiosis generated the tree of life complements the Darwinian competitive view. Margulis (1998) describes her views using a tree metaphor:

> Family trees usually are grown from the ground up: a single trunk branches off into many separate lineages, each branch diverging from common ancestors. But symbiosis shows us that such trees are idealized representations of the past. In reality the tree of life often grows in on itself. Species come together, fuse, and make new beings, who start again. Biologists call the coming together of branches—whether blood vessels, roots, or fungal threads—anastomosis. Anastomosis, branches forming nets, is a wonderfully onomatopoetic word. One can hear the fusing. The tree of life is a twisted, tangled, pulsing entity with roots and branches meeting underground and in midair to form eccentric new fruits and hybrids ... Symbiosis, like sex, brings previously evolved beings together into new partnerships. Like sex, too, some symbioses are protracted unions with stable, prolific futures. (pp. 51–52)

Lynn Margulis's view of how mitochondria, the energy factories in every cell of our body, were once free living organisms that learned to cooperate in the evolution of higher forms of life, has a profound lesson for us in our relationship with nature—Gaia at large. Margulis and Sagan (1986) describe the need to recognize the cooperative dynamics in the creative replays of the continuing coevolution of nature, gene, mind, and behavior:

> The ancestors to the mitochondria of our cells were probably vicious bacteria that invaded and killed their prey. But we are living examples that such destructive tactics do not work in the long run: mitochondria peacefully inhabit our cells, providing us with energy in return for a place to live. *While destructive species may come and go, cooperation itself increases with time.* People may expand, plundering and pillaging the Amazon, ignoring most of the biosphere, but the history of cells says we cannot keep it up for long. To survive even a small fraction of the time of the symbiotic bacterial settlers of the oceans and Earth, people will have to change. Whether we move into space or not, we will have to dampen our aggressive instincts, limit our rapacious growth, and become far more conciliatory if we are to survive, in the long-term, with the rest of the biosphere. (p. 18, italics added)

Learning to cooperate with Gaia is, of course, fundamental for our survival. The DNA within all cells of our bodies are continually exposed to trauma and toxins that can lead to genomic instability, illnesses, and death. Wood et al. (2001) have explored the entire human genome sequence and listed the locations of 130 genes (and more to come) involved in repairing damaged DNA. Current studies that focus on how the protein products of these repair genes coordinate their activity, hold the promise of developing fundamentally new insights on how cellular dynamics go awry in cancer and the aging process as well as a variety of other problems.

This deepening understanding of the coordination between Gaia and genes will be of central significance in our future approaches to facilitating the creative dynamics of individuals coevolving in families, communities, societies, nations, and now the global village. The cooperative dynamics explored at the level of gene expression in human experience in this book are the most recent development in the coevolution of Gaia, gene, body, and mind. We are now, for the first time, confronted with numinous mysteries of the role of consciousness and the human spirit in interacting with its own molecular-genomic sources in ways that previous generations could not have conceived. What sort of new tools can we develop to explore these currently evolving hyperlinks between mind and gene?

PATHWAYS NINE AND TEN:
DNA MICROARRAYS FOR EXPLORING BRAIN DEVELOPMENT AND PSYCHOBIOLOGICAL STATES OF CONSCIOUSNESS

Pathways nine and ten can be thought of as "hyperlinks" between mind and gene that are made possible by new developments in the union of molecular biology and gene-chip technology. It is a truism in science that the invention of a new instrument, technique, or technology can revolutionize our understanding of nature. The telescope, microscope, EEG, and now, in our own time, the Human Genome Project and the development of DNA microarrays, expand human perception of our own nature far beyond what could have been imagined previously. DNA microarrays (commonly called gene chips) consist of wafers of glass or other bonding surfaces about 1.5 centimeters square that appear to be analogous with the silicon chips of computer technology. Each chip is lined with thousands of microscopic wells or sites, on which are attached short bits of DNA that can bond with any matching genes in a biological sample being studied. DNA microarrays can be used to assess the expression and coordinated activity of as many as 10,000 genes in a single experiment.

The new DNA microarray technology follows naturally from the basic structure and replication dynamics of the DNA molecule and the genetic code, as described by the Nobel prize-winning work of Watson and Crick. Genes are arranged in pairs that are zippered together on long strands of DNA and positioned around each other in a double helix in the chromosomes. During cell replication, the double helix formed by the base pairing of nucleotides comes unzipped momentarily, until new bonds are formed. Using highly specialized robots, molecular biologists have discovered how to arrange thousands of carefully identified pieces of the unzipped DNA in a known order on a glass slide or chip. Each microscopic piece of DNA acts as a molecular "probe." Each probe is "sticky"—and just waiting to attach or zipper itself to a corresponding piece of DNA, which is a "target gene" in any solution that is incubated or "hybridized" with the probes on the chip.

We are now on the threshold of creating DNA microarrays that contain the entire human genome, recently found to number about 30,000 genes that are comprised of about 2.91 billion pairs of nucleotides (Venter et al., 2001). Such DNA microarrays open the prospect of being able to identify the activity patterns of gene expression, at any given moment, of any condition or state of health or illness. For example, we can compare the differences in gene expression during different patterns of learned behavior (Bućan & Abel, 2002). We can explore the differences in gene expressions between normal and cancerous cells. A variety of technologies for manufacturing DNA microarrays has been developed for these different applications and are known by many different names (biochips, genome chips, cDNA arrays, DNA chips, gene chip arrays, oligionucleotide arrays, high density microarrays, etc.). Similar technologies are being developed to analyze proteins that make up the proteome—the entire population of proteins contained in cells of the body. Typical experiments, using DNA technology at the National Human Genome Research Institute, identify patterns of gene expression in varying states of the organism on the cellular-genomic level. If we believe that sleep, dream, arousal, and hypnosis are different states, then we can expect that DNA microarrarys will identify the patterns of gene expression associated with these different states. This means that eventually we may be able to define more precisely exactly what we mean by arousal, sleeping, dreaming and the various states of healing and therapeutic hypnosis using these DNA microarrays.

Classical Mendelian Genetics and the New Functional Genomics

To fully understand the significance of the DNA microarray revolution, it is important to review the difference between *classical Mendelian deterministic genetics* (that most people think of when they read about genes) and *the functional genomics of modern molecular biology* that is generating the new DNA technology. Early classical Mendelian concepts were focused on genetic determinism—the laws of inheritance—how dominant and recessive genes determine the transfer of phenotypic traits (easily observed physical characteristics such as eye, hair, or skin color) from one generation to the next. The basic theory of Mendelian genetics was that one, or at most a few, genes determined one biological or behavioral trait. According to this early view, *nature* alone determined such physical traits; *nurture,* or life experience, supposedly had nothing to do with it. *The nature-nurture controversy* arose when Mendelian genetics was extended to the study of human behavior. For example, many statistically-oriented studies on animals and human twins documented how intelligence was related, in a deterministic manner, to the inheritance of genes. This view led to the belief that intelligence—and, by implication, many other human traits and behaviors—were determined entirely by *nature* and could not be changed (Plomin et al., 2001).

While such traditional Mendelian studies of one-gene/one-function have demonstrated that genes and behavior are related, these studies are slow, tedious, expensive, and difficult to replicate. What is worse, such studies have given rise to the misconception that we can identify one or a few genes as the ultimate cause of every human illness and behavioral problem: Fix the gene with some sort of chemical manipulation, and the problem is solved. Most human behaviors, however, are not turned on by a single gene (McGuffin et al., 2001). This oversimplified and erroneous point of view ignores the broader truth that most genes are expressed (that is, turned on and off) in coordinated patterns of activity in response to signals from the extracellular environment (Brown, 1999). The new functional genomics of molecular biology focuses on these broad patterns of gene expression rather than the one-gene/one-function or trait approach of early Mendelian genetics. The microarray technology enables us to formulate a dynamic picture of the coordinated activity of hundreds of genes in a single experiment exploring the fluctuating states of the organism as it interacts with its environment in an adaptive and creative manner. In a recent technological symposium, for example, Conklin (1999) identified 600 genes with DNA microarray technology that were differentially expressed in heart disease. Similarly, several hundred genes are involved in mediating the effects of chronic high-sugar diets (Johnson, 1999). The study of varying patterns of gene expression associated with human choice and behavior is one aspect of the newly emerging field of psychobioinformatics (Bassett et al., 1999; Rossi, 1999a, 2000b).

TEMPORAL WAVES OF GENE EXPRESSION
IN BRAIN DEVELOPMENT

In mapping CNS development in the rat, Wen et al. (1998) found *temporal waves of gene expression* that well illustrate how DNA microarrays can be used in future studies of how gene expression and human experience are related. They summarize their work as follows:

> The data provide a temporal gene expression "fingerprint" of spinal cord development based on major families of inter- and intracellular signaling genes. . . . We found five basic "waves" of expression that characterize distinct phases of development. The results suggest functional relationships among the genes fluctuating in parallel. We found that genes belonging to distinct functional classes and gene families clearly map to particular expression profiles. The concepts and data analysis discussed herein may be useful in objectively identifying coherent patterns and sequences of events in the complex genetic signaling network of development. Functional genomics approaches such as this may have applications in the elucidation of complex developmental and degenerative disorders. (p. 334)

Figure 1.3 (see color insert) illustrates the temporal waves of gene expression in CNS development identified by Wen et al. These waves of gene expression are illustrated as computer printouts on the top of Figure 1.3. The names of the genes border each of the four major waves depicting CNS development, along with a family of genes labeled "constant" (housekeeping genes), and another group simply called "other." In the constant group we find our old friend, the ODC gene, which can be turned on by human touch, as previously described. The constant group also contains the brain-derived neurotropic factor (BNDF) associated with the growth, development, and survival of neurons. The dynamics of BDNF gene expression are related to a wide range of psychobiological processes, from memory and learning to the stress response and psychiatric conditions such as depression (Duman & Neslter, 2000). The simultaneous expression of the ODC and the BNDF genes suggests the hypothesis that turning on the ODC gene via human touch and associated psychosocial experiences may simultaneously evoke the expression of the BNDF gene and many of its psychobiological dynamics. We hypothesize that such associations between gene expression and psychosocial experience provide the bridge between nature and nurture that is central to mind-body communication and healing in the therapeutic arts. The experimental assessment of this hypothesis could provide a new database for psychosomatic medicine in the future (Chin & Moldin, 2001).

An examination of the family of genes expressed together in wave 1 indicated that they are representative of an early stage of rapidly dividing neuroglial progenitor cells. The authors recommend this gene family as containing candidates for further study of spinal cord disease and injury. The reactivation of this gene, which is associated with early CNS development, could lead to the healing of physical injury and, from our perspective, entirely new approaches to functional psychobiological disorders.

Wave 2 in Figure 1.3 represents a family of genes associated with *neurogenesis,* indicated by the simultaneous expression of neuronal markers in this group. In Chapter 3 we will review current research in neuroscience that indicates how experiences involving novelty, environmental enrichment, and voluntary physical exercise can turn on gene expression that leads to neurogenesis in the networks of the brain, where new memory, learning, and behavior are encoded. Wave 2 may be the clearest marker of functional relationships and information transduction between mind and body. From our psychobiological perspective wave 2 may be conceptualized as a marker of how (1) *networks of gene expression* in (2) *neural networks of the CNS* are related to (3) *networks of state-dependent memory, learning, and behavior (SDMLB).* These three steps model how the mind-body communication between gene and psychological experience may be bridged in the new approaches to therapeutic hypnosis, psychotherapy, and the healing arts discussed in Part II.

Wen et al. (1998) describe the dynamics of wave 3 in gene expression as follows:

> Wave 3, although exclusively covering neurotransmitter signaling genes and neuronal markers, is distinguished from wave 2 by a characteristic *low-high-low pattern of developmental gene expression* and a slower rise time. A fundamental phenomenon in the development of the brain is overproduction of cells, many of which are later eliminated during maturation and cementing of synaptic connections. . . . The pattern of gene expression in wave 3 emphasizes that neu-

ronal-signaling gene expression is not a gradual linear process in which genes asymptotically approach their mature tissue levels but that there is a transient phase of high expression, analogous to the transient overabundance of cells in neurodevelopment. (p. 338, italics added)

The *"characteristic low-high-low pattern of developmental gene expression"* is another clue linking the complex adaptive system of (1) gene expression, (2) neurogenesis, and (3) psychological experience; it is identical to the inverted U-pattern that is highly characteristic of many psychobiological processes of behavioral state-related gene expression and activity-dependent gene expression, as well as human performance and the creative process. Furthermore, the inverted U-pattern is highly characteristic of the nonlinear dynamics of most complex adaptive systems of physiological and psychobiological experience. Low levels of psychobiological arousal and motivation are usually associated with low levels of psychological performance in memory, learning, and behavior. An optimal level of arousal is associated with peak psychological performance. Any further arousal after this peak, however, leads to a decrement or low level of psychological performance. In addition, the inverted U-pattern is highly characteristic of the nonlinear dynamics of the circadian map of consciousness and the four stages of the creative process illustrated in Chapter 2 and described in detail in Part II. Finally, the same inverted-U pattern will be found in the complex dynamics of psychotherapy, hypnosis, and the healing arts illustrated throughout this book and others (Rossi, 1996a, 2000a, 2000b, 2000c, 2001).

Wen et al. (1998) describe how wave 4 of gene expression in the development of the brain and nervous system may be a marker for the final maturation of neuronal tissue. They comment that "we now need a simple model of conceptualizing how large numbers of genes interact to generate a complex but robust system. Threshold levels of gene expression are a possible mechanism by which the genetic program makes decisions about the timing of development" (p. 338). What Wen and colleagues describe is essentially a modern expression of the search for understanding how life and mind co-evolve in nature. They go on to note the common perception in neuroscience that Boolean networks, based on binary information processing of the gene expression, are replayed in the aperiodic, circular loops of self-organizing complex adaptive systems of all life processes.

Pioneers in the study of the neurobiology of human cognitive and emotional development now recognize how the origin of the psychological sense of the self may be traced to gene expression in neurons of the brain and cells of the body that interact with the pre- and postnatal environments as well as psychosocial processes throughout life (Damasio, 1999; Schore, 1994). While it is recognized that all humans are about 99.9% identical in the composition of their 30,000 or so genes, it is now known that there are nearly six billion base pairs of DNA in the human genome (Kucherlapati & DePinho, 2001). Within these there are about 3 million small differences in the arrangements of the four nucleotide bases (adenine, thiamine, guanine, and cytosine) that make up the composition or code of our genes. These natural variations in our genes are called "single nucleotide polymorphisms" (SNPs). SNPs are a major source of individuality out of which we co-create ourselves in our psychosocial interactions within ourselves and with each other and nature. This is the essence of the emergent science of psychosocial genomics (Rossi, 2000c). Current research, theory, and practice suggest that our uniqueness on all levels, from the physical to the psychological, may be traced to the complex adaptive interactions between nature and nurture involving these SNPs (Sullivan et al., 2000; Wong & Licinio, 2001). In the following chapters we will explore this idea by integrating current research on the relationships among human experiencing, gene expression, neurogenesis, and healing with a new view of their role in the ever-evolving creation of meaning in culture, the humanities, and the therapeutic arts.

SUMMARY

Now that the structural phase of the Human Genome Project is coming to a successful conclusion, attention is shifting to the process of annotation: How does the society of our genes interact with external events to creatively replay the dramas of our daily experiences? We now know that significant life events can turn on genes that lead to the synthesis of proteins—which, in turn, generate new neurons and connections in our brain. Our daily and hourly life experiences, thoughts, emotions, and behavior can modulate gene expression and neurogenesis in ways that actually can change the physical structure of the brain.

Emerging from the Human Genome Project, this new worldview of the relationships between gene expression and human experience is setting the stage for a profound expansion of our understanding of life. The essential mission of the humanities and sciences as well as psychotherapy and the healing arts in the new millennium is to explore and utilize this new worldview to facilitate health and well-being. We can now document the existence of a society of genes that cooperate to create, maintain, and replay the natural dynamics of psychosocial evolution in daily life. The concept of a cooperative society of genes currently emerging from the work of Lynn Margulis and others complements the concept of selfish genes in the neo-Darwinian views of evolution. The general evolutionary principle that ontogeny recapitulates phylogeny can now be extended to include the psychobiological development of the individual's entire life cycle. Many of the essential dynamics of gene expression involved in the formation of the brain and body in embryology are now recognized as a continuing creative development throughout an individual's lifetime.

Psychobiological research on the many types of *behavior-state* and *activity-dependent gene expression* provides a new database for understanding the connecting links between *nature and nurture* as well as *body and mind*. In the following chapters of Part I, we will review the essential research that documents how the creative replay of gene expression modulates human experience, and vice versa, in the circular dynamics of all complex adaptive systems of emergent life processes. In Part II we will utilize this encompassing psychobiological perspective to develop a new class of activity-dependent approaches for facilitating the creative replay of gene expression and neurogenesis in psychotherapy and the healing arts.

Chapter 2

BEHAVIORAL STATE-RELATED GENE EXPRESSION

How Gene Expression Modulates Human Experience

"The study of Immediate Early Genes [IEGs] indicates that sleep and wake, as well as synchronized and desynchronized sleep [dreaming], are characterized by different genomic expressions, the level of IEGs being high during wake and low during sleep. Such fluctuation of gene expression is not ubiquitous but occurs in certain cell populations in the brain. . . . IEG induction may reveal the activation of neural networks in different behavioral states. Although stimulating, these finding leave unanswered a number of questions. Do the areas in which IEGs oscillate during sleep and wake subserve specific roles in the regulation of these physiological states and in a general 'resetting' of behavioral states? Is gene induction a clue to understanding the alternation of sleep and wake, and REM and non-REM sleep? . . . could behavioral state-related IEG induction underlie, at least in part, learning mechanisms? The oscillation of IEGs affects the expression of target genes, and thus brings about other questions: May the transcriptional cascade explain the biological need and the significance of sleep? Does this explain the molecular and cellular correlates of arousal, alertness, and, more in general, of consciousness? We do not have final answers to these questions."

—M. Bentivoglio & G. Grassi-Zucconi
"Immediate Early Gene Expression in Sleep and Wakefulness"

Important advances in research on gene expression and the molecular biology of waking, sleeping, and dreaming have profound implications for the deep psychobiological dynamics of psychotherapy and the healing arts. *"Behavioral state-related gene expression"* refers to research documenting how a wide range of behavioral states, such as sleeping, dreaming, consciousness, vigilance, stress, emotional arousal, and depression, is associated with different patterns of gene expression. The essence of this new view is that *immediate early genes* and *clock genes* play a central role at the cellular and molecular levels in the dynamics of waking, sleeping, dreaming, and healing.

This chapter explores how immediate early genes and clock genes function as mediators of information transduction among environment, psychological experience, behavioral states, and biology in the new discipline of *psychosocial genomics.* These interactions between genes and the deep psychobiological processes of consciousness addressed in psychotherapy and the healing arts are complex adaptive systems engaged in a constant replay of creative adaptation between Gaia and gene that can be modeled by the nonlinear dynamics of deterministic chaos theory (Rossi, 1996a).

A FOUR-STAGE MODEL OF MIND-BODY COMMUNICATION

The mystery of the relationship between mind and the matter of the brain has led to unending controversy throughout human history. In this chapter we seek relief from casual philosophical speculation by reviewing neuroscience research that traces the natural pathways of mind-body communication. Here we have a great advantage over previous generations, for we have new brain-imaging tools—such as positron emission tomography (PET) and functional magnetic resonance imaging—(fMRI) that can tell us which *areas of the brain's matter* are active during various activities of the *brain's psychological experiences of mind.*

Damasio et al. (2000) have explored how subjective human experiences of emotion, such as happiness, sadness, anger, and fear, are associated with different areas of heightened brain activity. They found that the subjective experience of emotional states during the voluntary recall and reexperiencing of personal life episodes is associated with, or "mapped on to," brain regions (such as the somatosensory cortices and the upper brainstem nuclei) that are involved in the regulation of psychobiological states of homeostasis. Damasio and his team concluded that their results support the view that the subjective process of emotions is grounded in ever-changing dynamic neural maps, which represent different aspects of the organism's continuously changing internal state.

Modern neuroscience is a practical exploration of relationships between subjective experience and objective measurements that gives us interesting and potentially useful psychobiological perspectives on what has been called "the spectrum of consciousness" (Deikman, 1980a, 1980b; Wilber, 1993). The implication of this view is that consciousness is a *complex adaptive system* that shares the same basic pathways of creative adaptation as all other dynamics of life. This idea gives new meaning to the ancient notion that "All is One" and provides practical clues about how we can facilitate mind–body communication in health and illness.

Let us begin by tracing the nonlinear, circular pathway between the subjective states of personal memory and emotions, with their deepest psychobiological roots in gene expression (Rossi, 1986/1993, 1996a). The four-stage model of mind–body communication and healing illustrated in Figure 2.1 outlines some of the major relay stations that are typically engaged in normal daily life as well as psychotherapy and the healing arts. This general map of mind–body communication as a complex adaptive system will warrant careful study; it is fundamental for a full understanding of the major concepts of this book and will be referred to frequently in the following chapters.

STAGE ONE: THE MIND-GENE CONNECTION AND PSYCHOSOCIAL GENOMICS BRIDGING THE CARTESIAN GAP

Stage one of mind–body communication is well illustrated by the dynamics of the limbic–hypothalamic–pituitary–adrenal (LHPA) system in Figure 2.1. Papez (1937) originally described this system as the major information transducer between mind, brain, and body. The Scharrers (1940) and Harris (1948) then confirmed how the secretory cells within the hypothalamus mediate information transduction via the flow of hormones between brain and body. We now recognize their work as an initial step in bridging the so-called Cartesian gap between mind and body on the cellular–genomic level that is the essence of psychosocial genomics. Cells within the hypothalamus (stage one of Figure 2.1) transform the essentially electrochemical impulses of the neurons of the cerebral cortex that transduce the phenomenological experiences of "mind" (such as thoughts, imagery, and emotions, documented by Damasio et al., 2000) into the hormonal messenger molecules of the endocrine system. These hormones (also called "primary messengers" or "informational substances") flow through the bloodstream in cybernetic replays of circular loops of information transduction between brain and body (Edelman & Tononi, 2000).

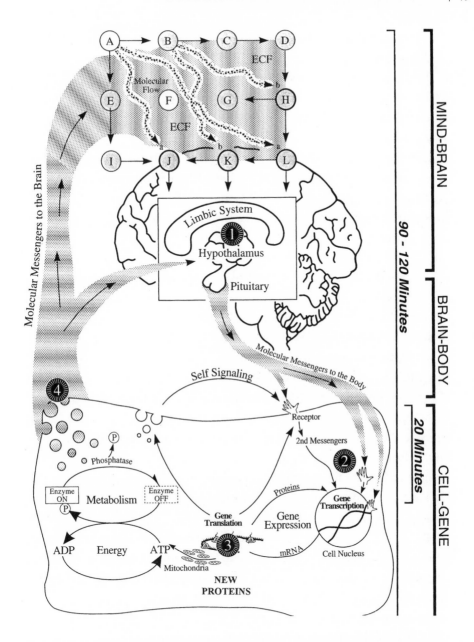

FIGURE 2.1 | Psychosocial genomics illustrated as a four-stage model of the complex adaptive system of mind-body communication and healing. This circular loop of information transduction illustrates the flow of hormones (messenger molecules) between the mind-brain, brain-body, and cellular level of gene expression. The brain's neural networks at the top represented by the rectangle of neural units, A through L, that receive molecular signaling through the extra cellular fluid [ECF] are regarded as a complex adaptive system: a field of self-organizing communication processes that are the psychobiological basis of mind, memory, learning, and psychosomatic medicine.

This complex adaptive system modulates the action of neurons and cells at all levels, from the basic pathways of sensation and perception in the brain to the intracellular dynamics of gene expression throughout the body (Rossi, 1986/1993, 1996a). These molecular messengers of the endocrine, autonomic, and immune systems mediate stress, emotions, memory, learning, personality, behavior, and symptoms in sickness as well as their healing (Rossi, 2001a). This circular communication loop between mind and gene has been described as a *"psychosomatic network"* (Pert et al., 1985, 1989), a two-way street by which (1) mind can modulate the physiology of the brain and body, and (2) biology, in the form of hormonal molecular messengers, can modulate mind, emotions, learning, and behavior.

The limbic-hypothalamic-pituitary-adrenal (LHPA) axis is the broad system of mind-body communication that integrates stress inputs throughout the body, which then converge on the final common neuronal pathway of the medial paraventricular nucleus in the hypothalamus. There neurons synthesize corticotrophin-releasing hormone (CRH) and arginine vasopressin (AVP) that coordinate mind-body communication. The current consensus is that CRH is a central psychobiological coordinator of the generalized arousal and stress response system (Akil & Morano, 2000; Felker & Hubbard, 1998). In humans the CRH gene is located on chromosome 8. There is a high degree of homology in the base sequences of the variations in the CRH gene, which indicates that it has been highly conserved throughout evolution because of its central importance in coordinating life processes on many levels. There is a wide distribution in the expression of the CRH gene throughout the central nervous system (CNS) and the peripheral nervous system (PNS). CRH synthesizing and secreting neurons are found throughout the neocortex, but their highest densities are found in the prefrontal, insular, and cingulate areas, where they mediate cognitive and behavioral processing (De Souza & Grigoriadis, 2000). CRH signals the anterior pituitary to produce proopiomelanocortin (POMC), which is cleaved into about a dozen messenger molecules, including adrenocorticotropic hormone (ACTH), which facilitates states of arousal, and beta-endorphin, which facilitates relaxation. These hormones then flow through the bloodstream as primary molecular messengers that have varying effects upon the psychobiology of arousal, stress, memory, learning, and relaxation at the cellular-genomic level (Brush & Levine, 1989). Let us now turn to this cellular-genomic level of mind-body communication and healing.

STAGE TWO:
SECONDARY MESSENGERS AND IMMEDIATE EARLY GENES

Current theory and research in molecular biology and neuroscience conceptualizes living cells as information processors and problem-solving systems (Brey, 1995; Brodsky, 1992). Marijuan (1996) regards the flow of "bioinformation" through the cell as the basic dynamic that mediates "work-cycles" of fundamental life processes involving self-organization, self-synthesis, and self-reconfiguration. Stage two in Figure 2.1 illustrates how secondary messengers within the cell (such as cAMP, cGMP, Ca, InsP3, diacylglicerol, ceramide, arachidonic acid, nitric oxide) convey the extracellular signals from the environment (including psychosocial cues) to the nucleus of the cell, where they initiate gene expression and the synthesis of proteins.

Classical Mendelian genetics focuses on genes as the units of biological heredity that are transmitted from one generation to another through some form of reproduction. Today, by contrast, we know that there are many classes of genes that are activated or turned off, from moment to moment in everyday life, to carry out the important life functions of homeostasis, adaptation, learning and healing. We have already met one major class of such genes—immediate early genes—which are expressed continually in response to hormonal messenger molecules mediating psychobiological adaptation. Many of these extracellular stimuli ultimately come from outside of the organism, such as temperature, food, sexual cues, psychosocial cues (both stressful and pleasing), physical trauma, and toxins. It is now known that there are persistent alterations in immediate early gene expression in the process of adaptive behavior on all levels, from the sexual and emotional to the cognitive (Lydic, 1998; Merchant, 1996; Tölle et al., 1995).

As noted, immediate early genes (IEGs) have been described as the newly discovered mediators between nature and nurture (Rossi, 1998a). Immediate early genes transduce signals from the external environment to regulate the transcription of other "target genes" that code for the formation of proteins, which then carry out the metabolic and adaptive functions of the cell. Immediate early genes therefore can initiate series of molecular-genomic transformations that transduce relatively brief but meaningful signals from the environment into enduring changes in the physical structure of the developing nervous system throughout life (Dragunow, 1995; Richardson, 2000). C-fos, for example, is an immediate early gene that is turned on by arousing or stressful environmental stimuli to activate neurons of the brain, which trigger the production of a protein called "Fos." In turn, Fos can then bind onto the DNA molecule where it can turn on—that is, initiate the transcription of—other target genes. C-fos, in combination with other IEGs (such as those in the "jun family"), can regulate and be regulated by other genes that are involved in the material, energetic, and informational dynamics of the cell in response to psychosocial cues as well as physical stress and trauma. This complex process is illustrated in Figure 2.1 as the major circular pathway of mind-body communication that is constantly replayed, with creative variations, in the normal dynamics of adaptation on all levels, from mind to gene, in everyday life.

The most fundamental experimental evidence that IEG expression is associated with behavioral states of arousal, waking, sleeping, and dreaming (REM sleep) in the ultradian time frame (90–120 minutes) noted in Figure 2.1 comes from the observation of limbic-hypothalamic-pituitary oscillations during the normal circadian replay of the sleep-wake cycle (Kelner & Bloom, 2000). Bentivoglio and Grassi-Zucconi (1999) describe this fundamental link between behavior and gene expression as "behavioral state-related gene expression." Behavioral state-related gene expression is a fundamental link between psychology and biology; indeed, it is the essence of psychobiology. Bentivoglio and Grassi-Zucconi (1999) summarize their findings:

> The expression of immediate early genes oscillates in the brain during the sleep/wake cycle. . . . We observed that Fos immunoreactivity increased in the brain in parallel with the increase in the proportion of wake in the 1.5 hours preceding sacrifice. In other studies, rats that had been awake in the dark were compared to animals that had been spontaneously awake during the light period for about 30 minutes before sacrifice, and c-fos [gene] expression was found to be higher in the former group than the latter. These studies confirmed an increase of Fos [protein] expression after a sustained (1.5-hour) period of spontaneous wakefulness in respect to a short (0.5-hour) period, which, however, induced c-fos mRNA expression but was not sufficient for adequate protein synthesis. (p. 241)

The research team of Cirelli, Pompeiano, and Tononi (1998) confirm the presence of this ultradian time frame in the behavioral states of arousal, waking, and sleeping associated with immediate early gene expression, as follows:

Over the last few years it has become clear that the activation or deactivation of the expression of specific genes can occur in a matter of hours or even minutes. This time frame is compatible with the duration of sleep-wake states and with the time constants of their regulation. Thus, it becomes relevant to ask whether gene expression in the brain changes across the sleep-wake cycle and after sleep deprivation. . . . A parallel series of studies indicated that after a few hours of sleep deprivation the patterns of IEG expression were remarkably similar to those observed after spontaneous wakefulness, suggesting that such patterns are associated with waking, per se, rather than with circadian or stress factors. Recently we showed . . . that the expression of c-fos during waking is strictly dependent on the level of activity of the noradrenergic system. . . . In our studies, the most consistent increase in Fos expression during waking was found in the preoptic area (POA) of the hypothalamus, a region that has been previously implicated in sleep regulation. . . . Thus, Fos protein expression in the POA during waking may be an integral part of mechanisms that assess the duration and intensity of prior waking and/or the homeostatic or executive mechanisms that bring about sleep. (p. 46)

This is a technical way of saying that the behavioral states of consciousness, waking, and sleeping are intimately associated with gene expression and protein synthesis—the essence of mind-body communication and healing.

Gene expression is also used to identify neuronal networks that are associated with REM sleep during dreaming. Merchant-Nancy et al. (1992), for example, found changes in c-fos proto-oncogene expression in relation to REM sleep duration. The brain distribution of c-fos expression was found to include brainstem regions, the basolateral amygdala, and the lateral hypothalamic area, which are all part of the neuronal network associated with the dynamics of REM sleep (Merchant-Nancy et al., 1995). This gene expression was also found in the locus coeruleus, dorsal raphe, medial pontine reticular formation, and mesopontine nuclei that contain REM-on and REM-off neurons. In another study (Yamuy et al., 1993), the gene expression was evident in the pons and medulla, the medial and lateral vestibular nuclei, and the motoneurons of the nucleus abducens involved in the generation of pontine-geniculo-occipital (PGO) spikes associated with REM sleep.

The quotation at the beginning of this chapter by Bentivoglio and Grassi-Zucconi (1999) summarizes the current implications of these and other studies that raise fundamental questions:

IEG induction may reveal the activation of neural networks in different behavioral states. Although stimulating, these findings leave unanswered a number of questions: Do the areas in which IEGs oscillate during sleep and wake subserve specific roles in the regulation of these physiological states and a general "resetting" of behavioral states? Is gene induction a clue to the understanding of the alternation of sleep and wake, and of REM and non-REM sleep? (p. 249)

Bentivoglio and Grassi-Zucconi (1999) emphasize the profound role of immediate early genes in mediating the constant creative replay of nature and nurture as well as mind and body in all adaptive life processes:

The inducibility of transcription factors indicates that external cues can modulate cell function through regulation of gene expression. The variation of IEG expression during sleep and wake seems to indicate that this could also be true for internal cues. The high expression of IEGs during wake could be related to the animals activity, to a momentary excitation of single neurons in the course of transferring physiological "everyday information." . . . In view of the role played by transcription factors in neural plasticity [that is, changes in neurons of the brain during memory and learning], could behavioral state-related IEG induction underlie, at least in part, learning mechanisms? . . . The oscillations of IEGs affects the expression of target genes, and this brings about other questions: May the transcriptional cascade explain the biological need and significance of sleep? Does this explain the molecular and cellular correlates of arousal, alertness, and more in general, of consciousness? (p. 249)

These fundamental questions raised by researchers on the cellular-genomic level surely indicate that a profound rapprochement is taking place between biology and psychology. There is, as yet, no research exploring the effects of psychotherapy on the expression of immediate early genes or their role as transcription factors in regulating a wide variety of target genes and their protein products that modulate mind-body communication and healing. Richardson (2000) estimates that approximately 90% of all genes are engaged in self-regulatory, adaptive responses in cooperation with signals from the environment. The implication of current research is that a deep psychobiological model of how psychotherapy operates to facilitate healing must involve these molecular-genomic levels (Lockhart & Winzeler, 2000; Rossi,

2000a, 2000b;Vukmirovic & Tilghman, 2000). Let us now review the creative replay of the gene expression cycle that is the essence of this molecular-genomic view of consciousness and behavior.

STAGE THREE:
CLOCK GENES, GENE EXPRESSION, AND BEHAVIOR

The third stage in the mind-gene communication loop is illustrated in Figure 2.1 as the gene expression and protein synthesis cycle via the production of messenger RNA (mRNA). This gene expression cycle is the heart of metabolism, homeostasis, and behavior in the complex adaptive systems of life. Timing, via clock genes, regulates the daily rhythms of gene expression that generate physiology, behaviorial states, and the psychological experiences of consciousness, sleeping, and dreaming (Lloyd & Rossi, 1992, 1993). In all higher organisms the central circadian "clock," which resides in the suprachiasmatic nuclei of the hypothalamus, coordinates the individual clock genes in cells throughout the body. Recently, it has been found, however, that there are many clocks that operate in many tissues and organs of the body independently of the central suprachiasmatic nucleus (Stokkan et al., 2001).

Next we will briefly introduce a number of areas of special interest for a new understanding of the role of time in the psychobiology of gene expression in psychotherapy and the healing arts.

Clock Genes Modulating Time, Love, and Memory

Clock genes mediate a surprisingly wide variety of adaptive processes of mind-body communication and healing. The history of the discovery of these time-processing genes is described in an accessible book on *Time, Love, and Memory* (Weiner, 1999). In brief, clock genes are needed to turn on the target genes that code for the production of proteins, which, in turn, generate the hormones coordinating the psychobiological dynamics of time, sexual behavior, memory, and learning. A major breakthrough in the modern understanding of gene expression and behavior took place when Seymour Benzer (1967, 1971, 1973) and his colleagues discovered a series of mutant flies with abnormal time patterns. Studying the abnormal mutations in flies served as an effec-

tive model for gaining insight into the normal role of clock genes in all higher organisms. This area is highly significant for the study of behavior because
clock genes closely interface with the outer environment, on the one hand,
and the inner molecular biology of the cell, on the other. In an important
review, Dunlap (1999) noted that, "clocks represent at once both a nearly
ubiquitous aspect of cellular regulation and also a molecular regulatory process
that has clear and immediate effects on organismic behavior" (p. 286).

Figure 2.2 (see color insert) illustrates a number of the common elements and functions of the gene expression and protein synthesis cycle that
generates the rhythms of metabolism, physiology, behavior, and cognition. The
fundamental dynamic of a cybernetic process is that whenever a system moves
away from a setpoint or equilibrium, a signal is produced that feeds back to
speed the process up or slow it down (the negative elements in Figure 2.2)
so that it returns to equilibrium. Biological oscillators need a source of energy, activation, or excitement (the positive elements of Figure 2.2) that keeps
the cybernetic cycle from winding down.

There has been a great deal of controversy about the location and coordination of these nonlinear biological oscillators that regulate virtually all
physiological and behavioral processes in an adaptive manner (Lloyd & Rossi,
1992, 1993). It was originally thought by some researchers, for example, that
the suprachiasmatic nuclei in the hypothalamus, which receives light signals
from the retina of the eye, was the major circadian oscillator that entrained
or regulated the rest of the body. It is now known, however, that, "all known
circadian oscillators use loops that close within cells. . . . They rely on positive and negative elements in oscillators in which transcription of clock genes
yields clock proteins (negative elements) which act in some way to block the
action of positive element(s) whose role is to activate the clock gene(s)"
(Dunlap, 1999, p. 273). This finding suggests that the ultimate sources of psychobiological rhythms—which may be modulated by creative work, psychotherapy, and the healing arts—are to be found within cells in the adaptive
replay of the gene expression and protein synthesis cycle. This is the molecular-genomic nucleus of what has been called "the mind-gene connection"
(Rossi, 1990a, 1994c, 1996a, 1996b, 2001a, 2001b).

One of the most interesting aspects of clock genes is that, to a certain
extent, they can be entrained (that is, reset or modulated) by environmental
stimuli. *Many biological clocks could be described as coordinating genes that entrain*

and integrate, in a timely manner, creative adaptation to outer world signals such as light, temperature, food availability, performance demands, and psychosocial stress. Changing environmental and psychosocial signals leads to the expression of clock genes that generate the proteins and messenger molecules that modulate many adaptive systems in the rhythmic replays of psychobiology. The fact that psychosocial cues entrain and synchronize immediate early genes and clock genes opens the door to understanding how psychotherapy and the healing arts can facilitate optimal performance and healing.

Figure 2.3 (see color insert) illustrates the time dynamics of how light and temperature entrain (modulate, regulate) the oscillations in the expression of the clock gene called *frequency* (frq) in the relatively simple system of the bread mold *Neurospora*. Many studies have found that the same genes are responsible for the same function in many organisms on the evolutionary scale (Greenspan et al., 1995). This is certainly true of the clock-controlled genes (*ccg's* illustrated in part D of Figure 2.3) that modulate the timing of many fundamental biological processes, ranging from development to energy production and adaptive responses to stress. To date, at least 17 timing genes have been identified along the phylogenetic scale, from molds and microbes to mammals such as mice and men (Dunlap, 1999; Kelner & Bloom, 2000). The sinusoidal rhythm of FRQ protein production between dawn and dusk is well illustrated in section C of Figure 2.3. It is a prototype of how the essentially circular cybernetic loop of the gene expression and protein synthesis cycle, illustrated in section D of Figure 2.3, gives rise to the sinusoidal rhythm that is typical of most psychobiological processes. I propose that these rhythms at the level of gene expression and protein synthesis are the psychobiological basis of the ultradian rhythms in behavior, the creative process, and healing that are developed in Part II.

Figure 2.4 (see color insert) illustrates some of the more complex adaptive dynamics of how clock genes operate in mammals, including humans. It emphasizes how the light-sensitive neurons of the suprachiasmatic nucleus (SCN) can function as a pacemaker in the release of neuropeptides that entrain the overt psychobiological rhythms of the entire organism (Dunlap, 1999; Young, 1998, 2000). Researchers remind us that "there is a circadian rhythm for virtually every homeostatic function" (Kupfermann et al., 2000, p. 1012). Section D of Figure 2.4 summarizes how these clock-controlled processes of gene expression and protein synthesis regulate how the stress response is medi-

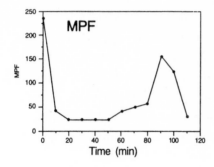

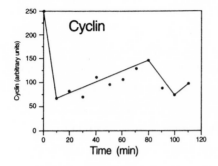

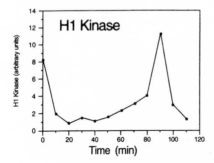

FIGURE 2.5 | The cellular-genomic level of a typical 90–120 minute ultradian rhythm (with permission from Murray et al., 1989). In this series of graphs it can be seen that a typical 90–120-minute ultradian rhythm is fundamental in cell growth and replication. The approximately 20-minute peak in maturation promoting factor (MPF), the protein cyclin and the enzyme H1 kinase act in concert to signal the final stage of genetic replication and cell division (mitosis). Some researchers believe this may be the basic ultradian pacemaker that sets all other levels such as the metabolic, neuroendocrinological, cognitive-behavioral and the socio-cultural as illustrated below.

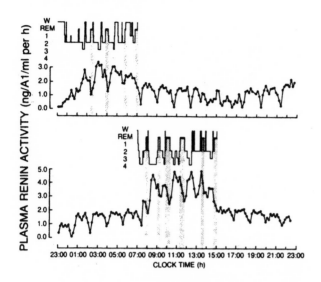

FIGURE 2.6 | The endocrine-behavioral level of a typical 90–120-minute ultradian rhythm (with permission from Brandenberger, 1992). An example of the interaction between the cognitive-behavioral and hormone levels. The typical 90–120-minute ultradian rhythms in the pulsate expression of the endocrine renin-angiotensin-aldosterone system have their higher amplitude peaks shifted to the right by an 8-hour delay of the sleep-wake cycle in the lower graph. The small inset on the upper part of each graph illustrates the relationship between plasma renin and the stages of waking consciousness, rapid eye movement sleep (REM dreaming state) and the four major levels of sleep depth.

ated at the cellular level. In Chapters 5 and 6 we will discuss, in greater detail, how this stress response at the cellular-genomic level is manifest at the behavioral and psychological levels of human experience.

A wide range of psychobiological rhythms that have their source in gene expression on all levels, from the cellular-genomic to the cognitive-behavioral and social, are illustrated in Figures. 2.5–2.10. All these examples of the creative replay of circadian and ultradian rhythms are responsive to psychosocial stress; they are all entrained and modulated by psychosocial cues (Lloyd & Rossi, 1992, 1993). Figure 2.10 (p. 58) illustrates complex adaptive rhythms in the ultradian healing response and self-hypnosis in a pilot study that now requires replication with larger groups of subjects.

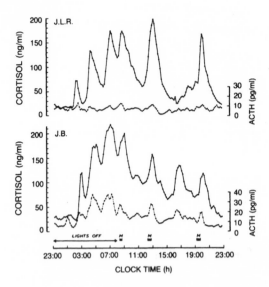

FIGURE 2.7 | The neuroendocrine level of a typical 90–120-minute ultradian rhythm (from Brandenburger, 1992). Two profiles of the individual ultradian rhythms in ACTH and Cortiol in two subjects when blood samples were taken at 10-minute intervals over a 24-hour period. While there are obvious differences, they both illustrate 90–120-minute pulsate rhythms with varying amplitudes. The lower profile, which is perhaps the more typical, illustrates how these two hormonal messenger molecules that mediate states of arousal tend to their highest peaks in the early morning hours and gradually dwindle in the afternoon and evening when our energy levels are lower.

The Creative Replay of Complex Adaptive Systems of Mind-body Communication and Healing

Any study of the history of science reveals that the dynamics of change over time are fundamental—as fundamental as Galileo's measurement of the time required for the movement of a ball down an inclined plane, to the development of calculus relating time and motion by Newton and Leibniz (Devlin, 1999). The nonlinear time parameters of the mind-body communication cycle, illustrated on the right side in Figure 2.1, are fundamental in the sense that they integrate the generative dynamics of the gene-protein cycle with the major pathways of physiology, psychobiology, and psychosocial genomics into a single complex adaptive system.

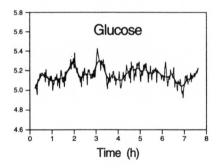

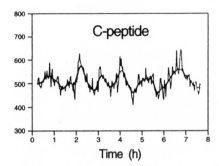

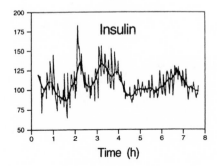

FIGURE 2.8 | The endocrine-energy-metabolic level of a typical 90–120-minute ultradian rhythm (with permission from Sturis et al., 1992). These three graphs illustrate how the blood glucose (top), C-peptide (middle), and insulin (bottom), obtained at 2-minute intervals during an 8-hour fasting period, have rapid 10–15-minute oscillations that are shown superimposed on the approximately 90–120-minute ultradian rhythms, which were obtained by a best-fit curve with a regression algorithm.

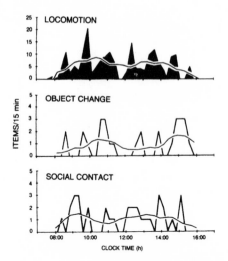

FIGURE 2.9 | The behavioral-social-cultural level of a typical 90–120-minute ultradian rhythm (with permission from Meier-koll, 1992). Time series illustrating the ultradian rhythms of the locomotion (about 80 minutes), object change (about 65 minutes), and social contacts (about 103 minutes) for the Indian child "Ram" in a naturalistic setting in a small village in India. Similar ultradian rhythms are found in a child of Ko Bushmen and social sychronization in a village community of Columbian Indians as well as free play in highly urbanized children in Germany.

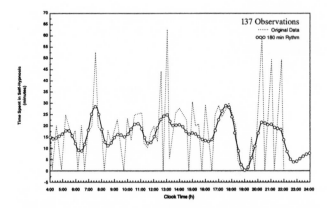

FIGURE 2.10 | The cognitive-behavioral level of a typical 90–120-minute ultradian rhythm (Rossi, 1992). An illustration of the approximately 180-minute ultradian rhythms (averaged data) of self-hypnosis that were recorded by 16 subjects in their diaries. Similar rhythms are found in the time series of subjects who were instructed to simply enjoy an "ultradian healing response" whenever they felt a need to throughout the day.

Figures 2.5–2.10 illustrate a half dozen interrelated life processes, from the cellular-genomic level to the psychosocial (including self-hypnosis). A superficial glance suggests these systems are all very different. A more careful examination, however, indicates that these apparent differences are due, in large part, to the fact that (1) the systems were sampled with different numbers of data points on (2) different time scales that (3) are not strictly regular in their periodicity but manifest nonlinear behavior that is sometimes called "quasi-periodic." Quasi-periodic behavior is found when two or more rhythms with different periods interact to produce oscillations that never repeat and replay themselves exactly. Many physical systems, from the two-coupled pendulum to most biological and brain processes, are described in the technical literature as quasi-periodic (Rossi, 1996a).

Quasi-periodicity is the signature of the nonlinear dynamics of communication in the *complex adaptive systems of life* that are replayed with creative variations on all levels, from the psychosocial to the cellular-genomic, which we describe here as *psychosocial genomics* (Lloyd & Rossi, 1992; Newtson, 1994; Vallacher & Nowak, 1994). What all psychobiological systems have in common is that they can be modeled by the nonlinear, quasi-periodic replays of circadian (about 24 hours) and ultradian (less than 20 hours) rhythms. The value of this mathematical perspective is that it enables us to create models of how biological and psychological processes modulate each other. What seems "mysterious" about the Cartesian gap between mind and body can be effectively modeled by the mathematics of cybernetic replay, even when it is difficult to understand in ordinary language.

Time and Complex Adaptive System of Mind-body Communication

Table 2.1 provides an overview of how nonlinear ultradian rhythms fit into the broader context of complex adaptive biological systems, from the milliseconds of enzymatic and neural action (Goldbeter, 1996) to the typical temporal dynamics of gene expression in Kleitman's (1963, 1969, 1970; Kleitman & Rossi, 1992) 90–120-minute basic rest–activity cycle (BRAC).

One of the most significant conclusions of these numerous studies is that the chronobiology of waking, sleeping, and dreaming does not operate with regular periodicity, like a clock (Lloyd & Rossi, 1990; Rossi, 1996a). *All psychobiological processes are manifest as the creative replay of complex rhythms of adap-*

TABLE 2.1 | EXAMPLES OF COMPLEX ADAPTIVE SYSTEMS OF SIGNIFICANT ULTRADI-
AN RHYTHMS REPLAYING CREATIVE HUMAN EXPERIENCE ON ALL LEVELS FROM THE
MOLECULAR–GENOMIC TO THE COGNITIVE–BEHAVIORAL.

Rhythm	Period	Reference
Neural transmission	0.01 sec.	Johnston et al., 1995
Cardiac (heart beat)	1.0 sec.	Stupfel, 1992
Respiration cycle	4–5 sec.	Brodsky, 1992
Immediate early gene activation	1–2 min.	Eriksson,1998; Lydic, 1998
Energy (glycolytic ATP cycle)	2–3 min.	Lloyd & Rossi, 1993
Typical biochemical cell oscillations	1–20 min.	Goldbeter, 1996
Human incubation stage of creative work	5–15 min.	Smith, 1996
LTP neural synapse induction initiation	10–30 min.	Lüscher et al., 2000
Human panic attacks	15–20 min.	NIH Report, 1991
Human nap time	15–20 min.	Stampi, 1992; Reuters, 2002
Typical self-hypnosis period	15–20 min.	Sanders, 1991
Epiphany experiences	20 min.	Yeats, 1934
Light-brain entrainment	20 min.	Barinaga, 2002
Creative and spiritual inspiration	20 min.	Smith & Tart, 1998
Personal journal therapy	20 min.	Pennebaker, 1997
Typical meditation period	20 min.	Deikman, 1980
Typical ultradian rest period	20 min.	Rossi & Nimmons, 1991
Typical bacterial cell cycle	20 min.	Rossi, 1992a
Psychosocial stimulus to mRNA	20 min.	Lloyd, 1992
Salivary IgA response	20 min.	Crabtree, 1989
Mother milk relaxation response	20 min.	Green & Green, 1987
Human cell cycle (mitotic cycle)	20 min.–1 day	Murray et al., 1989

Rhythm	Period	Reference
Enzyme and protein metabolism	20–90 min.	Brodsky, 1992
Basic rest–activity cycle (BRAC)	90–120 min.	Kleitman, 1969
Cerebral hemisphere cycle	90–120 min.	Werntz, 1981
Nasal breath cycle	90–120 min.	Shannahoff-Khalsa, 1991
Appetite and gastrointestinal	90–120 min.	Hoppenbrouwers, 1992
Memory and learning	90–120 min.	Bailey & Kandel, 1985
Dream rhythm, REM Sleep	90–120 min.	Aserinsky & Kleitman, 1953
Burn treatment by hypnosis	90–120 min.	Ewin, 1986
Immune cell DNA synthesis	90–120 min.	Crabtree, 1989
Hormonal, ACTH, cortisol, etc.	90–180 min.	Veldhuis, 1992
Sleep/wake endocrines	90–180 min.	Brandenberger, 1992
Human social rhythms	90–180 min.	Meier-Koll, 1992
Early response genes	1–2 hours	Incyte, 1999
Late response genes	2–48 hours	Incyte, 1999
Posttraumatic stress	2–8 hours	Wallace et al., 1998

tation to the varying contingencies of everyday life. Stress, motivations, demands, and expectations in normal living can shift the normal nonlinear ultradian and circadian dynamics of arousal, stress hormones, and their associated psychobiological processes on all levels, from the behavioral to the cellular-genomic (Lloyd & Rossi, 1992, 1993). The highly complex adaptive dynamics of all psychobiological and chronobiological processes are best described mathematically by the nonlinear dynamics of modern chaos and complex adaptive systems theory (Rossi, 1996a, 2001).

Solé & Goodwin (2000) have described the dynamics of chaos theory and complex adaptive systems in psychobiology and brain function in the following manner:

The observation that brain function is chaotic is puzzling. One might think that ordered dynamics would be the state of "normal" brain dynamics, but in fact, as we saw with the heart, *a shift to more ordered dynamics is typically associated with pathological states.* . . . Understanding the origins of neural complexity requires new model approaches to exploit the still unknown laws of brain dynamics. Why is the brain chaotic? Perhaps because the computational constraints involved in memory dynamics and perception require a large amount of flexibility. *Chaos provides dynamics that are at once ordered and innovative.* Some experiments suggest that in fact, our brains might be operating on the edge of instability. (p. 139, italics added)

Solé and Goodwin describe health as an emergent dynamic property of the whole organism. They note, for example, how healthy human subjects share the same nonlinear chaotic variability in their heartbeats that is known as a "self-similar" or fractal pattern. People with heart problems, by contrast, have less variability in the interbeat interval of their heart rates. *This loss of variability means that they have a reduction in their normal capacity for coping with the constant challenges of change in their inner and outer environments.* The easiest way to lose the natural pattern of Darwinian variation and selection that is needed to function as a complex adaptive system is for the heart (or brain or any other organ) to get locked into one frequency in preference to others that may be more adaptive at any particular moment.

In sleep apnea, for example, the heart becomes locked into a periodic breathing pattern that interferes with the normally adaptive variability of interbeat intervals. This leaves the heart more vulnerable to dysfunction when attempting to adapt to stress. Dysfunctions of the heart have been found to be intimately associated with human emotions in general and the experience of loneliness in particular (Lynch, 2000). From this perspective we can infer that *deterministic chaos* provides a mathematical model for the operation of the *optimal sensitivity of complex adaptive systems* in response to important signals from the psychosocial environment. Box 2.1 presents one such mathematical model of how we can conceptualize the emergent dynamics of psychosocial genomics as creative replays of complex adaptive systems of life on all levels, from gene expression (*genomics*) and protein synthesis (*proteomics*) to sleep, dream, and waking consciousness behavior (Rossi, 1996a).

BOX 2.1 | TOWARD MATHEMATICAL MODELS OF PSYCHOSOCIAL GENOMICS AS CREATIVE REPLAYS OF THE COMPLEX ADAPTIVE SYSTEMS OF MIND–BODY CIRCADIAN AND ULTRADIAN COMMUNICATION AND HEALING.

$$\frac{dX}{dt} = \frac{k_1}{(Z^n+1)} \, k_4X,$$

$$\frac{dY}{dt} = k_2X - k_5Y,$$

$$\frac{dZ}{dt} = k_3Y - k_6Z.$$

This set of differential equations could illustrate the emergent dynamics of a complex adaptive system engaging in gene expression (*genomics*), protein synthesis (*proteomics*), and the *behavioral levels* of psychosocial stress and healing (Goodwin, 1965; Rossi, 1996a; Ruoff & Rensing, 1996; Solé & Goodwin, 2000). There are many possible interpretations of such sets of equations to describe the multilevel dynamics of psychobiological systems from mind to gene. For example, X could represent the *genomics* of gene expression in the form of mRNA. Y could represent the translation from mRNA into a protein (*proteomics*) such as an enzyme, while Z could be a metabolite from the catalytic activity of Y or a more phenotypic *behavior*.

Described more generally, X could represent the *genomic level* (dotted line below) via the concentration of mRNA from gene transcription in response to extracellular signals that ultimately come from a need from the environment. Y could represent the *proteomic level* (dashed line below) that emerges from the translation of mRNA into the synthesis of proteins that function as the molecular machines of life on the levels of matter, energy, and information. Z could represent how the material, energetic, informational, and *behavioral* (solid line below) systems of life are emergent from the genomic and proteomic levels.

An analogous set of differential equations could model the emergent dynamics of a self-organizing cycle of complex adaptive behavior, such as the Gaia, gene, body, and mind system illustrated in Figure 1.2, or the gene, protein, hormone, SDMLB system proposed later in Chapter 5. The ultimate purpose of such mathematical models of mind-body healing is to explore and validate how psychosocial interventions on the cultural-behavioral-cognitive-spiritual levels could entrain, modulate, and facilitate the psychobiology of healing via psychotherapy, meditation, hypnosis, and therapeutic rituals.

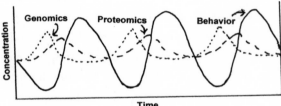

Solé and Goodwin (2000) describe the different influences to which the heart must adapt simultaneously, on many time scales, ranging from the *millisecond* intervals of neural impulses (which include emotional shifts); *seconds* for shifts in respiration; *minutes* for hormonal pulses; *hours* for the basic rest-activity cycle that is expressed in behavioral states such as waking, sleeping, and dreaming; *daily and monthly* hormonal sifts in the seasons of the year as well as others listed in Table 2.1. Here we will briefly summarize some of the psychobiological implications of the nonlinear temporal dynamics and wave nature of the gene expression and protein synthesis cycle that modulates human experience; more detail is forthcoming in the following chapters.

Psychosomatic Medicine, Stress Hormones, and Circadian Clocks

Excessive psychosocial stress can lead to changes in gene expression that result in the production of stress proteins that are associated with many malfunctions (Morimoto et al., 1990; Taché et al., 1989). Recently, for example, it has been found that the phase of circadian gene expression in tissues of the heart, liver, and kidney can be reset by stress hormones such as the glucocorticoids (Balsalobre et al., 2000). This is a major psychosomatic pathway by which chronic psychosocial stress can modulate gene expression, leading to the excessive production of stress proteins in the heart, liver, kidneys, and probably many other tissues and organs as well. The current challenge for research using DNA microarrays to investigate stress, healing, and gene expression, is to assess more directly the effects of psychotherapy, placebos, and the healing arts in modulating such stress–induced alterations in gene expression.

Psychoimmunology, Stress, and Healing

The research team of Ronald Glaser (1990, 1993) reported a series of experimental studies that clearly demonstrated how psychosocial stress can down-regulate gene expression, so that the immune system becomes compromised and more vulnerable to opportunistic infections. More recently, Castes et al. (1999) demonstrated the reverse: how emotionally supportive experiences of children in therapeutic groups optimized the functioning of their immune systems and, by implication, facilitated the up-regulation of gene expression. Details of research in these and other areas of psychoneuroimmunology associated with psychosocial stress and healing are presented in Chapter 5, which explores the possible role of gene expression in therapeutic hypnosis and the healing arts.

Socialization Can Modulate Gene Expression

An elegant example of the interconnectedness of environmental stimuli, socialization, gene expression, and behavior was reported recently by entomologists Toma et al. (2000). When bees start to forage, they must synchronize their internal clocks with the cycle of daylight and peak nectar-producing periods in flowers. It was found that the clock gene named "period" (per) turned on, at appropriate times of the day, to generate higher levels of per mRNA and protein synthesis in the brains of worker bees. When the younger bees were isolated to prevent touch and socialization with older, mature workers, it was found that expression of the gene per mRNA changed. Analogous dynamics between gene expression and human touch experience (Schanberg, 1995), together with their implications for psychiatry, have been described as the role of social experience in transforming the genotype into the phenotype (Eisenberg, 2000).

Novelty, Stress, Neurogenesis, and Brain Growth

Current research on how gene expression and protein synthesis (stage three of Figure 2.1) lead to neurogenesis in a wide range of responses to stress and drug-induced neural plasticity has been reviewed, in a very revealing manner, by Duman and Neslter (2000). Details of the experimental conditions, such as the experience of novelty and environmental enrichment, that can facilitate the growth and development of new neurons associated with new memory and learning (Eriksson et al., 1998; Gage, 2000a, 2000b) will be reviewed in the next chapter.

Creative Replays of Gene Expression and Neurogenesis During Dreaming

Recent research (Ribeiro et al., 1999) has documented how a day rich in memorable experiences leads to creative replays of gene expression, protein synthesis, and neurogenesis (literally, brain growth) during dreaming (REM sleep). Details of this research are presented in Chapter 5, along with practical implications for developing a variety of new approaches to psychotherapy and the healing arts. We will now turn our attention to how the gene expression and protein synthesis cycle can modulate human experience at the psychological level via state-dependent memory, learning, and behavior.

STAGE FOUR:
MESSENGER MOLECULES MEDIATING MEANING AND STATE-DEPENDENT MEMORY, LEARNING, AND BEHAVIOR

Stage four of Figure 2.1 illustrates how the messenger molecules that have their origins in the processing of the larger protein "mother-molecules" (in stage three) are stored within the cells of the brain and body as a kind of "molecular memory." These molecular messengers are released into the bloodstream, where they can complete the complex cybernetic loop of information transduction by passing through the "blood-brain barrier" (Davson & Segal, 1996) to modulate the brain's neural networks (illustrated by the block of letters A–L at the top of Figure 2.1). These localized neuronal networks are modulated by a complex field of messenger molecules that can reach the limbic-hypothalamic-pituitary system as well as certain areas of the cerebral cortex. This completes the nonlinear loop of information transduction and communication via the messenger molecules of the psychosomatic network between mind and body.

Since the neuronal networks of the brain are modulated by changes in the strengths of synaptic connections, we could say that *meaning* is continually modulated by the complex, dynamic field of messenger molecules that continually replay, reframe, and resynthesize neuronal networks in ever-changing patterns. Most of the sexual and stress hormones that have been adequately tested, for example, have state-dependent effects on our cognitive and emotional states. Freeman (1995) has pointed out how oxytocin, a hormonal messenger molecule released during lactation and sexual arousal, encodes state-dependent memory and learning. Oxytocin in concert with other messenger molecules in the extracellular fluid (*ECF* at the top of Figure 2.1) of the brain make up the constantly changing field of meaning expressed in the ever-emergent phenomenological experience of "mind" and behavior.

Research indicates that most forms of learning (Pavlovian, Skinnerian, imprinting, sensitization, etc.) are now known to involve these hormonal messenger molecules in the body that can reach the brain to modulate the neural networks that encode mind, memory, learning, and behavior. Insofar as these classical forms of learning use messenger molecules, they *ipso facto* have a state-dependent component (Erickson, Rossi, & Ryan, 1992; Rossi, 1986/1993, 1987, 1990a, 1990b, 1996a; Rossi & Cheek, 1988).

It is important to reiterate what state-dependent memory, learning, and behavior (SDMLB) actually entail. When subjects are given memory and learning tasks while under the influence of stress hormones (such as ACTH, epinephrine, sex hormones, or even psychoactive drugs that mimic these natural hormonal messenger molecules), there is a varying degree of amnesia for these tasks once the stress hormone or drug has been metabolized out of the system. That is, when memory and learning are encoded under conditions of high emotional arousal, sex, stress, or trauma, they tend to become state-dependent (or state-bound) to that psychobiological condition. Memory and learning are dependent on the original psychophysiological state when they were first encoded. State-dependent memory and learning become dissociated or apparently "lost" after the person recovers and the stress or sexual hormones have been metabolized and returned to normal levels. Reactivating stress or sex in another context, however, has a tendency to reestablish the original encoding condition, and with it, the emotions and varying degrees of memory of the task, experience, or trauma. This is the psychobiological basis of much psychopathology related to early sexual and stressful life events, as originally discovered by the early pioneers of therapeutic hypnosis and classical psychoanalysis more than a century ago (Edmonston, 1986; Zilboorg & Henry, 1941). SDMLB is a focal concept in completing the psychobiological loop of information transduction between molecules and mind that will be utilized throughout this book. State-dependent memory, learning, and behavior brings the so-called "mystery" of the connection between mind and body into the arena of scientific research (outlined in Part I) and the psychotherapeutic arts (outlined in Part II).

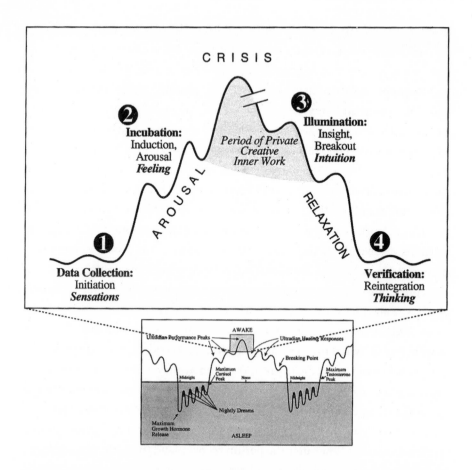

FIGURE 2.11 | The four-stage creative process in psychobiologically oriented psychothera-py. The lower diagram summarizes the alternating 90–120-minute ultradian rhythms of wak-ing and sleeping for an entire day in a simplified manner. The ascending peaks of rapid eye movement (REM) sleep characteristic of nightly dreams every 90–120 minutes or so are illus-trated along with the more variable ultradian rhythms of activity, adaptation, and rest in the daytime. This lower figure also illustrates how many hormonal messenger molecules of the endocrine system such, as *growth hormone*, the activating and stress hormone *cortisol* and the sexual hormone *testosterone*, have typical circadian peaks at different times of the 24-hour cycle.

The upper diagram outlines the basic psychobiological unit of psychotherapy as the cre-ative utilization of one of the natural 90–120-minute ultradian rhythms of arousal and relax-ation illustrated in the lower diagram. The classical four stages of the creative process—(1) data collection, (2) incubation, (3) illumination, (4) verification as documented by Wallas (1926) are described in detail in Chapter 6. The four basic psychological functions of sensations, feeling, intuition, and thinking originally described by Jung (1923) are discussed and illustrated throughout Part II.

A New Map of the Psychobiology
of Consciousness and Behavior

Current explorations of the nonlinear dynamics of the chronobiology of consciousness, sleep, and dreams is leading to an integration of all previous work on the creative dynamics of psychotherapy and healing in both theory and practice. An overview of this chronobiological approach is illustrated in Figure 2.11. When the 90–120-minute ultradian rhythms of mind-body communication are replayed over the daily circadian cycle, they can be viewed as graphs of the alternating rhythms of activity and rest (illustrated on the lower part of Figure 2.11). A new map of the psychobiological foundations of consciousness is presented in the lower part of Figure 2.11, which summarizes the alternating ultradian rhythms of the awake, sleep, and dreaming states of an entire day in a simplified manner. The ascending peaks of rapid eye movement (REM) sleep characteristic of nightly dreams every 90–120 minutes or so are illustrated along with the more variable ultradian replays of activity, adaptation, and rest in the daytime. The lower part of Figure 2.11 also illustrates how many hormonal messenger molecules of the endocrine system (such as *growth hormone*, the activating and stress hormone *cortisol*, and the sexual hormone *testosterone*) have a typical circadian peak at different times of the 24-hour cycle. Because the nonlinear chronobiological release of many of these hormones is recognized as having profound state-dependent effects on memory, learning, emotions, and behavior throughout the day, it is important to consider their relevance for a psychobiological model of psychotherapy and the healing arts.

The Psychobiological Foundation
of Psychotherapy and the Healing Arts

The upper part of Figure 2.11 illustrates the hypothesis that the natural unit of psychobiologically oriented psychotherapy is the utilization of one of the 90–120-minute ultradian cycles of activity and rest illustrated in the lower part (Lloyd & Rossi, 1992; Rossi, 1986a, 1996a, 2000c, 2000d). In support of this hypothesis, we can cite endocrinal research by Iranmanesh et al. (1989), Felker and Hubbard (1998), and others. These researchers document how the ultradian peaks of cortisol secretion that lead to psychophysiological states of arousal every 90–120 minutes or so throughout the day (labeled "Ultradian Performance Peaks" in the lower part of Figure 2.11) are typically followed, after about 20 minutes, by ultradian peaks of beta-endorphin that lead to rest and relaxation. The natural periods of rest and relaxation every hour and a half or so are labeled "Ultradian Healing Responses" in the lower part of Figure 2.11 (Rossi, 1982, 1986a; Rossi & Nimmons, 1991). It appears as if nature has built-in natural but flexible (highly adaptive) psychobiological work cycles of activity, rest, and healing every 90–120 minutes throughout the 24-hour day. A detailed discussion of the four stages of the creative process illustrated in the upper portion of Figure 2.11 is presented in Chapter 6.

What, exactly, is the "work" that is done in each 90–120-minute ultradian cycle? The essence of such psychobiological work is a replay of the gene expression and protein synthesis cycle that initiates a creative response to changing environmental conditions in embryonic development (Palmeirin et al., 1997) as well as to the conditions of stress and healing throughout the life cycle. Previously cited research by Wen et al. (1998; Figure 1.3), describing four fundamental waves of gene expression during the embryonic development of the central nervous system, may be a prototype of the four stages of the creative process that occur throughout the lifetime. Replays in the cycles of gene expression and protein synthesis underlying the fundamental dynamics of healing at the cellular level have been described in detail by Todorov (1990).

The similarity in the time parameters of psychotherapeutic sessions and the chronobiological dynamics of gene expression leads to a new hypothesis about the role of suggestion in psychotherapy, therapeutic hypnosis, and the healing arts (Rossi, 1981, 1982, 1986a, 1996a). *From this perspective, what has been traditionally called "therapeutic suggestion" may be, in essence, the accessing, entrainment, and utilization of ultradian/circadian replays of mind-body communication on all levels, from the cellular-genomic to the behavioral, that are responsive to psychosocial cues* (Rossi, 1982, 1987, 1989b, 1996a). The classical phenomena of therapeutic hypnosis may be conceptualized as extreme manifestations and/or preservations of state-dependent psychobiological processes that are responsive to psychosocial cues (described in greater detail in Chapter 5).

THE THERAPEUTIC UTILIZATION OF
BEHAVIORAL STATE-RELATED GENE EXPRESSION

In this section we integrate the creative dynamics of self-organization and adaptive complexity theory with new psychobiological approaches to the healing arts in what has been called "the symptom path to enlightenment" (Rossi, 1996a). The case studies presented here have all been updated from Rossi, Lippincott, and Bessette (1994, 1995) to illustrate how we can utilize our new psychobiological awareness of behavioral state-related gene expression and nonlinear dynamics with a variety of clinical problems. These cases were selected to highlight the types of family processes, psychodynamic issues, and performance demands that may lead to a disruption of the normal "wave nature" of being. Repeated disruptions of this basic psychobiological process generate stress and consequent psychosomatic problems.

One caveat: These abbreviated case examples should not be taken to imply that the psychobiological aspects of circadian and ultradian behavioral state-related gene expression are more important than traditional psychodynamic and interpersonal aspects. To the contrary, the ultradian healing response is probably a period during which many traditional approaches to the resolution of clinical problems can be optimized. Indeed, the natural psychobiological replays of ultradian cycles is hypothesized to be a missing link between the social-psychological and the biological levels of gene expression, neurogenesis, and healing in the traditional approaches to psychotherapy.

The case histories presented here are of heuristic value only; they suggest how the ultradian/circadian dynamics of behavioral state-related gene expression can be used as a "window of opportunity" for accessing and resolving many mind-body problems. We freely speculate about the possible role of gene expression and a variety of hormonal messenger molecules that may be involved at various critical phase transitions of consciousness, mood, and memory in psychotherapy. These cases also suggest the types of mind-body research that need to be done at the level of gene expression (Christopher et al., 1997). At the present time, the most obvious application of the ultradian/circadian dynamics of gene expression is to those stress syndromes wherein the presenting problem involves a clear desynchronization of the patient's typical psychobiological rhythms of activity and rest.

CASE 1
Jane: Narcolepsy, Catalepsy, and Depression

Presenting Problem

"Jane" is a 36-year-old woman who was referred by her family physician because of depression brought on by her current divorce. Jane also was suffering from narcolepsy and cataplexy and was receiving medical treatment from a nearby university sleep disorder clinic. Extensive research has documented the complex adaptive system that engages the dynamics of gene expression, neurotransmitters, and circadian factors in narcolepsy and cataplepsy (Okura et al., 1999). Jane was prescribed a high daily dose of Ritalin to keep her awake while driving. For the first year of treatment at the clinic, she and her therapist focused on the depression associated with her divorce. During the second year, deep-seated issues related to growing up in a dysfunctional alcoholic home became the focus of treatment. Jane's *DSM-R* diagnosis at the clinic was as follows:

> Axis I: 296.23 Major Depression with Melancholia
> Axis II: 301.81 Narcissistic Personality Disorder
> Axis III: Narcolepsy and Cataplexy

Narcolepsy is a psychobiological disorder characterized by abnormal REM patterns, cataplexy, sleep paralysis, and hypnogogic hallucinations. Observations in sleep laboratories show that narcoleptics experience early REM periods within the first 15 minutes of sleep. Cataplexy, by contrast, is a brief, sudden loss of skeletal muscle tone that is often observed after the person has had an emotional reaction such as laughter, anger, or excitement. The cataleptic remains aware of the surroundings during the episode, which differentiates this condition from epilepsy. About 60% of narcoleptics experience cataplexy. Narcoleptics usually begin to experience periods of excessive daytime sleepiness in their early teens or twenties. It is currently thought that this disorder is due to an imbalance in the psychobiological periodicity of waking, REM, and REM sleep. Spouses, friends, and employers may misinterpret the narcoleptic's frequent naps and sleepiness during the daytime as lethargy, sloth, and lack of motivation.

Treatment

After being referred from the clinic, Jane was interested in exploring the thoughts and feelings of the past and present that seemed to be associated with her narcolepsy and cataplexy. Since she was interested in trying meditation to clear her mind, Jane was given an opportunity to practice "a comfortable, natural, hypnotic meditative trance" during the first therapy session, while her inner mind reviewed and dealt with whatever issues came up all by themselves.

After a quiet period of *about fifteen minutes* of apparently rapt inner focus, Jane stirred a bit. This was taken as a natural behavioral indication that she would be awakening soon, so she was asked, **"Will your eyes open as you awaken, when your inner mind knows you have done as much healing as is appropriate at this time?"**

Notice how this question is an implied directive (Erickson et al., 1976; Rossi, 1996a), or what we call an *implicit processing heuristic* in Chapter 9. *Implicit processing heuristics are highly permissive and nondirective approaches to facilitating a patient's psychodynamics on an implicit or more unconscious level.* They allow patients to experience and explore the natural process of Darwinian variation and selection that takes place in the creative replays of their inner life. The implicit processing heuristic typically presents the patient with a problem that neither the patient nor the therapist may know how to solve. The implicit processing heuristic provides the patient with an opportunity to solve a problem on an unconscious level and then give an observable behavioral signal to indicate when the problem has been solved (Rossi, 1996a).

An evaluation of this initial experience, using the Indirect Trance Assessment Scale (Rossi, 1986b), suggested that Jane had easy access to a very rich inner life that could be used for exploring and utilizing her own inner resources for problem-solving. After she awakened, for example, she spontaneously reported feeling "refreshed, alive, and energized . . . even the colors in the room now seemed brighter." She wondered if this "energizing experience could be used to support the effects of Ritalin at those times of the day when the drug seems to wear out?"

Jane was encouraged to explore this intuition by reviewing whatever possible patterns might be evident in her symptoms in the course of her typical day. In the context of this review, we discussed the normal ultradian variations in consciousness and energy level that most people experience throughout the day. She quickly developed an intuitive appreciation of how she was "pushing herself into further stress" when she did not let herself have these normal breaks because of her fears that any feeling of tiredness or depression meant that she was falling into narcolepsy. Jane's perceptive intelligence led her to wonder whether some of her fatigue, depression, and even the narcolepsy itself could be manifestations of her personal "ultradian stress syndrome" (Rossi & Nimmons, 1991). She was intrigued by the therapist's suggestion that it might be interesting to explore the possibility that her ultradian stress syndrome could be transformed into an "ultradian healing response" by letting herself experience "a comfortable meditative trance" several times a day whenever she felt her energy level waning (Rossi & Nimmons, 1991).

The following week Jane reported some success with potentiating her "consciousness and energy levels" by enjoying an ultradian healing response several times a day when she felt herself getting tired. Most important, she said, was the relief she felt now that she no longer had to fear that taking such breaks meant that she and her medicine were failing. After a month of continuing success, in cooperation with her medical doctor's advice, she was allowed to substitute a 20-minute ultradian healing response in the mid-afternoon for her usual 10 mg dose of Ritalin.

Two months later, with her medical doctor's permission, she substituted an ultradian healing period for another 10 mg dose of Ritalin after lunch. She reported: "This really works for me—I'm more productive now because my mind is clear and alert throughout the workday." When she was indirectly queried about what was clear in her mind, she discussed a number of personal issues and relationships that she was handling better since she was "taking time out to mull them over during ultradian healing periods."

The final stage of her therapy, when she stopped taking any Ritalin, came a few months later when she added another ultradian break at about 10:00 A.M. This first healing period of the day "rejuvenates" her and allows her to keep her energy up until the after-lunch ultradian break. When she occasionally forgets to give herself an ultradian healing response, she finds she can use the first mild signs of waning enthusiasm and tiredness to remind herself that it is time to take a break. It is possible that a participating factor in Jane's depression, narcolepsy, and cataplexy was the stressful overriding of her natural ultradian rhythms of rest and rejuvenation during the several years when she was unable to resolve the many familial and psychodynamic issues that she was trying to cover up with obsessive concern about her "productivity."

Jane's clinical progress is consistent with the view that, for some patients, practicing the ultradian healing response when they experience chronic stress may facilitate a therapeutic synchronization of waking and sleeping patterns of behavior (Dement, 1972; Kleitman, 1970). That the resolution of Jane's clinical depression was, at least in part, related to this ultradian/circadian synchronization is consistent with the research literature on biological rhythms and mental disorders (Kupfer et al., 1988).

At least two important mind–body questions are raised by this case: (1) Which psychobiological syndromes can be ameliorated by using the ultradian healing response to support and even replace medication? (2) To what extent can the early, mild symptoms of stress disorders be used as a signal to initiate an ultradian healing response to thereby avert the course of more severe ultradian stress syndromes (Rossi, 1990d; Rossi & Cheek, 1988; Rossi & Ryan, 1986)? An extensive scientific literature already documents how administering drugs at appropriate phases of ultradian and circadian rhythms can reduce dosage levels for many clinical conditions, including asthma, cancer, and depression (Klevecz, 1988; Klevecz & Braly, 1987). The case of Jane presented here, however, may be the first report of the use of the ultradian healing response to completely replace the need for a drug.

We strongly suspect that Jane's personal belief system in the efficacy of meditation, her extraordinary sensitivity to her inner states, and her lively interest in "converting my ultradian stress syndrome into an ultradian healing response" were contributing factors to her success. We speculate that the utilization of this interest in her inner states enabled her to go through a critical phase transition in converting the symptoms of stress into signals that she needed a period of rest and recovery. From the psychobiological perspective, Jane learned to practice *low-phase hypnosis* (discussed in Chapter 5), "whenever she felt herself getting tired." She was doing very important inner work in relation to herself and others during this period, however, as was indicated by her report of "taking time out to mull things over." Apparently useful therapeutic insights were more accessible to her during the ultradian healing response. The appropriateness of using the ultradian approach with many different types of personality and life situations (even when psychodynamic insights are not easily available) is illustrated in a number of the following cases.

CASE 2
Millie: Insomnia, Clock Genes, and Ultradian Synchronization

Presenting Problem

"Millie" is a 68-year-old woman who was referred by her internist for treatment of persistent insomnia. Current research on insomnia documents how the loss of psychobiological synchronization via clock gene expression may be disruptive to circadian rhythms that are at the source of a variety of general health problems (Seppa, 2001; Winstead, 2001). Millie had a sleep onset latency of 60–90 minutes almost every night. For the past nine months, she had slept for less than four and a half hours per night, and daytime fatigue had led to depression and irritability. This mood disturbance had become so severe that it was affecting her marriage of 38 years, and there was potential for divorce.

Although it became apparent during the initial stages of her psychotherapy that there was a connection between Millie's mood disturbance and her marital problems, her husband was unresponsive to insight therapy and just kept telling her to "try harder to relax" and "it's all in the mind." It got so bad that he accused her of getting more sleep than she claimed, and he began to sleep in another bedroom. She had been prescribed benzodiazepine drugs, the antidepressant Desyrel, and the barbiturate Seconal, but she did not like the side effects of these medications.

Now that both she and her husband were retired, Millie felt she "had nothing important to do since the kids are gone." She began lying in bed until late morning. She also stopped eating, meeting friends, shopping, and walking, which she had formerly enjoyed. During the first session, Millie talked about fears of her husband having a heart attack, their home being burglarized, or her and her husband dying in a fire.

Treatment

Our first impression is that Millie is facing a crisis, an obvious *critical phase transition* in her life, having to do with retirement and a new relationship that was now required with her husband. To facilitate an initial phase of rapport in therapy and to reassure Millie that her worldview was fully accepted, we had her contact the police and fire inspectors to check the locks and smoke detectors in her home. She was taught a progressive approach to relaxation to help reduce anxiety during the day, but it did not help with her insomnia. Millie was told of the sleep research that documented how older individuals sleep less due to a decrease in delta stage sleep (deep sleep) and an increase in light sleep, which makes them easier to awaken.

Millie was taught to go into a natural ultradian healing response in an early therapy session with these words. **"When your eyes close all by themselves in a moment or two, will it be a signal that your inner mind is ready to help solve your sleep problem? I wonder if you can recall a time when you slept very well? And will your unconscious help you go comfortably to sleep before you even realize it? When you feel like waking up and stretching, will it be a signal that your inner mind has learned something useful to help you find comfort and healing when you lay down to rest?"** Notice how all of these permissive and open-ended questions function as implicit processing heuristics that utilize common psychobiological associations between sleep, rest, and healing to facilitate an the ultradian healing response.

After 13 minutes, Millie spontaneously opened her eyes, stretched, and talked of her inner experiences: vague, fragmented thoughts about her husband, seeing colors with no form, and being very relaxed. Millie was assured that everyone normally experiences 90-minute basic rest–activity cycles throughout the day, during which healing as well as solutions to problems occur in a natural manner. She was told to continue her usual schedule but to allow herself to enjoy an ultradian healing response whenever she felt a need for one. It was emphasized that her mind and body could use this time to solve "their own problems in their own way." Problems with marriage and family, for example, sometimes could be "sorted out" in ways that were surprising, or even unknown, to the conscious mind.

The next week she reported much less irritability and "a lot more energy throughout the day, but I only slept a little better." She said that her ultradian healing periods took 20–30 minutes, and she noticed that she seemed to have "a lot going on inside during these times." She was encouraged to discuss whatever relevant ideas and fantasies were going on during her ultradian rest-rejuvenation periods. She reviewed a number of family problems and then summarized them by saying it was important to "live and let live in order to get along in this world." The depth of this insight certainly was not profound from a traditional psychodynamic point of view, but it apparently was enough so that she could recover her emotional equilibrium and sense of well-being.

For the next week, she was instructed to keep her bedtime the same but to get up at 8:00 A.M. instead of later in the morning, and to keep up the ultradian healing breaks when it felt right. She reported that this routine gave her energy during the day so that she was not irritable, but she still had trouble getting to sleep. She stated: "The energy is so good I have my husband doing ultradian healing too. It helps pace our day, and we have been eating regular meals and bike-riding together again." This was a hint that although her husband still remained refractory to psychological insight, he was amenable to changes that synchronized their activities and improved the quality of their relationship. She reported that her husband now feels rested enough to enjoy baby-sitting their grandchildren with her two evenings per week, while their son and daughter-in-law attend night-school courses. That is, by enjoying comfortable, guilt-free, low-phase ultradian healing responses throughout the day, they were both able to enjoy better ultradian phases of peak activities and family relatedness as well.

It is very important that senior citizens in retirement have adequate rest so that they can more fully participate in the 20-minute peaks of the basic rest-activity cycle. Getting optimal exercise in the daytime, such as bicycle riding for this couple, facilitates better sleep at night. Ideally, sufficient exercise leads to deeper sleep and the release of growth hormone (during the first 90–120 minutes of sleep) that, in turn, tends to optimize the immune system as well as a host of other natural healing functions.

One of the more significant unresolved issues of this case involves the patient's communication problems with her husband, which had become acute due to their spending more time together after retirement. There is at least a hint in this case, however, that their *learning how to synchronize their mind-body states by eating, resting, bike-riding, and baby-sitting together may lead to further rapport on an emotional-cognitive level.* This view is supported by empirical studies that document the importance of the synchronization of ultradian and circadian rhythms—for harmonious interaction between couples (Rossi, 1996a; Rossi & Nimmons, 1991) and the social process in general (Dement, 1992; Moore-Ede et al., 1982)—that have their sources in the dynamics of gene expression (Singer & Ryff, 2001).

Addictions: Alcohol, Sugar, Cocaine, Marijuana, Caffeine, Nicotine

The identification of psychobiological dynamics of gene expression and their modulation of neurogenesis and the long-term plasticity (changeability in response to environmental signals) of the brain is leading to generating a new view of how mind and body are modulated by addictions (Nestler, 2001). The relationship between stress and addiction (Gottheil et al., 1987) has led us to explore the utilization of ultradian dynamics in treating a variety of chemical dependencies and their associated behavioral and psychosomatic dysfunctions. It has been proposed (Rossi, 1990b; Rossi & Nimmons, 1991) that when stress-related disruptions of ultradian rhythms lead to an experience of excessive fatigue, many people will attempt to override their need for rest by taking an "upper" (caffeine, nicotine, cocaine, etc.). When overly stimulated by those natural messenger molecules that mediate heightened activation (the stress-associated hormones such as ACTH, cortisol, and epinephrine), people will attempt self-medication with "downers" (alcohol, marijuana, opiates, etc.). Such artificial methods of chemical self-regulation ultimately fail, however, and leave a tell-tale trail of many associated habit, behavioral, and psychosomatic disorders. The following cases illustrate how a variety of ultradian approaches can facilitate recovery.

CASE 3
James: Reframing Smoking and Stress into Ultradian Healing

Presenting Problem
"James" was self-referred because he wanted to use hypnosis to quit smoking. James felt he was constantly under stress working as a high-school teacher in a school where violence, drug abuse, and suicide were almost daily occurrences. His wife was pregnant with their first child, and he wanted to quit smoking before the baby was born. The relationships between gene expression, smoking, and stress are well documented (Siegfried et al., 2000).

Treatment
James was administered Spiegels' Hypnotic Induction Profile (Spiegle & Spiegel, 1978), on which he scored in the highly hypnotizable zone. Even in his first session, he appeared to experience a deep trance that was ratified by

arm levitation, amnesia, and other behavioral cues. Direct suggestions that utilized his desire to be a healthy role model for his child were enough to motivate him to stop smoking in one session. By utilizing his desire to be a healthy role model, the therapist recognized how the anticipated birth of his child was a *critical phase transition* in his life that could generate creative adaptive responses.

Three weeks later James called and reported that he had not smoked, but that *"several times a day"* he noted having an intense craving for a cigarette. He resisted the impulse but didn't know how long he could hold out. An immediate follow-up appointment was made. During the session, we talked about the smoking problem for half an hour. Then James yawned, his eyes teared, and he stated, "I really need a cigarette, I really want one right now. I have a lot of stress and need to smoke." This synchronization between the typical behavioral cues for ultradian rest-rejuvenation (yawning, tearing), the experience of stress, and the symptomatic need to smoke was recognized by the therapist as an opportunity to convert an "ultradian stress syndrome" into an "ultradian healing response" (Rossi & Nimmons, 1991).

From a psychobiological perspective, his yawning suggested that he was beginning to experience what Milton Erickson called "the common everyday trance" (Erickson, Rossi, & Rossi, 1976; Erickson & Rossi, 1979). James's ongoing experience was utilized to facilitate a permissive naturalistic trance induction with the following words: **"That's okay, just let yourself take a comfortable break right now. That's right, just allow yourself to sit back and be with the craving and enjoy not having to do anything right now about your stress. When your inner mind is ready to let the relaxation deepen so that it can relieve the stress and craving while you rest, will you notice your eyes closing naturally all by themselves?"**

James's eyes closed immediately and remained closed for about 12 minutes, during which he appeared to be in rapt inner focus; his body remained absolutely still. When he finally stirred a bit, giving a little sigh, he was asked: **"When your conscious and unconscious minds know they can continue to allow you to enjoy periods of rest and comfort like this *several times a day* so you don't have to experience stress and smoking, will you find yourself awakening refreshed and alert?"** About three to four minutes later, James's eyes opened and he stretched. He reported visualizing himself holding his baby and feeling proud that he did not smoke. He stated that his craving for smoking was gone.

James was instructed to let the craving to smoke *several times a day* become a signal that he needed an ultradian break; whenever he felt a craving, he was to take the next available break at school to sit down and explore the deeper and more satisfying comfort of an ultradian healing rest. Two weeks later James reported that the urge to smoke had been converted into comfortable ultradian breaks *"several times a day,"* during which he enjoyed positive fantasies. The school brunch break was between 10:20 and 10:40, and he had developed a new habit of using it to enjoy an ultradian healing break. This also prevented him from eating the two or three doughnuts and drinking the cup of coffee he usually had during the break—which helped him avoid the weight gain that is feared when one stops smoking.

By reframing his need to smoke into a need to do an ultradian healing response, he was able to reduce stress and give up his smoking addiction. An important clue to the appropriateness of this ultradian reframe was his exact way of phrasing how he needed to smoke *"several times a day."* While this phrase can be understood as being merely a common cliché from a linguistic point of view, it acquires deeper significance from a chronobiological perspective, where it often serves to identify an ultradian rhythm of behavior.

This brief example illustrates the integration of at least four approaches to the ultradian psychobiological model proposed in this chapter:

1. The utilization of the patient's exact words, behavior, and worldview (Erickson, 1958/1980);
2. The casual and permissive utilization of the patient's ongoing behavior to initiate an indirect, permissive, naturalistic hypnotic induction (Erickson, 1959/1980; Erickson & Rossi, 1979);
3. The reframing of symptoms of stress and converting them into signals for therapeutic self-care (Erickson & Rossi, 1989; Rossi, 1996a);
4. Associating a posthypnotic suggestion with a future behavioral inevitability (Erickson & Erickson, 1941/1980; Erickson & Rossi, 1976; Rossi, 1982).

The posthypnotic suggestion to enjoy a comforting ultradian healing response several times a day, instead of smoking, was associated with the behavioral inevitability that he would experience periods of stress several times a day. Taken together, these four permissive approaches work particularly well as a mini-model for facilitating many types of critical phase transitions associated with addiction problems.

CEREBRAL HEMISPHERE DOMINANCE
AND ULTRADIAN RHYTHMS

One of the most interesting mind-body ultradian relationships is the contralateral shift that takes place between cerebral hemispheric dominance and nasal breathing every 90–120 minutes (Werntz et al., 1983). In general, when the right nostril is more open (right nasal dominance) so that air enters the nose primarily on that side, the left-hemispheric functions manifest more dominance. When the left nostril is more open, the right-hemispheric functions appear to be more evident (Klein & Armitage, 1979, Klein et al., 1986). A number of researchers are now accumulating experimental data that support the view that this may be the psychobiological process that has been used for centuries as a yogic method of breath-brain-autonomic nervous system regulation (Kennedy et al., 1986; Rama et al., 1976).

It is possible to shift the nostril through which more air is entering the nose by either blocking one nostril manually or simply by lying down on one side or the other (Rao & Potdar, 1970; Werntz et al., 1987). Lying down on the right side, for example, will open the left nostril within a few minutes for most individuals, reflexively activating right-hemispheric functions and sympathetic system dominance on that side of the body. In this way it is possible to gain some control over the autonomic nervous system and associated mental states (Shannahoff-Khalsa, 1991; Werntz et al., 1981, 1983, 1987). The associations found between hypnosis and laterality of cerebral hemispheric dominance (Levine, Kurtz, & Lauter, 1984; Rossi, 1982, 1986a, 1990c) suggest that naturalistic hypnotic states may be facilitated by allowing patients to explore and creatively replay and reframe their personal experiences of this "brain-breath-autonomic nervous system link" (Rossi, 1986a; Rossi & Cheek, 1988).

CASE 4
Rex: Alcoholism and the Ultradian Brain-Breath Link

Presenting Problem
"Rex" is a 44-year-old construction worker who has been involved in long-term treatment for alcoholism. After going through a traditional 30-day inpatient treatment program, he was referred for ongoing alcoholism counseling.

From the beginning, he complained of cravings for alcohol *"throughout the day but especially after work."* Because Rex's psychodynamically oriented therapy was very complicated, only the specific ultradian intervention for alcohol cravings will be presented here.

Recent research has documented the important role of genetics and immediate early genes (IEGs) in determing the effects of alcohol on the central nervous system (CNS) (Reilly et al., 2001). As reviewed earlier in this chapter, IEGs are expressed within a minute or two in the nucleus of cells in the CNS in stage two of the complex adaptive system of mind-body communication (Figure 2.1). It is now known that high doses of acute ethanol administration are able to suppress stress-induced c-fos mRNA expression as well as immunoreactivity in the hippocampus (locus of learning) and piriform cortex of the rat (Ryabinin et al., 1995). Lower doses of acute ethanol decreased stress-induced c-fos expression in the cerebral cortex, the hippocampus, and the hypothalamus (a central locus of hormonal psychoendocrine modulation of mind-body communication). On the other hand, alcohol (ethanol) increased the expression of c-fos mRNA in the central nucleus of the amygdala (locus for stress, fear, and trauma-encoding in the brain). After experiencing stress, acute alcohol administration suppresses the gene expression of zif-268 and fos-B in these animal models of how alcohol may relieve experiences of environmental stress. In Chapter 4 we will review research on the expression of zif-268 (which facilitates neurogenesis in the encoding and updating of experiences) during dreams (REM sleep) after a day of salient life experiences.

It is interesting to note that the famous Swiss psychiatrist C. G. Jung is reported to have commented that even a small glass of wine during the day could banish the recall of his dreams. Researchers now report that *IEG activation during the stress phase of alcoholic withdrawal* induces the expression of a variety of other target genes that may be responsible for withdrawal syndromes and the relapse back into addiction (Beckmann et al., 1997). The following treatment protocol for reducing the stress of alcohol withdrawal via the ultradian healing response is proposed as a model for easing withdrawal syndromes for the addictions in general. More research is now needed to assess the value of the ultradian healing response for modulating the expression of IEGs, their target genes, and stress syndromes during the withdrawal phase of a variety of addictions.

Treatment

Rex was first taught traditional craving control techniques, such as how to achieve a state of general comfort using a progressive relaxation method. Along with this, he received training in developing "alcohol refusal skills" and self-assertiveness. Rex was taught the decision delay technique, whereby he would delay for 20 minutes before deciding to drink or not. These techniques eventually failed, however, and Rex relapsed after work one day when his "buddies" asked him if he wanted a beer. Rex intellectually understood the techniques, but "at certain times of the day the craving was just too much."

Rex's association of drinking with "certain times of the day" suggested that there may have been an ultradian component to his alcoholism. Because of this possibility, the therapist decided to help Rex explore the ultradian brain-breath link in his cravings for alcohol (Rossi, 1990b; Rossi & Cheek, 1988). During the next session while he was discussing his craving for alcohol, he was asked to check for nasal dominance. He found that his right nostril was open, suggesting that he was experiencing left-hemispheric dominance at that moment. Rex was instructed to lie on his right side to "turn on his right-brain hemisphere" and explore whether his feelings and cravings changed. A permissive approach to suggestion was facilitated with these words: **"It's curious how just lying down on one side or the other can allow us to feel differently. If your unconscious is really ready to see pictures, have ideas, thoughts, or feelings about how to handle your problem with greater comfort and ease [pause], will your eyes close comfortably all by themselves?"**

Rex's eyes closed after a few moments, and the therapist followed up with a *basic accessing question* (Rossi, 1986/1993; Rossi, 1996a): **"When your inner mind has learned some interesting ways of dealing with those problems, will you awaken feeling refreshed and alert?"**

Fourteen minutes later, Rex's body moved a little in what seemed to be a prelude to awakening. The therapist then administered a posthypnotic suggestion: **"When you feel like moving and opening your eyes, will that be a signal that your inner mind will continue to help you deal comfortably with your problems** *throughout the day but especially after work?"*

Three minutes later Rex opened his eyes and sat up. He reported visualizing some friends from Alcoholics Anonymous saying, "One day at a time, one minute at a time." Rex was instructed to continue using the ultradian healing response to explore shifts in his cerebral-hemispheric dominance at appropriate times throughout the day. He chose after work as a consistent time to do the exercise, since this was the time when he usually began drinking. When he checked which nostril was open at that time, he found it was usually the right. This suggested that he was still in a "work mode," when his left cerebral hemisphere was more dominant. We speculated that this might be one reason why he felt a need to drink after work: Perhaps the alcohol was facilitating a shift to a right-hemispheric mode of relaxation. He therefore was encouraged to continue experimenting with doing an ultradian healing response by lying down on his right side to shift reflexively his hemispheric dominance to the right after work.

This therapeutic approach seemed to be just what he felt he needed to navigate safely through the dangerous time of temptation after work. In association with his continuing attendance at Alcoholics Anonymous and further insight-oriented therapy, he has been able to stop drinking completely for over a year.

It is not known whether the purported cerebral-hemispheric shift during an after-work ultradian healing period was the effective mind-body dynamic in this case, or whether it was simply another permissive suggestion with a pseudo-rationale of the type sometimes used by Erickson (1966/1980). Current research indicates that the right hemisphere is associated with "the vigilance aspects of normal human attention to sensory stimuli" (Levine et al., 1984) rather than with simple relaxation and imagery. Further controlled research studies are required to determine whether the focus on internal sensory stimuli, emphasized in the self-exploratory brain-breath shift, can actually enhance certain psychobiological aspects of brain laterality. The use of a comforting ultradian approach to relaxation and self-hypnosis as a therapeutic substitute for alcohol in this case, however, does appear to be an effective therapeutic strategy that has much in common with many of the other cases of addiction reported in this chapter.

CASE 5
Mary: Headaches, Nausea, and the Brain-Breath Link

Presenting Problem
"Mary" was referred by the County Health Department because she was experiencing extreme headaches and nausea with no explainable physical cause. She was a 17-year-old high school senior who had been addicted to

cocaine for one year (from ages 15 to 16) and had gone through a hospital treatment program. During the time she was using drugs, she was involved with several older boyfriends who abused her sexually.

She had received more than a year of weekly psychotherapy sessions with a psychiatrist who prescribed antidepressant medication. In this period of psychotherapy she worked through many of her past problems and eventually was taken off the medication, but the headaches and nausea continued. Hypnosis was considered as a last resort. Mary's present life situation was stressful: She was working 30 hours a week while attending high school. She was almost constantly on the move. In her words, *"I never seem to have a moment to take a breather."*

Treatment

During the first session, the Hypnotic Induction Profile (Spiegel & Spiegel, 1978) was administered to Mary, the results of which suggested that she was not a good subject for hypnosis. Nonetheless she appeared responsive to a traditional eye closure technique, and an arm levitation was used to ratify trance. Mary was given several suggestions using auditory (relaxing music), kinesthetic (warmth), and visual (breathing in light) modalities. Prior to trance she reported her headache at 70 on a subjective scale of 1–100. After a therapeutic trance experience, she reported only a slight improvement to 68— on her subjective experience even though she also said, "I did have more relief while I was under." During the next eight sessions, several more hypnotic approaches were used, including progressive relaxation, age regression, autogenic training, and metaphor. All were without much success.

At the ninth session Mary reported a severe headache (90 on her subjective scale). She reported that she was in a "terrible state of tension right now because I'm so much on the go today." We wondered together whether her current feelings were typical of the way she habitually forced herself into too much activity. When she was asked to check, she found that her right nostril was open, suggesting that her left cerebral hemisphere was dominant at this moment when she had a headache. Mary was instructed to lie comfortably on her right side to allow gravity to clear her left nostril and to induce a shift in cerebral-hemispheric dominance to "the right emotional hemisphere." She was told that **"even though your conscious mind may not know what is happening, your unconscious can help find the source of your current stress and symptoms."**

As Mary focused inwardly in apparently deep concentration, her eyelids began to flutter occasionally with a slight but high frequency as expressions of frowning, fear, and discomfort crossed her face. In a few moments the facial cues of distress and restlessness became more and more pronounced. She was encouraged to explore these negative feelings with the words, **"That's right, you can have the courage right now to continue receiving all those feelings of distress that tell the real story of your troubling symptoms."** She began to whimper softly with a few tears and then cried loudly, restlessly tossing about on the couch.

Since it was now evident that she was going into a strong cathartic reaction that threatened to get out of control, she was supported with a *therapeutic dissociation*: **"That's right, as one part of you really feels that fully, another part of you can watch comfortably and calmly from the sidelines, so you can give an accurate report about the source of your problems later."** From the new perspective of nonlinear dynamics theory, this type of *hypnotherapeutic dissociation* would be called a *bifurcation of consciousness* that facilitates the patient's path to self-organization (Rossi, 1996a).

Mary then went through what appeared to be a series of emotional "flashbacks," occasionally shouting words and phrases that suggested that she was reliving some of the negative sexual encounters she had experienced while under drugs. After about 20 minutes of couch-pounding and kicking off imaginary assailants, Mary appeared exhausted and began to calm down. She apparently dozed for about five minutes, her face calm, then she coughed and seemed about to awaken. Taking this cue, the therapist facilitated her awakening by giving a therapeutic posthypnotic suggestion regarding her ultradian healing response periods throughout the day: **"When your unconscious mind knows it can continue this healing work all by itself whenever it's entirely appropriate [pause], and when your conscious mind knows it can cooperate by helping you recognize those moments throughout the day when it is right to *take a breather* [pause], will you find yourself awakening feeling refreshed, alert, and as aware as you need to be of the meaning of your experience here today?"**

With this question, awakening from trance was made contingent on the healing continuing on an inner, implicit, unconscious level, while the conscious mind focused on cooperating by recognizing those ultradian mind-body cues that she needed to enjoy to break the stress cycle. When she awakened, Mary reviewed some of the negative memories she had just reexperienced with an attitude that is rather characteristic of many young patients dealing with flashbacks to traumatic drug experiences: A part of her always seemed to know that this was the source of her continuing distress, but she was not able to help herself other than by trying to deny it and keeping herself overly busy all the time to avoid coping with it.

This highly permissive psychotherapeutic work allowed both parts of her to come together: (1) the side that experienced the painful, emotionally-laden memories, and (2) the more rational side that needed time and support in a safe therapeutic milieu to assimilate the traumatic experiences in a meaningful manner. The next few sessions were spent exploring her deepening appreciation of psychodynamic issues associated with her traumatic drug experiences, being "too much on the go," and helping her recognize how both were related to her symptoms of headache and nausea. During a few of her ultradian rest-rejuvenation periods, she became aware that, even before she had a bad headache, she would feel a heat in her stomach and tension around her eyes as the first symptoms that she was under too much stress. She was instructed that whenever she felt these initial *"signaling symptoms,"* she was to check which nostril was open and to lie down on that side so that her **"inner mind could work through whatever it needed to in a safe way."** Arrangements were made with the school nurse so that Mary would be allowed to go to the nurse's office to do this therapeutic exercise.

During the next three weeks, Mary was able to prevent all but one headache and nausea attack, and this single episode reached only 40 on her subjective scale of pain. Mary reported needing to use the ultradian healing response two or three times per day. This exercise has developed into a positive part of her lifestyle as a means of dealing with inner issues and coping with outer sources of tension before they become full-blown physical symptoms.

"Mary" is a composite case containing dynamics from a number of typical cases seen by the author, wherein catharsis and insight therapy were facilitated by the purported brain-breath link to access another side of the personality that might facilitate healing. Again, we stress that these cases do not prove that such a link is actually responsible for the therapeutic result. The so-called brain-breath link may be merely a new metaphor for utilizing the current belief system in "the other side of the brain" that has been popularized in the press. Current research on the psychobiological dynamics of the brain-breath link would be an appropriate approach to investigating this area.

CASE 6
Lucy: Gene Expression, Stress, and Breast-Feeding

Presenting Problem

"Lucy" was a 21-year-old patient in marriage counseling when her first child was born. Shortly after the birth she phoned the therapist at 11 P.M. and reported anxiety so severe that she could not breast-feed her daughter. She had become depressed and felt guilty that her marriage problems had affected her ability to bond with the baby. At night, Lucy would lie down while facing her daughter and attempt to breast-feed her, but sometimes the "milk would just not come." Then the baby would cry and a vicious cycle of guilt, anxiety, and further upset would take place between husband and wife. It is now known that there are many psychobiological connections between the circadian rhythms of gene expression, hormonal regulation, breast-feeding, and stress (Tay et al., 1996; Tortonese et al., 1998).

Treatment

During a previous session with the couple, they were both taught to recognize which nostril was open. Then they were taught to lie on the side of the open nostril "to change cerebral-hemispheric dominance and mood," when they were having a bad time together. This procedure was initially suggested to provide a strategic intervention that would disrupt the pattern of escalating argument and mutual recrimination that had developed between them concerning their relationship issues and the breast-feeding problem.

On the phone the therapist asked Lucy to check which nostril was open; she reported the right nostril, adding that she had been lying on her left side trying to feed the baby. The therapist instructed Lucy to lie down on her right side and take 15 minutes "to allow your inner mind-body to make the necessary changes that would enable you to nurture the baby." Then her husband was to bring her the baby to feed while she was still lying on the right side.

The next morning, a follow-up call determined that this intervention had indeed helped Lucy's milk to resume flowing. She was instructed that whenever she had problems feeding her daughter, she was to check which nostril was open and lie on that side until she could "feel her cerebral-hemispheric dominance change," and then to feed the baby lying on that same side. She reported that this was very effective and greatly reduced her guilt and self-derogation.

This clinical presentation does not enable us to distinguish whether the purported nasal-hemispheric dominance shift associated with a change in body position was the real therapeutic mechanism. It could, for example, simply be another illustration of a strategic family intervention facilitated by stochastic resonance (a type of noise or nonrelevant stimuli that can potentiate important psychobiological signals). Milton Erickson sometimes used suggestions that were laden with "false contingencies and false relationships" (Erickson & Rossi, 1980, 1981, 1989; Erickson, Rossi, & Rossi, 1976). Suggestions containing false relationships and contingencies might have been effective simply because the patient and her husband believed in them. While we do not recommend the intentional use of suggestions with false relationships and contingencies, human history is rich in documentation of their effectiveness when they are believed.

This case does provide a model, however, of a new class of ultradian approaches that could be used in controlled clinical research to explore the possibility of a complex cybernetic loop of mind-gene communication. It is now well known, for example, that prolactin is a peptide messenger molecule (hormone) released from the pituitary in typical ultradian rhythms (Veldhuis & Johnson, 1988) to mediate lactation at the cellular-genomic level (Veldhuis, 1992). Recent reports have documented how a 20-minute relaxation/imagery tape can increase milk flow in mother's of premature infants (Feher et al., 1989). Research is now needed to assess the hypothesis that many methods of relaxation, imagery, hypnosis, and holistic healing may be effective because they can facilitate a shift in the cerebral-hemispheric brain-breath link which may, in turn, entrain our natural ultradian rhythms of mind-body communication at the molecular-genetic level (Singer & Ryff, 2001).

GENE EXPRESSION, DIET, AND GASTROINTESTINAL DISORDERS

Eating and gastrointestinal disorders are another group of mind-body problems that has its sources in psychosocial genomics with many ultradian components. Current research on the relationships between diet and gene expression in health and disease represents a profound paradigm shift in the study of nutrition and the optimal functioning of mind and body (Berdanier, 1996; Hargrove & Berdanier, 1993; Moustaïd-Moussa & Berdanier, 2001). There are well documented 90–120-minute ultradian rhythms in gene expression associated with eating (Hiatt & Kripke, 1975), diet, insulin, glucose levels (Mejean et al., 1988; Van Cauter et al., 1989), and urination (Lavie & Kripke, 1977). Friedman has outlined a psychophysiological model of psychosomatic illness based on the relationships among stress, psychodynamic factors, and the desynchronization of ultradian/circadian rhythms (Friedman, 1972, 1978; Friedman & Fischer, 1967). The neuroscientist Walter Freeman (2000) has outlined new views of mind-body interactions in the exercise of free will and choice as an essential and inalienable property of human life. We believe that the heightened states of self-sensitivity and self-reflection that are available during the quiet periods of the ultradian healing response may engage our natural psychogenomic processes, wherein the critical but often fragile power of choice may be optimized for the co-creation of human identity and behavior.

CASE 7
Debbie: Stress, Diabetes, and Gene Expression

Presenting Problem

This 37-year-old client requested help with her "addiction for sugar." "Debbie" is a severe diabetic (diagnosed at age 11) and is currently taking 65 units of insulin daily. Her father died in 1969 from diabetes. Debbie had a long history of denying the seriousness of her condition and routinely bingeing on sugar and sweets. She had a particular weakness for the icing on birthday and wedding cakes and was hospitalized once for such sugar bingeing. At the present time a beginning exploration is underway of the psychosocial genomics of how sugar (glucose, fructose, etc.) in the diet modulates gene expression in health and diabetes (Hwang et al., 2001). In this case history we will focus on the practical, clinical aspects of how the ultradian healing response in the form of 20-minute naps can ameliorate diabetic problems.

The younger of two children, with a half-sister five years her senior, Debbie grew up in turmoil. Her father was physically abusive, and both her mother and sister had histories of alcoholism and suicide attempts. Debbie attempted suicide at age 15, then became pregnant and got married the same year in order "to get away from home." She was divorced four years later and raised her son on her own.

Treatment

Debbie was seen a total of seven times over a two-month period. It was learned that she routinely took "20-minute naps" during her lunch break. The therapist talked to her about the ultradian healing response and suggested that she increase the number of her healing breaks to about four to six times a day. She never was able to do that, but she did average at least one or two extra healing periods daily. Since she apparently had an ability to do lucid dreaming, we utilized that skill during the ultradian healing response to enhance her feeling of comfort and well-being. Her ability to relate creatively to her dreams was continually linked with her growing ability to cope with her sugar addiction.

Debbie reports that she has not eaten any sugar for the past four and a half months. She described how a friend had her over for a visit this week and offered her a piece of pie with coffee. She casually refused, saying, "I don't do pie!" During this period, she has attended three birthday parties where she was "exposed" to the type of cake she formerly found irresistible. She ate none of it. She is still aware of her psychological/emotional desire for sugar, but she now feels she can continue to control it. She describes herself as being "inner-motivated by the deep comfort" she feels during her ultradians— which is a new experience for her. She still takes her two or three rest breaks daily. She describes them by saying: "They take the cake! Really, they relieve the anxiety that, in the past, would have caused me to eat cake or sweets. They replace the cake!" When asked if it was difficult to find time for the ultradian breaks each day, she laughingly replied, "Are you kidding? It's a piece of cake!"

The humorous word play Debbie used to describe her experience is reminiscent of Erickson's use of humor as a significant dynamic in the healing process (Erickson & Rossi, 1989). Her spontaneous remarks, equating her ultradian comfort with the relief that she formerly received from cake ("They replace the cake. . . . It's a piece of cake!") sound like what Erickson might describe as the "literal and concrete" crossover or overlapping of two essentially different types of experience (rest and eating) to facilitate the healing process (Erickson & Rossi, 1976). She was able to access the comfort of the ultradian rest-rejuvenation state and utilize it rather than cake to cope with anxiety.

The psychobiological mechanisms involved in sugar addiction are a good illustration of psychosocial genomics and the complex adaptive system illustrated in Figure 2.1. We know that many messenger molecules with similar relaxing and comforting effects (e.g., the endorphin family) are released by eating and rest (Tache et al., 1989). That may be the reason why eating and ultradian rest feel so right together. Further research is needed to assess the hypothesis that there is an actual release of similar messenger molecules and gene expression (e.g., for the constitutive glucose transporter, GLUT1, Hwang et al., 2001) in the ultradian approaches to facilitating psychotherapy. Such research would provide a psychobiological communication link at the molecular level for the complex adaptive processes whereby the inner resources of one type of life experience (e.g., the ultradian healing response) are used to facilitate healing of another life experience (e.g., anxiety problems, sugar addiction).

CASE 8
Fred: Stress, Obesity, Nutrition, and Gene Expression

Presenting Problem

After a year of rather unsuccessful conventional hypnotic approaches to diet control (and a lifetime of unsuccessful dieting by most advertised methods) a 67-year-old lawyer had success for the first time with the ultradian diet. The ultradian diet consists of six small meals of 200–300 well-balanced calories, spaced throughout the day to coincide with our natural 90–120-minute ultradian hunger rhythm (Rossi & Nimmons, 1991). The lowered cholesterol levels, the reduced need for insulin, and the generally health-promoting aspects of such an ultradian eating pattern are now well-documented (Jenkins, 1989).

The patient is usually advised to **"enjoy the meal and then reward yourself with a 20-minute ultradian healing response, during which your body will make an efficient beginning in digesting the food to the point where you will feel completely satisfied."** This process is carried out first in a hypnotherapy session, wherein the patient actually eats 200–300 calories and then experiences a natural ultradian therapeutic trance under the direction of the therapist. Posthypnotic suggestions are given to support the same type of ultradian eating/rest rejuvenation experience every few hours in daily life.

At the present time a remarkable series of discoveries relating psychosocial stress, nutrition, and gene expression in health and illness are transforming our understanding of the relationships between diet and a number of dysfunctions associated with obesity, ranging from diabetes, hypertension, and cardiovascular disease to immune system problems and cancer (Christensen, 2001; Morrison & Farmer, 2001). Within the past decade we have learned that the adipocyte (fat cell) is much more than a passive storage of fat. Adipocytes and fatty acids play an active and focal role in the dynamics of energy formation and utilization as well as body composition at the level of gene expression. Kersten et al. (2000) describe how some fatty acids can act as hormones by binding to, and activating, nuclear factors, thereby regulating gene expression. These researchers emphasize that fatty acids are not just passive energy-providing molecules, but active participants in metabolic regulation. They describe how PPARs (perixosome proliferator-activated receptors) function as hormone receptors within the nucleus of the cell, where they can modulate gene expression in response to signals from the environment (such as diet, psychosocial stress, etc.).

Accumulating evidence suggests that the natural ultradian healing response that accompanies the eating cycle can reduce psychosocial stress by lowering stress-related hormones (catecholamines and glucocorticoids). The dietary and hormonal regulation of the fatty acid synthase gene (FAS), the insulin-dependent glucose transporter (GLUT4), leptin, and other patterns of gene expression that are involved in the regulation of appetite and weight control (Morris et al., 2001; Morrison & Farmer, 2001), are all subject to modulation by psychosocial stress (Jordan, 2001). This emerging data base of research on the relationships between psychosocial stress, diet, hormones, and gene expression supports the therapeutic approach of utilizing the natural ultradian healing response to ameliorate obesity and its related dysfunctions.

Fred was able to use the natural ultradian healing periods after each meal to reduce a lifetime pattern of psychosocial stress by what he called "deep meditations" on the course and meaning of his life. He "spontaneously" received insights into the early life sources of his hunger in a series of remarkable "flashbacks," wherein he felt himself to be a baby who continually spit out food because he was overfed by his mother. During a naturalistic and permissively oriented hypnotherapy session, he would spontaneously (with no suggestions from the therapist) lay down on the couch and go through an apparent age regression, wherein he made sucking and eating movements as if he were an infant. He believed that this forced overeating as an infant accounted for the fact that, even now, when he overeats, he really does not feel hungry and does not want to eat.

He was able to use this insight to relate to his compulsion to overeat by following the posthypnotic suggestion to have a small meal every two or three hours, followed by what he called his "deep ultradian meditation." The effectiveness of this posthypnotic suggestion in helping him steadily lose 30 pounds over a two-year period may be accounted for by Erickson's method of reinforcing posthypnotic suggestions by "associating them with behavioral inevitabilities" (Erickson & Erickson, 1941/1980; Erickson & Rossi, 1979). It is a normal behavioral inevitability that (1) most people experience hunger every few hours throughout the day (Freidman & Fischer, 1967; Hiatt & Kripke, 1975), (2) most people need to experience rest, relaxation, and sometimes sleep after eating. Associating posthypnotic suggestions for small meals with both of these behavioral inevitabilities thus provides a double reinforcement.

CASE 9
Gary: Encopresis, the Ultradian Moment, and Emotional Rapport

Presenting Problem
A family was referred by the school counselor because "Gary's" older brother, "Andy," age 14, had developed a severe behavioral problem. In the group sessions, Andy discussed some of the problems in his family that were associated with both parents being "high-powered executives" under terrible stress. These parents apparently had very high expectations that both Andy and his little brother, Gary, age 10, did not feel they could uphold.

Family therapy was indicated and the family complied, probably because of the guilt the parents felt about their overworking. Initially Andy improved, supposedly in response to the group and family sessions. At one point, about three months into the treatment, Gary replaced Andy as the identified patient when the issue of Gary's soiling was raised. Gary had been toilet-trained by the age of four but occasionally soiled his pants until age six, when his mother went back to work.

For the past three years, however, one or two times per week Gary would have an "accident" and pass feces while playing during the day. He would then call his mother to come and take care of him. This could be very shaming for him at the school, day care, or even at home if he was playing with friends. Sometimes his mother could not get off work, and the father would have to bring Gary clean clothes.

The family dynamics showed an overly strong coalition between Gary and his mother, with Gary's father remaining distant and aloof from him. The father enjoyed Andy more because "he's more into sports and cars, like I am." This structure indicated that more bonding was needed between Gary and his father. An adequate presentation of the psychodynamics of this entire family therapy is beyond the scope of this section; only the utilization of the ultradian healing response as a strategic maneuver to cope with the encopresis will be discussed. A variety of gene expression profiles has been associated with gastrointestional dysfunctions and muscular control that might be associated with encopresis (Lei, 2001; Moustaïd-Moussa, 2001), but specific data on this clinical problem are not yet available.

Treatment

A few sessions were conducted with Gary and his father to help them relate better with one another. In one of these sessions Gary's father brought up the soiling issue. While we were discussing it, Gary's eyes teared up, his breathing deepened, and he began staring out the window. The therapist believed this withdrawal behavior was related, at least in part, to his need to turn inward to cope with the shame he was experiencing. Since some of these withdrawal behaviors are similar to those that are naturally experienced when entering an ultradian rest-rejuvenation period—what we might call "the ultradian moment"—the therapist reframed his developing state of discomfort and withdrawal into an ultradian healing response, by saying: **"I notice you seem to be getting quieter in the last few minutes, and you're not saying much while you're looking out the window. I wonder if that means you would really like to just rest a bit and be comfortable with your father. Just continue as you are, Gary, and see if it feels good to take a rest or maybe** nap with your dad." Gary eyes fluttered for a few minutes and then closed. I motioned the father to put his arms around Gary as he snuggled close to his father. It was evident that they both enjoyed participating in this "rescuing hug" (see Chapter 1) that may have generated healing at many levels, from family dynamics to gene expression (Singer & Ryff, 2001).

The father then closed his eyes as the therapist said, **"I wonder just how aware you are that you're both doing just what you need to do to feel good together and solve your problems? Take some time to enjoy it."** After about 20 minutes they stirred, so the therapist said: **"Continue to enjoy your good feelings together and know that these good feelings of being together can continue in the whole family. When something within you feels it has dealt with your problems as much as possible right now, will you find yourselves waking up refreshed and alert with a good stretch?"**

After a while they both opened their eyes and stretched. They commented how they had not really rested together like that since before Gary started school. Gary said, "Even though it seemed kind of kooky, I really feel good now." A few minutes later Gary asked if he could use the bathroom, as he could now feel his "tummy moving." He successfully passed stools under complete voluntary control and was obviously pleased with himself as he came out of the rest room and faced his father and the therapist with a broad smile.

Gary and his father were told to rest comfortably together this way, once or twice a day, when it "felt right." They agreed. In addition, the school nurse was asked to let Gary lie down in her office during recess or lunch-time if he requested permission to do so. His teacher was also asked to let him go to the nurse's office upon request. Gary later reported that the "napping" helped him feel what was "going on" inside him. He has not had an accidental bowel movement for the past six months.

Recognizing the optimal therapeutic moment when the young boy's shameful behavior overlapped the typical withdrawal aspects of the normal rest-rejuvenation phase of ultradian behavior served as the critical transition to a deeply healing naturalistic experience that brought Gary and his father together in a comfortable mind-body state of nonverbal synchrony. This was clearly an example of how the natural ultradian healing response can be used in parent-child relationships (Rossi & Nimmons, 1991). The use of appropriate touch and nonverbal communication for healing via "the common everyday trance" has been a central theme in much of Erickson's work (Erickson, 1964/1980; Erickson, Rossi, & Rossi, 1976). The ultradian approach to problem- solving and healing makes it clear that there are particularly auspicious periods throughout the day when creative replays of mind-body synchrony and emotional rapport can be facilitated in an optimal manner.

ULTRADIAN SENSITIVITY
AND PSYCHOLOGICAL INSIGHT

From a clinical point of view, one of the most significant aspects of experiencing the ultradian healing response is the heightened sensitivity many people develop for their previously unnoticed mind-body sensations during subtle shifts in consciousness throughout the day (Broughton, 1975; Dinges & Broughton, 1989). This is particularly true of women undergoing the hormonal shifts in the many mind-body messenger molecules associated with their monthly cycles.

CASE 10
Barbara: Ovulation, Sensation, and Insight

"Barbara" wrote the following in her diary.

"I was very intent on finishing the final chapter of the book I had been reading for the past few hours when my eyes began to droop and my concentration waned. I forced myself to keep reading, yet realized an ultradian was coming on—so maybe I'd best let my eyes close for just a moment. Just as I sighed and closed my eyes, I was aware of feeling a module-like substance in my fallopian tube with an accompanying familiar feeling of sensitivity and receptivity. I thought, Oh, I must be ovulating.

"As I relaxed more, I became aware of an immense feeling of warmth and lubrication in my genital area. I decided to feel down there. Sure enough—a warm, lubricated, and pulsating vagina—all ready to make a baby—if only I were 26 years old instead of 47 years old—and didn't already have all the beautiful children I could handle! Yet how wonderfully and simply this body responds and communicates with consciousness moment by moment. We need only to 'tune in' to the rhythm by paying attention to the natural ultradian moments throughout normal everyday life. If I were 26, this would be the exact time to become impregnated."

This unusually perceptive ultradian diary report supports the view that the ultradian healing response can be used as a window of heightened inner awareness for accessing subtle mind-body signals and inner resources for dealing with problems (Rossi & Nimmons, 1991). The following case reports another excerpt from a patient's ultradian diary, in which she apparently receives "spontaneous insights" into psychological as well as dietary sources of her sinus condition.

CASE 11
Andrea: Sinus Congestion and Ultradian Insights

"Andrea" wrote the following in her diary.

"I laid down, but my mind seemed resistant to taking an ultradian break. I listened to sounds around me (birds, children, classical music). Thoughts were of things I 'have to do' or 'should be doing.' I just let these thoughts be, but also chanted a mantra silently to myself. Breathing became more shallow and I worried about falling asleep. *One thought that kept recurring that seemed significant: Sinus congestion occurs more frequently when I have something to say that might not be agreeable to my partner.*

"Next day: Lots of long, luxurious stretching. But, would you believe it, also feelings of guilt for tuning into quiet ultradian time (my partner would call this 'the Catholic-girl syndrome'). After a while these guilt feelings /thoughts went away and I felt extremely relaxed. *Another thought that came to mind regarding my sinus congestion is wine. I've noticed that if I have wine with my dinner, I am much more congested the following morning.*"

These spontaneous insights into the possible interpersonal (difficulty talking with her partner) and dietary sources of her sinus condition certainly are not evidence, in themselves, of causation in a scientific sense. They do, however, provide her with hints about what changes she might explore to cope with her sinus symptoms. Further research is needed to determine the similarities and differences between perceptions received during the ultradian healing response, classical psychoanalytic free association, and what has been traditionally called "hypnogogic" and "hypnopompic" states on the border of going to sleep and waking up (Rossi, 1972/1985/1998, 1990d, 1996a).

SHOCK AND INSIGHT IN THE RESOLUTION
OF PSYCHOSOMATIC PROBLEMS

One of the most dramatic examples of how the ultradian healing response can facilitate psychological insight and the resolution of traumatically induced psychosomatic problems is presented in the following diary record of a psychotherapist, in his sixties, who had previously tried many forms of psychotherapy and hypnosis, with no success, in attempts to resolve a chronic problem of "throat constriction" that had bothered him for 30 years. This person attended a professional presentation on "The Ultradian Healing Response" and a few months later mailed me an ultradian diary that included the following entries from the third and fourth day of exploring his personal experiences. This case is a particularly striking example of how a psychosomatic symptom can become manifest as a state-dependent encoding of memory generated via activity-dependent gene expression (see Chapter 3) due to a psychological shock (Rossi, 1973a).

CASE 12
John: Throat Constriction, Electric Shock, and Memory

John wrote the following in his diary.

"Relaxed almost immediately, letting go of thoughts, trying not to concentrate on anything. Memory of an electrical shock in 1954 (one phase 220V of a 440V circuit), the sulfuric taste in my throat again just as it was then. Strange, I haven't thought of that in years. Throat relaxes, taste fades, no sense of constriction. As I arose from rest, my legs tingled with the increased flow of blood to them as I stood up.

"Next day: Relaxed almost immediately, became aware of my body, tension in my throat. Throat begins to relax, all feeling of stricture leaves throat. Thoughts of the afternoon's plans and activities. Up relaxed and refreshed."

Follow-up calls six months and a year after these journal entries indicate that he is still using the ultradian healing response about once a day to maintain a satisfactory resolution of his throat constriction. This apparently "spontaneous" resolution of a significant, traumatically induced psychosomatic symptom during the ultradian healing response calls to mind the discovery of the role of trauma and stress in the early history of psychopathology and psychoanalysis (Ellenbeger, 1970; Zilboorg & Henry, 1941). Breuer and Freud (1985/1957) reported how they gradually replaced an authoritan approach with hypnotic induction and then the use of "free association" to access and trace the sources of traumatically induced psychological problems and psychosomatic symptoms.

Freud believed that he had given up the use of hypnosis. From our point of view he had only given up a direct and obvious form of hypnotic induction; his procedure of allowing patients to lie on a couch and free associate is actually an indirect form of hypnotic induction that has much in common with what we now call the "ultradian healing response." In the ordinary course of everyday life most people experience an "ultradian rest-rejuvenation deficit": They have skipped one or more of the natural 20-minute rest-rejuvenation phases of the basic rest-activity cycle. Because of this ultradian deficit, most people will immediately and automatically experience an ultradian healing response whenever they are given an opportunity to sit or lie down and rest a bit.

We could hypothesize that the mild stress of lying down and "free associating" about one's problems can occasionally reactivate the state-dependent encoding of trauma- and stress-induced psychosomatic problems that lead to a recovery of memories about their sources (Rossi, 1986a; Rossi & Cheek, 1988; Rossi, 1990c, 1990d; Rossi, 1996a). I outlined *the neuropeptide hypothesis of consciousness and catharsis* to account for this phenomenon (1990a). The arousal and relaxation phases of cathartic psychotherapy are mediated by the time-linked release of ACTH (for arousal) and beta-endorphin (for relaxation) from their mother molecule, POMC (proopiomelanocortin), in the limbic-hypothalamic-pituitary system over a 20–30-minute period. Carefully controlled research monitoring the "real time" release of these informational messenger molecules during psychotherapy is required to assess this hypothesis and this proposed use of the 20-minute ultradian rest-rejuvenation period as a "window of opportunity" for accessing and facilitating many natural mind-body healing processes.

The following report, contributed by Jeffrey Auerbach, Ph.D., a psychotherapist in Los Angeles, documents a vivid experience of this hypothesized relationship between shock, trauma, stress, symptom formation, and the ultradian healing response. It illustrates how a spontaneous "hypnotic phenomenon" like catalepsy (italicized below) can occur in response to a shock (Erickson & Rossi, 1981) that can modulate gene expression (Kaufer et al., 1998).

CASE 13
Jeffrey: Shock, Trauma, and Ultradian Dynamics

"Does a shock to the body trigger an ultradian rest state? One morning at the office, I fell off a chair and crashed to the ground on my right hip. Stunned, I lay there without moving and thought that I had badly injured myself.

"My secretary rushed in, asked if I was all right, and said, *"You can put that chair down."Without realizing it, I was holding a chair that I had bumped into, up in the air with one hand.* I found that I had no injuries. After a few moments I stood up and prepared to see patients. I felt unfocused, a lack of ability to concentrate, tired, and in general, 'spacey.' I realized that I felt as if I were in the 'rest' phase of an ultradian cycle—which I normally did not feel at that time of the day. I wanted very much to just lay down and close my eyes, which I did.

"It was as if the sudden physical shock to my body produced an alteration in consciousness very similar to the feeling of being in the rest phase of the ultradian cycle. Cheek reports that a sudden movement to the body of a hospitalized patient often produces a hypnotic state (Rossi & Cheek, 1988). Anyway, 20 minutes later I felt alert and returned to work."

UTILIZING THE ULTRADIAN EXPERIENCE OF BEHAVIORAL STATE-RELATED GENE EXPRESSION

Taken together, these case studies suggest a number of new ways of utilizing the psychobiological dynamics of ultradian/circadian rhythms for facilitating behavioral state-related gene expression to optimize problem-solving and healing in psychotherapy.

How to Recognize Ultradian/Circadian-Related Problems

Therapists can recognize a possible ultradian/circadian etiology whenever patients describe the *periodic occurrence* of their symptoms and problems with commonly-used phrases whose significance as signals of the ultradian periodic nature of the reported symptoms is usually not recognized by patients or therapists. For example: "These symptoms keep coming up **several times a day** . . . every once in a while throughout the day . . . when I get tired . . .

when I get stressed . . . a few times in the afternoon and evening . . . when I get hungry . . . usually after eating . . . when I need to go to the rest room . . . when I start to fall behind in my work . . . every few hours throughout the night."

The circadian periodicity of problems can be recognized by the patient's statements about the *daily* occurrence of a symptom at a particular time of the day: "It usually feels worse in the morning . . . afternoon . . . evening . . . night." Typical circadian issues take place in the middle or late afternoon. This time period has been called "breaking point" (Tsuji & Kobayashi, 1988), and it is precisely the time when the *per* gene turns on to initiate another circadian cycle at the cellular level (Takahashi & Hoffman, 1995). By this point in the day, many people have already skipped two or three ultradian rest periods, producing an accumulated ultradian deficit and stress syndrome expressed with these common complaints:

"I'm exhausted by mid-afternoon."
"I get stressed, tense, and irritable toward the end of the workday."
"I need a drink after work."
"My addiction gets worse later in the day when I have to **have** something."
"I get sleepy in the afternoon.
"The worst time is when I have to go home after school and I'm too tired to do homework."
"Just before dinner everybody is irritable and that's when arguments start."

Many of these acute and chronic problems can be ameliorated by taking one or two ultradian breaks earlier in the day or taking a nap after lunch. Other periodic psychobiological problems with longer rhythms, such as those associated with the monthly menstrual cycle and the yearly seasonal affect disorders (SAD), have been dealt with elsewhere (Kupfer et al., 1988; Rossi & Nimmons, 1991).

Utilizing Minimal Cues and Heightened Sensitivity to Mind-body Signaling

Our hypothesized association between the normal behaviors of the rest phase of the basic rest-activity cycle and the indicators of Erickson's common everyday trance (Erickson & Rossi, 1976) can help therapists become more aware of the many minimal cues that signal patients' changing states of consciousness, which can be utilized to facilitate creative moments of inner work. These moments can be optimal times to use therapeutic approaches such as imagery, free association, meditation, and emotional reexperiencing. These approaches were documented in most of the cases in this chapter at the point when patients learned to become more aware of the first subtle mind-body signals of their need to take a break to avoid the type of excess stress that could exacerbate their symptoms.

Symptom-scaling is a very useful approach for heightening patients' sensitivity to their symptoms so that they can better appreciate those periods throughout the day when those symptoms get either worse or better, as well as the just noticeable therapeutic improvements that can motivate and reinforce their progress. The associated therapeutic approach of *symptom prescription* to help patients learn to recognize "how to turn on their symptoms to make them momentarily worse" is useful for accessing the state-dependent encoding of mind-body problems so that they can more readily learn "how to make the symptom better" (Rossi, 1986/1993, 1996a).

The Ultradian Entrainment and Synchronization of Inner Resources for Problem-Solving and Healing

Although Milton Erickson and his students have extensively documented how to help patients access their personal repertoire of social and psychological life experiences as "inner resources" for problem solving, there has been relatively little updating of what Erickson called the "psycho-neuro-physiological" basis of psychotherapy (Erickson, 1980c; Rossi, 1999a, 2000b; Rossi & Cheek, 1988; Rossi & Ryan, 1986). A variety of the cases presented in this chapter documents how it is precisely those mind-body processes that have an ultradian or circadian periodicity that is sensitive to distortion by the types of psychosocial stress that may modulate gene expression in a way that leads to psychosomatic symptoms. Just as acute and chronic aversive psychosocial stimuli can desynchronize our normal ultradian rhythms of psychobiological health, so too can the constructive use of psychosocial stimuli be used to entrain and synchronize creative replays of mind-body communication for healing within individuals (cases 1, 3, 4, 5, 7, 8, 12, and 13) as well as healing between people (cases 2, 6, 9, 11). Many chronic dysfunctions associated with psychosocial stress and diet, ranging from addictions, obesity, diabetes, cardiovascular problems and hypertension, the immune system, and cancer, could be ameliorated at the level of gene expression by using the ultra-dian healing response in the highly individualized process of self-discovery in psychotherapy (cases 4, 7, 8).

Converting Symptoms into Signals, Problems into Resources

The psychobiological perspective presented in this chapter interprets many mind-body problems as symptoms of the ultradian stress response and the behavioral state-related patterns of gene expression associated with it. Many psychosomatic symptoms are a response to the stress engendered when patients chronically ignore and override their natural need to have a rest-reju-venation period every few hours throughout the day. A useful therapeutic approach in such cases frequently involves the utilization of patients' current feelings of distress and experiences of psychosomatic symptoms in the ther-apy session to access the sources of their mind-body problems and motivate appropriate changes in their lives.

This approach may involve utilizing and reframing many stress symp-toms of apparent resistance in psychotherapy into an ultradian healing expe-rience. This was demonstrated in case 9, when Gary's feelings of shame and withdrawal were converted into an ultradian healing response wherein he could experience a warm, nonverbal bonding experience with his father. A careful review of Erickson's methods of "utilizing the patient's resistance" from this perspective documents how he frequently channeled the patient's withdrawal behavior into a creative replay, reframe, and resynthesis of inner experience during therapeutic trance (Erickson, 1958/1980, 1959/1980.)

Evolving Consciousness in Psychotherapy and Everyday Life

From the broadest point of view, the ultradian healing response can be under-stood as a window for accessing and facilitating creative moments in every-day life as well as in psychotherapy. Erickson demonstrated how psychotherapy can facilitate profound shifts in, and broadening of, an individual's con-sciousness during the common every day trance (Erickson, 1980; Erickson & Rossi, 1989; Rossi & Ryan, 1991). In Part II we will greatly extend the range of possibilities for utilizing the ultradian healing response to facilitate the creative replay and resynthesis of experience in individuals as well as in society as a whole.

Summary

This chapter outlined an evolving view of how the creative replay of complex adaptive systems of psychosocial genomics mediate mind-body communication across all levels, from the psychological and social to the cellular-genomic. This new field of psychosocial genomics generates a unified psychobiological theory of the creative dynamics of consciousness, sleeping, and dreaming in psychotherapy and the healing arts. The essence of this new view is that immediate early genes and clock genes play a central role in the deep psychobiology of waking, sleeping, dreaming, stress, and healing. Behavioral state-related gene expression is a genomic source of behavior that can be modulated with psychosocial cues. Behavioral state-related gene expression and immediate early genes are bridges between body, brain, and mind that can be accessed to facilitate creative replays of psychotherapy and the holistic healing arts.

A variety of problems, ranging from stress and trauma to the addictions and psychosomatic problems, were presented to illustrate how familiar psychotherapeutic approaches can be used to facilitate the ultradian–circadian dynamics of behavioral state-related gene expression. Current research suggests that we may have gone as far as we can with today's therapeutic approaches that lack the capacity for direct measurement of gene expression profiles during the psychotherapeutic process. Further progress requires new clinical-experimental research assessing gene expression during the experiences of psychotherapy and the healing arts. We propose that this assessment of the psychosocial dynamics of mind-body therapy could be made with tools such as brain imaging, DNA, and protein microarrays. In the following chapters we propose new activity-dependent approaches to the psychosocial genomics of psychotherapy and the healing arts that will be suitable for this future assessment on the level of gene expression.

| CHAPTER 3 |

ACTIVITY-DEPENDENT GENE EXPRESSION

How Human Experience Modulates Gene Expression

"One might say, metaphorically, that consciousness is the tutor who supervises the education of the living substance, but leaves its pupil alone to deal with all those tasks for which it is already sufficiently trained. But I wish to underline three times in red ink that I mean this only as a metaphor. The fact is only this, that new situations and the new responses they prompt are kept in the light of consciousness; old and well practiced ones are no longer so."

—Erwin Schrödinger
Mind and Matter

"Presentations of surprising or unexpected reinforcers generate positive prediction errors, and thereby support learning. . . . Expected reinforcers do not generate prediction errors and therefore fail to support further learning even when the stimulus is consistently paired with the reinforcer. . . . Dopamine neurons also respond to novel, attention-generating, and motivating stimuli. . . . Modeling studies have shown that neuronal messages encoding prediction errors can act as explicit teaching signals for modifying synaptic connections that underlie associative learning."

—P. Waelti, A. Dickinson, W. Schultz
"Dopamine Responses Comply with Basic Assumptions
of Formal Learning Theory"

It required a bit less than half a century for this flash of insight by Schrödinger, one of the founders of quantum physics, to become well established in the laboratories of modern neuroscience. The basic idea is that conscious human experience emerges from a complex interaction between the environment and *activity-dependent gene expression and protein synthesis.* We review the facts of how the conscious experiences of novelty, environmental enrichment, and physical exercise optimize neurogenesis: the growth of new neurons and their interconnections throughout the brain. Current research illuminates how the hippocampus is the part of the brain that encodes new experiences. The replay of new experience between the hippocampus and the cortex consolidates new memory, learning and behavior.

We utilize this creative replay as a basic mechanism to evoke activity-dependent gene expression and protein synthesis for generating growth factors and neurotransmitters in the dynamics of neurogenesis in psychotherapy and the healing arts. Evidence now suggests that a reduced rate of gene expression and neurogenesis may be linked to psychological depression, whereas an increase may be associated with happiness and well-being. Integrating this new research leads to a new concept of the relationships between gene expression, neurogenesis, and conscious human experience, described as the *novelty-numinosum-neurogenesis effect*: Highly motivated states of consciousness can turn on and focus gene expression, protein synthesis, and neurogenesis in our daily creative work of building a better brain.

NEUROSCIENCE, NOVELTY, AND NEUROGENESIS

Remarkable research in neuroscience over the past decade, dedicated to studies of the brain, has been summarized by two pioneering workers, Kempermann and Gage (1999):

> Contrary to dogma, the human brain does produce new brain cells in adulthood. . . . With continued diligence, scientists may eventually be able to trace the molecular cascades that lead from a specific stimulus, be it an environmental cue, or some internal event, to particular alterations in genetic activity and, in turn, to rises or falls in neurogenesis. Then they will have much more of the information needed to induce neuronal regeneration at will. Such a therapeutic approach could involve administration of key regulatory molecules or other pharmacological agents, delivery of gene therapy to supply helpful molecules, transplantation of stem cells, *modulations of environmental or cognitive stimuli,* alterations in physical activity, or some combination of these factors. . . . Compilation of such techniques could take decades. (p. 53, italics added)

Decades, indeed, or perhaps an entire millennium! With the words *"modulations of environmental or cognitive stimuli,"* Kempermann and Gage are intuiting a new neuroscientific worldview of what we would regard as the essential goal of psychotherapy and the healing arts. From a broad historical perspective we could say that the mission of the perennial philosophies, the humanistic arts, spiritual practices, and consciousness studies of all ages and cultures has always been a search for effective "modulations of environmental or cognitive stimuli" to better the human condition. It is only with the recent rich harvest of research on the continuous interplay among environment, gene expression, and psychological experience, however, that we can finally build a meaningful bridge between neuroscience and our humanistic heritage. In this chapter we survey the new findings of neuroscience for their potential contributions to a deeper understanding of the relationships between consciousness and the creatively-oriented psychotherapies and healing arts of the past, present, and future.

EARLY EVIDENCE OF NEUROGENESIS AND BRAIN GROWTH

The current era of research on neurogenesis and growth of the brain actually began more than two generations ago with the neuropsychological theory of the Canadian psychologist Donald Hebb (1949) and the pioneering microscopic observations of Joseph Altman (1962; Gross, 2000). Hebb proposed a molecular mechanism whereby connections and growth could take place between neurons during the acquisition of new memory, learning, and behavior. Hebb theorized that when the axon of neuron A stimulates neuron B repeatedly or persistently, a growth process takes place in one or both cells so that neuron A's efficacy in firing B is increased. The connecting links between neurons that replay this function—manifesting *an activity-dependent increase in their associative strength—are now called Hebbian synapses.* The early reports that active and intense experiences of new memory and learning in experimental animals were associated with more proteins and heavier brain weight supported Hebb's theory that *increased mental activity can lead to a growth process in the brain.*

Early research on the neurochemistry of memory and learning in this area was summarized (Rossi, 1972):

> During the past decade innovations in biochemical analysis that permit the isolation and measurement of extremely small amounts of macromolecular RNA

and enzyme systems, such as cholinesterase, have enabled researchers to determine how they are utilized in the neurochemistry of memory and learning. In essence, the evidence suggests that "changes in behavior may produce concomitant changes in the rate of synthesis and/or the structure of RNA and proteins," (Russell, 1966). In a clinical series of experiments (Hyden & Egyhazi, 1963) demonstrated that changes in the coding of RNA are associated with new learning in rats. Rats exposed to wide variations in environmental complexity and behavioral experience have significant differences in the acetylcholine systems of their brain tissues (Krech, Rosenzweig, & Bennett, 1964). (pp.143–144)

This early research was the precursor to what we now call "activity-dependent gene expression" in neurogenesis and the dynamics of memory, learning, and behavior that is the psychobiological core of psychotherapy and the healing arts.

ACTIVITY-DEPENDENT GENE EXPRESSION IN NEUROGENESIS, MEMORY, AND LEARNING

Three factors relate psychology to gene expression and neurogenesis: (1) *novelty* (Erikson et al., 1998), (2) *environmental enrichment* (Kempermann et al., 1997; Van Praag et al., 2000), and (3) *physical exercise* (Van Praag et al., 1999). Activity-dependent gene expression facilitates the generation of functional neurons in the hippocampus of the brain, which encodes new memory, learning, and behavior (Gould et al., 1999a, 1999b; Van Praag et al., 2002).

In a series of pioneering papers, Kandel (1983, 1989, 1998, 2000) proposed how molecular-genomic mechanisms could account for many of the classical phenomena of memory and learning as well as clinical psychopathology such as chronic anxiety, neurosis, and schizophrenia. Kandel's theoretical and experimental work is the clearest expression of the possibilities of activity-dependent gene expression as the molecular basis of psychosomatic medicine, psychotherapy, and the healing arts (Bailey & Kandel, 1995; Rossi, 1986/1993, 1996a). The role of activity-dependent gene expression and neurogenesis in the way we express ourselves with language (using measures of "idea density" and grammatical complexity) from an early age is evident in the quality of human experience as well as organic brain dysfunctions such as Alzheimer's disease. A major finding of Snowdon's (2001) research in his study of nuns suffering from Alzheimer's is that exercising whatever mental capacity one has with positively stimulating intellectual activities and emotions, offers some protection against age-related mental decline (Fredrickson, 2001; Lyubomirsky, 2001; Schneider, 2001).

The Anatomical Locus of Neurogenesis: The Hippocampus

Figure. 3.1 illustrates a cross section of the hippocampus, which is located in the medial temporal lobe of the brain. The human hippocampus is a small structure about the size of a child's thumb; it is enlarged in Figure 3.1 to illustrate the horseshoe area of the dentate gyrus that is involved with encoding new memory and learning. Clinical and experimental studies confirm that damage to the hippocampus in humans and animals interferes with the acquisition and storage of new memories and learning (Squire & Kandel, 1999). The current view of the function of the hippocampus and associated parts of the medial temporal lobe is that they receive input from many areas of the cortex during memorable stimulation and life events. *The hippocampus serves a binding function by converting short-term memory into long-term memory.* The hippocampus is only a temporary storage site for memories that are being processed, however. The final storage sites for memory are distributed throughout the cortex of the brain. Long-term memories find their ultimate localization in the same areas of the brain that initially processes incoming information about events, people, places, and things. Sets of neurons in the hippocampus can encode space and directions in activity-dependent firing patterns that generate internal cognitive maps as representations of outer space as well as time (Shors et al., 2001).

The Molecular-Genetic Locus of Neurogenesis: The Gene Expression and Protein Synthesis Cycle

Short-term memory, called *primary memory* by William James (1890), is a kind of *immediate or working memory* lasting a few minutes to an hour or so. Long-term memory, by contrast, can last weeks, months, years, or a lifetime. The difference between short- and long-term memories on a molecular level is illustrated in Figure 3.2 (see color insert). *Short-term memory utilizes existing*

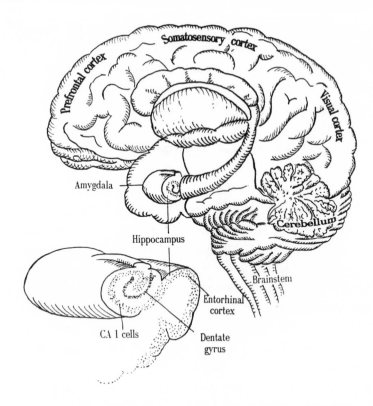

FIGURE 3.1 | Human brain with enlarged view of the hippocampus and associated areas such as the dentate gyrus and entorhinal cortex that encode new memory, learning, and behavior. Neurons in the CA1 region serve as "gating switches" in the construction and reconstruction of memory, particularly during the recall and replay of memories associated with fear. Fear memories processed via the amygdala return to a liable state when recalled and reframed in the gene expression and protein synthesis cycle. The zif-268 gene is involved in the fundamental dynamics how mind and meaning are updated during the replay of memory in dreaming, psychotherapy, and the healing arts.

pathways of neurotransmitters and molecular communication only. Long-term memory requires activity-dependent gene expression and new protein synthesis leading to the growth of new synapses and connections between neurons. The activation of the gene expression and protein synthesis cycle that engages synaptic growth to encode long-term memory, learning, and behavior requires a typical ultradian basic rest–activity cycle of one (Kohara et al., 2001) to several hours (Martin et al., 2000; Squire & Kandel, 1999). Memories are encoded and stored by increases in the strength, number, and patterns of synapses in the neural networks of the brain.

A great deal of research has been focused on the molecular "switch" that converts short-term to long-term memory and learning that can lead to lasting changes in behavior. Kandel's (2000) research on the gill-withdrawal reflex in the marine mollusk *Aplysia* provides an excellent model for elucidating the molecular dynamics of learning-related growth in the synaptic connections between neurons in the brain. The fundamental insight that the conversion from short- to long-term memory requires the activation of the gene expression and protein synthesis cycle connects traditional psychological approaches to memory with the emerging molecular biology of the Human Genome Project (Waelti et al., 2001). Figure 3.2 illustrates the bridge between traditional paradigms of memory and learning that emerged a century ago with the early work of Russian physiologist Pavlov and American behaviorists such as Watson and Skinner, and current research on neurogenesis at the molecular genetic level by Kandel and others.

Figure 3.2 illustrates the temporal dynamics of memory and learning at four loci in the synaptic bridges between sensory and motor neurons. Memory, learning, and extinction on the molecular level are illustrated at the first locus labeled *Extinction/Habituation*. Kandel's molecular model of the acquisition and loss of a learned association corresponds, in part, with what Pavlov and the classical behaviorists described as the "conditioning and extinction" of a learned response; that is, the building and loss of an association between a stimulus and response (Squire & Kandel, 1999). Because the marine snail *Aplysia* has a relatively simple nervous system, researchers were able to quantify the acquisition of new memory and learning by microscopic examination of the synaptic terminals between neurons. They determined that there is a doubling of the number of synaptic terminals in each sensory neuron from about 1,300 to 2,600 during new memory and learning. Without further reinforcement, this molecular memory decays over a period of three weeks, during which synaptic terminals gradually return to their original number.

There was an intense controversy about whether these growth processes took place in the presynaptic sensory neuron or in the postsynaptic motor neuron. It is now known that the molecular growth process is replayed, in a coordinated manner, via structural change on both the pre- and postsynaptic sides of the bridge between neurons. Long-term memory and learning may become manifest on the molecular level by the loss as well as the gain of synaptic terminals. In the process of habituation (extinction) in *Aplysia,* for example, the number of synaptic terminals may be reduced from the original 1,300 to about 800 per neuron.

Short-Term Memory: Neurotransmitters Signal Secondary Messenger Molecules

Short-term memory, illustrated at locus 2 on Figure 3.2, allows us to hold something in memory for about 1–20 minutes. Neurotransmitters such as serotonin, norepinephrine, dopamine, and acetylcholine are released from a modulatory neuron to stimulate receptors on the target neuron. A *secondary messenger molecule,* cyclic AMP (cAMP), is released to initiate a number of memory processes. The fastest and most direct molecular route for short-term memory occurs when cAMP leads to a temporary activation of the protein kinase PKA which acts at the synapse to enhance neurotransmitter release for a period of minutes. Kandel's research, for which he received the Nobel Prize in Medicine in 2000, documented how short-term memory is created by means of phosphate addition to proteins that make up the pores (or gates) in the cell membrane that calcium and other ions flow through in the energy dynamics of nerve transmission (Balter, 2000).

Long-Term Memory: Gene Expression and Protein Synthesis

Kandel and his research team found that with repeated stimulation from the environment, which is transmitted via modulatory interneurons (illustrated in Figure 3.2), genes receive signals to turn on the mechanisms that form the new proteins involved in memory and learning (Mayford et al., 1996). "Persistent PKA" is involved in a circular loop of communication from the environment to the nucleus of the neuron where the genes are located. Figure 3.2 illustrates how PKA initiates a genetic switch to turn on *immediate early genes* (CRE:EARLY) that then activate other *late response genes* (CRE:LATE) leading to the formation of proteins for the growth process of converting short-term to long-term memory. This activation of gene transcription, which is the molecular basis of neurogenesis and new growth, is illustrated at locus

3 in Figure 3.2. This source of psychobiological growth is called the *genomic action potential* (Clayton, 2000). As is typical of the gene expression and protein synthesis cycle throughout the brain and body illustrated earlier in figure 2.1, it requires an ultradian periodicity of about 90–120 minutes.

Long-Term Potentiation, Gene Expression, and Neurogenesis

Recent research is clarifying another mechanism contributing to long-term memory and learning, *long-term potentiation* (LTP), which is illustrated at locus 4 of Figure 3.2. In LTP nitric oxide gas functions as a messenger molecule that can diffuse a few cell diameters in all directions from the stimulated postsynaptic neuron, to mediate the encoding of long-term memory in a neural network. Long-term potentiation was originally described by Bliss and Lomo (1973), who found that when a living slice of brain tissue from the hippocampus was strongly stimulated, the neurons become more sensitive than usual to much smaller stimuli for several hours. The initial strong stimulus sensitized the brain neurons so that their response to much smaller stimuli was increased for an unusually long period of time; thus the name "long-term potentiation."

The role of LTP in new memory and learning has remained controversial since its original discovery. A series of studies by a variety of researchers using many different approaches, however, has recently confirmed the role of learning in inducing LTP in the mammalian neocortex. Rioult-Pedotti et al (2000), for example, have summarized their work in this area.

> The hypothesis that learning occurs through long-term potentiation (LTP)—and long-term depression (LTD)—like mechanisms is widely held but unproven. This hypothesis makes three assumptions: Synapses are modifiable, they modify with learning, and they strengthen through an LTP-like mechanism. . . . Here we investigated whether learning strengthened these connections through LTP. . . . [Our] results are consistent with the use of LTP to strengthen synapses during learning. (p. 533)

We will now overview those states of psychological arousal and LTP that are of significance for gene expression, neurogenesis, and a new theory of psychological growth in psychotherapy (Rossi, 1967, 1968, 1972, 1999c, 2000a).

Long-Term Potentiation, Psychological Arousal, and Neurogenesis

Since its initial discovery, LTP has been a major focus of interest for neuro-scientists exploring the cellular-genomic basis of memory, learning, behavior, and mood states. It is now believed that LTP takes place in many regions of the cerebral cortex. Johnston and Maio-Sin Wu (1995) define LTP "as an enduring, activity-dependent increase in synaptic efficiency . . . *which persists on the order of an hour or more . . . initiating gene expression*" (p. 459, italics added).

Another type of neural plasticity that appears to be the reverse of LTP is long-term depression (LTD). Computer simulations indicate that more information can be stored in a system that permits both increases (LTP) and decreases (LTD) in synaptic communication. Psychobiological processes appear to optimize memory and learning by maintaining a balance between arousal and depression. Under certain conditions, stimulation of presynaptic neurons can lead to an enduring depression in synaptic communication in the hippocampus, cerebellum, and neocortex. LTD (depression), in associa-tion with LTP (arousal), appears to play a complementary role in sculpting memory and learning along the lines of classical conditioning in the cere-bellum. Although the cerebellum has long been regarded as a "motor learn-ing machine," it is now known to contribute to cardiovascular conditioning, spatial memory, and complex cognitive processes (Beggs et al., 1999) that are surely involved in psychosomatic medicine and the therapeutic arts.

A model of the dynamic process by which LTP and LDP generates the formation and growth of new synapses to encode activity-dependent mem-ory, learning, and behavior has been proposed recently by Lüscher et al. (2000). As illustrated in Figure 3.3 (see color insert), LTP generates an increase in the number of synapses connecting one neuron with another within an hour. Within the first 10 minutes of LTP induction, there is an activation of the receptors that are involved in synaptic communication via neurotransmitters. Within the first 30 minutes, the size of the synaptic spine increases and the stimulated AMPA receptors move to the postsynaptic membrane; this leads to an increase in the size of the postsynapse. Within an hour, some synapses divide in two, as illustrated in Figure 3.3. This leads, in turn, to further growth in presynaptic multiplication and remodeling to create new neural networks for encoding the memory, learning, and behavior change that is of essence for psychotherapy and the ultradian healing response.

It is somewhat startling to realize how such experimental research serves as the molecular-genetic model for psychotherapy within the typical ultradian time period of about an hour. Does this imply that a shorter period is not sufficient for the molecular-genomic dynamics of psychotherapy? Not really. Common experience indicates that a brief but memorable life experience of a few seconds or minutes can change one's perspective and behavior for a lifetime. Figure 3.3 illustrates that a memorable experience can initiate changes at the molecular–genomic level that we can measure within the limitations of current laboratory techniques. We often "think about" and replay brief but significant life experiences. Such reviews and replays are the essence of many of the new approaches to creative experience and psychotherapy developed in Part II.

By analogy with the process of LTP, Lüscher and his colleagues propose that LTD leads to a removal of AMPA receptors so that the postsynaptic density decreases and there is a loss of synaptic strength and neurogenesis. The role of AMPA receptor-mediated communication, LTP, and LTD has been highlighted by the newly recognized interaction between neurons and the major type of glial cells, the astrocytes, in the brain (Gallo & Chittajallu, 2001). Previously it was thought that glial cells were only structural elements of the brain that physically supported the neurons. It is now known that the long, thin extensions of the astrocytes cover the synaptic connections between neurons and modulate the process of neurotransmission via the release of the neurotransmitter glutamate (Kandel, 2000; Oliet et al., 2001). Since the formation of neurotransmitters is itself an activity-dependent process of gene expression, at least in part, we can expect that future studies in this area of psychosocial genomics will uncover many individual differences in human experiencing that may be traceable to variations in single nucleotide polymorphisms (SNPs), described earlier in Chapter 1.

Activity-Dependent Dendritic Growth in Neurogenesis

A recent review of the details of the actual growth process at the dendritic level of neurogenesis by Matus (2000) fills in some of the details of the mechanisms proposed by Lüscher et al. Matus begins by noting that during brain development, neurogenesis proceeds in a number of distinct phases. The first phase, which takes place during embryogenesis, involves the initial creation of neural networks under the guidance of neurotropic messenger molecules.

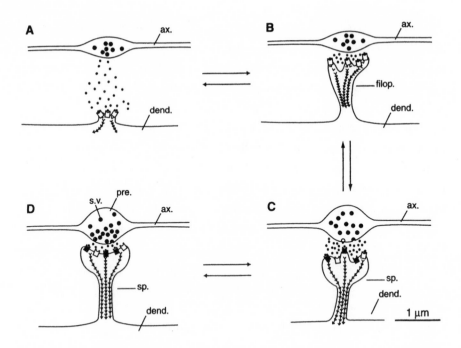

FIGURE 3.4 | Reversible neurogenesis at the level of dendritic growth as proposed by Matus (2000). Phases A and B illustrate the initial steps of dendritic growth stimulated by the activation of NMDA receptors that are involved in memory and learning. Phases C and D illustrate the activity-dependent acquisition of AMPA receptors that stabilizes the growth of dendrites. The reversible arrows illustrate a fundamental psychobiological mechanism for the creation and maintenance of memory and learning as well as its change, reframing and transformation with further experience. Research suggests that the transition states (B, C, & D) in this sequence are reversible depending on the strength and maintenance of activity-dependent gene expression and neurogenesis. (With permission from Matus, 2000).

The second phase takes place after birth, during a relatively short critical period, when synaptic connections are refined by activity-dependent sensory and motor experience. The third phase, which is active throughout the life span, engages activity-dependent memory and learning.

Figure 3.4 illustrates neurogenesis at the level of dendritic growth as proposed by Matus. Phases A and B in the figure illustrate the initial steps of dendritic outgrowth, which appears to be stimulated by the activation of NMDA (n-methyl-d-asparate) receptors that are known to be involved in memory and learning. Phases C and D illustrate the activity-dependent acquisition of AMPA receptors that stabilizes the growth dendrites.

It is of greatest importance to recognize the reversible arrows between the four phases of this growth process of neurogenesis at the level of dendritic spine development and stabilization. *This reversibility illustrates a fundamental psychobiological mechanism for the creation and maintenance of memory and learning as well as its change, reframing, and transformation with further experience.* Current research suggests that the four transition states (A, B, C, and D in Figure 3.4) in this sequence of development are reversible *depending on the strength and maintenance level of activity-dependent stimulation at the presynaptic terminal.* Matus (2000) summarizes the interaction between the psychological level of memory and learning and the molecular level of neurogenesis at the dendritic level in this way:

> Such a scheme may help explain experience-dependent shaping of neuronal circuits because it would make sense to require a mechanism that depends on strong stimulation, like that of NMDA receptor activation, for the initiation of new connections but to then support them with a mechanism sensitive to low rates of neuronal activity, like that of AMPA receptor activation. Learning responses could thus be maintained during periods when they are not actually being used. Similarly, should synaptic stimuli fall below a certain threshold indicative of disuse, it would be appropriate for synaptic connections to be broken and re-formed in new configurations. (p. 757)

This process of neurogenesis and neural networks being "re-formed in new configurations" appears to be the psychobiological foundation of the "re-framing of psychological states" during psychotherapy and the healing arts that we will illustrate in detail of Part II. For now, we will illustrate this mind-body interaction by looking at a new view of the growth process, or the lack of it, in the relationship between the psychological state of depression and neurogenesis.

A NEW THEORY OF PSYCHOLOGICAL DEPRESSION: A LACK OF NEUROGENESIS

In clinical states of depression patients are caught in cycles of emotional gloom with accompanying loss of appetite, fatigue, mental confusion, memory problems, absence of motivation, and sleep disturbances. The neuroscientific approach to clinical depression has generally focused on an imbalance of neurotransmitters in the brain. The presence or absence of neurotransmitters, however, is itself fundamentally dependent on the adequacy of the gene expression and protein synthesis cycle that generates the proteins that make the neurotransmitters. With the growing recognition of the role of gene expression in human experience, many neuroscientists are now exploring the relationship among gene expression, neurogenesis, and psychological depression. Although it is still controversial, this new theory emphasizes that the varying states of depression are associated with "a lack of new cell growth in the brain" (Vogel, 2000a, p. 258).

Evidence of the association between psychological depression and a lack of neurogenesis comes from the finding that depressed patients have 12–15% less volume in their hippocampus (Sheline et al., 1996). Follow-up studies have confirmed this finding, prompting neurobiologist and stress researcher Robert Sapolsky to conclude, "It is absolutely clear that really prolonged major depression is associated with loss of hippocampal volume" (quoted in Vogel, 2000a, p. 259). Decades of experimental studies with animals have documented how chronic states of stress lead to heightened levels of glucocorticoids, which cause the death of cells associated with new memory and learning in the hippocampus (Sapolsky, 1990, 1992, 1996). Trauma and psychosocial stress can lead to a suppression of neurogenesis and growth processes in primate brains (Gould et al., 1998) that may be manifest as the psychological experience of depression in humans (McEwen, 2000; Vogel, 2000a).

Further support for this new theory of the association between depression and a lack of neurogenesis comes from the molecular mechanisms by which antidepressant medications reverse depression. Antidepressants lead to increases in the availability of neurotransmitters that facilitate communication between neurons as well as to actual increases in neurogenesis, as measured by increases in brain volume. Prozac (fluoxetine) and similar drugs that increase the availability of the neurotransmitter serotonin, for example, are well-known promoters of cell growth during fetal development as well as in adult animals. The mood stabilizer lithium has recently been shown to increase brain gray matter volume after four weeks of treatment, as measured by magnetic resonance imaging. It is reported that three different classes of antidepressant drugs as well as electroconvulsive (ECT) therapy all lead to the generation of significantly more dividing cells in the hippocampus (Vogel, 2000a). This suggests that increased neurogenesis is at least one basic mechanism for reversing psychological depression by antidepressant medication as well as ECT.

As is typical of all biological processes, there is a range of responses between too little, too much, and just the right amount of stimulation. Too little stimulation can result in a lack of activity-dependent gene expression and neurogenesis that leads to depression. The right amount of stimulation, surprise, and pleasant shock leads to optimal gene expression and neurogenesis to support memory, learning, and healing (Rossi, 1973a, 1973b). Excess stimulation leads to changes in gene expression associated with posttraumatic stress disorder (Kaufer et al., 1998). An appropriate dose of ECT can simulate gene expression and neurogenesis to facilitate recovery from depression. Too much electroconvulsive shock, as dramatically illustrated in a series of illustrations in Figure 3.5 (Steward, et al., 2000; Wallace et al., 1998), however, leads to the typical ultradian parameters of PTSD. The amount of gene expression taking place after excessive ECT in the rat can be seen as a dark pigmentation that is a measure of gene expression (mRNA accumulation). Gene expression is clearly evident within the dentate granule cells of the hippocampus in the first 15 minutes after shock. Gene expression continues to increase to an ultradian peak of intensity between 2 to 4 hours, and then it returns slowly to basal levels of recovery between 8 to 12 hours. There could hardly be a more vivid demonstration of the effects of severe shock and trauma on the ultradian time parameters of PTSD. Let us now turn our attention to some of the molecular-genomic processes of recovery from trauma and stress by the reconstructive dynamics of growth-oriented psychotherapy.

THE NEUROSCIENCE OF GROWTH THERAPY: THE RECONSTRUCTIVE DYNAMICS OF TRAUMATIC MEMORY, LEARNING, AND BEHAVIOR

Relationships among psychobiological arousal, depression, and stress have important implications for understanding the reconstructive dynamics of psychotherapy. The time parameters of activity-dependent gene expression and neurogenesis in generating the dynamics of memory, learning and behavior

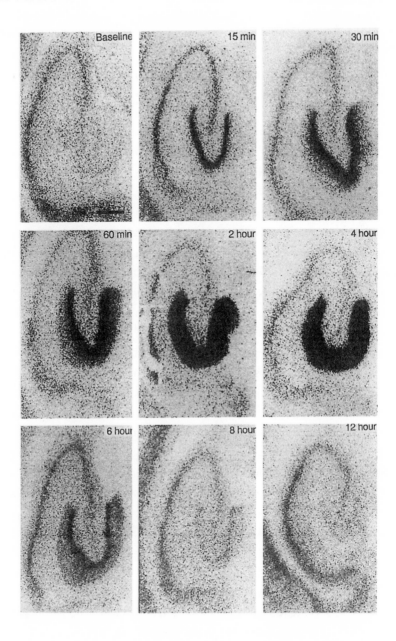

FIGURE 3.5 | A vivid demonstration of the ultradian time parameters of posttraumatic stress in the hippocampus in response to a single electroconvulsive shock. The darkened areas indicate a peak in diving cells within 2–4 hours and a gradual return to baseline in about 12 hours. (With permission from Wallace, Lyford, Worley, & Steward, 1998. Differential intracellular sorting of immediate early gene mRNAs depends on signals in the mRNA sequence. *Journal of Neuroscience, 18*, 26–35. Copyright 1998 by the Society of Neuroscience.)

are currently being discussed by many investigators (Gross, 2000; Shimizu et al., 2000). There is evidence that many adult-generated neurons have a time-limited existence that is correlated with the growth and decay of memory (Milner et al., 1998) and its transformations via psychotherapy (Rossi, 1972, 1985, 1996a, 2000a, 2000b, 2000c, 2000d).

Memory is not a static or reliable record written "in stone." Living memory is a process of construction and reconstruction over time (Izquierdo et al., 1988), a highly individualized and subjective process colored by personal history before and after the original event. As the ancient Greek philosopher Heraclitus said, we cannot step into the same river twice. This malleability or plasticity of memory may be the bane of oral history and the legal profession, but it contains the essential therapeutic seed of psychotherapy and the healing arts.

The recall of a memory in everyday life invariably involves a mingling of the old representations of the past with the new perceptions of the present. This mingling of past and present results in new associations that lead to a natural process of change and transformation in memory. Current research documents how the classical process of Pavlovian stimulus-response conditioning requires the activation of a conditioned response before it can be extinguished (Dudai, 2000). *This process of reactivating a memory in order to extinguish it has profound implications for the practice of psychotherapy and, by extension, many of the therapeutic arts.*

Research by Shimizu et al. (2000), for example, clarifies how a particular region of the hippocampus, called "CA1," is crucial for converting new memories into long-term memories. This psychobiological process may continue for weeks after the initial learning and memory event. They found that the NMDA receptor in the CA1 region serves as a "gating switch" in the construction and reconstruction of memory. They summarize their findings:

> Our results indicate that *memory consolidation may require multiple rounds of site-specific synaptic modifications, possibly to reinforce plastic changes initiated during learning,* thereby making memory traces stronger and more stable. Recent studies report that *the learning-induced correlation states among CA1 neurons are reactivated spontaneously in a post-learning period.* Such a *co-activation* of these neurons might suggest the existence of the natural condition within the hippocampus by which recurrent synaptic strengthening can occur during memory consolidation. We

hypothesize that such a *synaptic re-entry reinforcement* (SRR) process can also be applied to explain how the hippocampus transfers newly created memories to the cortex for permanent storage. As the hippocampus undergoes *reactivation during consolidation, it may also act as a coincidence regenerator for activating neurons* in the cortical area such as the association cortex. *This would allow cortical neurons previously corresponding to the different sensory modalities to be reactivated together,* leading to the strengthening of the connections between them through SRR. Indeed, such a *coordinated reactivation of hippocampal-cortical neurons after learning* has been observed recently. . . . Once these cortical connections are fully consolidated and stabilized, the hippocampus itself becomes dispensable for the retrieval of the "old memory." . . . Therefore, we postulate that the hippocampus, by serving as a *coincidence regenerator,* may induce the reinforcement of synaptic connection within the cortex during memory consolidation as the cellular means to convert short-term memories into long-term memories. (pp. 1172–1173, italics added)

This summary outlines a neuroscientific approach to dealing with many of the mysteries of memory consolidation that are significant for psychotherapy and the healing arts. One is the so-called "binding problem of consciousness": How are the qualia of sensory and cognitive associations bound together to make up a single, holistic perception we call consciousness? The answer on the neural-brain level is implied when Shimizu et al. state: "This would allow cortical neurons previously corresponding to the different sensory modalities to be reactivated together . . . [through] the coincidence regenerator." One wonders if this explanation could account for our delight in, and sense of mystery about, the "meaning" of coincidence and numinous experience (Jung, 1960; Richardson, 2000). The work of Shimizu and colleagues is consistent with the synthetic, constructive, and growth approach to understanding dreams as developed previously (Rossi, 1972, 1973b, 2000a, 2000d) and updated in the next chapter. The *reactivation process* that takes place *spontaneously* in normal everyday life as we *review and replay* significant life experiences is inherent in the development of wisdom. On the negative side, the reactivation process could be the source of obsessive and compulsive ruminations when we are experiencing overwhelming effects of trauma and stress and stuck in stage two of the four-stage creative process (the incubation stage of Figure 2.11). In Part II we discuss and illustrate strategies for reducing trauma and stress by transforming symptoms and problems into signals and creative resources.

An ingenious research design by Nader et al. (2000a, b) illustrates how this malleable, plastic, and reconstructive aspect of memory contains the neurobiological seed of psychosocial genomics as well as the growth paradigm of psychotherapy (Rossi, 1967, 1968, 2000a-d, 2001). These researchers demonstrated how fear memories that require protein synthesis in the amygdala of the brain could be disrupted and made to disappear when they are reactivated under special experimental conditions. *Their research illustrates how consolidated fear memories return to a labile state when reactivated by the gene expression and protein synthesis cycle during recall.* When the rat brain is infused with anisomycin (an inhibitor of protein synthesis) shortly after the reactivation of a long consolidated memory (from one to 14 days after conditioning), the memory is extinguished. The same treatment of the brain with anisomycin but without reactivating the consolidated memory leaves the memory intact. *This means that the gene expression and protein synthesis cycle is reactivated when important memories are recalled and replayed.* As these researchers phrased it:

> Consistent with a time-limited role for protein synthesis production in consolidation, delay of the infusion until six hours after memory reactivation produced no amnesia. Our data show that consolidated fear memories, when reactivated, return to a labile state that requires *de novo* protein synthesis [and gene expression] for reconsolidation. These findings are not predicted by traditional theories of memory consolidation. (Nader et al., 2000a, p. 723)

It is certainly true that the traditional academic theories of classical conditioning, based on stimulus-response learning, do not predict the natural modification and/or disappearance of memories upon reactivation. However, most paradigms of psychotherapy over the past few centuries, ranging from hypnosis and classical psychoanalysis to the behavioral and cognitive therapies, involve a combination of the same two-step process that is the essence of Nader et al.'s research design. The central dynamic of classical and modern psychotherapies typically involves (1) a reactivation of old traumatic memories, which is (2) immediately followed by some form of therapeutic intervention designed to heal the old hurt one way or another (Rossi, 1996a). Many traditional, spiritual, and shamanistic healing rituals, wherein this two-step dynamic has long been evident, involve (1) an initial phase of heightened emotional arousal (sometimes described as a hysteria-like "crisis" in the psychiatric literature), which is followed by (2) a relaxing ritual that is experienced as healing within the patient's cultural belief system (Edmonston, 1986; Ellenberger, 1970; Greenfield, 2000; Zilboorg & Henry, 1941). This two-step process of arousal and relaxation, illustrated in the top image of Figure 2.11, is the psychobiological foundation of the four-stage creative process described in greater detail in Chapter 6.

In the field of therapeutic hypnosis this two-step paradigm was systematized by Pierre Janet (1925) in his method involving suggestion and reeducation: Patients were (1) first led into an emotional reliving of traumatic experiences, which was then (2) followed by suggestions that reorganized the significance of the patient's experiences. This two-step approach became the basis for Freud's classical psychoanalysis, wherein the patient is encouraged to (1) "free associate" whatever comes to mind; when the patient stumbles, apparently by accident, onto old traumatic memories that evoke fresh tears and affect, the psychoanalyst then (2) intervenes in the reactivated memory by adding a new insight or interpretation that reframes its meaning. In the best of circumstances this insightful reinterpretation of the old memory alters it by placing it into a new context (this is the "reframe") so that the patient feels better and symptoms seem to disappear of their own accord.

In classical behavior therapy (Wolpe, 1969), the patient is (1) reexposed in current, real life to a traumatic or phobic situation with the reassuring and supporting presence of the therapist, which leads to (2) a therapeutic reframe and change in the way the patient experiences the old trauma or phobia.

In recent innovations of the cognitive-behavioral approach, such as EMDR (eye movement desensitization and reprocessing), there is a carefully scripted plan to (1) reactivate and reprocess the past problematic experience, which is then (2) infused with the positive thoughts, images, and feelings that constitute the therapeutic change the patient would like to realize (Bjick, 2001; Hollander & Bender, 2001; Shapiro, 1999; Shapiro & Forrest, 1997). Likewise in AMT (anxiety-anger management training), "The fundamental principle is the importance of (1) exposing individuals to anxiety arousal and then (2) deactivating the emotional arousal through relaxation skills" (Suinn, 1990, 2001, p. 32; Suinn & Richardson, 1971). Notice how these two-step cognitive-behavioral approaches mirror the general psychobiological process of *arousal and relaxation* in the *basic rest-activity* and *creative cycle* illustrated in Figure 2.11. What is of great value in these cognitive-behavioral approaches is that they are highly structured processes that are well researched, easily learned, short-term (about six sessions), and applicable to wide range of problems.

Nader et al.'s (2000a, 2000b) research gives us a new understanding of the molecular dynamics underlying the gene expression and protein synthesis cycle in the recall and therapeutic modification of traumatic experiences. In association with the current neuroscience reviewed previously, we could now hypothesize that Nader et al.'s research also illustrates an essential molecular mechanism of creativity and the transformation of consciousness in everyday life. In the common parlance of psychotherapists today, it is said that "every recall is a reframe." That is, every recall of a memory automatically reframes or changes it in some way. This research provides strong experimental validation of this psychotherapeutic experience on the molecular level of the gene expression and protein synthesis cycle.

Further support for the "every recall is a reframe" principle comes from more recent research on the molecular-genomic dynamics of novel learning and memory versus relearning with familiar cues. Berman and Dudai (2001) describe their research in this area, as follows:

> Experimental extinction is the decline in the frequency or intensity of a conditioned response following the withdrawal of reinforcement. It does not reflect forgetting due to the obliteration of the original engram, but rather "relearning," in which the new association of the conditioned stimulus with the absence of the original reinforcer comes to control behavior. . . . If extinction is indeed an instance of learning, why does it not share with the original learning the activation of the muscarinic receptor and of the MAPK cascade? [See Figure 5.2 for a partial illustration of the role of MAPK in the psychosocial genomics of communication at the cellular level]. . . . The reason could be that in CTA [conditioned taste aversion] training, the taste is unfamiliar [that is, novel], whereas in extinction it is already familiar. *This is in accordance with the suggestion that the cortex contains molecular "novelty" switches, which are turned on only on the first highly salient [numinous] encounter with the stimulus.* . . . In conclusion, we show that extinction of long-term CTA memory is subserved by the same brain region that subserves the acquisition and consolidation of that same memory. Further, we identify in the IC [insular cortex] essential core elements of the molecular

machinery of learning, with other obligatory elements added according to the stimulus dimension and content. *This means, among other things, that the cortex honors at the molecular level the distinction between learning anew and learning the new.* (pp. 2417–2419, italics added)

We can recognize in the research of Nader et al. (2000a, 2000b) and Berman and Dudai (2001) the essential mechanisms of psychosocial genomics in the cognitive-behavioral psychotherapies, therapeutic hypnosis, and the healing arts. This new psychobiological understanding can function as an Occam's razor to eliminate everything that is proprietary and irrelevant in the plethora of contending schools of psychotherapy and the healing arts. In the following sections we do just that: We focus on the "bare bones" of a neuroscientific approach to understanding the role of consciousness in the creative growth process that is of essence in all the therapeutic arts.

The Role of Subjective Consciousness in Focusing Gene Expression and Neurogenesis to Build a Better Brain

We are now in a position to better appreciate Schrödinger's (1944) metaphor "that consciousness is the tutor who supervises the education of the living substance . . . new situations and . . . new responses . . . are kept in the light of consciousness; old and well practiced ones are no longer so" (p. 103). Current consciousness studies, however, have repeatedly challenged the value and significance of consciousness in human experience. Rather than celebrating and continuing the exploration of all possible opportunities for facilitating the human experience of consciousness, the 20th century seems to have taken a perverse delight in baiting what has been called "the conceits of consciousness." Many of the most famous paradoxes of our century, such as Bertrand Russell's paradoxes of logic, Gödel's incompleteness theorem, Turing's halting problem, and Heisenberg's uncertainty principle, have been interpreted as degrading the significance of consciousness as a supreme value (Chaitin, 2000; Robertson, 2000). The mathematician Morris Klein (1980) summarized the history of this provocatively doubting zeitgeist of our century in his book, *Mathematics: The Loss of Certainty.*

Current neuroscientific research emphasizes the view that human experience seems to be determined more by what are called unconscious or "implicit" processes than by conscious or "explicit" processes. For example, during a lifetime of research, the neurobiologist Benjamin Libet (1993) determined experimentally that human consciousness is usually about half a second behind the brain's "readiness potential" that determines, on an implicit level, how we will behave in an apparently voluntary manner that we experience as free will. Tor Nørretranders' (1998) popular book, *The User Illusion*, recounts recent arguments from the scholarly and famous that imply how Libet's research supports the view that consciousness is a thin, fragile thread that is riddled with error and illusion. It is easy to make this argument when reviewing the fearful consequences of the flimflam foibles and fallacies of human history. One wonders whether consciousness can get anything right (Rossi, 2000a, d). Indeed, some have asked: Why has consciousness evolved at all? Why aren't we zombies with no consciousness (Koch & Crick, 2001)? What good is consciousness? What does consciousness do (Hameroff et al., 1996)?

Conscious Explicit Memory
Versus Unconscious Implicit Processing

A ray of hope for understanding the significance of consciousness comes from Squire and Kandel's (1999) analysis of the implications of current research on the molecular-genomic level of neurogenesis. A new understanding of the role of consciousness follows from theory and research on the distinction between declarative or explicit versus nondeclarative or implicit memory. Declarative memory is the conscious recollection of information about places, objects, and people that we can talk about. Nondeclarative memory encodes the *unconscious performance or implicit information* about perceptual, motor, and cognitive skills and habits that we cannot explain in detail. Conscious, voluntary, explicit, declarative memory is used to tell a psychotherapist the story of one's life. We use unconscious, involuntary, implicit, nondeclarative memory, on the other hand, to automatically balance and peddle a bike without thinking too much about it. Nondeclarative memory is not necessarily the Freudian unconscious that involves psychologically-motivated suppression of unpleasant memory. Rather, nondeclarative memory is more like the automatic stimulus-response learning of Pavlov's classical conditioning that guides behavior after it is well learned and no longer requires conscious monitoring.

Now comes the important part for an updated understanding of the significance and role of consciousness in gene expression, neurogenesis, and psychotherapy as well as creative work in the arts and culture. Both declarative and nondeclarative memories are located throughout the brain in the same sensory, perceptual, and motor systems that originally encoded them. Musical memory, for example, is encoded in the auditory pathways of the brain, while visual memory is encoded in the visual pathways. *Conscious declarative memory in humans and higher animals, however, requires an engagement of the hippocampus in order to accomplish the transition from short- to long-term memory. It is this engagement of the hippocampus in the creation of declarative memory that may account for many of the special qualities of human experience that could provide us with clues about why consciousness evolved.* As described above, the hippocampus is the site where short-term memory (a few minutes) is converted into the typical long-term memory (hours) that can be maintained and transformed throughout a lifetime.

The hippocampus and its associated components in the medial temporal lobe of the brain bind together the sources of our scattered life experiences into a single fabric of long-term conscious declarative memory that becomes the perpetually changing thread of our personal identity. This suggests that the hippocampus and its pathways to the prefrontal lobes of the brain (associated with thinking, planning, and foresight) are intimately involved in facilitating the dynamics of consciousness, self-reflection, self-awareness, and an integrated view of life (Rossi, 2000a, 2000b, 2000c, 2000d, 2001). This view of the important role of the hippocampus in consciousness, mind, memory, and meaning is well expressed by Shors et al. (2001):

> It has been suggested that hippocampal-dependent learning processes, such as those involved in trace conditioning, involve *conscious awareness or recollection, and that the hippocampus performs such functions* (Clark & Squire, 1998; LaBar & Disterhoft, 1998). (p. 374)
>
> The vertebrate brain continues to produce new neurons throughout life. In the rat hippocampus, several thousand are produced each day, many of which die within weeks. Associative learning can enhance their survival; however, until now it was unknown whether new neurons are involved in memory formation. Here we show that a substantial reduction in the number of newly generated neurons in the adult rat impairs *hippocampal-dependent trace conditioning, a task in*

which an animal must associate stimuli that are separated in time. (p. 372)

On the basis of the relative immaturity of the neurons and their apparent sensitivity to both temporal and spatial associations, it could be hypothesized that the cells are used to maintain stimulus representations of newly experienced stimuli in the event that these representations are associated with subsequent events in time and space (Shors et al., 2000, 2001; Wallenstein et al., 1998).

Although it seems that the new cells are associated with the formation of hippocampal-dependent memories, we note that the rats treated with saline learned the delay task faster than the trace. Thus, *newly generated neurons may not be used for learning under more lenient conditions, but become involved as task demands increase.* A reduction in the number of newly generated neurons in the dentate gyrus of the hippocampal formation impaired hippocampal-dependent, but not hippocampal-independent forms of associative memory formation. These results suggest *that immature neurons in the adult brain are involved and perhaps necessary for the acquisition of new hippocampal-dependent memories about temporal relations or the accurate timing of learned responses, such as during the acquisition of trace memories.* (p. 375, italics added)

Current research on the consolidation and reconstruction of memory conceptualizes the replay between the hippocampus and cortex as a central dynamic in the constant process of reframing and updating consciousness (McGaugh, 2000). This concept has been well summarized by Lisman and Morris (2001):

. . . newly acquired sensory information is funneled through the cortex to the hippocampus. Surprisingly, only the hippocampus actually learns at this time— it is said to be online. Later, when the hippocampus is offline (probably during sleep), *it replays stored information, transmitting it to the cortex.* The cortex is considered to be a slow learner, capable of lasting memory storage only as a result of this *repeated replaying of information by the hippocampus.* In some views, the hippocampus is only a temporary memory store–once memory traces become stabilized in the cortex, memories can be accessed even if the hippocampus is removed. *There is now direct evidence that some form of hippocampal replay occurs. . . .* These results support the idea that the hippocampus is the fast online learner that "teaches" the slower cortex offline. (pp. 248–249, italics added)

Note the haunting similarity of this idea to Edwin Schrödinger's (1944) prescient view quoted at the head of this chapter:

> One might say, metaphorically, that consciousness is the tutor who supervises the education of the living substance, but leaves its pupil alone to deal with all those tasks for which it is already sufficiently trained. . . . The fact is only this, that new situations and the new responses they prompt are kept in the light of consciousness; old and well practiced ones are no longer so. (p. 103)

This replay between the hippocampus and the cortex in the construction and reconstruction of memory is related to the basic Hebbian paradigm of activity-dependent learning at the synaptic level by Lisman and Morris (2001):

> . . . a connection between two cortical neurons might be formed and strengthened offline [that is, during hippocampal replay] in a two-step process. The first step would be an ongoing process involving the perhaps random growth of new synapses and the breaking of older, non-stabilized ones. We would assume that this so-called "random growth" actually involves Darwinian natural variation and selection. *As the replay occurs repeatedly, when two cells do by chance become physically connected, they would then be sensitive to LTP if they were simultaneously excited as a result of hippocampal replay.* . . . LTP is thought to be a molecular mechanism underlying learning, and requires the protein alpha-CaMKII (Frankland et al., 2001). This protein [generated as a result of the activity-dependent gene expression and protein synthesis cycle described above] is greatly enriched at many excitatory neuronal connections in the central nervous system.
>
>   The observed development of new synapses in adult brains is consistent with this proposal. When animals are raised in mentally stimulating environments, the number of synapses increases. . . . These studies show that changes in the input activity to the cortex can lead to increases in axonal branching and in the number of synaptic boutons [synaptic terminals]. *This suggests that new connections can be made in adult brains, and that these connections are determined by processes that depend on neuronal activity.* (p. 249, italics added)

What, then, is the role of consciousness? *Consciousness focuses the replay of gene expression, protein synthesis, and neurogenesis on important time-related associ-*

ations, memory, and learning when "task demands increase." What attracts consciousness? As detailed above, *novelty, environmental enrichment, and physical exercise.* Squire and Kandel (1999) review research with newborn human infants that documents how a colored image—say, a blue square, for example—will attract and hold their attention until the novelty of it wears off. Then substitute a red square and their attention is again immediately fixated—until the novelty of it wears off. We now know that novelty, environmental enrichment, and physical activity can activate psychobiological arousal that evokes the gene expression and protein synthesis cycle of neurogenesis. Anything that is experienced as new, different, changing, shocking, or surprising in any way immediately attracts consciousness, which facilitates the transforming reframes of declarative memory via the hippocampus. We could now propose that this type of transformation at the cellular-genomic level is the psychobiological basis of what many constructively oriented thinkers have emphasized as the pathways of creativity, psychotherapy, and the healing arts. These pathways range from the Socrate's method of creative questioning in ancient Greece, to the educational therapy of Pierre Janet (1925), to Carl Jung's (1966) "synthetic or constructive method," to current innovative psychotherapists (Corsini, 2001).

Making an *active effort to consciously rehearse and replay* any behavior, whether an outer activity like sports or an inner focus as in memorizing a poem, generates activity-dependent gene expression, protein synthesis, and neurogenesis. *Consciousness may be the fragile thread of civilization from an outer perspective, but it is more like an intense laser beam focusing and fusing the psychobiological basis of human experience from the inner subjective perspective.* Is this why we have an inner life of psychological experience? Sensitive but highly focused consciousness facilitates inner rehearsal and creative dialogue with the emergent, novel, and numinous images of our dreams and fantasies that Jung called "active imagination." Active imagination is a dramatic and numinous experience of arousing, engaging, and facilitating the psychobiological processes of healing and individuation. From a psychobiological perspective this is what consciousness does best: *Highly motivated states of consciousness turn on and focus gene expression, protein synthesis, and neurogenesis in our daily creative work of building a better brain.*

Creative Replay in the Dynamics of Consciousness

Experimental evidence for the role of consciousness in the inner rehearsal and creative replay of experience between the hippocampus and the cortex is consistent with the Nobel Laureate Gerald Edelman's view of "how matter becomes imagination." Edelman & Tonini (2000) have described the role of "reentry" in the neural dynamics of consciousness:

> Finally, if we consider neural dynamics (the way patterns of activity in the brain change with time), the most striking special feature of the brains of higher vertebrates is the occurrence of a process we have called reentry. Reentry ... depends on the possibility of cycles of signaling in the thalamocortical meshwork and other networks [of the brain]. . . . It is the ongoing, recursive interchange of parallel signals between reciprocally connected areas of the brain, an interchange that continually coordinates the activities of these area's maps to each other in space and time. This interchange, unlike feedback, involves many parallel paths and has no specific instructive error function associated with it. Instead, it alters selective events and correlations of signals among areas and is essential for the synchronization and coordination of the area's mutual functions. (p. 48)

One of the most important characteristics of the reentry process is the widespread synchronization of the activities of many different neural groups distributed over many area of the brain that mediate sensory, perceptual, emotional, and cognitive processes. Edelman and Tonini (2000) comment further that reentry is the most unique feature of higher brains: "There is no other object in the universe so completely distinguished by reentrant circuitry as the human brain" (p. 49). In his theory of neuronal group selection, Edelman (1987) proposed and documented how Darwin's theory of natural selection can be extended to the developmental, neuronal, cellular, and molecular levels. Cirelli et al. (1996) have supported and extended Edelman's neural Darwinism theory to include the process whereby changes in neuronal gene expression in the waking state are associated with areas of the brain connected with consciousness, such as the locus coeruleus. Tononi et al. (1995) described how change in gene expression during the sleep-wake cycle provides a new view of activating systems in the brain. *Taken together, these neuroscientific developments, ranging from gene expression to the replay and reentry dynamics of neural activity, set the stage for the role of consciousness as a novelty-seeking modality of experience.*

CONSCIOUSNESS AS A NOVELTY-SEEKING MODALITY

Developmental Studies of the Co-Creation of Intelligence

The most salient characteristic of consciousness is that it is sensitive and responsive to novelty. Consciousness is context-sensitive, highly flexible, and adapts to nuances of change. These features are consistent with current research that documents how novelty, along with environmental enrichment and exercise, are important conditions for facilitating gene expression and neurogenesis. Consciousness, novelty, gene expression, and neurogenesis comprise a mutually supportive, complex adaptive system. Consciousness is necessary for memory and learning when large amounts of information need to be assimilated and integrated by selecting whatever is relevant and new in any situation. *This leads us to view consciousness as a novelty-seeking modality that evolved as a sensitive detector to facilitate rapid and creative adaptation to environments manifesting constant change and Darwinian variation* (Zeki, 1999, 2001).

It is only now, after several generations of research on the developmental aspects of cognition in children (Rossi, 1964), that we can appreciate the role of consciousness, attention, and awareness in the novelty, gene expression, and neurogenesis cycle. This research began with the now legendary ideas of Vygotsky (1962) on the early development of thought and language as an internalization of physical activity in children, and the constructivist views of child development by Piaget (1954). While there has been an engaging controversy about the interpretation of this early research, over the generations methodology in child development has gained in scientific sophistication by exploring the connections among consciousness, novelty, and attention span.

As any mother knows, a child's attention span and learning is intimately associated with what is novel and interesting. In a variation of the research on novelty and attention span involving blue and red squares described above, Starkey et al. (1980, 1983) showed 16- to 30-week-old babies a visual slide with two bright dots that immediately caught their visual gaze and evident conscious attention. With repeated showing of the same slide, the novelty effect wore off and the babies spent less and less time gazing at it. When the slide was replaced with a new one containing a slightly novel variation in the arrangement of the dots, the babies' attention was recaptured for a longer span of time. With each repetition, attention span again grew less and less until the slide suddenly showed the novelty effect of a shift from two to three dots that again rekindled the babies' consciousness and extended their attention span. Antell and Keating (1983) used this same experimental methodology with infants just a few days old to show that they could distinguish between the numbers two and three.

Bijeljac-Babic et al. (1991) extended this association between consciousness and novelty from the visual to the auditory modality with an ingenious experimental design that explored the association among novelty, attention span, and the sucking reflex in four-day-old infants. The infants were given a nipple connected to a pressure transducer and a computer that recorded response times to the delivery of two or three nonsense syllables over a loudspeaker. Initially the babies showed great interest and attention span by sucking vigorously on the nipple. When the novelty wore off the sucking rate decreased. This led the computer to shift the number of syllables between two and three, which immediately reawakened a novelty-interest-behavior-effect, which was replayed as a renewed high rate of sucking.

These developmental studies of consciousness and cognition in infants took place before the current neuroscientific research on the relationship among novelty, gene expression, and neurogenesis. Taken together, however, these independent lines of research have profound implications for a theory of consciousness, novelty, gene expression, and neurogenesis as a complex adaptive system that bridges the Cartesian gap between mind and body. A more comprehensive theory, however, requires an understanding of how the conscious experience of novelty generates the interest and motivation that evokes psychobiological arousal and its attendant dynamics of gene expression, protein synthesis, and neurogenesis.

Consciousness, Surprise, and the Human Quest

A recent paper by Waelti et al. (2001) reports the discovery of a new mechanism bridging the gap between consciousness and the molecular dynamics of motivation, memory, and learning in this way:

> Classical theories assume that predictive learning occurs whenever a stimulus is paired with a reward or punishment. However, more recent analyses of associative learning argue that simple temporal contiguity between a stimulus and a

reinforcer is not sufficient for learning and that a discrepancy between the reinforcer that is predicted by a stimulus and the actual reinforcer is also required. This discrepancy can be characterized as a "prediction error." *Presentations of surprising or unexpected reinforcers generate positive prediction errors, and thereby support learning,* whereas omissions of predicted reinforcers generate negative prediction errors and lead to reduction or extinction of learned behavior. *Expected reinforcers do not generate prediction errors and therefore fail to support further learning even when the stimulus is consistently paired with the reinforcer.* Modeling studies have shown that *neuronal messages encoding prediction errors can act as explicit teaching signals for modifying synaptic connections that underlie associative learning.* (p. 43, italics added)

This is a radical finding! Classical theories of memory, learning, and behavior change were based on the continuity of stimulus, response, and reward as described by Pavlovian theory, behavior theory, and especially Hebb's activity-dependent theory of how connections are formed between neurons. There are no surprises in these classical theories: There is no essential role for novelty, consciousness, and the subjective experiences of motivation and reward. Waelti and colleagues, however, introduce these very elements as "teaching signals for associative learning:"

Current research suggests that one of the principal reward systems of the brain involves dopamine neurons. Both psychopharmacological manipulations and lesions of the dopamine system impair reward-driven behavior of animals, and drugs of abuse, which provide strong artificial rewards, act via dopamine neurons. Neurobiological investigations of associative learning have shown that dopamine neurons respond physically to rewards in a manner compatible with the coding of prediction errors, whereas slower dopamine changes are involved in a larger spectrum of motivating events. . . . *During initial learning, when rewards occur unpredictably, dopamine neurons are activated by rewards. They gradually lose the response as the reward becomes increasingly predicted. . . . Dopamine neurons also respond to novel, attention-generating, and motivating stimuli, indicating that attentional mechanisms also contribute.* Results from neuronal modeling suggest that the responses of dopamine neurons to primary rewards and conditioned stimuli may constitute particularly effective teaching signals for associative learning and embody the properties of the teaching signal of the temporal difference reinforcement model. (p. 43, italics added)

Students of the arts, humanities, and culture have long felt bored with the classical psychological theories of human memory and motivation. There was nothing that could account for the sense of surprise, fascination, and wonder that is so motivating in human experience. There was no satisfying understanding of soul, the human quest, spiritual striving, and the experience of the numinosum—no place, really, for art, beauty, and love. The research of Waelti and colleagues, however, opens the window to a new worldview of the relationship between the most interesting and motivating experiences of consciousness and the novelty-numinosum-neurogenesis effect at the cellular-genomic level. *It is not so much what is expected and easily predictable in human affairs that is motivating but the exact reverse! That which is surprising, unknown, and unpredicted garners our attention and sets us forth on the human quests for adventure in the novelty-numinosum-neurogenesis experiences of love and life as well as the arts and sciences.*

Consciousness and the Novelty-Numinosum-Neurogenesis Effect in the Humanities and the Healing Arts

Consciousness and its associations with novelty, gene expression, and neurogenesis have important implications for a new appreciation of the role of the creative arts and culture for facilitating the growth of the brain. As illustrated in Figure 3.6, consciousness plays an important role in the novelty-numinosum-neurogenesis effect as a complex adaptive system wherein all levels, from mind to gene, belong to a single co-creative, generative system.

Novel interactions between the organism and the environment, which evoke gene expression, protein synthesis, neurotransmitters, and neurogenesis, operate via the Darwinian principles of natural variation and selection on all levels from the molecular-genomic to the subjective states of consciousness. Zeki (1999, 2001) has described how all forms of artistic creativity are related to the dynamics of Darwinian variation and the selection of subjective states of consciousness and experience. This evolutionary perspective on the role of consciousness in focusing and optimizing the growth of the brain has deep implications for understanding the relatedness of many approaches to creative experience in all times and cultures. In the West, the ancient Greek concept of *ekstasis* (ecstasy), the breakthrough moment of creative con-

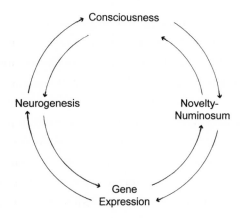

FIGURE 3.6 | The novelty-numinosum-neurogenesis effect is a generative cycle of con-
sciousness, novelty, numinosum, gene expression, neurogenesis, and healing. It is mutually reinforcing complex adaptive system that bridges the Cartesian gap between mind and body.

sciousness, was described as the activating emotive force of a *daemon* (demon)
driving human experience, whether we liked it or not. In the Buddhist tradition, the Zen koan was developed as a way to activate and intensely focus
meditative consciousness to facilitate heightened states of arousal called
"satori" or a milder "kensho." Ritual, music, dance, drama, and storytelling in
all cultures are means of focusing the alternating states of attention, arousal,
and relaxation to facilitate the dynamics of curiosity and wonder that nur-
tures imagination and psychological transformation (Figure 2.11).

Rudolph Otto (1932/1950) summarized the human experience of *fascination, mystery,* and the *tremendous* by coining the concept of the "numinosum" or "the numinous," derived from Greek roots. Otto believed that the
experience of the numinosum was highly characteristic of the peak moments
of spiritual insight and transformation (Rossi, 1972, 1985, 2000a). From the
perspective of the basic rest-activity cycle developed in Chapter 2, Otto's
numinous experiences are characteristic of the alternating psychobiological
states of arousal and relaxation in the dynamics of creativity and the transformations of consciousness. For Carl Jung (1960, 1966), the experience of
the numinosum became the essential driving force of human motivation in
personal development and individuation.

The alternation between heightened states of consciousness, deep relaxation, sleep, and dreaming of the basic rest-activity cycle are all phases in the generation of neurogenesis and healing. The novelty-numinosum-neurogenesis effect is a core dynamic of psychobiology: It integrates experiences of mind (sensory-perceptual awareness of *novelty* with the arousal/motivational aspects of the *numinosum*) with biology (gene expression, protein synthesis, *neurogenesis,* and healing). As indicated in Figure 3.6, the novelty-numinosum-neurogenesis effect is a two-way street: experiences of mind can modulate gene expression (e.g., activity-dependent expression of the CREB genes in memory, learning, and behavior reviewed earlier in this chapter), and body states can modulate mind (e.g., behavioral state-related gene expression reviewed in Chapter 2).

Activity-dependent creative experiences in the arts, cultural rituals, humanities, and sciences as well as the peak experiences of everyday life are all manifestations of the novelty-numinosum-neurogenesis effect. When viewing awesome art or architecture, when moved by cinema, music, and dance, when enchanted by drama, fantasy, fairytale, myth, or poetry, we are experiencing mythopoetic transmissions of the numinosum from the life of the artist to our own. In fact, we might say that *the evolutionary function of the creative arts and sciences is to evoke experiences of novelty and numinosum that drive gene expression, protein synthesis, neurogenesis, and healing.* From this perspective we may regard the heightened experiences of the novelty-numinosum-neurogenesis effect as facilitating the evolution of consciousness.

A more general recognition of the common ground shared by both the arts and the sciences in the novelty-numinosum-neurogenesis effect could liberate the creative arts and sciences to pursue their exploratory quests. We could be free from the misunderstanding and apparent opposition between the two cultures of the arts and sciences (Snow, 1993). In Part II we will explore a variety of practical approaches for utilizing subjective consciousness to focus and optimize the creative work of gene expression and neurogenesis in psychotherapy and the healing arts. For now we will review how the late Milton H. Erickson, the founder of the American Society of Clinical Hypnosis and one of the most innovative psychotherapists of our time, anticipated and utilized the novelty-numinosum-neurogenesis effect to facilitate healing.

THE NOVELTY-NUMINOSUM-NEUROGENESIS EFFECT IN ERICKSON'S NEURO-PSYCHO-PHYSIOLOGICAL DYNAMICS

Today it seems as if we all have our own favorite Milton H. Erickson, depending on what we read and the anecdotes we hear about him. My favorite Erickson comes from a few of his early papers where he writes about the role of *arousal, reassociation, and resynthesis as the essence of hypnosis and psychotherapy.* Here is one of my favorite excerpts that gains new significance in the light of the current research on novelty and neurogenesis. In a basic paper titled "Hypnotic Psychotherapy," Erickson (1948/1980) writes the following about the role of suggestion in hypnosis:

> *The induction and maintenance of a trance serve to provide a special psychological state in which patients can re-associate and reorganize their inner psychological complexities and utilize their own capacities in a manner in accord with their own experiential life.* Hypnosis does not change people nor does it alter their past experiential life. It serves to permit them to learn more about themselves and to express themselves more adequately.
>
> Direct suggestion is based primarily, if unwittingly, upon the assumption that whatever develops in hypnosis derives from the suggestions given. It implies that the therapist has the miraculous power of effecting therapeutic changes in the patient, and disregards the fact that *therapy results from an inner re-synthesis of the patient's behavior achieved by the patient himself.* It is true that direct suggestion can effect an alteration in the patient's behavior and result in a symptomatic cure, at least temporarily. However, such a "cure" is simply a response to the suggestion and does not entail that *re-association and reorganization* of ideas, understandings, and memories so essential for an actual cure. *It is this experience of re-associating and reorganizing his own experiential life that eventuates in a cure, not the manifestation of responsive behavior, which can, at best, satisfy only the observer.*
>
> For example, anesthesia of the hand may be suggested directly, and a seemingly adequate response may be made. However, if the patient has not spontaneously interpreted the command to include a realization of the inner need for reorganization, that anesthesia will fail to meet clinical tests and will be a pseudo-anesthesia.
>
> An effective anesthesia is better induced, for example, by initiating a train of mental activity within the patient himself by suggesting that he recall the

feeling of numbness experienced after a local anesthetic, or after a leg or arm went to sleep, and then suggesting that he can now experience a similar feeling in his hand. *By such indirect suggestion the patient is enabled to go through those difficult inner processes of disorganization, reorganization, reassociating, and projecting of inner real experience to meet the requirements of the suggestion and thus the induced anesthesia becomes a part of his experiential life instead of a simple, superficial response.*

The same principles hold true in psychotherapy. The chronic alcoholic can be induced by direct suggestion to correct his habits temporarily, *but not until he goes through the inner process of reassociating and reorganizing his experiential life can effective results occur.*

In other words, hypnotic hypnotherapy is a learning process for the patient, a procedure of reeducation. Effective results in hypnotic psychotherapy, or hypnotherapy, derive only from the patient's activities. The therapist merely stimulates the patient into activity, often not knowing what that activity may be, and then guides the patient and exercises clinical judgment in determining the amount of work to be done to achieve the desired results. How to guide and judge constitute the therapist's problem, while the patient's task is that of learning through his own efforts to understand his experiential life in a new way. *Such reeducation is, of course, necessarily in terms of the patient's life experiences, his understandings, memories, attitudes, and ideas; it cannot be in terms of the therapist's ideas and opinions.*

For example, in training a gravid [pregnant] patient to develop anesthesia for eventual delivery, use was made of the [direct] suggestions outlined above. . . . The attempt failed completely even though she had previously experienced local dental anesthesia and also her legs "going to sleep." Accordingly, the suggestion was offered that she might develop a generalized anesthesia in terms of her own experiences when her body was without sensory meaning to her. *This suggestion was intentionally vague since the patient, knowing the purpose of the hypnosis, was enabled by the vagueness of the suggestion to make her own selection of those items of personal experience that would best enable her to act upon the suggestion.*

She responded by reviewing mentally the absence of any memories of physical stimuli during *physiological sleep,* and by reviewing her dreams of walking effortlessly and without sensation through closed doors and walls and floating pleasantly through the air as a disembodied spirit looking happily down upon her sleeping, unfeeling body. *By means of this review she was able to initiate a process of reorganization of her experiential life. As a result she was able to develop a remark ably effective anesthesia,* which fully met the needs of the subsequent delivery. Not until sometime later did the therapist learn by what train of thought he had initiated *the neuro-psycho-physiological process* by which she achieved anesthesia. (pp. 38–39, italics added)

Here, in a nutshell, are the essential dynamics of what Erickson called the *naturalistic and utilization approach* to the *neuro-psycho-physiological* dynamics of therapeutic hypnosis (Erickson, 1958/1980, 1959/1980). Notice how Erickson distinguishes between the *direct and relatively superficial and short-acting suggestions* of traditional hypnosis that attempt to program the patient from the outside, versus the indirect suggestions that evoke, prompt, and stimulate patients to co-create the dynamics of their own inner process. In the language of modern neuroscience, we would say that Erickson facilitated the growth of consciousness, identity development, and problem-solving by integrating explicit (conscious) and implicit (unconscious) processes via the so-called "indirect suggestions" that functioned as *"implicit processing heuristics"* that are summarized in Table 3.1 and discussed in greater detail in Chapter 9.

Erickson's patients often became emotionally excited and aroused; they might weep or rage or became hot and actually sweat during the *psychobiological work* of their hypnotherapeutic sessions (Erickson, 1954/1980). Table 3.1 lists some of Erickson's approaches that were designed to heighten states of *neuro-psycho-physiological* arousal in hypnotherapy. Indeed, my very first published paper on Erickson's work was titled "Psychological Shocks and Creative Moments in Psychotherapy," wherein I was myself shocked by the novelty of the fantastic extent to which Erickson would go to arouse his patients on deeply challenging levels so that they could break out emotionally and resynthesize a new experiential reality for themselves (Rossi, 1973a, 1981, 1989b).

Erickson described his approach to me in an informal manner as a "yo-yoing" of the patient's consciousness and expectancies "to initiate response readiness and the hypnotic process" (Erickson & Rossi, 1989, p. 1). In our first book (Erickson et al., 1976), we initially described the secret of Erickson's success as the facilitation of the patient's "response tendencies":

We witness a simple secret of the effectiveness of his approach: he *offers* suggestions in an open-ended manner that admits many possibilities of response as acceptable. Suggestions are offered in such a manner that any response the patient makes can be accepted as a valid hypnotic phenomenon. These open-ended suggestions are also a means of exploring the patient's response tendencies (the "response hierarchy" of learning theory and behavior therapy). The therapist can utilize these response tendencies to effect the therapeutic goals. (pp. 27–28)

Particularly noteworthy in Table 3.1 is the common denominator of how to utilize each patient's personal patterns of arousal and responsiveness. Psychobiological arousal is evident in Haley's early interpretation of Erickson's approach as "arduous therapy" as well as in current views of expectancy and response set theory by Kirsch (2000) and in socio-cognitive models of hypnosis (Lynn & Sherman, 2000). Research documenting how Erickson's innovative neuro-psycho-physiological approaches can provide a road map for optimizing the novelty-numinosum-neurogenesis effect should be our highest priority.

A careful study of Erickson's case reports indicates that he was always engaging his patients in their own deeply motivated neuro-psycho-physiological work rather than effecting "miracles" (Erickson et al., 1976; Erickson & Rossi, 1979, 1981, 1989; Siegel, 1986, 1989, 1993). Many patients seek hypnosis hoping the therapist will have the magic to heal them. Erickson was constantly teaching, to the contrary, that patients must become engaged in their own numinous and creative healing work, just as Erickson did in his own struggle with polio all his life. Notice Erickson's emphasis on patients' *personal responsibility* for reorganizing and resynthesizing the neuro-psycho-physiological foundations of their own experience. This is contrary to the common misunderstanding of therapeutic hypnosis as a technique of programming people where the ultimate locus of control is in the therapist rather than the patient.

This focus on each person's ultimate responsibility for guiding his or her own neuro-psycho-physiological inner work is entirely consistent with the pioneering concepts of Thomas Szasz (1997) in re-visioning the patient's *response-ability* in the medical model of psychotherapy. We are all ultimately responsible for recognizing and facilitating the novelty-nunimosum-neurogenesis effect in the co-creation of our intelligence, personal development, and healing (Richardson, 2000). While the social and cultural environment may support and reinforce many aspects of human behavior, only the individual can recognize the novelty-numinosum-neurogenesis effect as a sub-

TABLE 3.1 | ERICKSONIAN APPROACHES TO THERAPEUTIC HYPNOSIS THAT ANTICI-
PATED MANY OF CURRENT FINDINGS OF NEUROSCIENTISTS IN THEIR NEWLY INSPIRED
SEARCH FOR THE CREATIVE COGNITIVE-BEHAVIORAL-ENVIRONMENTAL–PSYCHOSOCIAL
REPLAYS THAT FACILITATE NOVELTY, NEUROGENESIS, AND HEALING.

1. **Behavioral state-related gene expression:** Suggestions for hypnotic induction via comfort, relaxation, sleep, dream, dissociation, somnambulism, emotional, and cognitive arousal (Erickson, 1958/1980).

2. **Immediate early gene expression:** Facilitating surprise and psychobiological shock (Erickson & Rossi, 1979; Rossi, 1973a).

3. **Entrainment and modulation of clock gene expression:** Time distortion and post-hypnotic suggestion (Cooper & Erickson, 1959; Erickson & Erickson, 1958/1980).

4. **Activity-dependent gene expression and neurogenesis:** The naturalistic and utilization techniques (Erickson, 1958/1980,1959/1980; Erickson et al., 1976; Langton & Langton, 1983; Zeig, 1999), and others such as:

• Activity-dependent activities for learning, memory, arduous work, behavior change, and organic symptom resolution (Erickson, 1963/1980, 1965/1980b; Haley, 1963,1985)

• Experience-inducing anecdotes (Zeig & Geary, 2000)

• Yo-yoing consciousness, expectancy, and response sets (Erickson & Rossi, 1976a, 1976b, 1981; Lynn & Sherman, 2000; Kirsch, 2000)

• Numinous enchantment, fascination, wonderment, stories, metaphor, and humor to facilitate problem-solving and healing (Braid, 1855/1970; Erickson et al., 1983; Rosen, 1982)

• Physical exercise in Erickson's self-recovery from polio at age 18 and his recommenda tions to patients to climb Squaw Peak Mountain (Erickson & Rossi, 1977/1980)

• Utilizing ultradian dynamics of two-hour therapy sessions (Rossi, 1986a, 1996a)

• Creative edge therapy and strategic focus on most important life issues (Haley, 1963, Rossi, 1989; Zeig, 1985, 1990, 1994)

• Activity-dependent learning, education, puzzles, breaking-out-of-the-box thinking, posthypnotic suggestion, and pantomime techniques (Erickson, 1964a/1980; Erickson & Rossi, 1979, 1981)

• Provocation and utilizing emotional arousal in rehabilitation (Erickson, 1965/1980b)

• Questions, Erickson's healing version of Socrates (Erickson et al., 1976; Erickson & Rossi, 1976)

• Therapeutic double binds (Erickson et al., 1976), multilevel communication (Erickson & Rossi, 1976/1980), and indirect suggestion (implicit processing heuristics) (Erickson et al., 1976; Erickson & Rossi, 1980)

• Implication and the implied directive (Erickson et al., 1976, 1992; Erickson & Rossi, 1979, 1981, 1989; Rossi & Cheek, 1988)

• Replaying memory and re-dreaming a dream to explore new possibilities (Erickson, 1952/1980) and recreate identity (Erickson, 1980a, Erickson & Rossi, 1989)

• Surprise, my-friend-John, and resistance techniques (Erickson, 1964b, 1964c/1980;

Erickson
& Rossi, 1981, 1989; Erickson et al., 1992)

• Replaying memory and sensory-perceptual experiences to facilitate the creative process (Erickson, 1965/1980a)

• Activity-dependent family therapy for schizophrenics, and social encounters for the lonely (Dolan, 1985)

---

jective experience providing intuitions of what is really important for his or her own inner life. This is the essence of the individual human sense of "soul" and "quest" that is played out over the destiny of a lifetime. James Kirsch (1966) has illustrated how the sense of the numinous may be experienced as the essence of drama in a number of Shakespeare's plays.

A transcript of one of Erickson's hypnotic demonstrations, filmed at a 1964 meeting of the American Society of Clinical Hypnosis (available for professional use through The Erickson Foundation in Phoenix, Arizona), is particularly revealing of how Erickson attempted to use novelty, fascination, and surprise to motivate the subject, a woman, to take responsibility for her own inner work and to co-create the conditions for the neurogenesis that neuroscientists are only now beginning to explore on the molecular level. Following are a few excerpts from the first part of that 1964 film, when he was introducing a naive subject to the experience of therapeutic hypnosis (Haley, 1993). I have italicized Erickson's use of the concepts of *novelty, curiosity, not knowing, work, fascination, surprise,* and other aspects of psychobiological arousal that anticipated the finding of current neuroscience regarding the significance of novelty and neurogenesis. Notice how his very first words to her are a series of open-ended questions that immediately evoke a sense of *novelty, curiosity, not knowing, expectation,* a focus on *her own inner work,* as well as the possibility of experiencing *fascination and surprise.*

> ERIKSON: Tell me, have you ever been in a hypnotic trance before?
> SUBJECT: No.
> ERIKSON: Have you ever seen one?
> SUBJECT: No.
> ERIKSON: Do you know what it's like to go into a hypnotic trance?
> SUBJECT: No.
> ERIKSON: Did you know that *you do all the work,* and I just sit by and enjoy watching you *work*?
> SUBJECT: *No, I didn't know that.*

Having first facilitated this self-acknowledged state of *not knowing, curiosity, expectation, and wonder* within the subject, Erickson now proceeds with what will be *a novel experience* for her—a *fascinating hypnotic induction with a hand levitation approach,* as follows.

> ERIKSON: Now the first thing I'm going to do is this. I'm going to take hold of your hand and lift it up. It's going to lift up, like that.

Erickson appears to lift her hand in a slow, hesitant manner, but he is actually only providing very light sensory cues to her arm and wrist. The subject is doing the actual lifting in this *novel experience of implied or involuntary movement* (Erickson & Rossi, 1981).

> ERIKSON: And you can look at it.

Erickson lets go of his tactile contact with her arm and it remains up, as *she now stares at it in wide-eyed fascination.* Only now, when her consciousness is fixed and focused in fascination and arousal, does Erickson offer the first direct suggestion.

> ERIKSON: And close your eyes and go deeply, soundly asleep. So deeply, so soundly asleep, so deeply so soundly asleep that you could undergo an operation, that anything legitimate could happen to you.

We now recognize that this classical suggestion for sleep in hypnotic induction may be understood from a neuroscientific perspective as a way of facilitating the *behavioral state-related gene expression* associated with sleep, as discussed previously in Chapter 2.

> ERIKSON: Now *I'm going to surprise you*, but that's all right. I will be very careful about it.

Erickson takes her ankle and uncrosses her leg—*a surprise, indeed!* This certainly is not a recommended procedure in our litigious society today, but notice how Erickson reassures her, at the same time, with the words "that's all right" and "I will be very, very careful." The large professional audience observing this demonstration as well as the presence of a video-recorder, may constellate the sense of safe shelter for a socially acceptable ritual since ancient times. On an archetypal level this setting has aspects of a healing temenos, a cultural ritual, for people going through significant life transitions in vulnerable situations (Jung, 1950).

> ERIKSON: And are you comfortable? And you can nod your head again. Do you know how the ordinary person nods her head? You really don't, but they nod it this way.

Erickson nods his head up and down rapidly in a typical, everyday fashion.

> ERIKSON: And you nod it this way.

Erickson nods his head very slowly, as is typical of the head-nodding that occurs in the automatic and dissociated manner of the hypnotic state. Erickson is modeling one of the classical behavioral characteristics of the hypnotic state wherein subjects manifest a slowing down and time lag in their responding. He is teaching her how to recognize her own hypnotic state, via *her own expe-*

riencing and behavior. This is the first indication that he is facilitating experiential learning. From the perspective of functional genomics, he may be facilitating *activity-dependent gene expression.*

> ERIKSON: You don't know what I'm talking about, but that's all right. And now your hand is going to lift up toward your face.

Erickson provides very light, subtle, tactile cues to her arm to facilitate hand levitation.

> ERIKSON: And you didn't really know it was that easy, did you? And when it touches your face, you'll take a deep breath and go way deep sound asleep.

In suggesting that she will "go way deep sound asleep," it seems likely that Erickson is facilitating the *behavioral state-related gene expression* associated with sleep, as discussed in the previous chapter.

> ERIKSON: And you didn't know it would be that easy, did you? And it is so far different from the show–off stage hypnosis, isn't it? *Because you realize that you're the one who is really doing it. You know that, do you not?*

In suggesting "you're the one who is really doing it," is Erickson facilitating *activity-dependent gene expression?*

> ERIKSON: I am going to ask you to open your eyes. [She opens her eyes]. Hi, have you been in a trance?
> SUBJECT: I don't know.
> ERIKSON: You don't. You really don't know. I'll tell you the way to find out. Watch your eyelids to see if they start closing on you. And if they start closing on you, that will mean that you've been in a trance. And down they go, that's beautiful, down they go, down they go. That's right. All the way now. And all the way. All the way until they stay shut. *And now all the proof came from within you, did it not? . . . Isn't it charming? Isn't it interesting?*

Notice the *attention-arresting paradox* in Erickson's approach that tends to facilitate *fascination and wonder* within the subject. When she is asked to watch for the subtle differences between her voluntary behavior (*conscious explicit level*) and the involuntary, dissociated behavior (*unconscious implicit level*) of her eyelids, her own observation will constitute her *activity-dependent* proof that she is in a special state called trance.

From the very beginning to the end of this hypnotic induction, Erickson facilitates experiences of fascination, novelty, not knowing, curiosity, charm, interest, and wonder. That is, Erickson is evoking *numinous experiences* by using many words that have suggestive connotations of the numinous. He continually emphasizes the point that it is the subject who is doing the important activity-dependent inner work. In this video he repeatedly suggests little experiential exercises that enables his subject to recognize, explore, and validate her own hypnotic experience by finding a behavioral *activity-dependent validation for it—within her own experience.* Erickson's innovative approaches appear to involve the modulations of environmental or cognitive stimuli, alterations in physical activity, or some combination of these factors that neuroscientists describe as novelty, life enrichment, and physical activity that are associated with neurogenesis (Kempermann et al., 1997; Kempermann & Gage, 1999).

This leads us to propose that many of Erickson's unique approaches to facilitating therapeutic hypnosis and psychotherapy may be conceptualized as *implicit processing heuristics* that stimulate immediate early, behavioral state–related, and activity-dependent gene expression to facilitate problem-solving and healing at the molecular level. When Erickson suggests "comfort" and "deep sound sleep," as is done in classical hypnosis, he may be facilitating *behavioral state-related gene expression* that is associated with sleep. When Erickson startles us with his innovative approaches that evoke psychological shock and surprise, however, he may be activating *immediate early gene expression*; when he prompts people to engage in challenging, novel, and numinous experiences, he may be facilitating *activity-dependent gene expression and neurogenesis.* Where can we turn for a new look at Erickson's neuro-psycho-physiology of therapeutic hypnosis and psychotherapy in the new millennium? Why, to brain imaging and the new DNA microarrays, of course—where else?

Summary

This chapter began with Schrödinger's intuitive idea that consciousness is somehow the tutor who supervises the education of the living substance. Current neuroscience has identified how novelty, enriching life experiences, and physical exercise can optimize gene expression, neurogenesis, and the growth of the brain. Surprising and unexpected experiences activate the neurotransmitter dopamine in neurons that respond to novel, attention-generating, and motivating stimuli. During initial learning, when rewards occur in a surprising and unpredicted manner, dopamine neurons are activated by rewards. This activation response gradually decreases as the rewards lose their surprise, fascination, and numinous value.

This pattern implies that consciousness is a novelty-sensitive modality that has an essential role in focusing gene expression and neurogenesis in the flowering of intelligence and personality. We describe this as *the novelty-numinosum-neurogenesis effect: Highly motivated states of consciousness can turn on and focus gene expression, protein synthesis, neurotransmitters, and neurogenesis in our daily creative work of building a better brain.* Activity-dependent creative experiences in the arts, cultural rituals, humanities, and sciences as well as the peak experiences of everyday life are manifestations of this novelty-numinosum-neurogenesis effect. A lack of optimal gene expression and neurogenesis is now believed to be associated with psychological depression. We hypothesize that happiness and positive psychological attitudes, by contrast, are associated with optimal gene expression that leads to neurogenesis, healing, and well-being.

The inner rehearsal and creative replay of novel and surprising life experiences between the hippocampus and cortex of the brain plays a significant role in optimizing activity-dependent gene expression and neurogenesis in the creation and recreation of memory, learning, and behavior. *This leads us to propose that creative replay is the fundamental dynamic of all forms of growth-oriented psychotherapy. Creative replay is an inner theater wherein natural Darwinian variation on all levels from the molecular-genomic to the psychosocial interacts with the conscious possibility of selecting positive experiences and life choices to facilitate healing transformations of memory, learning, and behavior.* Creative replay is a complex adaptive system that provides a new neuroscience foundation for bridging the Cartesian divide between mind and body in the deep psychobiology of alternative and complimentary medicine, therapeutic hypnosis, and psychotherapy that will be illustrated in practical detail in Part II.

| CHAPTER 4 |

DREAMING, GENE EXPRESSION, AND NEUROGENESIS

Self-Reflection and the Co-Creation of Experience

"Within the past decade one of the most exciting research breakthroughs in psychology has been in the understanding of memory and learning as an organic growth process. Experimental animals raised in more complex and stimulating environments appear to synthesize more complex proteins within their brain cells. Specific life experiences can be encoded in the synthesis of specific protein molecules. If we extrapolate these findings to the area of dreams, we realize that dreams are frequently just as vivid as any real life experiences and, as such, dreaming must lead to the synthesis of new protein structures within the brain. These new protein structures, of course, can then become the nuclei of new developments in the personality."

—Ernest Rossi
Dreams and the Growth of Personality:
Expanding Awareness in Psychotherapy

"The present study constitutes a first demonstration that the expression of an activity-dependent gene is up-regulated during REM sleep that follows exposure to an enriched environment. . . . Given the association between zif-268 [gene] expression and neuronal plasticity, the phenomenon we describe may represent a window of increased plasticity during REM sleep that follows a rich waking experience. Our results thus provide a possible mechanism whereby previous waking experiences can contribute to long-lasting changes in the brain."

—Sidarta Ribeiro et al.
"Brain Gene Expression During REM Sleep
Depends on Prior Waking Experience"

Controversy over the significance of dreams began with the earliest written records of the nature of human experience in myths, epics, and spiritual treatises. Even today, opinions range from those who follow Freud in believing that dreams are the royal road to the unconscious, to those who believe that dreams are the phenomenological detritus of a brain that is somehow offline in sleep.

This chapter begins with an overview of current neuroscience research on the role of gene expression in dreams as a psychobiological process of adaptation to novelty and significant life experiences. *Dreaming is viewed as a creative process integrating behavioral state-related gene expression with activity-dependent gene expression, and neurogenesis.* This creative psychobiological process is illustrated by a particularly vivid series of dreams that traces the development of a young woman's consciousness, identity, and personality. *Our conscious experience and worldview are continually updated with implicit processing during the creative replays of the drama of the dream.* A series of hypotheses summarizes the dynamics of dreaming as a co-creative psychobiological process of self-reflection and self-development throughout a person's lifetime.

THE DREAM-PROTEIN HYPOTHESIS AND ACTIVITY-DEPENDENT GENE EXPRESSION

Early research on the molecular dynamics of memory and learning led to the original formulation of the dream-protein hypothesis: "Dreaming is a process of psychophysiological growth that involves the synthesis or modification of protein structures in the brain that serves as the organic developments in the personality" (Rossi, 1973b, p. 1094). These new proteins were regarded as the biological basis for the synthesis of new developments in consciousness and personality during the psychobiologically aroused state of REM dreaming. In 1972 we only dimly understood the deeper implication of the finding that RNA was associated with the new protein formation that occurs during dreaming and new memory and learning (Russell, 1966): *Genes are turned on in everyday life during significant life events* to make the messenger RNA (mRNA) that serves as the blueprint for synthesizing new proteins and neurogenesis to encode memory and learning.

Such an idea seemed too preposterous at the time, however. Most researchers believed the early dogma about the strict separation between nature (genes) and nurture (life experience) in the generation of human behavior. Researchers did not understand the exciting concept of *activity-dependent gene expression in everyday life:* Novel, stimulating, and challenging life experiences can activate neurons to release neurotransmitters and other messenger molecules (such as hormones and growth factors) to activate gene expression and protein synthesis to facilitate neurogenesis in the brain. Proteins are the molecular machines managing the dynamics of matter, energy, and information in all living systems. Proteins and the new science of *proteomics* currently emerging from the Human Genome Project are clarifying the molecular dynamics of adaptive responses to challenges from the outside world in health and optimal performance, as well as stress, illness, and healing.

NOVELTY, GENE EXPRESSION, AND NEUROGENESIS DURING DREAMING

The novelty-neurogenesis process that takes place during waking activities apparently continues during REM sleep after a particularly exciting day with many new experiences and opportunities for new learning. Ribeiro et al. (1999) have demonstrated how exposure to a "memorable" life experience turns on an immediate early gene, called zif-268, during subsequent REM sleep. Zif-268 (also known as nerve growth factor-induced A as well as egr-1, krox-24, and tis-8) is turned on within a few minutes when the organism is stimulated by exciting or novel situations (Bentivoglio & Grassi-Zucconi, 1999). Ribeiro et al. report how the activation and transcription of the zif-268 gene leads to the formation of new proteins that can encode new memory and learning:

> We studied brain expression of the plasticity-associated immediate-early gene (IEG) zif-268 during SW [slow-wave or deep delta sleep] and REM sleep in rats exposed to rich sensorimotor experience in the preceding waking period. Whereas non-exposed controls show generalized zif-268 down-regulation during SW and REM sleep, zif-268 is up regulated during REM sleep in the cerebral cortex and the hippocampus of exposed animals. *We suggest that this phenomenon represents a window of increased neuronal plasticity during REM sleep that follows enriched waking experience.* (p. 500, italics added)

This research documenting the role of gene expression in encoding new memory and learning during dreaming receives support from two recent studies that reveal a more general role for the zif-268 gene in long-term memory formation as well as emotionally encoded state-dependent memories. Jones et al. (2001) documented how the absence of the zif-268 gene in mice leads to an impairment of long-term potentiation and memory for a number of learning tasks even while short-term synaptic plasticity and memory remained intact. Hall et al. (2001) demonstrated still another central role for this particular gene: Contextual fear and emotionally associated memory retrieval was marked by zif-268 expression in CA1 neurons of the hippocampus. Given that we now know that *fear memories return to a labile state when recalled and reframed in the gene expression and protein synthesis cycle* (Nader et al. 2000a, 2000b), the zif-268 gene may be involved in the fundamental dynamics whereby mind, memory, and meaning are changed during the replay of memory in dreams, psychotherapy, and the healing arts.

Let us now turn our focus to research demonstrating how the creative replay of salient life experiences during dreaming (Chicurel, 2001a, 2001b) leads to an updating, resynthesis, and reframing of memory that is at the heart of the growth approaches to psychotherapy, therapeutic hypnosis, and the healing arts (Rossi, 1996a, 2000a).

Creative Replay During Dreaming

Maquet et al. (2000) used positron emission tomography (PET) to elucidate how humans replay novel and enriching daytime experiences during dreaming (REM sleep). These researchers made three-dimensional images of the daytime patterns of brain activation in seven human subjects while they were engaged in learning a computerized program of hitting certain keys in response to signals from different areas that flashed on the screen. They found that the brain activation patterns of their subjects during REM sleep were similar to their novel daytime activation patterns when engaged in this computerized learning task. The reactivation of the same brain areas in REM sleep that were activated in novel daytime learning strongly supports the idea that dreaming is associated with gene expression, protein synthesis, and neurogenesis in the brain. Maquet and his team note that the levels of acetylcholine, a neurotransmitter known to be associated with new memory consolidation, surge in the brain during REM sleep. Their research in brain imaging during REM sleep is entirely consistent with the Ribeiro et al. (1999) study that demonstrated how dreaming can consolidate new memory and learning. This combination of findings linking the level of gene expression with neuroanatomy and novelty is the strongest evidence, to date, of the heuristic value of the dream-protein hypothesis.

<h2 style="text-align:center">Replay Integrating Past
and Present During Hypnagogic Sleep Imagery</h2>

Further evidence connecting the levels of (1) gene expression, (2) neurogenesis, and (3) brain anatomy to (4) conscious human experience is being explored by Strickgold et al. (2000a). These researchers designed an experimental study that explored the relationship between learning Tetris, an interesting video computer game, and hypnagogic imagery. Tetris is a colorful and attention-absorbing game involving spatial reasoning that requires the player to arrange spinning and falling blocks to form layers on the bottom of the screen. After an initial ultradian training period of two hours, verbal reports were collected from the subjects during the hypnagogic pre-sleep period and after intervals of 15–180 seconds of objectively measured sleep during the first hour after they had retired for the night.

The researchers found that the probability of the subjects reporting a replaying of the striking visual images of Tetris was inversely related to initial performance, suggesting a relation to the learning process itself. That is, less experienced subjects for whom Tetris was more novel, dreamed more about it than more experienced subjects for whom Tetris was no longer novel. In all cases the reports of imagery were focused on the novel aspects of learning Tetris, such as the blocks rotating, falling, and fitting into empty spaces on the bottom of the screen, rather than the more familiar imagery of the computer keyboard. This important finding appears to confirm Schrödinger's prescient intuition of the role of consciousness in "tutoring" the brain. *Hypnagogic imagery appears to be a replay or rehearsal of the new imagery and learning of the novel aspects of life.*

The role of consciousness in new learning of the Tetris game must be interpreted with caution, however, in light of the responses of the amnesic

subjects who had extensive bilateral medial temporal lobe damage (hippocampal damage). These organically amnesic subjects reported the typical
Tetris hypnagogic imagery, even though they were unable to recall playing
the game itself! This suggests that the conscious declarative memory processed
by the hippocampus is not needed to report such hypnagogic imagery. The
preservation of hypnagogic images in amnesic subjects means that such
imagery is processed on *implicit or unconscious levels* that do not require conscious declarative memory mediated by the hippocampus. Strickgold and colleagues believe that this lack of hippocampal involvement in image
construction may explain many of the experiences of bizarreness, disconti-
nuities, incongruities, and uncertainties reported during the *implicit level* of
REM sleep, when there is evidence that hippocampal output is blocked. This
is an example of the type of experimental data that motivates our utilization
of *implicit processing heuristics* to facilitate psychotherapy and the healing arts
in Part II.

Another striking finding was that the experienced Tetris players had
incorporated imagery from memories of experiences playing Tetris one to
five years previously. This means that their recent experience of playing Tetris
during this controlled experiment called up past memories. What could be
the reason for this apparent replay and integration of past memories with
present experience? Strickgold et al. (2000a) provide an answer that summarizes the implications of their experiment:

> One possible function for the activation of these memories would be to alter
> their strengths, structures, or associations in ways that are adaptive to the organ-
> ism. With such predictable associations being made and presumably strength-
> ened at sleep onset, one can only surmise that during REM sleep more
> unpredictable, potentially valuable, but frequently useless, associations are tested
> and, when appropriate, similarly strengthened. (p. 352)

We may interpret this conclusion as providing further support for the
dream-protein hypothesis first proposed 30 years ago (Rossi, 1972, 1973b,
1985, 2000a).

Strickgold et al. (2000b) also reported that long-term learning on a visual discrimination task required at least six hours of post-training sleep prior
to retesting; they found that learning was proportional to the amount of sleep

in excess of 6 hours. Using 18–25-year-old subjects recruited from the college campus, they found that overnight improvement was proportional to the amount of slow-wave sleep experienced in the first quarter of the night as well as the amount of REM in the last quarter. This finding indicates that both slow-wave and REM sleep are required for optimal learning in humans—this means that the humor about sleepyhead kids going to school in the morning is really no joke. The last 90–120-minute ultradian period of sleep in the morning is very important for the consolidation of new memory and learning. As we know, circadian rhythms such as the daily wake-sleep cycle take about 24 hours, and ultradian rhythms such as the basic rest-activity cycle of REM and slow-wave sleep take about 90–120 minutes (Lloyd & Rossi, 1992, 1993). If we miss that final hour-and-a-half cycle of sleep that allows us to wake up feeling refreshed in the morning, we are losing a lot of what we learned yesterday as well as limiting how much we can learn today.

Such experimental findings provide a psychobiological perspective on the significance of the 90–120-minute rhythm of the basic rest–activity cycle when we are awake as well as when we are in deep sleep or dreaming. Insofar as creative experience in psychotherapy and the healing arts can contribute to psychobiological arousal, it may be possible to help people find optimal levels of mental stimulation to facilitate the genesis, growth, maturation, and interconnections of new neurons while awake as well as during sleep (Frank et al., 2001). Facilitating new frames of reference (reframing), memory, learning, and behavior is the essence of the growth paradigm in psychotherapy as well as the humanistic arts. Let us now take a deeper look at how the molecular dynamics of gene expression and neurogenesis may become manifest on the experiential level in dreaming during important periods of life transition.

SELF-REFLECTION AND CO-CREATION OF PERSONALITY IN A DREAM SERIES

In this section we explore the dreams of a young woman, "Davina," who was going through the critical life transitions from adolescence to marriage (Rossi, 1972, 1985, 2000a). She reported a total of 110 dreams, visions, and fantasies during her 10-month therapy consisting of 42 sessions at weekly intervals. Seven of Davina's dreams and the hypotheses they generated are presented here to illustrate how gene expression and neurogenesis may be experienced on a phenomenological level in dreams. Davina was a graduate student in the humanities who was in the habit of writing down her dreams in her personal journal every morning when she awakened. Eighteen months before she decided to embark on a course of psychotherapy for problems associated with her recent marriage, she recorded the following dream.

A Captive Maiden (18 months before psychotherapy)

I am a captive maiden, dressed in a long frilly gown, on a pirate vessel sailing out to sea. The captain is an ugly pirate with a long, black, scraggly beard. He wears a black patch over one eye and a blue and white soiled tee-shirt over his stocky, ape-like frame. Hair is tangled on his arms and, all in all, he seems to be a beast. I am tied up beside him, and he is steering the wheel of the ship.

All of a sudden he gets ill and says he is dying. He grabs his stomach and says that this is his end. He tells me that now I have to do the steering for the whole ship. He unties me and shows me how to steer. I grab the wheel and then he tells me the secret of the ship. There is a printing press that must be kept going if the ship is to stay afloat, and ink and paper must be fed constantly into this press as I turn the wheel. Soon I am doing it very well, and just as soon as I can handle everything, the giant-captain says that now "you must carry on," and then he dies right there on the floor. I feel afraid but strangely reassured and newly strengthened.

The *captive maiden* theme describes the state of Davina's mind—her own personal and unique point of view—as being in captivity. This is a dream she had a year and a half before applying for therapy. She is a captive, but the inner capacity for freedom and self-direction appears evident from the fact that the ugly pirate sets her free and even gives her control of the ship as he dies. What is unique in this dream is the unusual detail about a "printing press that must be kept going if the ship is to stay afloat." A sense of the meaning of the printing press becomes obvious when we learn that she is a graduate student in literature interested in writing. The dream indicates a dramatic change is already taking place in her development on an implicit level: from captive maiden to guiding her own ship via the creative process of feeding ink and paper into the printing press. The important part played by the unique function of the printing press in this dream suggests a general hypothesis about the self-organizational dynamics of the creative process in dreams in general. The printing press is the first indication of how a new pattern of awareness, understanding, or meaning can become manifest in consciousness.

HYPOTHESIS 4.1. That which is novel, unique, odd, strange, fascinating, numinous, or intensely idiosyncratic in a dream is a manifestation of gene expression and the neurogenesis of original psychological experience, out of which new consciousness and individuality evolve.

Whatever is subjectively experienced as surprising, novel, numinous, unique, odd, strange, fascinating, interesting, or intensely idiosyncratic is usually an indication that the person is becoming aware of something new. We take this to mean that something new really is being created within the brain on the levels of gene expression, protein synthesis, and neurogenesis—the novelty-numinosum-neurogenesis effect—which constitutes the psychobiological bases of new experiences of consciousness, meaning, and personality development.

The Broken Record (18 months before psychotherapy)

I buy myself some recordings of music and go home. When I get home and close the door of my room, I hear my parents, brother, and sister-in-law talking behind my back, saying that I am ugly, fat, sloppy, crude, etc. I start to cry, and suddenly the record I was listening to on the phonograph breaks into a million pieces. I feel dashed, panicked. I call up Joyce, who in the dream is working at a record store. She tells me to stop calling her about my problems. She tells me to go to sleep, to get some rest, and in time, with rest, the old record that broke will be replaced. Then I go to sleep.

It is evident from this dream that Davina is overidentified with the views of others, as is typical of many young people in the early stages of finding themselves. Her self-image is caught in the criticisms of others: "I hear my parents, brother, and sister-in-law talking behind my back, saying that I am ugly, fat, sloppy, crude, etc." In the first few sessions of psychotherapy she confirms the fact that her relatives are very critical of her appearance and behavior in just this way. This aspect of the dream is thus a true reflection of her relationship with her family: She is a captive of the unfortunate stereotypes her family has of her.

This dream portrays the current condition that is unsatisfactory in her life and which causes her to cry. It is from this condition that she is trying to escape, and it certainly appears that this is the central problem of the dream. She has bought new records for herself; she is trying to listen to her own music, her own inner world. But under the impact of their criticism—the old, stereotyped image of herself as ugly—her old music record shatters into a million pieces. She then says, "I feel dashed, panicked." The new sounds of her precariously developing individuality are obliterated when her family imposes their voices on her. Yet the dream ends on a hopeful note. Her records, her new sounds, will be replaced. The great significance associated with the words overheard by Davina in this dream suggests a second hypothesis about the self-organizational dynamics of consciousness and personality development in dreams.

HYPOTHESIS 4.2. Words, messages, or any other form of self-description in a dream are usually important aspects of personality that are in a process of change.

When the dreamer reacts *negatively* to the words or behavior of others, it usually points to places where the dreamer's consciousness has been frozen in old, rigid patterns that block its free development. We will see in later dreams how Davina responds positively to voices that tell her something new about herself. *Positive* responses of the dreamer are indications of new developments in the personality, whose growth represents a significant step toward individuation and self-actualization.

The positive voice of her friend Joyce in this dream seems mildly cautionary yet also promising. Davina needs to stop calling for help from the outside and recognize that healing will take place in the dramas of her dreams on the inside as she goes to sleep and gets some rest. This seems like merely homely advice, yet it contains a deep psychobiological truth. Various patterns of behavioral state-related gene expression are turned on and off during the stages of rest, sleep, and dreaming, leading to the synthesis of new proteins that are the basis of physical and emotional healing as well as new memory, meaning, learning, and behavior.

Baby Clothes Don't Fit (one year before psychotherapy)

Suddenly as a dreamer, I see myself *inside the store* looking at all kinds of baby clothes, *and I* wonder why *I look at baby clothes, because* they wouldn't fit me any-more. *In the aisles I see all sorts of people I dislike and have known in the past. I'm shocked to see them. They are salespeople and try to sell me baby clothes to wear. But I insist that the clothes wouldn't fit me, and look at them as though they were crazy or something.*

Davina had this dream about a year before entering psychotherapy. Is not the theme of baby clothes an obvious indication that her mind is still occu-pied with the problem of growing up? She does not like the people of her past who try to sell her baby clothes. Here, again, we observe that the dream-er's negative reaction to characterization by others is an indication of a devel-opmental transition—in this case, obviously a critical phase transition from childhood to maturity.

Davina makes a very important observation about herself in this dream: "I wonder why I look at baby clothes, because they wouldn't fit me any-more." This indicates that within the dream she experiences two aspects of herself: one that looks at baby clothes and another that recognizes it as inap-propriate to look at baby clothes anymore. In the language of the new dynam-ics of self-organization, this indicates that a division or *bifurcation* is taking place in her state of being and in her consciousness about herself. We hypoth-esize that during this stage of the drama of the dream, gene expression and neurogenesis are generating molecular computations at the phase transition between the identity states of childhood and adulthood (Rossi, 1996a, 1999a, 1999b, 1999c). This is the current crisis in her identity: She is delicately poised between the realm of childhood and that of chaos and a new order in her psychological development. We could say that the conflict between these two sides of her bifurcating self-image is the developmental problem that the drama of the dream attempts to resolve by replaying it in a new way. Davina currently exists in this phase transition between two phenomenological worlds: She can experience herself as a child interested in looking at baby clothes and also as an adult who recognizes that this behavior is no longer appropriate. The experimental drama of the dream is replaying the conflict between these two states on the levels of gene expression, neurogenesis, and phenomenological experience to generate a new path of identity and personality development.

HYPOTHESIS 4.3. When two or more states of being interact in a dream, it means that a process of psychobiological change—a critical phase transition in consciousness, identity, and personality development—is in progress on the levels of gene expression, neurogenesis, and psychological experience.

Notice that Davina also experienced a change in her state of being in her two previous dreams. In the Captive Maiden dream she existed first as a captive and later as master of the ship. In the Broken Record dream she got caught in her family's negative view so that her own record, her own view of herself, was dashed. There is, however, a subtle but very important difference between these two previous dreams and this one. In the former dreams things happened to her; she played an essentially passive role. In the Captive Maiden dream the developmental block represented by the pirate seemed to dissolve by itself to give her control of the ship. In the Broken Record dream the developmental block represented by her family's criticism took a more severe turn and dashed her. In these two previous dreams Davina is a rather passive participant who could only respond to whatever forces were active in her. To use the new language of neuroscience, her state is primarily a manifestation of the preconscious, implicit, self-organizational dynamics of behavioral state-related gene expression within the drama of her dream. She does not yet know how to use her consciousness and will to facilitate activity-dependent gene expression to co-create herself. That is, she does not yet know how to recognize novel developments within her experiencing and use them to facilitate the novelty-numinosum-neurogenesis effect to recreate herself. She does not know how to use her explicit, conscious awareness to interact cooperatively with her implicit, unconscious processes in the activity of co-creating her evolving identity.

In this third dream, however, Davina is no longer in a passive role. She is not merely a recipient of sensations, emotions, images, thoughts, and conflicts within the drama of the dream. She now sees herself experiencing *and* she evaluates the experience—*she is self-reflecting.* She is not simply in a store looking at baby clothes; rather, "Suddenly as a dreamer, I see myself inside the

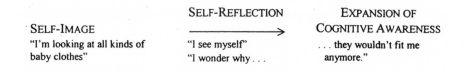

SELF-IMAGE | SELF-REFLECTION ⟶ | EXPANSION OF COGNITIVE AWARENESS
"I'm looking at all kinds of baby clothes" | "I see myself" "I wonder why . . . | . . . they wouldn't fit me anymore."

FIGURE 4.1 | Phenomenological equation proposing how self-reflection via images, thoughts, emotions, or behavior in a dream are mediated by gene expression and neurogenesis to facilitate an expansion of consciousness. (From Rossi, 2000a, p. 30.)

store looking at all kinds of baby clothes, and *I wonder why I look at baby clothes. . . .*" Wondering why represents a definite trend toward cognitive self-reflection. This develop-*mental* shift from the visual modality of looking at baby clothes to a cognitive one of reflecting on the implications of what she is doing is critical for her psychological growth. It is a *critical phase transition* from an essentially passive to a more active mode of being. Because she is able to see herself, the state of being she was experiencing in the dream develops into a new pattern of cognitive awareness: Baby clothes won't fit anymore.

We are all experiments of nature. The drama of the dream is a window through which we can view (self-reflect) and participate (co-create) in the interaction between behavioral state-related gene expression and activity-dependent gene expression that together generate neurogenesis and new identity. This suggests another hypothesis about the self-organizational dynamics of the dream and the critical phase transitions of co-creating consciousness.

HYPOTHESIS 4.4. Self-reflection—an examination of one's thoughts, images, emotions, or behavior in a dream—mediates critical phase transitions between nature (behavioral state-related gene expression that usually operates on an implicit level) and nurture (activity-dependent gene expression that is usually the outcome of an explicit, conscious effort).

We can express this psychobiological hypothesis as a phenomenological equation wherein a state of being is represented as a self-image (Figure 4.1).

Davina might have simply dreamed she was in a store looking at baby clothes. In this simple dream her state of being (looking at baby clothes) would be experienced as such, and that would be the end of the matter. Many dreams are experienced as one-dimensional dramas in which the dreamer simply responds to events as if they were true, without any self-reflection. Such simple experiencing on one level does not generate the drama of self-engagement that is necessary for activity-dependent gene expression, neurogenesis, and psychological change. In some dreams, however, an outside perspective develops; the dreamer sees herself in the dream as if from the perspective of a more objective observer.

In this dream Davina makes an important shift from a visual image of herself looking at baby clothes to an active cognition, "they wouldn't fit me anymore." This is the first example of how a critical phase transition from the visual modality to the cognitive modality can mediate the profoundly important and fundamental process of *intellectual insight in the dream*. Note that this significant developmental insight takes place as a co-creative act entirely within the privacy of Davina's dream. No interpretation by a therapist was required. We will see many more examples of such transitions from visual imagery to cognition mediated by self-reflection as an essential step in the evolution of her spirit.

What is the difference between a simple dream of experiencing oneself on only one level and the dream wherein two or more levels of identity are experienced? Neuroscience research, such as that of Ribeiro and Strickgold cited above, indicates that it is the challenge of novel experiences in everyday life that generates some of the conditions for activity-dependent gene expression and neurogenesis in the dramas of dreams. I propose that this neuroscience research offers a new perspective on why we are often spellbound by dramatic literature in the humanities, in general, and the case histories of psychoanalysis, in particular: This literature engages us in the novelty-numinosum-neurogenesis effect of self-reflection and self-creation.

It seems likely that the interactions between behavior-state-related and activity-dependent gene expression during the normal developmental stages of life (infancy, puberty, marriage, menopause, etc.) can generate the critical phase life transitions that are played out in dream dramas. These critical life transitions that prompt many people to enter psychotherapy are enough to generate dream dramas playing and replaying memories and the transference issues in the redevelopment of the personality. The novelty, polarity, and symmetry-breaking processes illustrated in Chapter 10 are designed to facilitate the creative interaction between behavior-state-related and activity-dependent gene expression in waking as well as dreaming.

A Hobo Leaves Home (three months before psychotherapy)

Dressed as a hobo with a knapsack on my back, I am leaving home. My husband doesn't know I have gone. I am walking down a dirt road but, in actuality, the scenery is moving while I appear to be stepping in place.

This beginning of a much longer dream that Davina had just three months before she entered therapy is presented here as a transparent expression of her developmental state just prior to her growth experience in psychotherapy. Psychologically, her state of being is that of a hobo as she leaves her home—her older, conventional, ordinary existence—to seek a new way in the outside world. This is the classical "call to adventure," celebrated in saga and myth, that is usually experienced in the developmental transition between adolescence and adulthood. As is typical of this early stage in the development of individuality, one's family and friends usually have no idea what is happening. Even Davina's husband does not know where she is going. She is embarking on an inner journey to find her own identity quite apart from her relationship with him.

Our hobo is now actively seeking the changes hinted at in the previous dreams. But as she does so, an unusual type of awareness takes place within her dream. She is not simply walking down a dirt road "but, in actuality, the scenery is moving while I appear to be stepping in place." What is happening here? She is obviously no longer having a simple dream limited to a one-dimensional awareness wherein she is entirely contained within the dream imagery and reacting to it in a naive manner. A second dimension of awareness is implied here as Davina watches herself with the perspective of an outside observer who realizes that the scenery is moving while she is only stepping in place. This outside perspective is a division or bifurcation of awareness from one to two levels in the dream. Its presence is very subtle and only implied here. A second, third, and even fourth bifurcation of awareness, each apparently evolving a new aspect of her consciousness, will be illustrated in a more obvious manner in the next few dreams as her development proceeds. These bifurcations of emergent consciousness are examples of the nonlinear self-organizational dynamics described by modern chaos theory (Rossi, 1996a, 2000a).

I Don't Know Where to Go (first psychotherapy session)

An ape climbs up a pole; suddenly it turns into me. I'm very high up. *There is a bright blue sky, and I look down on the earth far below. I* feel panic but I don't know where to go. *Then three things float by me in the sky: a typewriter, an old woman making flowers, and a lavender stalk of flowers. Suddenly all things come together. I cannot type but the flowers grow and the paper comes off the typewriter and floats down to earth. The paper has golden edges. I hold a little piece of it and try to type on it.*

The introductory theme of "I don't know where to go" is the typical situation of most people when they begin a new psychological process of inner development. In response to this dream, Davina reported a "fear of being abandoned" and an "everything will fall apart feeling." She was troubled about her strangely vivid dreams and the imaginary scenes that sometimes flashed before her open eyes, even during her waking hours. Because these images seemed to be something like what she supposed hallucinations must be, she feared she might be going psychotic. She was having problems getting along with her new husband, who was so preoccupied with his graduate studies that he seemed to have no time for her. She couldn't seem to get through to him any more. Love seemed lost. She couldn't ask her parents for advice. Their marriage was so stale and chilled that it made her shiver.

The most unique feature of this dream is how the ape climbing a pole suddenly turns into a visual image of Davina. It is precisely such apparently nonrational shifts in dream imagery that lead critically disposed observers to maintain that dreams are meaningless. Such *nonrational* transformations, divisions, or *bifurcations* in phenomenological experience, however, are typical examples of what is meant by the *nonlinear dynamics of self-organization*. The sudden visual transformation of the ape into Davina expresses an important *critical phase transition* in her state of being leading to new developments in her consciousness and personality. From our current genomics perspective, "climbing a pole" is *a phenomenological reflection of an activity-dependent process of gene expression and neurogenesis,* whereby the dreamer gets closer to a goal by a series of steps and successive approximations. This series of steps is a kind of *neurobiological calculation on an implicit level* that is called *iteration* and *recursion* in the mathematical language of *nonlinear dynamics*. Such iteration and

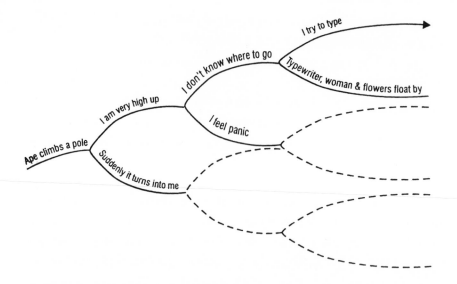

FIGURE 4.2 | Self-reflection and reality-creating choice points in Davina's "I don't know where to go" dream.

recursion are types of *replay or reframing* akin to what Jung described as the *circumambulation* of the goal of *individuation or self-organization* (Jung, 1966).

In this dream the transformation from the ape image to her self-image is immediately followed by a feeling of panic and self-reflection that generates her new awareness that "I don't know where to go." Implicit, unconscious processing intrudes in an apparently irrational way with the images of a typewriter, an old woman, and a lavender stalk of flowers. Davina responds adaptively to these seemingly nonrational presentations in the dream by trying to type. I propose that it is precisely such *active efforts during dreaming* generating novel images that are the phenomenological correlates of *activity-dependent gene expression and neurogenesis.*

The nonlinear dynamics of this dream can be diagrammed using a tree-branching or bifurcating pattern to illustrate the interaction or dialogue between conscious and unconscious processes (Rossi, 1999b, 1999c). Jung (1916) described this inner dialogue integrating the real and imaginary as the "transcendent function." In Figure 4.2 we adopt the convention of placing the representations of implicit, autonomous, and unconscious processing in the lower part of each bifurcation and her more conscious, explicit responses in the upper part. Each division or bifurcation illustrates a fork or "choice point" in the pathways of her psyche. Following Schrödinger's metaphor discussed in the previous chapter, such choices illustrate the phenomenology of how consciousness tutors the brain. The dashed branches represent potential pathways not taken.

The series of branching pathways in Figure 4.2 leads to an important idea about the evolution of consciousness that we witness over and over again in dreams. *Divisions or spontaneous bifurcating transformations in dream imagery are critical phase transitions replaying the important psychobiological choice points and developments that provide conditions for self-reflection, the expansion of awareness, and co-creation.* There are, of course, many more complex iterations and replays going on at an implicit level behind the conscious, explicit imagery of the dreams that we are able to recall. The next dream provides another instructive example.

New World Dream (sixth psychotherapy session)

Circumstances: *After a great deal of thinking about how my brother and I are different from any other member of our family, I have the following dream.*

Dream: *A terrible row rages upstairs, and I go up to the apartment above ours where an old actor lives, Willy Ford, who is forever screaming at his wife, drinking, and treating her brutally. I tell him off, lecture him, scold him, damn him, spit at him for his behavior, and tell him that I can't sleep at night because of his screaming and that I'm afraid that he'll kill his wife one of these days. I insist that he move out, that I don't want any part of his cruelty. Then he flies into a rage at me, tells me to stop meddling, and threatens my life. Both he and his wife scream at me and throw me out. It seems that both of them were unaware of their situation before, and now they are angry with me for making them conscious of their destructive relationship. I leave,* feeling very sad *and like a failure, as far as they were concerned.*

Then, as if by magic, the setting changes completely. I find myself on a ship in the middle of an ocean. There seems to be one main room and a few cabins. But deep in the ship are fighter planes and bombs. In the main room my mother, father, and all my past boyfriends are playing cards. I have a secret lover on the ship, however, who is a soldier-flyer on a special mission.

He is in my cabin, taking his farewell before he leaves. He promises to carry my scarf and flowers with him; he kisses me and promises to make the world a better place for me to live in. Before he flies off, though, a terrible explosion occurs. Everyone rushes outside to see great columns of smoke rise, and somehow everyone knows that the Chinese have set off a bomb. Then in reaction to the bomb, great earthquakes shake the land, sending the ship out of control. My soldier-flyer takes off into the sky to investigate. The rest of the group goes back into the main room, as though nothing had happened, and continues to play cards. The sky turns blood red.

I walk through the main room feeling very sad, wearing a strange dress made out of a red and pink sculptured bath towel. On one side of me, holding my hand, is a stark white figure that looks like me, but unfinished, vague, like a figure who is being painted into being. Holding my left hand is a child named Caroline, who looks exactly like me when I was four years old (the same picture of me that has appeared in previous dreams). We walk out onto the prow of the ship and sit in the very corner of the point of the prow. We are horrified to see that the ship is being drawn on by a mysterious power-force, like a strong current leading us through strange waters.

All the land around us is crumbling. To our left a series of bridges stretches like a tangled mass of wire, and autos keep driving at a terrific speed up the bridges, only to plunge off into the water. Fire burns the land to our right, and all around the earth keeps opening up.

I look around and discover that Caroline is missing. I look for her, the white figure of myself in panic, and find her hiding under a tarpaulin. I promise her that the world will be set straight again. We all go inside, and deep down I am terrified. I cry out that the world is being destroyed. But nobody in the main room believes me, and they all keep on playing cards. I ask where the captain is, but no one has seen him, and it is then that I realize that there is no captain on the ship. There is no force leading us that I can appeal to. I feel helpless and utterly alone.

It is then that my brother gets up from the group and sits down to play the piano, a particular Beethoven sonata. All the others complain and ask why he doesn't play a different piece, that he has played that piece of music ever since he was a little boy. He insists that he loves that piece of music and will play it as long as he likes.

Then he gets up from the piano and comes over to comfort me. He calls me aside and tells me that he believes me and realizes that the boat will meet disaster any minute, but that he has a plan for saving us.

He takes me out in the hall and shows me a lizard-skin suitcase that holds grains for planting, also seeds and food supplies, a tent, water—everything needed to survive and start a new world for us. He begs me not to worry, and we all go out to the side of the ship. (By all, I mean myself, my white self, Caroline, and my brother.)

Sure enough, soon the boat crashes into a chunk of earth and we jump off the ship *just in time. We watch it sink, and crash, and suddenly a huge gush of blood bubbles up from the area where the ship has sunk, and all the sea is bloody. As soon as this happens, the white figure melts into me and we become one, and the child Caroline, that is also me, flies up and becomes a bird who flaps away toward the sun that has just come into the sky for the first time in this dream.*

My brother holds my hand and we climb from rock to rock, just escaping death many times. Somehow we manage to keep our footing as the land keeps crumbling.

Then my brother sits down to rest, almost giving up. I insist that this is not the end, and continue climbing higher and higher. Suddenly I reach the top of the rocks, and instead of finding a slope leading down on the other side, a great green meadowland stretches into the distance, like a New World. White curly sheep are grazing nearby, and the mellow notes of a flute drift toward me, and a shepherd sits nearby. In the distance, a village shows itself, and I am overcome with joy. I race down to my brother and tell him of the New World that awaits us. He doesn't believe me, but I insist that it is there.

I pull him up, and the lizard-skin suitcase falls into the bloody water below. He cries out that now we will never survive, but I reassure him that we don't need that suitcase, as a New World awaits us at the summit of the rocks.

Finally, we reach the top ledge of rock, and lo and behold, the meadow stretches there as I had seen it, and we step firmly, though a bit fearfully, onto the New World. As we step on the grass, I wake up.

The circumstances that preceded this dream provide a clear context for understanding it. She is doing a great deal of thinking. She is becoming more aware of how she and her brother are different from other members of her family. In fact, she and her brother belong to a very different generation and culture. Their parents emigrated from Europe, where they had grown up with entirely different values and experienced many tragic hardships of family dislocation during World War II. Davina and her brother, on the other hand, were raised in America and identified almost entirely with its culture.

The dream begins by dramatizing a very unsatisfactory relationship between a husband and wife. In response to the dream Davina reports that the major source of anxiety in her early life (especially around the age of four) was witnessing the terrible arguments between her parents. As she grew older, she would try to intervene, but she never succeeded in reconciling them. Even now, in the dream, her parents and past boyfriends are nonchalantly playing cards while bombs are exploding and earthquakes are sending their

IMPLICIT PROCESSING

EMOTION $\longrightarrow$ IMAGE

"I feel very sad" ". . . unfinished, vague, like a figure who is being painted into being." "a stark white figure that looks like me . . . "

FIGURE 4.3 | A phenomenological equation illustrating the process whereby an emotion is transduced into an evolving self-image on an implicit level of processing within a dream. (From Rossi 2000a, p. 39.)

ship out of control. Only Davina and her secret soldier-flyer lover seem aware of the danger and try to do something about it.

Then a very unusual and unique thing happens within the dream imagery. Two other figures, both aspects of herself, appear by her side. On one side is "a stark white figure that looks like me, but unfinished, vague, like a figure who is being painted into being." Could we have a clearer expression of the emergence of a new aspect of her identity? A new being or a new level of awareness is now coming into existence within her consciousness. In the language of self-organizational dynamics, we would call this a *divergent bifurcation or critical phase transition* whereby two identities emerge from one. The new identity does not yet fully exist, even within the delicate fabric of the dream, but is just now in the process of being "painted into being" on an implicit level. We are witnessing the emergence of a new complexity, hopefully a new level of adaptation and coping skill (illustrated in Figure 4.3).

On Davina's left side is four-year-old Caroline "who looks exactly like me when I was four years old." Thus, within the replay of the drama of the dream, three aspects of Davina interact with one another: There is the frightened four-year-old child she once was and whom she now must care for; there is her uncertain self, as she is in reality today; and there is the yet unfinished white figure of herself. The four-year-old self clearly represents the first, earliest level of awareness and identity wherein she can only react to a frightening situation by crying. Davina's current identity, represented by her "strange dress made out of red and pink," suggests the current creation of a second level of identity in which she is frightened but still able to help her younger self. The unfinished white self, who appears to be the forerunner of a third level of emergent identity, provides a rare peek into the preconscious implicit level of processing. This identity does not get panicked by the dangerous situation but reacts with competence in looking for the four-year-old and stabilizing her current co-creative process of personality maturation.

While there could be many ways of describing the dynamics of this dream, the branching pathways of Figures 4.2 and 4.4 are consistent with a mathematical model—the Feigenbaum scenario (1980)—that has been used to illustrate the divisions and conflicts characteristic of human choice in the co-creation of consciousness (Rossi, 1997a, 1998b, 1999b, 1999d, 2000a). Box 4.1 at the end of the chapter outlines how the Feigenbaum scenario can be used as a mathematical model of psychobiological dynamics ranging from gene expression and neurogenesis to the relationships between implicit (unconscious) and explicit (conscious) processes we observe in Davina's dreams.

A new aspect of Davina is being created in the laboratory of her dream when she finds herself in the emotionally precarious position of losing her old world. A profound trauma, stressor, or life crisis can set the stage for a division, bifurcation, and critical phase transition in development. In Chapter 2 we reviewed how immediate early genes are turned on within a minute of experiencing strong shocks, whether physical or emotional. These immediate early genes then signal other genes to initiate protein production in brain cells. This protein production will form new neural connections on the pre-conscious implicit processing level (that is, the unconscious level) that become the psychobiological foundation for new experiences of self-reflection, consciousness, and healing (Rossi, 2000a. 2000b, 2000d).

I propose that this new loop of information transduction between the phenomenological experiences of mind and the gene-protein level is an example of what is commonly regarded as the "missing link" or the "bridge" between mind and matter. From the perspectives developed in Chapters 2 and 3, we could say that Davina's dream perception of "a stark white figure that looks like me, but unfinished, vague, like a figure who is being painted into being" is an intimation of a new aspect of her self-identity that she is co-creating in this dream. Applying Hypothesis 4.4, we can see this dream as a vivid example of the coordination of *behavioral state-related gene expression* (the behavioral state of REM sleep) and *activity-dependent gene expression* (Davina's strenuous efforts generate activity within the dream when she says, "I look for her, the white figure of myself in panic. . . .") on the phenome-

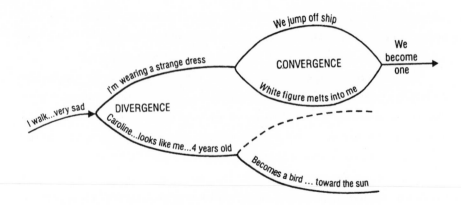

FIGURE 4.4 | The nonlinear dynamics of divergence and convergence in Davina's co-creation of consciousness and choice, which, we propose, integrates behavioral state-related and activity-dependent gene expression in the neurogenesis new identity formation and personality maturation.

nological or phenotypic level of psychological experience.

Davina's efforts in this dream bring to mind the concept of lucid dreaming, wherein the dreamer makes active efforts to replay the dream in a more constructive manner (LaBerge, 1985, 1990; LaBerge & Rheingold, 1990). From our perspective, lucid dreaming is a window through which we can glimpse the creative interaction between behavioral state-related and activity-dependent gene expression in the co-creation of consciousness, identity, and meaning. Psychological development rarely proceeds smoothly, however. The fruits of Davina's new identity are developed by further creative replays of old conflicts to update her emerging insights and personality, as we see in the next dream.

You Are Fruit of the Tree: The Hero Fight (nineteenth session of psychotherapy)

Circumstances: After a series of dreams in which she is frightened by two skeletons—one black, representing her father, and one white, representing her mother—Davina reports the following.

Hypnogogic experience: *At work I fall asleep for a few moments! Dozing, a dream fantasy takes me back to continue a dream I had last week—of being thrown into a room where my parents are bending in prayer before the religions of the world!*

In this continuation, I decide to venture forth and ask them why they are there, look-ing like monsters. I decide to ask why monsters come to terrify me!!! *At my ques-tions, my parents put on their skeleton clothes, shriek at me, go mad—with foaming mouths, bleeding eyes, etc.*

And suddenly a geyser of blood and blue colors shoots up from the center of the floor and they rush into it!! They scream at me: "Get out, get away from us! Go your own way now. We are a unit together, you don't belong!!"

Then the geyser turns into a weird tree, and the two skeletons form part of its trunk, and they shriek: "You are a fruit of the tree, but separate—you must build your own tree. Go away."

And a voice calls out: "The monsters are to frighten you away! They want to scare you out of their world, into one of your own!"

At this, horrid monsters rush out at me! I run out of the room into a little hallway. There, a staircase leads to an opening of light. I start to climb the stairs, but razor edges are on all the stairs and cut me horribly!! I bleed rivers of blood! The monsters scream: "Get out, get out—you don't belong here anymore!! Finally I reach the top and rush out into the light. I fall down on the grass and suddenly only a skeleton of me is left!! One side of my skull is black, the other side white!!! This skeleton rolls down the stairway and into the room of monsters. It sets the entire place on fire. Burning it all into nothingness.

Then the new-me comes out from behind a tree! I am fine and pretty. *I see the blaze, and the ashes that blow away in the wind! Then I run down a hill, get into our auto, and drive home to meet my husband. We go into our apartment and close the door!*

In adopting a questioning attitude toward the monsters, Davina initiates a self-reflective dialogue that leads to an activity-dependent process of gene expression and neurogenesis experienced as an expansion of awareness. This phase of her co-creative process of identity and personality maturation is illus-trated in the phenomenological equation of Figure 4.5a.

After achieving this new awareness, a profoundly important transforma-tion takes place in Davina's dream image of herself. She falls on the grass and suddenly only a skeleton of her is left: "One side of my skull is black, the other side white!!!" Her skull is comprised of her father (the black skeleton) and her mother (the white skeleton). That is, her identity up till now has been

SELF-REFLECTION

IMAGE	+	IMAGE	→	AWARENESS
Davina		monster aspect of her parents	Dialogue: "I decided to ask *why* monsters come to terrify me."	"The monsters are to frighten you away! They want to scare you out of their world, into one of your own!!"

FIGURE 4.5A | A phenomenological equation illustrating how a process of self-reflection on an implicit level can mirror activity-dependent gene expression and neurogenesis leading to a new consciousness and insight. (From Rossi, 2000a.)

PSYCHOSYNTHESIS

IMAGE	+	IMAGE	+	AWARENESS	→	IDENTITY
Davina's skeleton with black and white skull		Monster aspect of her parents		"The monsters are to frighten you away"	"This skeleton rolls down the stairway into the room of monsters. It sets the entire place on fire, burning it all into nothingness."	"Then the new-me comes out from behind a tree! I am fine and pretty."

FIGURE 4.5B | A phenomenological equation illustrating how Davina's new identity evolves out of her previous identity with her parents. The dynamics of psychosynthesis evolve out of the integration of behavioral state-related and activity-dependent gene expression and neurogenesis proposed in hypothesis 4. (From Rossi 2000a.)

half her father's world-view and half her mother's. The self-reflective and co-creative dynamics of this profound example of psychosynthesis, illustrated in Figure 4.5b, may be understood as a manifestation of the novelty-numinosum-neurogenesis effect described in the Chapter 3. Davina's use of three exclamation marks at the dramatic moments of this dream implies that she was experiencing the heightened motivation of novelty and emotional arousal characteristic of insight and new learning generated by gene expression and neurogenesis.

From a psychological perspective we say Davina was identified with her parents; she had a *symmetrical relation* with them. In mathematics the concept of "symmetry" is often used to keep track of "identity" during changing states of transformation. This connection between the psychological concept of

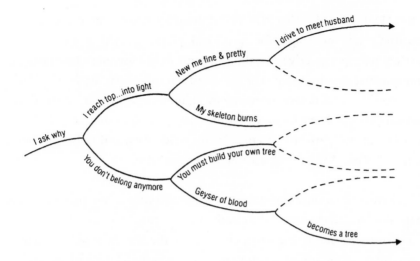

FIGURE 4.5C | A schematic representation of symmetry-breaking occurring in the identification between Davina and her parents when she asks, "Why?" Notice how the lower branch of each bifurcation or choice point is more closely associated with the implicit or autonomous aspect of the dream drama, while the upper branch expresses the outcome of a more conscious activity-dependent process of gene expression and neurogenesis. (From Rossi 2000a, p. 77.)

identity and the mathematical concept of symmetry is illustrated in the branching pathways of Figure 4.5c. Chapter 10 outlines a variety of new activity-dependent approaches that can be used to facilitate these dynamics in the practical, clinical work of conflict resolution, problem-solving, and healing.

A critical phase transition is taking place in this dream as a division emerges in her identification (symmetry) with her parents. In the inner work of the dream this identification with her parents is destroyed, burned to nothingness, along with the frightening skeleton aspect of her parents. After the destruction of this old identity, a "new-me comes out from behind a tree!" Here we observe the natural dynamics of *bifurcation, division, and conflict resolution:* The drama between Davina's skeleton and the monster aspect of her parents leads to the creation of a new identity ("the new-me comes out") and the possibility of a new phenomenal world (her life with her husband). This suggests another basic hypothesis about the phenomenological counterpart to gene expression and neurogenesis during dreaming.

HYPOTHESIS 4.5. Negative or frightening images in the dramas of dreaming set the stage for a critical phase transition that, from a mythopoetic perspective, is called a "hero fight" for one's psychological and spiritual development. The apparently nonrational dynamics of division, conflict, and destruction usually signifies a divergent bifurcation (a natural process Darwinian variation and selection) that may help one step out of old patterns of consciousness, identity, and behavior (selection or "choice") that are no longer appropriate.

Dramas of conflict and destruction often signify an essential early stage in the nonlinear dynamics of the creative process. Jung recognized the destruction, negativity, and darkness that characterize the *nigredo* stage in alchemy as a projection of the uncertain and often confused initial stage of the individuation process. In the language of mythology this stage is called the "night sea journey." In the spiritual literature it is called "the dark night of the soul." Henri Poincaré, the French mathematician who originated nonlinear dynamics around the turn of the century, described this as the stage of incubation and unconscious processing that characterized the creative process as he experienced it (Hadamard, 1954; Wallas, 1926). In Figure 2.11 we diagrammed incubation as stage two of the creative process, as it is experienced in daily life as well as in the arts, sciences, and psychotherapy, and which is illustrated in greater detail throughout Part II.

I propose that we can conceptualize the dynamics of creativity as a natural process of Darwinian variation and selection on all levels, from gene expression and neurogenesis to the co-creation of choice and consciousness that is reflected in the dramas of dreams, fantasies, fairy tails, myths, and sagas (Rossi, 1972, 1985, 2000a, 2000d). Being defeated by frightening images means that some aspect of one's emergent individuality is overwhelmed by previous patterns of life experience and identity. A successful confrontation with negative experiences is a manifestation of activity-dependent gene expression in the novelty-numinosum-neurogenesis effect, which in turn facilitates choice, consciousness, and the co-creation of new identity.

People who make the primitive mistake of acting out destructive patterns of behavior are often confusing inner and outer realities—they confuse the necessary inner work of the novelty-numinosum-neurogenesis effect with the unnecessary projection of conflicts that are then played out in the outer world with tragic consequences. When entire societies engage in warfare, they may be aborting important critical phase transitions in their own cultural development; they are enacting an outer drama of war rather than creating rituals of transition within the relatively safe havens of education, drama, dance, music, art, and spiritual practices. This is what Jung would call a failure of the *transcendent function* to integrate conscious (explicit) and unconscious (implicit) experiences that are still in a process of development. A successful working through of the seemingly negative but necessary process of symmetry-breaking in Davina's identification with her parents led to the generation of "the new-me [which] comes out from behind a tree. I am fine and pretty." This successful developmental process goes on!

A Red-Robed Woman of Art and Beauty (four years later)

Circumstances: After a year of psychotherapy Davina and her husband moved to a distant city where new opportunities beckoned. Four years later, out of the blue, I received a letter from her reporting all was well but, for whatever reasons, her dreams occasionally seemed to replay familiar themes wherein she continued to explore further growth possibilities. She reports that she is now able to handle her inner and outer life okay. She simply thought I would be interested in a follow-up dream that she reports in a poetic manner.

Trapped, in the midst of my parts
I cried for help
No one came.
No bird, no beast, no spirit.
My red room lay bare.
My red-robed self and all the other parts of me stuck together, staring at each other: the broom woman, innocence, the questioning one, mother, wife, sex, and the intellect. Seven parts stuck to my red-robed self. I put my head in my hands and fell into deep, deep thought, indescribable. I felt new courage, and the red-robed woman, me, glowed, illuminated from within. All the other parts became shadowy, but clung there still. I knew then that if I was to recreate my private room into a real room some day, I would have to reckon with these aspects of myself, somehow, some way! I had to struggle free! But how ????
    There beside the fire dosed a white-haired man, smoking a beautiful straight grain briar! Dressed entirely in green (odd color for HELL), he shuffled over. "Hi," he said—

> *"Blarney me, you're all split up, aren't you? Such an interesting one you are, too; quite pretty, but what conflicts you endure. Goodness! Hmm."*
>
> *"How can I become ONE?" I asked.*
>
> *"With words, words. You must write, speak and dream the words of Oneness!" he advised. He took out a golden brocade sack, then, and emptied it over me—millions of tiny printed words poured over me and disappeared into my skin.*
>
> *"Who are you?" I asked.*
>
> *"James Joyce," he replied.*
>
> *"The James Joyce?" I asked.*
>
> *"Yes," he answered and then he revealed to me private caves of words, wonderful golden words, stacks of paper, pens, all seeming to be alive and jumping. . . .*
>
> *I looked a strange sight; a RED-ROBED WOMAN OF ART AND BEAUTY, flowing hair, serious face, knapsack on my back. I ascended the first steps out of HELL.*

Davina's creative self is as active as ever, as expressed in this numinous image of a red-robed woman of art and beauty. This dream seems to sparkle with the novelty-numinosum-neurogenesis effect: New and novel inner experiences with strange, illuminated images and deep, indescribable thoughts motivate her to creative expression via painting and poetry. I propose that this creative expression in art and poetry is an activity-dependent effort that evokes gene expression and neurogenesis in her continuing process of self-development. Davina is still hard at work integrating at least seven aspects of her identity in this dream that comes four years after the formal termination of her therapy.

Davina's identity was initially in a process of splitting, dividing, or bifurcating in a *divergent* manner from her parents during the early phase of her therapy. In her New World Dream we observed how the process of diverging from identity with her parents was followed by the emergence of her new white self in a *convergent* process illustrated in Figure 4.4. Now we witness a further development in the convergence of *seven identities* that she struggles to integrate into one in this dream. In Box 4.1 I propose an application of the Feigenbaum scenario as a mathematical model of the dynamics of divergence and convergence in identity formation, such as those illustrated in this dream. The Feigenbaum scenario offers an intriguing visual picture of the alternating processes of *divergence* (symmetry-breaking, division, duality, or dissociation characteristic of all conflicts) and *convergence* (constructive integration, synthesis, or unification typical of the creative resolution of conflicts). From the perspective of modern nonlinear chaos theory, the Feigenbaum scenario provides a mathematical model of the infinite patterns of choice in the co-creation of identity and consciousness.

During this period of her life, Davina was deeply engaged in writing and painting. Words and James Joyce are clearly important modalities for her continuing development. Her poems, stories, and painting all pour forth in a creative manner that is gaining increasing professional recognition in the outer world.

The little detail in the very end of this dream, where she has a *"knapsack on my back,"* is sweet reminder of a dream she had three months before she began therapy five years ago (A Hobo Leaves Home). This recurring hobo metaphor of her developmental process is an example of the dynamics of replay in the self-reflective, co-creative imagery of her dreams. These are the natural dynamics of psychological development that are used to formulate and illustrate new approaches to facilitating the creative process of activity-dependent gene expression and neurogenesis in psychotherapy in Part II.

BOX 4.1 | THE FEIGENBAUM SCENARIO AS A MATHEMATICAL MODEL OF THE NOVELTY–NUMINOSUM–NEUROGENESIS EFFECT IN THE CREATIVE PSYCHOBIOLOGICAL REPLAYS OF CONSCIOUSNESS, DREAMS, PSYCHOTHERAPY, AND THE HEALING ARTS.

This mathematical model proposes how replays of the novelty–numinosum–neurogenesis effect could generate the symmetry-breaking bifurcations (divisions, duality, conflict, self-reflection, and choice) of consciousness that are the essence of dreaming, psychotherapy, and the healing arts. The word *bifurcation* means a sudden change or point of division in the number and pattern of solutions to an equation when an important factor or parameter is varied. In the equations below the letter **a** represents a control parameter that acts as a valve modulating the expression of the equations. The value of the control parameter at which a bifurcation takes place is called a *bifurcation point* or *bifurcation parameter value* (Nusse & York, 1998). A branching tree illustrates the essential dynamics of a Feigenbaum diagram. *Each branch represents an answer, behavior, or "choice" in the series of solutions to an equation generated by replaying a process of feedback and iteration.*

A mathematical model of human choice?
Could the Feigenbaum scenario be used as a simple mathematical model of consciousness and choice? This model has been used to illustrate many physical, chemical, and biological systems (Cvitanović, 1989; Feigenbaum, 1980). *I propose that it can also model many psychobiological systems—ranging from gene expression and neurogenesis to emotions and cognition—that are generated by replaying a continuous process of iterative feedback* (Rossi, 1996a, 2000a). Since mind and behavior obviously utilize information feedback on many levels, we wonder whether the Feigenbaum scenario can illustrate anything interesting about human choice points on conscious (explicit) or unconscious (implicit) levels.

Feedback and iteration in the logistic equation
The logistic equation was originally proposed as a model of population dynamics whereby feedback (environmental limitations of food supplies and space as well as the presence of predators [Peitgen et al., 1992]) prevents populations of bacteria, plants, and animals from growing infinitely. *Can the logistic equation also be used to model the creative replay of a population of ideas or states of consciousness?* In the logistic equation

$$x_1 = ax_0 (1-x_0)$$

the initial value (x0) is fed back into the equation to get the first solution (x1). This first solution is then fed back into the equation to get the second solution, x2, in a process called iteration.

$$x_2 = ax_1 (1-x_1)$$

Generating the Feigenbaum point on the path to deterministic chaos
Replaying this feedback process leads us to a series of solutions, choices, or states that generate the many pathways of the Feigenbaum bifurcation diagram illustrated on the next page. The first long stem coming down from the top could represent one choice or state of consciousness, which then branches (bifurcates or divides), as indicated. From each of these two branches we see two more bifurcating branches being generated, and so on. This is called the *period-doubling realm* on the path to *deterministic chaos*. In describing his discovery, Feigenbaum (1980) notes: "My first effort at understanding this problem was through the complex analytic properties of the generating function of the iterates . . ." (p.68). That is, Feigenbaum found that the feedback and iteration of many mathematical functions generate a series of solutions that, from a psychological perspective, we could call "choice points." From this mathematical perspective, *genes may be viewed as generating functions that process signals from the environment to transcribe (i.e., generate) messenger RNA, which is then translated into proteins and the dynamics of neurogenesis and psychological experience that is the essence of psychotherapy.*

Notice that the branches of the Feigenbaum diagram get shorter and shorter as they are generated down to the fourth bifurcation, after which they reach a threshold—when the parameter a = 3.5699+—that is called the *Feigenbaum point*. The Feigenbaum point splits the Feigenbaum scenario into two very different parts: the upper period-doubling realm wherein *periodic behavior* is clearly evident, and the lower dark realm of deterministic chaos where discrimination may be impossible.

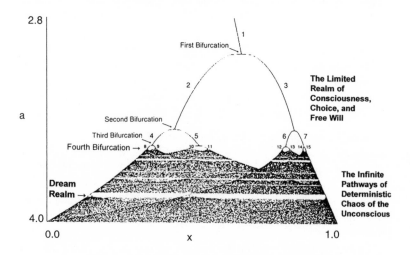

I propose that the Feigenbaum point marks the transition from the explicit realm of conscious choice and behavior to the implicate realm of deterministic chaos on the unconscious level (Rossi, 1996a, 1997a, 1998b, 2000a). The Feigenbaum point is the limit of *"subitizing"*—the ability to tll at a glance whether there are 1, 2, or 3 objects in conscious (explicit) perception (Lakoff & Núñez, 2000). Peitgen et al. (1992) comment on the *implicit* (unconscious) and previously unknown aspect of how numbers operate in the iterative replays that generate nonlinear dynamics:

The Feigenbaum diagram has become the most important icon of chaos theory. It will most likely be an image that will remain as a landmark of the scientific progress of this century. The image is a computer generated image and is necessarily so. That is to say that the details of it could never have been obtained without the aid of the computer. Consequently, the beautiful mathematical properties attached to the structure would definitely be still in the dark *and would probably have remained there if the computer had not been developed. The success of modern chaos theory would be unimaginable without the computer. (p. 587, italics added)*

Universality in the mathematical route from order to deterministic chaos

There is a ratio that quantifies the period-doubling path to chaos that is found to be true of many different equations when they are iterated. This ratio, made up of the lengths of any two successive branches of the Feigenbaum diagram, is called *Feigenbaum's constant,* and it converges to a value of 4.6692+. The Feigenbaum constant is now regarded to be as important to the nonlinear dynamics of chaos theory as the number pi is to geometry. Peitgen et al. (1992) comment on the surprise of this discovery:

Chaos theory began at the end of the last century with some great initial ideas, concepts, and results of the monumental French mathematician Henri Poincaré. . . . The more recent path of the theory has many fascinating success stories. . . . It is known as the route from order into chaos, or Feigenbaum's universality. . . . An even bigger surprise was the discovery that there is a very well defined "route" that leads from one state-order—into the other state-chaos. Furthermore, it was recognized that this route is universal. (pp. 585–86)

Route means that there are abrupt qualitative changes or bifurcations that mark the transition from the realm of order to the realm of deterministic chaos. Universal means that varying patterns of these bifurcations can be found in the mathematical equations that model many natural systems both qualitatively and quantitatively.

Deterministic chaos and the limits of consciousness

While the Feigenbaum point marks the onset of deterministic chaos, it is not really random. *Deterministic chaos only looks random because of the limitations of human perception.* If we zoom in on any small portion of the diagram with a computer, we will find that the overall picture is reproduced again in the smaller portion that we blow up (Peitgen et al., 1992). This is called the *fractal* or *self-similar* aspect of the equation on all scales.

Notice that when we look at the diagram, we can see seven paths clearly as labeled. If we look really closely, we can count 15 paths at the fourth bifurcation level. Above the Feigenbaum point, the choice between the bifurcations and points is clear. Below the Feigenbaum point, from about the middle to the bottom of the diagram, the distribution of paths seems to be a chaotic smudge with some vague structure and blank spaces that are difficult to discern. I propose that an association between the Feigenbaum point and psychobiological experience is suggested by research in memory and learning that indicates that the range of 7–15 items is a limit in many studies of learning. Kihlstrom (1980, 2000), for example, found that 15 items were the upper limit for memory even when it is highly focused with hypnosis. The number seven appears as a limit to a wide range of psychobiological processes that sustain human consciousness and performance. Hammond and Diamond (1997) ask:"Why are sustained energy budgets of humans and other vertebrates limited to not more than about seven times resting metabolic rate? The answer to this question has potential applications to growth rates, foraging ecology, biogeography, plant metabolism, burn patients, and sports medicine" (p. 457).

Further evidence from experimental psychology

A number of classical studies in psychology confirms that seven units (plus or minus two) is, in fact, the usual limit of human perception (Miller, 1956) and memory (Bower, 1997) illustrated at the second bifurcation level of the Feigenbaum scenario. Sperling (1960), for example, found that people could remember about seven letters over a one-second interval; he called this brief memory the *iconic trace*. Neisser (1967) found that the auditory trace, which he called *echoic memory,* had similar characteristics. Recall how telephone numbers also have seven digits. This again suggests that the number 7 is a typical limit in human consciousness that is modeled in the seven paths we can see easily in the upper part of the Feigenbaum scenario.

Another intriguing indication that seven is an important limit in memory (Bower, 1997) and human consciousness was observed by Hofstadter (1995), who described playing extensively with anagrams since his early youth. How many arbitrarily arranged letters, he asked himself, could he juggle in his mind simultaneously until they suddenly rearranged into a meaningful word? He found that he could regularly find meaningful rearrangements of six letters but never as many as 10. Similar limits are evident in other sensory-motor skills such as juggling physical objects. Ronald Graham, chief scientist at AT&T laboratories, is a gifted juggler who reports that he can juggle six balls consistently and sometimes seven (Horgan, 1997). Nine is the world record for juggling.

The practical significance of the number 7 as a typical limit in human choice and consciousness can be found in the extensive use of subjective rating scales of everything from sensation and performance to emotional experience. Research documents that most rating scales used in psychological assessment and personality testing, for example, can be optimized with a seven-point scale. Subjects are asked to judge how happy or satisfied they are with a particular situation or life experience on a five- to seven-point scale that runs something like: 1. Extremely Satisfied; 2. Satisfied; 3. Neutral; 4. Unsatisfactory; 5. Extremely Unsatisfied. Recently developed cognitive-behavioral approaches for assessing a patient's response to psychotherapy use a seven-point validity of cognition (VOC) scale to quantify a patient's conviction as well as a 10-point subjective units of discomfort (SUD) scale to quantify changes in the experience of pain or psychological discomfort during therapy (Rossi, 1996a; Shapiro, 1999). Further research is now needed to confirm the relevance of the Feigenbaum scenario as a model of the dynamics of human choice, perception, and performance in the range of 7 to 15 chunks of consciousness (explicit level) on the path from order to deterministic chaos (implicit or unconscious level).

The physicist John Archibald Wheeler (1994) described how limitations in the range of human consciousness may operate in our formulations of the *laws of nature* and how we can cope with these limitations. Hilgard (1977) conducted experimental research in hypnosis that provides evidence of a "hidden observer" that is able to report that pain is experienced on a deeper, more unconscious (or implicit) level, even when consciousness (the explicit level of conscious choice) does not complain of pain. Pain, symptoms, and problems are replayed in unconscious deterministic chaos at the cellular-genomic level to be selected by the more limited but rational range of consciousness at the neuronal-synaptic level. Some consciousness and potential choice operates even in *"Recalled Dreams"* (horizontal white areas within the implicit realm of the figure).

Explaining the hard problem of consciousness
The association between the Feigenbaum scenario and psychological experience leads me to propose *that consciousness evolved as a practical and rational method of coping with the apparently nonrational and unpredictable realm of deterministic chaos in living experience.* Consciousness, according to this view, is a psychological function that oper-

ates like an amplifying lens (on the self-similar or fractal aspects of nature) or mirror that allows us to sort out, rationalize, make linear and explicit the apparently nonrational, nonlinear, unconscious levels of deterministic chaos in our implicit experience that are necessary for survival (Rossi, 1996a, 1998b, 1999d). This could be an answer, in part, to the "hard question" of why consciousness has evolved.

Carl Jung (1953) speculated about the significance of the number 7 in alchemy, myths, mantic and mystical belief systems. When consciousness has to juggle more than seven items, dimensions, or levels, understanding seems to become chaotic, dark, fearful, unconscious, and unreal, even though we may gamble on the dim intuitions we have about the dynamics of deterministic chaos in the outside world and/or within ourselves. What are the mathematical implications and predictions of the Feigenbaum scenario regarding human experience that could be tested experimentally? How can we test the conjecture that the Feigenbaum point signals the transition from conscious, explicit choice to the implicit, autonomous, unconscious dynamics of deterministic chaos? Apart from the data summarized here, the connection between the Feigenbaum scenario and human experience is not well understood but remains an area of pioneering research (Guastello, 1995; Rössler 1992a, 1992b, 1994b; Vallacher & Nowak, 1994). We now need research that can identify what control parameters (plotted on the vertical axis above), ranging from gene expression and neurogenesis to cognition, can modulate the novelty-numinosum-neurogenesis effect, as illustrated by the many variations of the Feigenbaum diagram that are possible (Peitgen et al., 1992; Rossi, 1996a, 2000a). In Part II we invent new activity-dependent psychotherapeutic approaches to facilitate the novelty-numinosum-neurogenesis effect in the generation of creative choice via the dramatic replays of problem-solving and healing that are modeled here by the Feigenbaum scenario.

SUMMARY

Current neuroscience research is validating the dream–protein hypothesis I originally proposed 30 years ago. The immediate early gene—zif-268, expressed during dreaming—generates new protein synthesis and neurogenesis in a psychobiological process of adaptation to novelty and the consolidation of new memory and learning. We propose that dreaming is a complex adaptive system integrating *behavioral state-related gene expression* with *activity-dependent* gene expression in the natural Darwinian dynamics of variation and selection in the self-reflection and the co-creation of consciousness and choice. We illustrated this creative psychobiological process with a particularly illuminating series of dreams that depict the evolution of new developments in a young woman's consciousness, identity, and personality. This dream series illustrates how explicit, conscious, waking experience is continually integrated with implicit processing during the replays of significant life events in the dramas of the dream. We proposed an application of the Feigenbaum scenario as a mathematical model of symmetry-breaking in the dynamics of conflict resolution, self-reflection, and the co-creation of identity throughout a person's lifetime. In Part II we utilize these natural psychobiological processes to replay activity-dependent approaches to facilitate the novelty-numinosum-neurogenesis effect as the essence of psychotherapy in therapeutic hypnosis and the healing arts.

CHAPTER 5

PSYCHOSOCIAL GENOMICS AND
THE HEALING ARTS

Replaying Gene Expression via Therapeutic Hypnosis and the Humanities

"With the view of simplifying the study of reciprocal actions and reactions of mind and matter upon each other . . . the [hypnotic] condition arose from influences exist-ing within the patient's own body, viz., the influence of concentrated attention, or dominant ideas, in modifying physical action, and these dynamic changes re-acting on the mind of the subject. I adopted the term "hypnotism" or nervous sleep for this process . . . And finally as a generic term, comprising the whole of these phenomena which result from the reciprocal actions of mind and matter upon each other, I think no term more appropriate than 'psychophysiology.' "

—James Braid
The Physiology of Fascination and the Critics Criticized

"A successful hypothesis tells a story of a sort, one that takes multiple odd characters, strange events, unfamiliar settings, unlikely circumstances, and weaves them into a compelling narrative. The audience is left breathless with the sense that it must be true, since it makes order of the chaos of characters, events, settings, and circumstances that are otherwise unintelligible. The particularly elegant hypotheses accomplish all this with uncommon economy. . . . As happens so often in science, hypotheses multi-ply, as one creative idea leads to another and another. As knowledge advances, hypotheses propagate, and the new hypotheses lead to new knowledge. . . . Now the hard work of experimental verification, or falsification lay ahead."

—Ira Black
The Dying of Enoch Wallace: Life, Death, and the Changing Brain

This chapter adopts Ira Black's conception of the "compelling narrative" of hypotheses in science. The history of therapeutic hypnosis and healing certainly has "multiple odd characters, strange events, unfamiliar settings, unlikely cir-cumstances" that are compelling, indeed! The theory and practice of therapeutic hypnosis and the healing arts would be greatly advanced if we could "make order of the chaos of characters, events, settings, and circumstances that are otherwise unintelligible" and weave them together "with uncommon economy."

As yet, there is no general recognition of our basic idea about the creative replaying of the natural dynamics of gene expression as the psychobiological foundation of therapeutic hypnosis, the holistic healing arts, and the humanities. In previous chapters we reviewed well-documented research indicating that novel and enriching life experiences can initiate gene expression, leading to neurogenesis and the encoding of new memories, learning, and behavior. We explored how this is the psychobiological foundation of a new view of the essential role of consciousness, the arts, and the humanities in the evolution of personal identity and culture.

In this chapter we examine the more specific implications of activity-dependent gene expression for understanding the basic science underlying the therapeutic effects of hypnosis, psychotherapy, and healing within individuals. We use the same neuroscience principles to deepen our understanding of the role of the arts, humanities, and cultural rituals in the Gaia-gene-body-mind cycle that we introduced in Chapter 1 (Figure 1.2). We must emphasize, however, that this new vision is not yet a reality. We are simply making a beginning in exploring the possibilities for this future-science-and-therapy-in-waiting by proposing a series of hypotheses to guide research and clinical practice.

WHAT IS THERAPEUTIC HYPNOSIS?
AN ANCIENT QUEST FOR MIND-BODY HEALING

It usually comes as a bit of a surprise to modern students when they learn that it was James Braid, the same physician who popularized the term *hypnotism*, who apparently coined the word *psychophysiology* to describe the essential nature of therapeutic hypnosis. Indeed, there are many examples of how pioneers in hypnosis over the past two centuries struggled to understand the essence of hypnotherapeutic suggestion as a process whereby mind and body interacted and communicated with each other to achieve what they called "healing" (Edmonston, 1982; Tinterow, 1970; Zilboorg & Henry, 1941).

Because theory and experimental techniques were not sufficiently developed to explore the molecular parameters of mind-body communication and healing in those early years, however, the psychophysiological mechanisms proposed by the pioneers of therapeutic hypnosis remained a theoretical hope and clinical promise rather than a well-documented science and reliable method of therapy. In this chapter we formulate a series of hypotheses about mind-body communication and healing that could be assessed by current research techniques in neuroscience and functional genomics to establish a new psychobiological foundation for therapeutic hypnosis, psychotherapy, and the healing arts in general.

Following Braid's intuition, our psychobiological approach to therapeutic hypnosis and the healing arts requires an understanding and utilization of the circular pathways of mind-body communication and healing on all levels, from the subjective experiences of consciousness to gene expression as reviewed in the previous chapters. We begin by taking a look at a few revealing controversies in the history of therapeutic hypnosis to illustrate the typical problems and paradoxes that plague this field even today. Our guiding idea is that all the diverse approaches to holistic, complementary, alternative, and/or spiritual healing that have developed in many cultures over the course of human history are actually different paths to the same goal of what we today call "psychosomatic medicine" or "mind-body therapy." Healers in many different cultures developed and utilized many different worldviews and belief systems to deal with essentially the same question: How can we use human consciousness, psychological experiencing, and our perception of free will to communicate with our bodies in ways that facilitate healing and well-being?

Paradoxes of Hypnosis: The Illusions of Power, Programming, Manipulation, Compliance, and Control

In his initial perplexity over discerning the essence of the therapeutic efficacy of hypnosis, Braid noted his quandary concerning the role of fascination as well as voluntary and involuntary behavior in hypnosis with animals and humans:

> The power possessed by serpents to fascinate birds has always been a source of interest and admiration . . . by what means is this remarkable result effected? . . .

Is it a voluntary, or an involuntary process? . . . After due consideration, I feel satisfied that the approach and surrender of itself by the bird, or other animal, is just another example of the mono–ideo–dynamic, or unconscious muscular action from a dominant idea possessing the mind. (quoted in Tinterow, 1970, p. 365)

Braid's concept of the mono–ideo–dynamic action anticipated the modern view that it is the single-minded fixation of attention—the patient's intense focus on the words and ideas of the clinician using hypnosis—that is the stimulus for psychophysiological healing (Barber, 1969, 1984, 2000). Braid's questions about the role of active, conscious, and voluntary responsiveness versus more passive, involuntary, or unconscious processes in hypnosis, however, still remain central and unresolved in current theories. Until recently, Braid's emphasis on the significance of fascination in hypnosis fell into the oblivion of folklore about the "evil eye." As we saw in Chapter 3, however, fascination during novel and numinous life experiences plays a fundamental role in focusing our attention and engaging activity-dependent gene expression, neurogenesis, and healing in general.

Unfortunately, the ancient idea of the evil eye is still associated with the experience of fascination in the negative sense of gullibility, manipulation, and exploitation. Fascination, social dominance, and control are also associated with the shopworn ideas and illusions about power and compliance in human affairs. Modern studies document that there is, in fact, a small percentage of the general population—about 5–10%—who manifests high hypnotic susceptibility on standardized scientific scales (Hilgard, 1965, 1981, 1991). People scoring high on hypnotic susceptibility can respond to suggestions on a deep implicit level with little or no sense of explicit, conscious control over their behavior.

From the neuroscience perspective we could define *hypnotic suggestions* as *implicit processing heuristics.* The fact that activity-dependent gene expression and neurogenesis play a dynamic role in memory and learning means that hypnotic suggestions are *heuristics,* facilitators of creative experiences, not a means of programming. That is, highly hypnotizable subjects experience a high degree of dissociation between their behavior and their sense of conscious control over it in the hypnotic situation. A previously unrecognized aspect of this therapeutic dissociation is that it frees highly hypnotizable subjects to receive and creatively replay hypnotic heuristics (suggestions) primarily on an implicit level without too much interference from consciousness. For the great majority of people, however, hypnotic experiences are psychosocially inspired dramas of interesting self-exploration that engage them in an inner dialogue wherein hypnotic heuristics are creatively replayed for a healing reintegration and resynthesis of conscious and unconscious processes. They feel safe enough in the hypnotic situation to wonder about how they can learn to respond—on a more or less automatic (that is, implicit, unconscious, or involuntary) level—to hypnotic heuristics proffered by a therapist (Erickson et al., 1976; Erickson & Rossi, 1979, 1981; Rossi & Cheek, 1988; Zeig, 1985, 1990, 1994; Zeig & Geary, 2000).

The majority of people interested in self-exploration enjoy operating on two or more levels in this way. They enjoy watching—on the conscious, explicit, self-observer level—how a more implicit level of their experiencing self responds to the therapist's heuristic suggestions in a more or less automatic manner. Apart from the highly hypnotizable 5–10% of the general population, however, there is nothing in current research that supports the still popular misperception of hypnosis as a tool of power, programming, manipulation, control, and compliance (Nash, 2001). Quite to the contrary, in one of the earliest clinical-experimental studies in this area, Erickson (1932/1980) reported:

> Far from making them hypersuggestible, it was found necessary to deal very gingerly with them to keep their cooperation, and it was often felt that they developed a compensatory negativism toward the hypnotist to offset any increased suggestibility. Subjects trained to go into a deep trance instantly at the snap of a finger would successfully resist when unwilling or more interested in other projects. (p. 495)

Erickson later commented that programming "is a very uninformed way" of attempting to do therapeutic hypnosis (Erickson & Rossi, 1979, p. 288).

The Activity-Passivity Paradox of Hypnosis and Healing

The primitive state of psychophysiology during the first century of hypnosis meant that it could be conceptualized only as some sort of reflex or pathology. For example, Bernheim (1886/1957), the leader of the Nancy school in France described hypnosis as the "exaltation of the ideo-motor reflex excitability, which effects the unconscious transformation of the thought into movement, unknown to the will. . . . The mechanism of suggestion in general, may then be summed up in the following formula: increase of the reflex ideo-motor, ideo-sensitivity, and ideo-excitability" (p. 138). The idea that hypnosis involved an increase in "sensitivity" and "excitability" is in striking contrast to the view of the Salpêtrière School in Paris, led by Charcot, who maintained that hypnosis was a pathological condition of passivity that progressed from lethargy and catalepsy to somnambulism.

The conception of hypnosis as a state of passivity has its origins in many sources from ancient to modern (Ellenberger, 1970; Zilboorg & Henry, 1941). From the ancient healing rites in Greek temples to the current methods of inducing hypnosis (Edmonston, 1986; Hilgard, 1965, 1981), the proposed association between passivity and hypnosis has been a dominant theme (Weitzenhoffer, 1971, 2000; Weitzenhoffer & Hilgard, 1967). This passivity idea even became prominent in the school of self-hypnosis developed by Émile Coué. A French chemist and pharmacist who operated a free clinic, Coué reportedly taught the principles and practice of autosuggestion to as many as 40,000 patients annually. Coué (1923) described his approach as follows:

> All that is necessary is to place oneself in a condition of *mental passiveness,* silence the voice of conscious analysis, and *then deposit in the ever-awake subconscious the idea or suggestion which one desires to be realized.* Every night, when you have comfortably settled yourself in bed and are on the point of dropping off to sleep, murmur in a low but clear voice, just loud enough to be heard by yourself, this little formula: "Every day, in every way, I am getting better and better." Recite the phrase like a litany, twenty times or more. (p. 26, italics added)

Notice that in the first stage of this process, Coué recommended *mental passiveness* that was to be followed by a more active process of self-directed suggestion *in the ever-awake subconscious.* Coué recommended another more active ritual to enhance the effects of repetition: His patients were instructed to tie 20 knots on a piece of string and count off the repetitions just as one might say the rosary.

Coué (1922) offered two principles for the efficacy of his method: (1) It is impossible to think of two things at the same time; and (2) every thought that completely fills our mind becomes true for us and has a tendency to transform itself into action. The hidden problems and paradoxes within Coué's method have become apparent with the passage of time. If his method was so wildly popular, with many reports of its efficacy when he first introduced it, why is it no longer so popular today? Let us answer this question in the light of current neuroscience research on novelty, environmental enrichment, and exercise as the principle factors in facilitating activity-dependent memory, learning, neurogenesis, and healing, as reviewed in Chapter 3. Coué's method certainly enjoyed a novelty effect when it was first introduced. We can suppose that it contributed to an enrichment of the lives of patients who gained a great deal of psychosocial support from others who flocked to the most popular clinic of the day, where stories of miracles tended to evoke the novelty-numinosum-neurogenesis effect. It certainly was an exercise that practically anyone could pursue with self-reassuring comfort, and it was an exercise easily associated with the healing power of the spirit in the holy rosary. With the passage of time and the erosion of its reputation for mediating the miraculous, however, Coué's method lost its novelty-numinosum-neurogenesis effect.

Whenever a new and numinous method or psychosocial belief system is introduced, it will usually be able to boast a number of fast converts who will report marvelous results. It is likely that these immediate positive experiences come from that highly suggestible 5–10% of the general population who have a special talent for mind-body accessibility and healing. Only later, when the larger proportion of the population that does not have such talent complains of lack of success, does the novelty-numinosum-neurogenesis effect loose its psychogenomic potency (Rossi, 1990b). Nonetheless, the Coué phenomenon continues even today, with creative variations that reintroduce the healing aspects of novelty, environmental enrichment, and exercises in ever-new experiences that blend the passive and active aspects of self-hypnosis (Alman & Lambrou, 1991; Gillett, 2001; Simpkins & Simpkins, 2000). From our new neuroscience perspective, the central role of *repetition* in Coué's method could be understood as a primitive forerunner of what we now call the *creative replay, reframing, and resynthesis* of the dynamics of mind, memory, and psychological experience on all levels, from consciousness to neurogenesis and gene expression.

The basic paradox and question of whether hypnosis involves heightened activity, passivity, or both continued as a central theme for many early researchers. Pavlov (1927; Edmonston, 1986), for example, believed hypnosis was a state of cerebral inhibition, a kind of "partial sleep," whereas Clark Hull (1933/1986) maintained the opposite view that hypnosis is a state of arousal:

> We seem forced to the view that hypnosis is not sleep. . . . Thus the extreme lethargic state is not hypnosis, but true sleep: *only the alert stage is hypnotic.* Lastly, evidence has been presented which indicates not only that conditioned reflexes may be set up during hypnosis, but that this may perhaps be accomplished with even greater ease than in the waking state. This probably disproves Pavlov's hypothesis that hypnosis is a state of partial sleep in the sense of a partial irradiation of inhibition. (p. 221, italics added)

Hull's conclusion that "only the alert stage is hypnotic" is entirely consistent with the current neuroscience view of activity-dependent gene expression and neurogenesis in the creation and recreation of mind, memory, and consciousness.

Erickson was a student of Clark Hull and went on to develop innovative approaches to hypnosis that utilized either the passive, relaxed, and sleep-like tendencies of his patients to act on an implicit level ("You don't even have to listen to my voice") or their more explicit, active, conscious, "acting out" behavior, such as pacing around the therapy office in an agitated manner. Erickson taught that the appropriate choice of hypnotic induction should be a function of the patient's mood, attitudes, and behavior in the therapy session. Erickson described a great variety of innovative approaches to therapeutic hypnosis that made use of the full range of patients' ongoing behavior. Erickson (1958/1980, 1959/1980) used the terms "naturalistic" and "utilization" to describe the *neuro-psycho-physiological* process of therapeutic hypnosis (Erickson, 1948/1980; Rossi, 2001b).

One of the first efforts to understand the psychobiological basis of the activity-passivity paradox of therapeutic hypnosis was undertaken by one of Erickson's colleagues, Bernard Gorton (1949, 1957, 1958), in his papers on the physiology of hypnosis. A review of the existing literature at the time suggested to Gorton that "vasomotor activity" and the autonomic nervous system (ANS) with its two main branches, the sympathetic (arousal) and parasympathetic (relaxation), provided the major avenue of the physiological effects of hypnosis. More recent research supports Gorton's view that there is "a positive correlation between hypnotic susceptibility and autonomic responsiveness during hypnosis" (DeBenedittis et al., 1994, p.140) and that the nature of the "physiological responsiveness [arousal or relaxation] is dependent on the type of suggestions during hypnosis" (Sturgis & Coe, 1990, p. 205).

Even today, however, the use of hypnotic suggestion to modulate the active as well as the passive branches of the autonomic nervous system is not well understood by researchers who report the "paradoxical" nature of their results. Weinstein and Au (1991), for example, reported that norepinephrine levels were significantly higher in a hypnotized group of patients undergoing angioplasty than in the control group. Their findings were "unexpected and seemed paradoxical (p. 29). . . . One would expect that if hypnosis does cause relaxation, then those patients who were hypnotized would have a lower arterial catecholamine level than their controls. This was not the case. Just the opposite occurred and is hard to explain" (p. 35).

This so-called paradox is hard to explain only if one assumes, as these and other investigators do (Lazarus & Mayne, 1991), that hypnosis is essentially a state of relaxation. The importance of arousal in hypnosis, however, was emphasized by Amigo (1994), who presented evidence for "self-regulation therapy" as a "cognitive-behavioral approach to hypnosis" that involves the "voluntary reproduction of the stimulant effects of epinephrine" (p. 80). Likewise Harris et al. (1993) found that "both branches of the autonomic nervous system may contribute to hypnotic susceptibility" (p. 22). This accumulating research on how hypnosis can utilize both the active and passive (rest) branches of the autonomic nervous system, which operate in Kleitman's 90–120-minute basic rest–activity cycle suggests a resolution of the activity-passivity paradox of hypnosis and leads to an expanded model of the psychobiological domain of therapeutic hypnosis (Rossi, 1973a, 1982, 2000c).

EXPANDING THE DOMAIN OF THERAPEUTIC HYPNOSIS AND THE HEALING ARTS

The psychobiological model of therapeutic hypnosis and the healing arts is consistent with a generation of research that firmly established, contrary to popular belief, that there is no transcendence of normal abilities in hypnosis (Wagstaff, 1986). Researchers openly acknowledge, however, that they have no adequate theory of the source and parameters of hypnotic performance. Naish (1986), for example, has summarized this limitation of the cognitive-behavioral approach:

> As [hypnotic] susceptibility is normally assessed, a high scorer is one who produces the behavior [on measures of hypnosis], the reason for its production remains unknown. . . . The claim was frequently made that cognitive processes are involved in the production of "hypnotic" effects. However, the exact nature of these processes generally remained obscure. (pp. 165–166)

The 10 hypotheses developed in this chapter, which expand the psychobiological domain of therapeutic hypnosis and the healing arts, seek to clarify this obscurity. This psychobiological model proposes that what seems to be an extension of the normal parameters of mind-body performance skills, via hypnosis, is actually the optimization of the individual's normal range of abilities in response to the general process of adaptation on all levels, from gene expression to the central and autonomic nervous systems.

The limitations of cognitive-behavioral conceptualizations of suggestibility and hypnosis have been discussed by a number of researchers. Hilgard (1991) for example, summarized the current dilemma with these words:

> hypnotic behavior cannot be defined simply as a response to suggestion. . . . Although hypnotic-like behaviors are commonly responses to suggestion, the domain of suggestion includes responses that do not belong within hypnosis, and the phenomena of hypnosis covers more than specific responses to suggestion. (p. 45–46)

Hilgard seems to be saying that the domain of hypnotic suggestion, as currently defined on the cognitive-behavioral level, is not adequate or complete. Eysenck (1991) emphasizes the problem more pungently:

> There is no single, unitary trait of suggestibility, no one uniform type of reaction to different kinds of suggestion in human subjects. There are several, or pos-

sibly many different suggestibility's that bear no relation to each other. These are uncorrelated and, in turn, correlate differentially with other cognitive and emotional variables. This finding is of considerable interest and importance. . . . It does make books containing in their title the word "suggestibility" of rather doubtful value! (p. 87)

These statements imply that we are at a fundamental impasse in the current cognitive-behavioral theory of hypnosis. Researchers who began their careers by defining hypnosis as a cognitive-behavioral response to suggestion, reversed themselves by concluding that (1) hypnosis cannot be operationally defined simply as responsiveness to suggestion, and (2) there is no unitary human trait of suggestibility. Balthazard and Woody (1992), for example, stated: "Although a fair amount of factor analytic work has been done with the hypnosis scales, this work appears to be at an impasse methodologically, and it has failed to yield any consistent picture of the mechanisms that underlie performance on the hypnosis scales" (p. 22, italics added). Kirsch and Lynn (1995) continue to emphasize this impasse, as follows: "Is there a uniquely hypnotic state that serves as a background or gives rise to the altered subjective experiences produced by suggestion? Having failed to find reliable markers of trance after 50 years of careful research, most researchers have concluded that this has outlived its usefulness" (p. 854).

Paradox and impasse in science frequently indicate where theory and practice need to be revised and expanded in some fundamental way. This need motivates our first hypothesis about gene expression as the common language of both nature and nurture, thereby taking the first step in resolving the paradoxes in the psychobiological domain of therapeutic hypnosis.

GENE EXPRESSION AS THE PSYCHOBIOLOGICAL BASIS OF THERAPEUTIC HYPNOSIS AND THE HEALING ARTS

HYPOTHESIS 5.1. Therapeutic hypnosis and the healing arts engage the cybernetic pathways of information transduction that flow between the psychosocial environment and the psychosomatic network of the central and autonomic nervous systems and the neuroendocrine and immune systems, to the organ, tissue, and cellular levels of gene expression.

This psychobiological model of hypnosis and the healing arts expands the domain of therapeutic suggestion beyond the cognitive-behavioral level to include all systems of mindbody communication and healing at the molecular-genomic level that are responsive to psychosocial cues. It has been proposed that the major pathways of mind-body communication consist of the hormonal messenger molecule/cell receptor dynamics of the neuroendocrine, neuropeptide, autonomic, endocrine, and immune systems—all of which mediate stress, emotions, memory, learning, personality, behavior, and symptoms in a *"psychosomatic network"* (Pert et al., 1985, 1989; Rossi, 1986/1993, 1996a, 2000a, 2000b, 2000c, 2000d).

The four-stage model of mind-body communication, outlined previously in Figure 2.1, is reproduced with a sharper focus in Figure 5.1 to illustrate the psychobiological core of mind-gene communication in therapeutic hypnosis and the healing arts (Rossi, 1990a, 1990b). Figure 5.1 emphasizes how many processes of psychobiological arousal—such as pain, stress, novelty, the basic rest–activity cycle (BRAC), dreaming (REM sleep), and creative moments—can initiate (1) immediate early gene expression (IEGs) that, in turn, lead to (2) the expression of specific target genes, which (3) code for new protein synthesis that is the molecular basis of (4) state-dependent memory, learning, and behavior (SDMLB). Indeed, there are many genes that are turned on and off across the sleep-wake cycle that are of significance in modulating mind-gene communication via psychosocial cues and psychotherapeutic processes. These include genes such as galanin, which increases during REM sleep (Toppila et al., 1995), and many of the genes whose transcription and translation into proteins generate the messenger molecules of mind-body regulation and healing, such as tyrosine hydroxylase, somatosatatin, growth hormone releasing hormone (Porkka-Heiskanen et al., 1998), and nitric oxide (Brivanlou & Darnell, 2002; Williams et al., 1998).

Research on the changing molecular dynamics of prostate cancer by a research team lead by David Feldman at Stanford University (Zhao et al., 2000) presents an interesting example of how nature and nurture interact in the four-stage model of the psychosomatic network illustrated in Figures 2.1 and 5.1. In the early stage of prostate cancer, growth is modulated by testosterone and related hormones known collectively as the androgens. As prostate cells multiply, they release increasing amounts of a protein marker called prostate specific antigen (PSA). High levels of PSA in the blood are easily

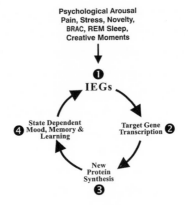

FIGURE 5.1 | A psychogenomic view of the psychosomatic network. Mindbody communication via the dynamics of psychobiological arousal that can initiate (1) immediate early gene expression (IEG) that, in turn, leads to (2) the expression of specific target genes, which (3) code for new protein synthesis that is the molecular basis of (4) state-dependent memory, learning, behavior (SDMLB) that is replayed on conscious and unconscious levels in the creation and recreation of human experience.

measured and are often the first sign of cancer. Medical treatment at this stage is oriented to lowering androgen levels in the blood so that androgen-dependent prostate cancers usually shrink and PSA levels fall.

In later-stage prostate cancer, two gene mutations take place so that androgens no longer bind tightly to the mutated receptor to turn on gene expression and further growth (stages two and three of Figures 2.1 and 5.1). Instead, the glucocorticoids *stress hormones* bind to the receptor to turn on gene expression and the out-of-control growth of prostate cancer:"Two mutations changed the androgen receptor structure so that now cortisol and cortisone could bind to the mutated receptor and act like 'pseudo-androgens,' " David Feldman said (Leslie, 2000). Molecular medicine now seeks to find other molecules that could block the mutated androgen receptor and prevent further proliferation of prostate cancer cells.

Note, however, that the newly discovered role of stress hormones in turning on the mutated androgen receptor changed what was initially a purely molecular problem into a psychosomatic problem: Excessive psychosocial stress processed in the limbic-hypothalamic-pituitary area (stage one of Figures 2.1 and 5.1) initiates a cascade of stress hormones throughout the body that turn on the pseudo-androgen receptor in stages two and three and exacerbate prostate cancer. Herein enters the possibility of using therapeutic suggestion to reduce psychosocial stress in the psychosomatic network to attenuate prostate cancer. Let us now explore the possibilities of this application of therapeutic hypnosis in facilitating the dynamics of psychosocial genomics in greater detail.

Integrating Nature and Nurture with Psychosocial Genomics

Nature and nurture are always interacting in the processes of psychobiological communication via the psychosocial dynamics of gene expression. Nature is more dominant in initiating the *behavioral state-related gene expression* that is associated with the changing states of waking, sleeping, the periodicity of the basic rest–activity cycle (BRAC), and the dream cycle (REM sleep). *The typical process of hypnotic induction is an exercise in facilitating behavioral state-related gene expression,* evident from the traditional approach to hypnotic induction that involves suggestions for relaxation, comfort, and sleep. Insofar as such suggestions are successful, they function as psychosocial cues initiating psychobiological processes associated with behavioral state-related gene expression (Rossi, 1999e).

Nurture is more dominant in initiating the psychological or experiential aspects of *activity-dependent gene expression,* which leads to neurogenesis and the concomitant changes in memory, learning, behavior, and healing. For example, when the hypnotized subject is given suggestions for optimizing performance involving the generation of new memory, learning, and behavior, we may infer that activity-dependent gene expression is being engaged to facilitate neurogenesis and healing. *Mind-body communication and healing in therapeutic hypnosis and the healing arts are inherent in the interaction between the two fundamental processes of (1) behavioral state-related gene expression, and (2) activity-dependent gene expression. Gene expression is the common language shared by nature and nurture in the dynamics of psychosocial genomics.*

Recent research documents how psychobiological mechanisms operate, via behavioral state-related gene expression, in the transitions between waking, sleeping, and dreaming (Bentivoglio & Grassi-Zucconi, 1999; Cirelli, 1998), as discussed previously in Chapter 2. How can we build a bridge between this level of behavioral state-related gene expression and the level of consciousness and free will in the subjective experiences of therapeutic hypnosis and the healing arts? Intriguing research by Born et al. (1999) suggests one possibility. These researchers investigated what we may call the "human alarm clock effect," which enables some people to awaken at a specific time in the morning without using an alarm clock. Awakening from a night of sleep is related to the daily (circadian) and hourly (ultradian) rhythms of behavioral state-related gene expression, which leads to the release of hormones along the central axis of the limbic-hypothalamic-pituitary-adrenal (LHPA) system. This daily circadian rhythm is comprised of a series of ultradian rhythms, wherein there is a peak in the release of ACTH and cortisol every 90–120 minutes throughout the day and night. Normally this release of ACTH and cortisol increases during the later stages of sleep and finally reaches a daily peak just before the time of awakening in the morning, as was illustrated in our new map of the psychobiology of consciousness in Figure 2.11.

Born et al. (1999) demonstrated that the conscious intentionality or anticipation of awakening at a specific time in the morning could shift the ultradian peak in the release of ACTH to that specific time. The implication of this research is that conscious intentionality, free will, and expectancy of awakening at a specific time can pervade sleep and modulate the expression of a normally involuntary rhythm of behavioral state-related gene expression and the flow of hormonal signals across the limbic-hypothalamic-pituitary system of mind-body communication (illustrated as locus one of Figures 2.1 and 5.1). Born's research is an example of how phenomenological experiences of consciousness—what is commonly called the "activity of mind" and "free will" in humans—can modulate physiology at the hormonal level, and ultimately, at the level of gene expression.

It would be of great interest to determine whether a posthypnotic suggestion achieves some of its mind-body effects by utilizing the same type of psychobiological mechanisms uncovered by Born at the hormonal level. Such a finding would imply that posthypnotic suggestions can operate as psychosocial cues to modulate the expression of "clock genes" and, in turn, their regulation of many ultradian processes of metabolism and development and neuronal activity associated with adaptation and stress (review Figures 2.2, 2.3, and 2.4). Such research would integrate therapeutic hypnosis with the vast database and methodology of neuroscience and the emerging disciplines of bioinformatics and psychosocial genomics. The complex dynamics of the many levels of interaction between mind and gene implied by Born's research motivates us to take a closer look at the mathematical modeling (Devlin, 1999, 2000; Solé & Goodwin, 2000) of the major systems of mind-body communication that may be operative in the psychosocial genomics of therapeutic hypnosis and the holistic healing arts (Rossi, 1996a).

THE MATHEMATICAL MODELING OF THERAPEUTIC HYPNOSIS AND THE HEALING ARTS

HYPOTHESIS 5.2. Therapeutic hypnosis and the healing arts utilize a wide variety of psychobiological states ranging from high arousal to low. The entire range of therapeutic states may be mathematically modeled as complex adaptive systems involving nonlinear, quasi-periodic, Yerkes-Dodson functions of arousal, creative work, healing, and relaxation.

The complex nonlinear relationship between performance and arousal illustrated in Box 5.1, called the Yerkes-Dodson function (1908), is one of the earliest and most well-established laws in psychology (Rossi, 1996, 2000c). All life processes go through an initial phase of arousal, reach a peak where performance and healing are optimized, and then relax back to a basal level. Guastello (1995) has updated the significance of the Yerkes-Dodson function in psychophysics and sensory-perceptual systems as well as the nonlinear dynamics of creativity, work, and social organizations. We hypothesize that this intensively researched Yerkes-Dodson function models the arousal (high-phase) and relaxation (low-phase) aspects of therapeutic hypnosis and Kleitman's basic rest–activity cycle. An outline of the parameters of this mathematical model, adapted from chronobiology, is presented in Box 5.2.

Box 5.3 illustrates how this mathematical model could integrate many of the apparently paradoxical phenomena and conflicting theories of high- and low-phase therapeutic hypnosis in a complementary manner on the level of gene expression. The continuum of therapeutic hypnosis could be assessed by DNA microarray technology (Rossi, 1999a, 2000a, 2000b). This continuum of therapeutic hypnosis ranges from the quasi-periodic ultradian performance peaks of (1) *high-phase hypnosis, with an active focus on outer effort and problem-solving,* as described by social-psychological theorists (Sarbin & Coe,

BOX 5.1 | THE CONTRAST BETWEEN LINEAR AND NONLINEAR DYNAMICS IN
PSYCHOLOGY.

The typical linear approach is illustrated on the left with the straight line cutting
through a cloud of data points is the "best linear fit" of the data. Traditionally, each
of the data points is said to be a combination of a measurable psychological factor
and experimental error or "noise." The nonlinear dynamics systems approach, however, recognizes that many apparently random deviations of so-called noise actually may be the signature of quasi-periodicity and deterministic chaos in the natural
ultradian rhythms of mind-body communication and healing.

DYNAMICS IN PSYCHOLOGY

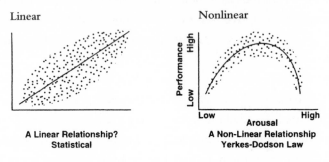

A Linear Relationship?
Statistical

A Non-Linear Relationship
Yerkes-Dodson Law

1972), to the (2) apparently *passive periods of deep inner absorption and ultradian
healing in low-phase hypnosis* emphasized by special state theorists (Hilgard,
1977, 1982, 1991; Hilgard & Hilgard, 1983).

It has been proposed that the typical stages of hypnotherapeutic work
follow the dynamics of the four-stage creative cycle (Rossi, 1996a, 2000c)
outlined in Chapter 2 (Figure 2.11). The first two stages (data collection and
induction) usually consist of an initial high phase of sympathetic system arousal as the subject becomes motivated and engaged in problem-solving. With
the resolution of a problem in stage three (illumination) and stage four (verification), the subject spontaneously slips into low-phase hypnosis wherein
parasympathetic system dominance is experienced as relaxation and healing
(Rossi, 1996a, 2000a, 2000c). Research is now needed to determine the value
of (1) identifying explicit, *conscious processing in high-phase hypnosis, which is associated with activity-dependent gene expression,* and (2) the more *implicit or uncon-*

BOX 5.2 | TOWARD A MATHEMATICAL MODEL OF THE CHRONOBIOLOGY OF THERAPEUTIC HYPNOSIS (FROM KRONAUER, 1984; STROGATZ, 1986).

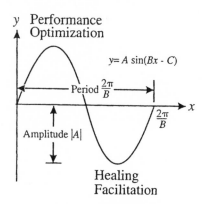

The parameters of a mathematical model of how the therapeutic applications of hypnosis may entrain and utilize the quasi-periodic psychobiological dynamics of adaptation to stress are as follows.

Amplitude refers to the absolute value of the height or depth of a quasi-periodic rhythm of behavior: it measures how far rhythm deviates from its mean level. The amplitude may correspond to the degree to which hypnosis may optimize performance variables or psychobiological healing.

Period is the time required for one complete cycle of adaption on all levels from the cognitive-behavioral to the cellular-genomic. The period is a quasi-periodic parameter of psychological rhythms that may be contracted, stretched or "phase locked" in the therapeutic applications of hypnosis. The *frequency* is the reciprocal of the period.

Phase identifies parts of natural quasi-periodic psychobiological rhythms of adaption that may be accessed, entrained, and utilized by hypnosis. The crest or peak phase is associated with the activation (sympathetic system) dynamics of "high phase hypnosis," while the trough phase of relaxation (parasympathetic system) may be entrained by "low phase hypnosis." The number C/B, called the *phase shift*, is a measure of the degree to which certain portions of chronobiological behavior can be entrained and modulated with hypnosis.

Entrainment or synchronization refers to the interaction of psychobiological rhythms (x and y below) with psychological cues such as hypnosis (z below). Many therapeutic applications of hypnosis may be conceptualized as the entrainment of portions of the basic rest–activity cycle (BRAC) that are utilized for enhancing performance of facilitating healing.

A mathematical model adapted from the field of chronobiology (Kronauer, 1984) is used to illustrate how the hypnotic entrainment of quasi-periodic psychobiological rhythms may operate. In the equations below F_{zy} is the "influence" or entrainment coefficient; w_x and w_y are the natural frequencies of quasi-periodic psychobiological rhythms x and y; k may be a constant associated with *hypnotic susceptibility*; that is, the entrainability of a particular psychobiological process by highly focused psychosocial cues.

$$k^2\ddot{x} + k\mu_x(-1 + x^2)\dot{x} + \omega_x^2 x + F_{yx}k\dot{y} = 0$$

$$(k = \pi/12)$$

$$k^2\ddot{y} + k\mu_y(-1 + y^2)\dot{y} + \omega_y^2 y + F_{xy}k\dot{x} = F_{zy}z$$

$$z = F_{zy}\cos(k\omega_z t + \theta_z)$$

scious processing in low-phase hypnosis associated with behavioral state-related gene expression.

What types of data are needed to assess this mathematical model of high- and low-phase hypnosis as a complex adaptive system (Callard et al., 1999; Rossi, 1996a; Solé & Goodwin, 2000) that may engage gene expression? While the therapeutic applications of hypnosis traditionally have focused on facilitating relaxation or low-phase hypnosis, a pair of research reports (Hautkappe & Bongartz, 1992; Unterweger et al.,1992) supports the idea that hypnosis involves a significant arousal or "work function" (our high-phase hypnosis illustrated here in Box 5.3). Unterweger et al. report that this work function operates differently in subjects with high versus low hypnotic susceptibility. They found that heart rate variability was a useful physiological parameter for discriminating high and low hypnotic susceptibility; high susceptible subjects have less heart rate variability. This may imply that "high susceptible subjects do not have to work as hard on passing a suggestion as do low susceptibles" (Unterweger et al., 1992, p. 87). This finding may mean that high hypnotic susceptibility is associated with a more efficient psychobiological use of information and/or energy on an implicit level.

What is the reason for this greater efficiency of highly hypnotizable subjects? Traditional hypnosis theory would emphasize that highly hypnotizable subjects can focus their attention to a greater degree. This view is consistent with Braid's early idea of defining hypnosis as a psychophysiological process that is focused by "mono-ideation." It is also consistent with research reviewed in Chapter 2 that indicated how a decrease in heart rate variability implied

BOX 5.3 | THE CONTINUUM OF THERAPEUTIC HYPNOSIS RANGING FROM
HIGH-PHASE TO LOW-PHASE (FROM ROSSI, 1996A).

The continuum of therapeutic of hypnosis to be assessed by DNA chip technology ranges from the quasi-periodic ultradian peaks of performance of (1) high phase
hypnosis with its active focus on problem solving as described by social-psychological theorists to (2) the apparently passive periods of deep inner absorption and ultradian healing with low phase hypnosis emphasized by special state theorists. A
complete cycle of hypnotic work usually consists of an initial high phase of sympathetic system arousal as the subject becomes engaged in problem solving. With
the resolution of a problem the subject spontaneously slips into low phase hypnosis
wherein parasympathetic system dominance is experienced as relaxation and healing.

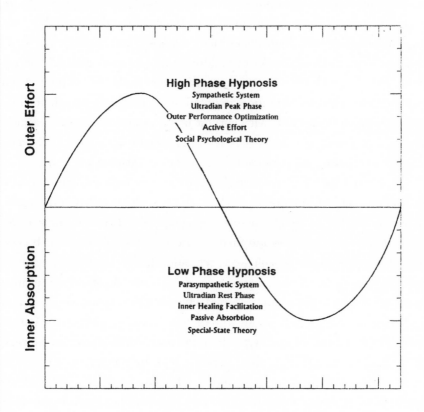

that *phase locking* was taking place somewhere in the complex adaptive psychobiological system. *This mathematical concept of phase locking is another way of understanding the essence of the psychobiological theory of therapeutic suggestion: how suggestions on a psychological level can modulate or "entrain" the biological dynamics of therapeutic hypnosis on the level of gene expression, neurogenesis, and healing.*

In Ericksonian terms, highly susceptible subjects have higher *response attentiveness* or *selective focus* on the implicit level. This focusing ability on the implicit level does the work more efficiently, so that an indiscriminate emotional arousal on the explicit level is not required (Erickson & Rossi, 1979; Lynn & Sherman, 2000). Erickson pioneered the use of psychological shocks and creative moments (Rossi, 1973a) to focus implicit-level activity in what we would now call *activity-dependent gene expression in high-phase hypnosis.* More recent clinical and experimental research by Barabasz and Barabasz (1996) has documented how such high-phase "alert hypnosis" can facilitate the neurotherapy (neural biofeedback) of children with attention-deficit/hyperactivity disorder (ADHD) by enhancing their skills in shifting from predominantly theta to beta modes of functioning, as measured by their EEG. Such findings are consistent with current neuroscience research that associates states of arousal, novelty, enrichment, and exercise with activity-dependent gene expression and neurogenesis in memory, learning, and behavior.

Further research is now needed to explore how high and low susceptible hypnotic subjects respond in the high and low phases of hypnotic works. One wonders, for example, whether high susceptible hypnotic subjects are more efficient in turning on and focusing the activating (sympathetic branch or high-phase hypnosis) as well as the relaxation (parasympathetic branch or low-phase hypnosis) of their neuroendocrine system. What are the relative merits of (1) the Stanford and Harvard Hypnotic Susceptibility Scales that tend to induce low-phase hypnosis with their emphasis on sleep; (2) the Barber Scale (2000), and the Spiegel Hypnotic Induction Profile (Spiegel & Spiegel, 1978) that may induce high-phase hypnosis with their emphasis on focused attention; and (3) Rossi's (1986b) Indirect Trance Assessment Scale (ITAS) that seeks to eliminate a bias toward either high- or low-phase hypnosis in the therapeutic applications of hypnosis?

The relative merits of these and future scales of hypnotic susceptibility will be found in relation to psychobiological variables on all levels from mind to gene. It has already been found, for example, that the essential difference, at the molecular-genomic level, between the greater efficacy of memory and learning in young versus older mammals is due to the length of time the NMDA receptor remains open in the CA1 region of the hippocampus (Tsien, 2000a). Genetically engineered mice that have the NR2B subunit of the NMDA receptor stay open almost twice as long (0.230 milliseconds) as those with the NR2A subunit (Tsien, 2000b). It is this greater amount of arousal time at the cellular level that allows more neurotransmitters to remain active, thereby strengthening the Hebbian synapse in memory, learning, and long-term potentiation on the molecular-genomic level (Malenka & Nicoll, 1999). These findings lead us to speculate that the importance of fascination in the fixation and focusing of attention in the classical descriptions of hypnosis (Edmonston, 1986) is due to its optimization of activity-dependent gene expression in states of arousal during high-phase hypnosis. From this perspective we can regard high-phase hypnosis as a work function that is the essence of Braid's original concept of the psychophysiology of fascination and mono-ideation in hypnosis.

MATCHING THE DOMAINS OF CHRONOBIOLOGY AND THERAPEUTIC SUGGESTION

HYPOTHESIS 5.3. The complex adaptive systems of mind-body communication and healing that are normally replayed during Kleitman's 90–120-minute basic rest–activity cycle (BRAC) can be entrained by therapeutic hypnosis and utilized in the healing arts.

The coordination of mind and body is synchronized by gene expression cycles that function as "biological clocks" (Buijs & Kalsbeek, 2001). At the limbic-hypothalamic-pituitary level, any powerful physical or psychosocial stimulus leads to the release, within minutes, of cortisol releasing hormone that, in turn, initiates an ACTH-cortisol-beta-endorphin cascade that coordinates adaptive psychobiological processes on the organ, tissue, and cellular-genomic levels (as illustrated in Figure 2.1). One of the most prominent of these is our now familiar 90–120-minute BRAC associated with the quasi-periodicity of hormonal systems of mind-body regulation during waking (Lloyd & Rossi, 1992, 1993), sleep, and dreaming (Kleitman, 1969; Kleitman

& Rossi, 1992; Rossi, 1986a, 2000a, 2000b, 2000c). Evidence summarized in Chapter 2 indicates that the highly adaptable quasi-periodic 90–120-minute ultradian cycle consists of an initial phase of psychobiological arousal, work, and performance mediated by the ACTH-cortisol phase of the BRAC. The relaxation phase of the BRAC is then mediated by beta-endorphin, which signals a molecular cascade at the cellular-genomic-protein levels to facilitate restoration and healing. Iranmanesh et al. (1989b), for example, reported that "cortisol was considered to lead ß-endorphin by 20 or 30 minutes. We conclude that ß-endorphin is released physiologically in a pulsate manner with circadian and ultradian rhythms and a close temporal coupling to cortisol" (p. 1019). This research finding is consistent with the hypothesis that arousal or high-phase hypnosis is associated with the cortisol peak of the BRAC, while relaxation or low-phase hypnosis is associated with the subsequent beta-endorphin response. Research is now required to monitor the release of ACTH, cortisol, beta-endorphin, and related messenger molecules during the quasi-periodic phases of hypnosis, as described previously (Rossi, 1988, 1990a & b, 1996a, b; Rossi and Cheek, 1988).

Well-replicated research on the psychobiological dynamics of the major systems of mind-body regulation and healing is evolving out of time parameter research conducted on many levels, ranging from behavioral state-related gene expression in the cell cycle to the neuroendocrine and behavioral levels discussed in Chapter 2 (Lloyd & Rossi, 1992, 1993). It now appears that many of the highly adaptive processes of psychobiological regulation that replay our natural circadian and ultradian quasi-periodic cycles are modifiable by psychosocial cues and hypnosis (Rossi, 1982, 1986/1993, 1996a). *This modifiability implies that what has been traditionally called "hypnotic suggestion" may be, from a chronobiological perspective, the accessing and utilization of the quasi-periodic variability of ultradian and circadian processes on all levels, from the cognitive-behavioral to the molecular-genomic.* Within this framework, many of the classical phenomena of hypnosis may be conceptualized as extreme manifestations and/or perseverations of the quasi-periodic psychobiological processes that are responsive to psychosocial cues. *What the clinician using hypnosis calls "therapeutic suggestion," the chronobiologist calls "the entrainment of biological processes by psychosocial cues."*

This matching of the domains of chronobiological and hypnotic phenomena leads to important falsifiability tests about the relevance of quasi-periodic parameters in hypnotic work. In brief, if one could find a single chronobiological phenomenon (aspects of memory, emotions, learning, behavior, or self-regulation, for example) that can be entrained by psychosocial cues but is not responsive to hypnotic suggestion, then one would have disproved the relevance of quasi-periodic parameters for the psychobiological model of hypnosis. The reverse is also true. If one could find a single psychobiological process modifiable by hypnotic suggestion that does not have a natural quasi-periodic ultradian or circadian rhythm, then one would have disproved the relevance of chronobiological parameters for the psychobiological model of therapeutic hypnosis.

Rossi (1996a) has discussed an unintended but highly informative example of this falsifiable test with research on the event–related P300 brain-wave potential that is a measure of the locus of neural arousal and activity. Initially the P300 wave was found to be modifiable by hypnosis (Barabasz & Lonsdale, 1983; Spiegel & Barabasz, 1988). Later, in unrelated research, the P300 wave was found to have a 90–120-minute ultradian periodicity (Escera et al., 1992). This is exactly what we would predict from our matching of those domains of chronobiological and hypnotic phenomena that are responsive to psychosocial cues.

Research findings consistent with the significance of quasi-periodic parameters in hypnosis have been reported by a number of investigators in the past two decades. Aldrich and Bernstein (1987) found that "time of day" was a statistically significant factor in hypnotic susceptibility. They reported a bi-modal distribution of scores on the Harvard Group Scale of Hypnotic Susceptibility by college students, who showed a sharp, major peak at 12 noon and a secondary, broader plateau around 5 to 6 P.M. Further research found a very prominent circadian rhythm with a peak between noon and 1 P.M. in self-hypnosis as well as an ultradian periodicity of about 90 to 180 minutes throughout the day that approximates Kleitman's 90–120-minute cycle (Rossi, 1992a). The pilot data for Figure 2.10 suggests that self-hypnosis usually lasts about 20 minutes. A replication of this research, using brain imaging and the new DNA microarray technology for measuring gene expression, would be needed to assess this hypothesis. New PET methods allowing assessment of gene expression throughout the human body in a noninvasive manner are currently being developed (Lok, 2001; Phelps, 2001). The adaptation of these imaging techniques for the assessment of therapeutic hypnosis would provide a definitive means of assessing the mind–gene connection in the healing arts.

It is interesting and probably not coincidental that much research assessing the therapeutic value of the various modalities of the holistic healing arts associated with alternative medicine such as acupuncture, biofeedback, imagery, meditation, music, therapeutic touch, etc., also use a core 20-minute therapeutic period (Bittman et al., 2001; Green & Green, 1987; Rossi, 1996a). While this core therapeutic period may be condensed or extended depending on practical exigencies, it is rarely extended beyond the Kleitman's typical 90–120-minute basic rest–activity cycle. Similar quasi-periodic parameters were found associated with hypnosis (Brown, 1991a, 1991b; Lippincott, 1992, 1993; Osowiec, 1992; Rossi, 1982, 1992a; Sommer, 1993; Wallace, 1993) and imagery (Wallace & Kokoszka, 1995).

Two studies (Mann & Sanders, 1995; Saito & Kano, 1992) were correct in emphasizing that a narrow interpretation of Rossi's original chronobiological hypothesis—that hypnotic susceptibility was a function of periodic or statistically uniform ultradian rhythms (Rossi, 1982)—needs to be amended. The experimental results of these two studies are entirely consistent with the current hypothesis about the quasi-periodic or nonlinear dynamics of therapeutic hypnosis as a complex adaptive system. As emphasized in Chapter 2 and elsewhere (Rossi, 1996a), the periodic life processes of chronobiology are actually complex adaptive systems (Glass & Mackey, 1988). In the terminology of complex adaptive systems, all life processes are balanced ("computing") on the "edge of catastrophe" between order and deterministic chaos (illustrated in Box 4.1) (Bélair et al., 1995; Morowitz & Singer, 1995). Psychobiological dynamics operate on the "edge of chaos" to optimize natural variation, selection, and adaptation on all levels, from the molecular-genomic to the psychosocial (Poon & Merrill, 1997; Solé & Goodwin, 2000). Further research utilizing more efficient mathematical algorithms for detecting the synchronization of psychobiological rhythms (Schäfer et al., 1998) as creative replays of complex adaptive systems is now required to examine how therapeutic hypnosis and the healing arts can entrain the natural quasi-periodic dynamics of mind-body communication and healing.

Are the similarities between the ultradian time parameters of (1) fundamental life processes at the level of gene expression, (2) Kleitman's basic rest–activity cycle, (3) stress, and (4) the therapeutic applications of hypnosis and the healing arts a simple coincidence? Or are they all associated as complex adaptive systems that are interconnected on many levels? The most recent research breakthrough in this area has established a connection between the chronobiology of gene expression, energy dynamics, and the light–dark circadian rhythm (Schibler et al., 2001). Researchers found that a transcription factor in the mammalian forebrain (neuronal PAS domain protein, NPAS2) modulates energy dynamics at the cellular level that are associated with states of alertness and tiredness (Reick et al., 2001). Rutter et al. (2001) have summarized the range of psychobiological states involved as follows:

> We believe that NPAS2 functions as a positive component of a molecular clock in the mammalian forebrain, equivalent to the function of Clock [gene] in the SCN [the suprachiasmatic nucleus of the brain that entrains mind-body rhythms to light]. The principle sites of NPAS2 [gene] expression include the somatosensory cortex, visual cortex, auditory cortex, olfactory tubercles, striatum, and accumbens nucleus. *These regions of the forebrain process sensory information, including touch, pain, temperature, vision, hearing, and smell, as well as emotive behaviors such as fear and anxiety.* We hypothesize that during wakefulness, these NPAS2-expressing regions of the brain receive and process stimulatory neuronal activity that entrains the molecular clock in the same manner argued for the retino-SCN pathway. *This entrainment could occur by three interlinked events:* (i) generation of extra-cellular glutamate by active neurotransmission, (ii) stimulation of glycolysis and lactate production [energy dynamics] by localized glutamate uptake into astrocytes, and (iii) fluctuations of intracellular redox potential by facilitated lactate transport and its consummation as a metabolic fuel in nearby neurons. If correct, *this hypothesis may ultimately help explain how the mammalian forebrain oscillates between states of alertness and tiredness.* (pp. 513–14, italics added)

These researchers do not relate their findings to current theories of therapeutic hypnosis and the healing arts. When they relate their research to "help explain how the mammalian forebrain oscillates between states of alertness and tiredness," however, we can interpret their results as consistent with the psychobiological foundation of therapeutic hypnosis and the healing arts developed in the hypotheses of this chapter. More direct tests of the engagement of the NPAS2 gene-expressing regions of the brain during conditions of chronic stress, trauma, illness, and healing are now needed to consolidate the chronobiological theory of therapeutic hypnosis at the genomic level proposed over the past two decades (Rossi, 1981, 1982, 1986a, 1990b, 1992a, 1992b, 1994c, 1995a, 1995b, 1995c, 1995d).

PSYCHOBIOLOGICAL STRESS, GENE EXPRESSION, AND PSYCHOSOMATIC PROBLEMS

HYPOTHESIS 5.4. Chronic stress engendered by traumatic and/or excessive work loads result in the desynchronization of circadian and ultradian dynamics of the 90–120-minute basic rest–activity cycle at the cellular-genomic level. This major etiological source of psychosomatic problems may be ameliorated by therapeutic hypnosis and the healing arts.

Research documenting the impact of acute (Kaufer et al., 1998) and chronic stress on gene expression and the desynchronization of human performance found heightened levels of cortisol and evidence of *spatial cognitive impairments* in pilots subjected to jet lag. A recent study by Cho (2001), for example, used MRI scans to assess the hypothesis that chronic cortisol elevations could induce the actual atrophy of the hippocampus in the brains of airline crewmembers. As we reviewed in Chapter 3, it is now known that chronic stress and elevated cortisol levels are associated with the death of cells in the hippocampus and a suppression of neurogenesis. Cho (2001) reported that airline personnel with short turnaround times between transmeridian flights have smaller temporal lobe structures (involving the hippocampus) than personnel whose schedules permit more recovery time. Higher cortisol levels associated with chronic stress were associated with a reduction in temporal lobe volume.

Since 1996, the National Transportation Safety Board has determined that a least 15 airline crashes could be attributed, in part, to pilot fatigue. When asked if pilots are well rested before flights, Captain Rand Harrell, a 21-year veteran and official of the Air Line Pilots Association (which represents pilots of most major airlines), answered, "There is a real possibility they are not [rested]" (Salant, 2001). He said his superiors tried to fire him twice for refusing to fly when tired. On one occasion he was so tired after three days with just four hours of sleep each day that he turned down the wrong taxiway. Turning down the wrong taxiway sounds like it could be related to the *spatial cognitive impairments* associated with jet lag reported by Cho. It is now believed that most, if not all, forms of physical and psychological stress are mediated by gene expression at the cellular level in the brain and body (Morimoto, 2001; Morimoto & Jacob, 1998). Here we will review a few examples of well-documented research in the areas of psychoimmunology, pain, and the aging process where new models of mind–gene communication, stress, and healing are currently being developed.

STRESS AND GENE EXPRESSION IN PSYCHOIMMUNOLOGY

The most revealing demonstrations of how psychosocial stress can modulate the actual mechanisms of gene expression in the immune system are described in a series of papers by Glaser and his colleagues (1990, 1993). Their research traces the effects of psychological stress (experienced by medical students during examinations) in down-regulating the transcription of the interleukin-2 receptor gene and interleukin-2 mRNA production. Since interleukin-2 is a messenger molecule of the immune system, its down-regulation in response to psychological stress is a fundamental demonstration of how optimal functioning can be impaired at the cellular-genomic level by psychosocial cues (Malarkey et al., 2001). Glaser's research gains even more profound significance for a general theory of mind–body communication and healing when we realize that other independent medical researchers (Rosenberg & Barry, 1992) found that interleukin-2, also known as "T-cell growth factor," is a messenger molecule of the immune system that "tells" white blood cells to attack pathogens and cancer cells. Rosenberg, of the National Cancer Institute, remarked in an interview (Newman, 2001), "I have seen cancers disappear. . . . What is clear is that if immune therapy is ever to work, the T-cells must learn to recognize antigens on the tumor and signal the immune system to attack. Theoretically, that's what happens in cases of spontaneous remission" (pp. 45–48). Traditional medical research represented by Rosenberg and mind–body medicine represented by Glaser have found the same interleukin-2 communication and healing link in psychoimmunology.

This molecular-genomic essence of communication in psychoimmunology is now known to be mediated by immediate early genes that typically turn on within a minute of the onset of psychological arousal and stress and operate for about an hour and a half or so (Schlingensiepen et al., 1995). Further evidence of how the immune system can be impaired at the cellular-genomic level was provided when Glaser et al. (1993) found that academic stress led to the down-regulation of two immediate early genes (also called proto-oncogenes), c-myc and c-myb, in peripheral blood leukocytes. The immediate early gene c-myc is part of an informational loop at the cellular level that activates oncogenes involved in breast cancer, stomach and lung cancers, and leukemia. Such research detailing how psychosocial cues can modulate immediate early gene expression provides a clue about the mechanisms of therapeutic hypnosis that could facilitate the so-called "spontaneous remissions" of cancer that are occasionally reported in a well-documented manner (Crasilneck, 1997).

The most obvious research frontier in this area of psychoimmunology is to document the reverse of the Glaser protocol outlined above. We need to design experimental protocols to assess whether a hypnotic intervention designed to reduce psychosocial stress could lead to a facilitation of the expression of the interleukin-2 receptor gene and mRNA. A step in this direction has been taken by Castes et al. (1999), who reported that a six-month program of relaxation, guided imagery, and self-esteem workshops with a group of asthmatic children led to a significantly reduced number of illness episodes and use of bronchodilator medication compared with a control group. The experimental group also showed an increase in the gene expression of the T-cell receptor for interleukin-2, precisely as predicted for a test of the mind-gene pathway of healing in therapeutic hypnosis and psychotherapy (Rossi, 1994a, 1994b, 1996a, 1998a), as well as a significant increase in natural killers cells and other immune system factors associated with psychosocial stress.

We now need to learn more about the actual pathways traveled by the positive and negative psychosocial stimuli that culminate in the activation or suppression of the interleukin genes (Rossi, 1986/1993). Sternberg (2000) has told the story of the discovery of more than 20 of these interleukin messenger molecules of the immune system (particularly interleukin-1 and interleukin-2) and their association with the stress, pain, and healing responses signaled by the brain (Samad et al., 2001). A summary view of some of these interleukin pathways at the cellular-genomic levels is illustrated in Figure 5.2 (see color insert). This molecular map, submitted by Walter O'Dell, is continually updated on the internet (www.biocarta.com/pathfiles/il2Pathway.asp). The psychosocial modulation of the secondary messenger cAMP at the cellular level has also been implicated in the dynamics of this interleukin-2 pathway (Glaser et al., 1990, 1993).

Positive and negative psychosocial stimuli can lead to the up- and down-regulation of pathways illustrated in Figure 5.2. Primary messengers (hormones) of the limbic-hypothalamic-pituitary-adrenal system that mediates psychosocial arousal and stress can signal (activate) the interleukin-2 receptor proteins that are located on the surfaces of many cells of the immune system (B-cells, T-cells, monocytes, natural killer cells, etc.). A series of recent psychoimmunological studies by Dhabhar and McEwen (2000) has demonstrated how ultradian periods of stress (two hours of restraint) in mice can lead to an apparently paradoxical enhancement of immune function in the skin by gamma interferon. Dhabhar and McEwen (1999) summarize their findings of the relationships between brain, stress hormones, and the immune system with the help of a colorful metaphor.

> In this manner, stress hormones may direct the body's "soldiers" (leukocytes), to exit their "barracks" (spleen and bone marrow), travel the "boulevards" (blood vessels), and take positions at potential "battle stations" (skin, lining of gastrointestinal and urinary-genital tracts, and draining lymph nodes). Moreover we hypothesize that, in addition to sending leukocytes to potential battle stations, stress hormones may also better equip them for battle by enhancing processes like antigen presentation, phagocytosis, cytokine function, and antibody production. Thus, a hormonal alarm signal released by the brain on detecting an *acute stressor may prepare and enhance the immune system* for potential challenges (wounding or infection) that may arise from the actions of the acute stress-inducing agent (e.g., a predator or attacker). In contrast, it is likely that *chronic stress suppresses immune function* by decreasing leukocyte redistribution and by inhibiting cytokine and prostaglandin synthesis and leukocyte function. (p. 1063, italics added)

A MODEL EXPERIMENT
ASSESSING GENE EXPRESSION USING DNA MICROARRAYS

Advances in imaging technology (Brown, 1999; Lok, 2001) place research on gene expression in psychoimmunology within the reach of clinicians and researchers in therapeutic hypnosis and the healing arts. DNA microarrays (gene chips) present up to 10,000 genes (including the interleukin-2 gene) that can be used to assess the patterns of gene expression associated with varying experimental conditions. A prototype of such research assessing interleukin-2 gene expression as well as the ultradian time dynamics of early-, intermediate-, and late-activated genes in the major cells of the human immune system is available on the internet (Incyte, 1999; gem.incyte.com/gem/data/unigemv/index.shtml).

This example of DNA microarray technology illustrates the ultradian time parameters of fundamental life processes, such as the immune system, that have been proposed as an important psychobiological modality of therapeutic hypnosis, placebos, and a host of other approaches to the healing arts (Rossi, 1986/1993, 2000a, 2000b, 2000c). Incyte researchers studied the peripheral blood mononuclear cells (PBMC)—which included many of the major cell types of the immune system, such as monocytes, natural killer cells, dendritic cells, T- and B-lymphocytes—in healthy human volunteers. These cells express the common housekeeping genes that are responsive to extracellular signaling, as described above, as well as many tissue-specific genes such as cytokines, transcription factors, and membrane receptors. These PBMCs were stimulated with the mitogens in vitro and gene expression was assessed after 0.5, 1, 2, 4, and 8 hours of activation. As illustrated in Figure 5.3, there were statistically significant changes in ultradian time in the expression of interleukin-2, which can be modulated by psychosocial stress as well as support. Interleukin-10, an antiinflammatory cytokine that suppresses allergic reactions, showed no changes in this experiment.

This model experiment in the use of DNA microarrays is particularly interesting because it confirms how this new technology can be used to easily and comprehensively replicate and greatly expand the one-gene/one-function approach of classical molecular genomics. Figures 5.4 and 5.5 illustrate the typical 90–120-minute ultradian time parameters of *early-activated genes* in response to extracellular signaling. Figures 5.6 and 5.7 illustrate a variety of ultradian time parameters in the *intermediate-* and *late-activated* ranges of gene expression in response to extracellular signaling. These are the fundamental ultradian time parameters that are associated with a wide variety of other biological processes including metabolism, homeostasis, cell division, growth, and healing (Rossi, 1999, 2000b, 2000c). These same ultradian time parameters are found in the psychobiological processes, involved in

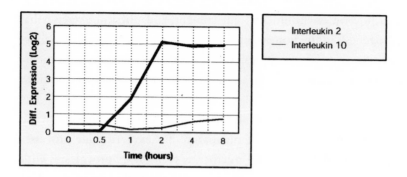

FIGURE 5.3 | A comparison of changes in gene expression of interleukin-2 (IL-2) and interleukin-10 (IL-10) in ultradian time. (With permission from Incyte corporation.)

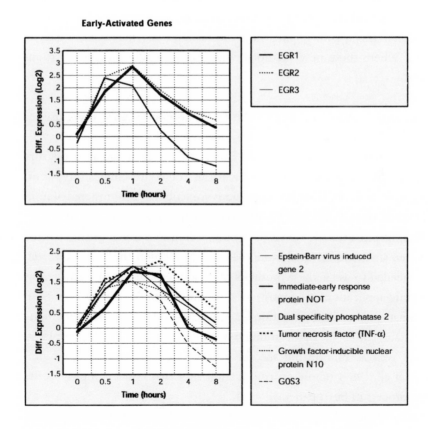

Figures 5.4 and 5.5 | The typical 90–120-minute ultradian time parameters of *early-activated genes* in response to extracellular signaling.

responses to novelty, environmental enrichment, and physical exercise that lead to neurogenesis in the human brain (Gould et al., 1999). Figure 5.8 illustrates how some families of suppressed genes are down-regulated immediately and continue downward for at least eight hours, unless conditions change.

The Incyte model experiments illustrate the rich potential harvest of the Human Genome Project and DNA microarrays for exploring life processes at the most fundamental level. DNA microarray technology provides us with opportunities to directly test the relevance of therapeutic hypnosis and all healing arts involving consciousness and psychosocial processes to psychosomatic medicine, healing, and health (Lloyd & Rossi, 1992; Rossi, 1996a). Using the Incyte model of assessing gene expression with DNA microarrays together with the noninvasive measurements of gene expression with PET imaging (Lok, 2001; Phelps, 2001), we could more directly assess the role of gene expression in therapeutic hypnosis and the healing arts. One interesting environment where these new methods could be used is in emergency room hypnosis.

Pain, Stress, and Gene Expression in Emergency Room Hypnosis

Based on a lifetime of clinical practice, Ewin (1986), an emergency room surgeon and a former president of the American Society of Clinical Hypnosis, documented how therapeutic suggestions for cooling administered within two hours (a typical ultradian BRAC) of incurring a severe burn can reduce pain and inflammation as well as facilitate healing to a much greater degree than when suggestion is used more than two hours after the burn. Research is now needed to assess the hypothesis that the mechanisms engaged by this therapeutic application of hypnosis modulate the patterns of gene expression that mediate pain and inflammation via the mind-body effects of interleukin-1ß (a signaling molecule of the immune system) (Samad et al., 2001; Sternberg, 2000) as well as minimizing the formation of excessive "stress proteins" (Pardue et al., 1989) whose overproduction after two hours complicates the healing process in burn patients.

Samad et al. (2001) documented the molecular mechanisms of inflammation factors, such as cyclooxygenase-2 (cox-2), that are associated with hypersensitivity and pain in areas of injury. Injured areas (such as the burns

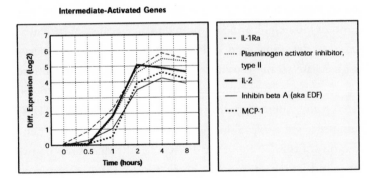

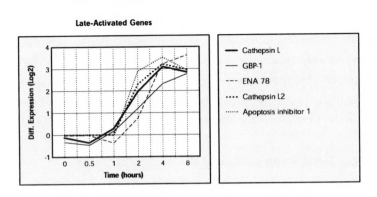

FIGURES 5.6 AND 5.7 | A variety of ultradian time parameters in the intermediate- and late-activated ranges of gene expression in response to extracellular signaling.

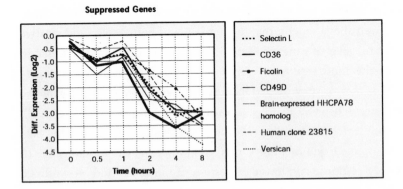

FIGURE 5.8 | How some families of suppressed genes are down-regulated immediately and continue downward for at least 8 hours unless conditions change.

Ewin treated via suggestions for cooling noted above) can generate secondary pain hypersensitivity in neighboring uninjured tissue because of increased neuronal excitability in the spinal cord and sensitization of the central nervous system. This gives rise to the typical feelings of being sick with diffuse muscle and joint pain, fever, lethargy, and anorexia. Samad et al. (2001) showed that cox-2 mediated pain by demonstrating the widespread induction of cox-2 mRNA (meaning that gene expression was involved) in spinal cord neurons and other regions of the CNS within the typical ultradian time frames of two to six hours and lasting through 24 hours. Cox-2 mRNA increased five-fold by two hours and 15-fold in six hours, and then descended to about eight-fold after 24 hours. They found that a major inducer of CNS cox-2 mRNA up-regulation in this ultradian time frame is interleukin-1ß. This research suggests another model by which we may assess the molecular-genomic dynamics of how therapeutic suggestion achieves its inflammatory and pain relieving effects, as demonstrated by Ewin in emergency room hypnosis and others in the laboratory (Kihlstrom, 2000; Montgomery et al., 2000).

Stress, Gene Expression, and the Aging Process

Ways of alleviating human stress have been explored on a multitude of levels leading to a cavalcade of approaches ranging from psychoanalysis, psychotherapy, hypnosis, meditation, shamanistic, and spiritual processes to biofeedback, body therapies, and the new medical model of functional genomics. Is there any common denominator to alleviating stress among all these different theories and approaches? Findings from the DNA array experiments reviewed above suggest that the dynamics of homeostasis and creative adaptation at the level of gene expression could be the common denominator.

One of the most interesting series of studies illustrating the cellular-genomic dynamics of stress used DNA array technology to investigate the changing patterns of gene expression during the aging process (Lee et al., 1999). These researchers found significant differences in gene expression patterns in the muscle tissue of adult versus aging mice, which are indicative of a marked stress response and lower expression of metabolic and biosynthetic genes (Table 5.1a). Further, they found that most of these stress-associated alterations of gene expression were either completely or partially prevented by caloric restriction, the only intervention known to retard aging in mammals (Table 5.1b). This research is especially informative about the relation between stress, gene expression, the aging process, and other aspects of adaptation at the cellular level. These include energy metabolism, protein metabolism, biosynthesis, and calcium metabolism as well as neuronal factors of primary importance for communication via psychosocial dynamics that are the essence of therapeutic hypnosis and the healing arts.

The relationship between stress and the aging process at the level of gene expression suggests a test of how therapeutic hypnosis may modulate stress-induced gene expression. It is significant to note that one of the stress-related genes expressed in aging mice translates into the production of Heat Shock 27 kDa protein (listed at the top of Table 5.1a). This suggests a possible pathway of mind-gene communication and healing by which Ewin's (1986) burn patients were aided via therapeutic suggestions. We predict that Heat Shock 27 kDa protein is one of the "stress proteins" whose overproduction is prevented when therapeutic suggestions are administered within two hours of receiving the burn trauma and stress.

Most current research using DNA chip technology has been carried out to assess purely biological variables at the molecular level in medical and pharmaceutical studies. We propose that in the near future, however, this DNA chip technology will be adapted as the ultimate test of the efficacy of psychosocial variables in facilitating mind–body healing and therapeutic hypnosis at the level of gene expression (Lee et al., 1999; www.Affymetrix.com; www.Gene-Chips.com).

STEM CELLS, STRESS, GENE EXPRESSION, AND HEALING

HYPOTHESIS 5.5. The modern molecular model of physical medicine as well as therapeutic hypnosis and the healing arts facilitate the expression of gene coding for the synthesis of proteins that function as the molecular machines of healing in the differentiation of stem cells into mature functioning cells throughout the brain and body.

A new view of how stem cells, stress, gene expression, and healing are interrelated is currently emerging in the health sciences. Stress on all levels, from the social and psychological to the physical and traumatic, leads to injury

TABLE 5.1A | AGE STRESS–RELATED CHANGES IN GENE EXPRESSION IN RAT MUSCLE. ONE OF THE STRESS-RELATED GENES EXPRESSED IN AGING MICE CODES FOR THE PRODUCTION OF HEAT SHOCK 27 KDA PROTEIN (TOP OF TABLE 5.1A). IT IS HYPOTHESIZED THAT HEAT SHOCK 27 KDA PROTEIN MAY BE ONE OF THE "STRESS PROTEINS" WHOSE OVER–PRODUCTION MAY BE PREVENTED WHEN THERAPEUTIC SUGGESTIONS ARE ADMINISTERED WITHIN TWO HOURS OF BURN TRAUMA AND STRESS.

ORF	Age (fold)	Gene	Function	CR Prevention
W08057	↑3.5	Heat Shock 27 kDa Protein	*Chaperone*	C
M17790	↑3.5	Serum Amyloid A Isoform 4	*Unknown*	N
AA114576	↑3.4	Heat Shock 71 kDa Protein	*Chaperone*	C
L28177	↑2.6	GADD45	*DNA damage response*	77%
M74570	↑2.4	Aldehyde Dehydrogenase II	*Aldehyde detoxification*	29%
AA059662	↑2.2	Protease Do precursor	*Protease*	C
L22482	↑2.2	HIC–5	*Senescence and differentiation*	C
X99963	↑2.2	rhoB	*Unknown*	87%
X65627	↑2.1	TNZ2	*RNA metabolism*	64%
X57277	↑1.8	Rac1	*JNK activator*	C
AA071777	↑3.8	Synaptic Vesicle Protein 2	**Neurite extension**	51%
X53257	↑2.5	Neurotrophin-3	**Reinnervation of muscle**	50%
X78197	↑2.2	Ap-2 Beta	**Neurogenesis**	N
X89749	↑2.1	mTGIF	**Differentiation**	C
AA014024	↑2.1	Dynactin	**Transport**	55%
X63190	↑2.1	PEA3	**Response to muscle injury**	C
AA106112	↑3.8	Mitochondrial Sarcomeric Creatine Kinase	*ATP generation*	C
AA061886	↑2.0	Dihydropyridine-sensitive L-type Calcium Channel	CALCIUM CHANNEL	67%

ORF	Age (fold)	Gene	Function	CR Prevention
AA061310	↓4.1	Mitochonrial LON Protease	*Mitochondrial biogenesis*	C
W55037	↓2.9	Alpha Enolase	*Glycolysis*	68%
V00719	↓2.6	Alpha-Amylase-1	*Carbohydrate metabolism*	N
M81475	↓2.5	Phosphoprotein Phosphatase	*Glycogen metabolism*	C
AA034842	↓2.1	ERV1	*mtDNA maintenance*	46%
AA106406	↓2.0	ATP Synthase A Chain	*ATO synthesis*	N
AA041826	↓2.0	IPP-2	*Glycogen metabolism*	C
L27842	↓2.0	PMP35	*Peroxisome assembly*	60%
Z49204	↓2.0	NADP Transhydrogenase	*Glycerophosphate shunt*	N
AA071776	↓1.9	Glycose-6-Phosphate Isomerase	*Glycosis*	C
M13366	↓1.9	Glycerophosphate Dehydrogenase	*Glycerophosphate shunt*	C
AA107752	↓2.9	EF-1-Gamma	*PROTEIN SYNTHESIS*	63%
U22031	↓2.6	20S Proteasome Subunit	*PROTEIN TURNOVER*	44%
AA061604	↓2.2	Ubiquitin Thiolesterase	*PROTEIN TURNOVER*	C
AA145829	↓2.1	26S Proteasome Component TBP1	*PROTEIN TURNOVER*	C
L00681	↓2.1	Unp Ubiquitin Specific Protease	*PROTEIN TURNOVER*	N
U35741	↓2.0	Rhodanese	*MITOCHONDRIAL PROTEIN FOLDING*	C
D83585	↓1.7	Proteasome Z Subunit	*PROTEIN TURNOVER*	C
D76440	↓2.9	Necdin	**Neuronal growth suppressor**	47%
X75014	↓2.7	Phox2 Homeodomain Protein	**Throphic factor**	65%
M32240	↓2.1	GAS3	**Myelin protein**	55%
M16465	↓3.4	Calpactin I Light Chain	CALCIUM EFFECTOR	C
L34611	↓2.3	PTHR	CALCIUM HOMEOSTASIS	N

ORF	Age (fold)	Gene	Function	CR Prevention
AA103356	↓2.2	Calmodulin	CALCIUM EFFECTOR	N
D29016	↓6.4	Squalene Synthase	CHOLESTEROL/FATTY ACID SYNTHESIS	52%
M21285	↓2.1	Stearoyl-CoA Desaturase	PUFA SYNTHESIS	C
U73744	↓2.1	HSP70	*Chaperone*	N

LEGEND |

Stress Response / **Neuronal factors** / *Energy Metabolism* / CALCIUM METABOLISM /

PROTEIN METABOLISM / **BIOSYNTHESIS**

and aging of the individual cells that make up the tissues and organs of the body. What is the most general mechanism of recovery from such stress, trauma, and injury? New answer: stem cells! Stem cells have been described as "Mother Nature's menders" because, like embryonic cells, they retain the ability to express whatever genes are needed to replace the injured cells (Vogel, 2000b). Stress, trauma, injury, and diseases of many types leave a trail of molecular signals that activate the gene expression/protein synthesis cycle in the stem cells still residing in malfunctioning tissues. *The molecular messengers generated by stress, injury, and disease can activate immediate early genes within stem cells, so that they then signal the target genes required to synthesize the proteins that will transform (differentiate) the stem cells into mature, well-functioning tissues.* These new tissue cells can then replace injured, aging, and dysfunctional cells that die by a process of apoptosis—the so-called cell suicide that takes place due to senescence, stress, injury, genomic mutations, etc. (McLaren, 2000; Temple, 2001).

The emerging evidence of stem cells as a general source of healing in stress-related physical and psychological medicine is implied in the existing scientific literature, even though many of the exact molecular mechanisms are not yet well known. Scores of growth factors that can activate the gene expression-protein synthesis cycle in stem cells have been isolated to date.

TABLE 5.1B | CALORIC RESTRICTION–INDUCED ALTERATIONS IN GENE EXPRESSION IN RAT MUSCLE. STRESS–INDUCED AGING PROCESSES ARE MODULATED BY CALORIC RESTRICTION. IT IS NOT YET KNOWN WHETHER STRESS REDUCTION VIA THERAPEUTIC SUGGESTION CAN MODULATE THE AGING PROCESS BY THE SAME OR SIMILAR MECHANISMS.

ORF	CR (fold)	Gene	Function
U5809	↑4.5	Transketolase	*Pentose phosphate pathway*
W53351	↑4.1	Fructose-bisphosphate Aldolase	*Glycolysis/Gluconeogenesis*
AA071776	↑3.5	Glucose-6-Phosphate Isomerase	*Glycolysis/Gluconeogenesis*
U34295	↑2.3	Glucose Dependent Insulinotropic Polypeptide	*Insulin sensitizer*
U01841	↑2.3	Peroxisome Proliferator Receptor Gamma	*Insulin sensitizer*
L28116	↑2.0	PPAR Delta	*Peroxisome induction*
D42083	↑1.9	Fructose 1,6-bisphosphatase	*Gluconeogenesis*
AA041826	↑1.9	Protein Phosphatase Inhibitor 2 (IPP-2)	*Inhibition of glycogen synthesis*
U37091	↑1.8	Carbonic Anhydrase IV	*CO_2 disposal*
M13366	↑1.8	Glycerophosphate Dehydrogenase	*Electron transport to mitochondria*
AA119868A	↑1.7	Pyruvate Kinase	*Glycolysis*
AA145829	↑2.3	26S Protease Subunit TBP-1	*PROTEIN TURNOVER*
AA107752	↑2.2	Elongation factor 1-gamma	*PROTEIN SYNTHESIS*
W53731	↑2.1	Signal Recognition Receptor Alpha Subunit	*PROTEIN SYNTHESIS*
U60328	↑2.1	Proteasome Activator PA28 Alpha Subunit	*PROTEIN TURNOVER*
X59990	↑2.0	mCyP-S1 (Cyclophilin)	*PROTEIN FOLDING*
W08293	↑1.9	Translocon-Associated Protein Delta	*PROTEIN TRANSLOCATING*
W57495	↑1.8	60S Ribosomal Protein L23	*PROTEIN SYNTHESIS*

ORF	CR (fold)	Gene	Function
X13135	↑4.7	Fatty Acid Synthase	**FATTY ACID SYNTHESIS**
X16314	↑2.5	Glutamine Synthetase	**GLUTAMINE SYNTHESIS**
AA137659	↑2.4	Cytochrome P450-IIC12	**STEROID SYNTHESIS**
L32973	↑2.0	Thymidylate Kinase	**dTTP SYNTHESIS**
X56548	↑2.0	Purine Nucleoside Phosphorylase	**PURINE TURNOVER**
AA022083	↑2.0	Huntingtin	**Unknown**
D76440	↑1.9	Necdin	**Growth suppressor**
AA062328	↓3.4	DnaJ Homolog 2	*Chaperone*
X63023	↓1.9	Cytochrome P-450-IIIA	*Detoxification*
U03283	↓1.8	Cyp1b1 Cytochrome P450	*Detoxification*
U14390	↓1.8	Aldehyde Dehydrogenase-3	*Detoxification*
X76850	↓1.8	MAPKAP2	*Unknown*
D26123	↓1.7	Carbonyl Reductase	*Detoxification*
L4406	↓1.7	Hsp 105-beta	*Chaperone*
U40930	↓1.5	Oxidative Stress-Induced Protein	*Unknown*
U66887	↓1.8	RAD50	**DOUBLE STRAND BREAK REPAIR**
AA059718	↓1.7	DNA Polymerase Beta	**BASE EXCISION REPAIR**
W42234	↓1.6	XPE	**NUCLEOTIDE EXCISION REPAIR**
D43694	↓1.8	Math-1	**Differentiation**
D16464	↓1.7	HES-1	**Differentiation**
W13191	↓1.6	Thyroid Hormone Receptor Alpha-2	*Thyroid hormone receptor*

LEGEND |

Stress Response / **Neuronal factors** / *Energy Metabolism* / CALCIUM METABOLISM /

PROTEIN METABOLISM / **BIOSYNTHESIS**

Stimulation by varying combinations of these growth factors and other signals from the environment lead to varying patterns of gene expression that generate the proteins that lead stem cells to differentiate into healthy, mature functioning tissues. Healing via gene expression is documented in stem cells in the brain (including the cerebral cortex, hippocampus, and hypothalamus), muscle, skin, intestinal epithelium, bone marrow, liver, and heart (Fuchs & Segre, 2000; Gage, 2000a; Kiger et al., 2001; Wasserman & DiNardo, 2001).

The most well-documented evidence of a relationship between psychosocial stress and a healing response via stems cells comes from research on neurogenesis and brain growth (Gage, 2000a & b; Temple, 2001). The inhibitory role of stress hormones (such as the glucocorticoids) on neurogenesis has been well replicated (Gould, et al., 1998). We now need to investigate the degree to which novelty, environmental enrichment, and physical exercise can evoke *activity-dependent gene expression and neurogenesis in stem cells of the body as well as the brain.* Hood (2001) has documented the activity-dependent gene expression in mitochondrial biogenesis and the dynamics of energy generation in skeletal, cardiac, and smooth muscle in response to physical exercise that is consistent with this hypothesis. *Hypothesis 5.5 implies, in essence, that many of the so-called miracles of healing via spiritual practices and therapeutic hypnosis (Barber, 1990), probably occur via this type of activity-dependent gene expression in stem cells throughout the brain and body.*

Our hypothesis of the stem cell-gene expression-protein synthesis cycle as the final common path to mind-body healing (modeled by the Feigenbaum scenario in Box 4.1) integrates current molecular medicine with the mind-body models of alternative, complementary, and holistic medicine. We propose that the ability to facilitate the creative replay of the gene expression cycle in stem cells throughout the body will become an important criterion, even a common yardstick, for evaluating all forms of therapeutic communication and holistic healing, including biofeedback, body work, emotional catharsis, cognitive-behavioral therapies, imagery, active imagination, hypnosis, meditation, prayer, ritual, ultradian healing response, yoga, or whatever. Researchers investigating the modulation of gene expression via acupuncture, for example, have developed promising research paradigms (Tsogoev et al., 2000)

All the holistic approaches may owe their therapeutic efficacy to the positive and creative replay of psychosocial genomics that are the deep psychobiological foundation of mind-body healing. Whatever the therapeutic method, we could test whether it has facilitated psychosocial genomics by using relatively simple DNA microarray assays to determine whether the *appropriate immediate early genes and their target genes* are expressed in the form of mRNAs that serve as "blueprints" for the synthesis of healing proteins. A dramatic color-coded positron emission tomography (PET) image of gene expression in the liver, intestines, and bladder of the human body is illustrated in Figure 5.9 (see color insert; Lok, 2001; Yaghoubi et al., 2001).

Screenings of medical patients with a variety of stress-related dysfunctions would reveal which patterns of gene expression are associated with the various psychosomatic problems. To determine which patients are most likely to benefit from therapeutic hypnosis, clinical research is now needed to match the patterns of gene expression enhanced in therapeutic hypnosis and the healing arts with those found in stress-related medical conditions. A Venn diagram representing the intersecting regions where gene expression and therapeutic hypnosis interact is illustrated in Figure 5.10.

Figure 5.10 illustrates how a Venn diagram could be used to assess which genes are accessed in common by (1) behavioral state-related gene expression, (2) activity-dependent gene expression, and (3) the varying psychogenomic states modulated by therapeutic hypnosis and the healing arts. Let us review this logic, as we have developed it in the previous chapters. Chapter 1 documented how *gene expression is related to human development and psychological experience in everyday life*—recall Schanberg's research that demonstrated how a mother's touch could turn on gene expression to facilitate her baby's physical growth and psychological development. Chapter 2 reviewed how *gene expression modulates human experience*—recall how circadian patterns of gene expression give rise to changing psychobiological states such as sleeping, dreaming, and being awake—this is called "behavioral state-related gene expression." Chapter 3 reviewed the reverse—*how human experience can modulate gene expression*—recall the research documenting how novelty, environmental enrichment, and physical exercise can turn on gene expression and neurogenesis to encode new memory, learning, and behavior in the process called "activity-dependent gene expression." Chapter 4 introduced neuroscience research that supported a series of hypotheses culminating in the idea that dreaming can be a creative psychobiological state that integrates behav-

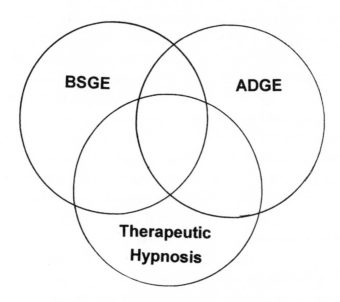

FIGURE 5.10 | A Venn diagram illustrating how overlapping regions of gene expression accessed by (1) behavioral state-related gene expression (BSGE) and (2) activity-dependent gene expression (ADGE) could be modulated by (3) the psychogenomics of therapeutic hypnosis, the healing arts, and cultural rituals.

ioral state-related gene expression and activity-dependent gene expression in a manner that updates an individual's sense of meaning and identity. Now, in Figure 5.10, we generalize and apply this psychobiological logic to account for the efficacy of therapeutic hypnosis in facilitating healing at the cellular-genomic level.

The psychobiological logic of the Venn diagram in Figure 5.10 could provide a new scientific criterion for assessing the therapeutic efficacy of hypnosis and related approaches to mind-body healing. Figure 5.10 illustrates how therapeutic hypnosis could facilitate a creative integration of behavioral state-related and activity-dependent gene expression. Whatever form of medical, psychological, or "alternative" approach to healing is used, the crucial question to answer remains the same: *Which patterns of gene expression are facilitated by which therapeutic procedures to optimize which pathways of mind-body heal-*

ing for which psychobiological dysfunction? The development of new technologies for assessing gene expression via DNA microarrays and positron emission tomography (PET) imaging to answer this question will lay the essential foundation for a genuine science of psychosocial genomics and healing.

For example, we could establish the reality of healing via therapeutic hypnosis on the genomic level by first using DNA technology to determine which patterns of gene expression can be modulated in a particular individual by which hypnotherapeutic procedure. Then we would determine which patterns of gene expression and protein formation are characteristic of the individual's stress-related psychosomatic disorder, using similar DNA technology. If an overlap were found between the first step (gene expression modulated by therapeutic hypnosis) and the second (gene expression modulated by stress), then by the logic of the Venn diagram in Figure 5.10 we would know there is at least a theoretical possibility that therapeutic hypnosis would be efficacious in modulating the stress-related dysfunction on a molecular-genomic level.

At this time, however, we do not know whether the gene expression patterns modulated by therapeutic hypnosis are similar for everyone or highly specific for each individual. Nor do we know, for that matter, how general or specific the gene expression patterns are for different people when they experience physical or psychosocial stress or any other mind-body dysfunction. It is evident that a great deal of fundamental research needs to be done in answering these questions before we can determine if therapeutic hypnosis and the healing arts facilitate genuine healing on the genomic and proteomic (protein) levels. State-dependent memory, learning, and behavior provide a bridge from the genomic to the psychological level that we now need to cross to fully appreciate these new possibilities for the psychotherapeutic arts.

STATE-DEPENDENT MEMORY, LEARNING, AND BEHAVIOR AS THE MODULES OF PSYCHOLOGICAL EXPERIENCE

HYPOTHESIS 5.6. The state-dependent pathways of mind-body communication and healing (SDMLB), which are encoded by the messenger-molecule receptor systems of the CNS, ANS, neuropeptide, and immune systems of the psychosomatic network, are two-way streets. Just as purely biological approaches to the messenger molecule/receptor system can modulate cognitive-emotional experience, the accessing, focusing, and creative replay of cognitive-emotional experiences via therapeutic hypnosis modulates the psychosomatic network to facilitate biological healing.

In Chapter 2 we outlined the dynamics of state-dependent memory, learning, and behavior as a bridge between the molecules and mental levels of psychological experience. Figure 2.1 illustrated how messenger molecules that have their origin in the processing of larger protein "mother-molecules" may be stored within cells as a kind of molecular memory. Messenger molecules from the peripheral cells of the body, such as epinephrine and norepinephrine from the adrenals, are released into the bloodstream. There they can complete the SDMLB loop of information transduction from all parts of the body to the brain's neural networks, illustrated by the block of letters A–L at the top of Figure 2.1. It has been theorized that messenger molecules can diffuse as much as 15 mm through the extracellular fluid (ECF) to any site in the cerebral cortex (Schmitt, 1984) to modulate memory, emotions, and behavior at the cellular-genomic level within neurons of the brain (Routtenberg & Meberg, 1998).

It is important to recognize how SDMLB completes the information transduction loop of mind-body communication so that the psychosomatic network that may be accessed and replayed by therapeutic hypnosis becomes a two-way street. One theoretical objection to the idea that the psychosomatic network is a two-way street is that the brain has long been regarded as a "privileged organ" with a "blood-brain barrier" that normally protects it from toxic substances that may be circulating throughout the rest of the body. The implication is that many messenger molecules from the body can be blocked from entering the brain by the blood-brain barrier. Recent research, however, has demonstrated that during highly stressful emotional conditions, such as physical and emotional trauma in the battle conditions of war, the blood-brain barrier is lowered so that many messenger molecules and other substances can enter the brain (Soreq & Friedman, 1997).

The two-way street hypothesis proposes a major mechanism of how the molecules of the body can modulate mental experience as well as how mental experience, such as therapeutic suggestion, can modulate the expression of genes and other molecules of the body. State-dependent memory, learning, and behavior is the quintessential psychobiological mechanism that bridges the so-called Cartesian dichotomy between mind and body. What is most significant about research in SDMLB is that it enables us to study the parameters of "reversible amnesia," which is the fundamental psychobiological phenomenon that theories of hypnosis and psychoanalysis have tried to explain (Rossi, 1996a, 1996b, 1996c, 1996d, 1996e). Most experiments in SDMLB demonstrate that reversible amnesia is only partial (that is, there is usually some memory/learning available in the dissociated condition after the messenger molecules return to normal levels). Likewise much hypnotic literature documents that hypnotic amnesia is usually fragile and partial in character. Since the earliest days of psychoanalysis, it has been noted that a sudden fright, shock, trauma, or stressor could evoke "hypnoid states" that were somehow related to amnesia and dissociated and "neurotic" behavior. A full amnesia that is completely reversible, however, is relatively rare in SDMLB research as well as in the literature of psychoanalysis and therapeutic hypnosis (Rossi & Cheek, 1988). The fragile and partial character of reversible amnesia has been responsible for many of the paradoxes of dissociation and memory described in the classical literature of hypnosis and psychoanalysis. State-dependent memory, learning, and behavior constitute the first experimental model that can account for many of the paradoxes of dissociation in a manner consistent with current neurobiological research on the effects of stress on mind, memory, and healing as well as the classical literature of psychopathology (Rossi & Ryan, 1986). We now urgently need research to assess how the creative replay, reframing, and resynthesis of SDMLB in therapeutic hypnosis and the healing arts on all levels, from mind to gene expression, may be the psychogenomic basis of many innovative approaches to psychotherapy (Corsini, 2001).

Creative Replays of Gene Expression in Hypnosis, Psychoanalysis, and State-Dependent Experience

HYPOTHESIS 5.7. Therapeutic hypnosis, psychoanalysis, dreaming, and the healing arts engage state-dependent experience that is encoded and

continually modulated by the creative replays of gene expression throughout the psychosomatic network.

The classical phenomena of depth psychology—conflict, dissociation, emotional complexes, reversible amnesia, and repression—are manifestations of state-dependent memory, learning, and behavior (Rossi, 1986a, 1987, 1988, 1990a, 1996; Rossi & Ryan, 1986, 1992). We generalize neuroscience research reviewed in the previous four chapters to hypothesize that replays of salient state-dependent life experiences during recall in everyday life (Berman & Dudai, 2001; Nader et al., 2000a, 2000b; Shimisu et al., 2000), dreaming (Ribeiro 1999; Strickgold, 2000a, 2000b), and psychotherapy (Rossi, 1996a) can engage creative interactions between behavioral state-related and activity-dependent gene expression in the psychosomatic network summarized in Figure 5.1. This psychogenomic perspective of state-dependent memory, learning, and behavior suggests a new research frontier for the psychobiological investigation of classical psychodynamic concepts as well as currently emerging theories of therapeutic hypnosis (Price, 1998; Rainville et al., 1997; Spiegel, 1998; Woody & Farvolden, 1998).

A new paradigm for evaluating these theories is implicit in the research of Cahill et al. (1994), who compared the effects of the beta-adrenergic receptor antagonist propranolol hydrochloride on subjects' long-term memory of an emotionally arousing versus an emotionally neutral short story. Their results were consistent with the hypothesis that the beta-adrenergic (fight or flight) dynamics of SDMLB mediate the enhanced memory associated with the emotionally arousing metaphors of a short story. Such research could become an effective psychobiological approach for investigating the relative merits of the naturalistic and utilization approaches to hypnosis (Erickson et al., 1976).

IMMEDIATE EARLY GENE EXPRESSION IN THE PSYCHOBIOLOGY OF THE HEALING ARTS

As reviewed above, many lines of research now indicate that immediate early genes (IEGs) are the newly discovered mediators between nature and nurture at the cellular-genomic level (Merchant, 1996; Rossi, 1996, 1997; Tölle et al., 1995). Immediate early genes act as transducers that allow signals from the external environment to regulate the adaptive transcription of "target" gene expression at the cellular level. Immediate early genes can initiate molecular-genomic transformations that transduce relatively brief signals of stress from the environment into enduring changes in the physical structure of the developing nervous system during childhood and the formation of new memory and learning throughout life (Morimoto & Jacob, 1998; Tölle et al., 1995). To coin a metaphor, IEGs seem to function as a "steering committee" mediating the stimuli from the outside world and the inner conditions of the cells in dynamics of creative adaptation at the genomic-protein level. For example, Puntschart et al. (1998) found activity-dependent gene expression of c-fos and c-jun, in human skeletal muscle after 30 minutes of physical exercise. In addition, Autelitano (1998) found that novel, arousing, or stressful environmental stimuli can induce the expression of immediate early genes, such as c-fos and c-jun and the proteins they code for within neurons of the brain. In response to conditions prevailing within the neurons, the Fos and Jun proteins associate to form a transcription factor that activates other genes that code for proteins mediating creative adaptations to signals from the outside world (Brivanlou & Darnell, 2002).

We now need research to document whether the class of environmental signals that activate IEGs includes the psychobiologically arousing psychosocial cues of therapeutic hypnosis and the healing arts. The complex range of interrelated biological and psychological functions that immediate early genes are already known to serve as illustrated in (Figure 5.11), however, recommends a central role for IEGs and their target genes in the deep psychobiology of therapeutic hypnosis that can now be explored with the new noninvasive gene expression technologies (Lok, 2001; Rayl, 2001). While more than 100 IEGs have been reported, many of their functions in activating the target genes associated with behavioral states in health and illness remain unknown. Most drugs dealing with pain and addictive drugs such as cocaine, amphetamine, and the opiates are mediated by immediate early genes. The implication is that immediate early genes are central in mediating human moods and behavioral addictions (Merchant, 1996). Indeed, immediate early genes are used as markers or indicators of changes in neuronal activity in psy-

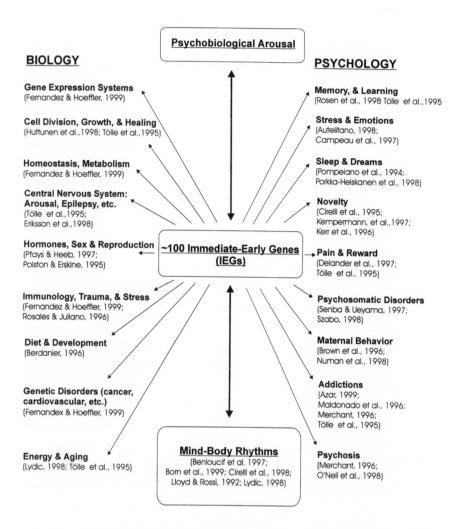

FIGURE 5.11 | The central role of immediate-early genes (IEGs) in psychobiological arousal that may be facilitated by therapeutic hypnosis and the healing arts. (From Rossi, 2000a, 2000b, 2000c.) Many arousing stimuli from the physical and social environment can signal IEGs, which, in turn, initiate a gene-protein cascade in neurons of the brain and other cells of the body to simultaneously modulate biological and psychological processes. Note the reciprocal relationships along the central axis between the levels of *psychological arousal, IEGs,* and *mind-body rhythms.* Together these three levels are windows into a continuum of mind-body communiction, adaption, and healing.

chopathological conditions such as schizophrenia. Antipsychotic drugs are currently being designed to modulate the effects of immediate early genes on pathways leading to the production and utilization of neurotransmitters such as dopamine, serotonin, and noradrenaline that are implicated in the "dopamine hypothesis" of schizophrenia.

It is the simultaneous mediation of both the biological and psychological functions—the psychobiological—that recommends a central role for immediate early genes in understanding the foundations of psychosomatic medicine and the healing arts. Hypnosis has been characterized as a continuum of mind-body states (DeBenedittis et al., 1994; Rossi, 1996a). Mass screenings of large samples of subjects of varying degrees of hypnotic susceptibility, using the new gene expression technologies, would be an ultimate way of precisely specifying what we mean by defining hypnosis as an altered state. We would expect that varying patterns of genes expression would be associated with the continuum of phenomenological states we call "therapeutic hypnosis."

Many arousing stimuli from the psychosocial environment can signal IEGs, which, in turn, initiate a gene-protein cascade in neurons of the brain and cells throughout the body that simultaneously modulate biological and psychological processes. Note in Figure 5.11 the reciprocal relationships along the central axis between the levels of psychobiological arousal, IEGs, and mind-body rhythms. Each of these three levels is a window into the continuum of mind-body communication in the creative replays of adaptation and healing that can occur in the experiences of everyday life as well as during the novelty-numinosum-neurogenesis effect in the psychotherapeutic encounter.

THE NOVELTY-NUMINOSUM-NEUROGENESIS EFFECT IN THERAPEUTIC HYPNOSIS

HYPOTHESIS 5.8. Therapeutic hypnosis and the healing arts facilitate the psychosocial genomics of encoding new experience, memory, learning, and behavior in physical rehabilitation by engaging long-term potentiation via the novelty-numinosum-neurogenesis effect.

As reviewed above, most psychologically arousing stimuli that have been studied can induce immediate early genes in the central nervous system within minutes; their concentrations typically peak within 15–20 minutes, and their effects usually continue for an hour or two. These are the same time parameters of the psychobiological model of memory and learning called "long-term potentiation" (LTP). It is now known that long-term potentiation takes place in many regions of the brain associated with stress and emotional learning (McKernan & Shinnick-Gallagher, 1997). Long-term potentiation operates in the same ultradian time frame of about 90–120 minutes (Bailey et al., 1996; Tully, 1996) that is typical of many dynamical processes of mind-body healing and therapeutic hypnosis associated with the novelty-numinosum-neurogenesis effect.

There are as yet no studies of the effects of hypnosis on immediate early genes and the novelty-numinosum-neurogenesis effect. Research on the role of the immediate early gene c-fos in the wake-sleep cycle (Bentivoglio & Grassi-Zucconi, 1999, outlined in Chapter 2), however, suggests that it may be related to the changing psychobiological states of therapeutic hypnosis as well. It has been found, for example, "that the expression of c-fos during waking is strictly dependent on the level of activity of the noradrenergic system . . . high levels of c-fos during forced and spontaneous waking and . . . low levels during sleep" (Cirelli et al., 1998, p. 46). While most research has been done with animals, it is tempting to hypothesize that stimulation of the noradrenergic system and immediate early gene expression may be associated with the arousal aspects of high-phase hypnosis illustrated in Box 5.3.

This is why an understanding of the activity-passivity paradox of hypnosis reviewed in the beginning of this chapter is so important. If hypnosis were nothing more than a passive state of relaxation, during which a therapist could program the patient, there would be no way to understand the rich phenomenology of mind-body communication and healing that Braid (1855/1970) described as the "psychophysiology of fascination." There would be no way to understand the entire continuum of psychobiological states, from arousal to relaxation, that invariably accompanies therapeutic hypnosis and the psychotherapeutic encounter (Corsini, 2001). The heightened states of psychobiological arousal that are associated with fascination and, more generally, the novelty-numinosum-neurogenesis effect as reviewed in Chapter 3, engage the dynamics of activity-dependent gene expression to facilitate the entire range of mind-body communication and healing.

A particularly vivid clinical case utilizing emotional provocation and psychobiological arousal to evoke the novelty-numinosum-neurogenesis effect was reported by Erickson (1965/1980b):

Karl was in his fifties, an energetic, hard-working man incapable of working for others because of his "German stubbornness," but fully competent to develop and successfully conduct his own business. Karl seldom wasted a moment. . . . *Self-reliance, as much as possible, was his guiding personal principle.* . . . Then unexpectedly an unbearable calamity struck Karl, a cerebrovascular accident that paralyzed him and rendered him physically a completely helpless bed patient, capable of understanding but unable even to read or talk. . . . As his wife explained, "Karl has always been so capable. He could do just anything, and if he couldn't, he would read up on it and then just do it. He just never let himself fail in anything. He is a proud, determined man, and now he is so pitifully helpless. He feels so ashamed because we lost our savings in medical bills and our shop because he couldn't run it and I couldn't. . . ." They kept him a whole year at the university on the neurological ward, trying to help him do a lot of things. But he was a "teaching case," and Karl would go half out of his mind when the medical students would come in and one after another examine him. Then they would hold clinics on him and talk about "irreparable damage," "hopeless prognosis," and talk about the parts of his brain he had lost because he can't talk and can't read, and Karl would get madder and madder and shake his head so furiously and sweat, so they told me they might have to put him on the psychiatric ward.

The purpose in seeing the author [Erickson], his wife explained [to Karl], was to have *hypnosis employed to reeducate new neural pathways so that he could learn new ways of functioning, new ways of using his arms and legs by employing newly developed neural pathways.* A family friend, a physician, had studied an article by the author (Erickson, 1963/1980) and had urged them to consult the author and have the possibilites of reeducation under hypnosis explored for Karl.

The patient listened to all of this, manifesting variously agreement, impatience, resentment, even anger, and boredom. More than once his wife commented, "look at him now. He is disgusted with all this talk. He wants to get started right now," to which remark Karl vigorously nodded his head in assent. . . . Karl listened with mounting impatience, breathing heavily, snorting, grunting, perspiring, and making many minor spasmodic movements with the one leg and arm over which he had gained some slight control. His wife stated in explanation, "Karl wants to start right now. He can't tolerate waiting until tomorrow for another appointment." Karl nodded a most emphatic assent. . . .

The author's reply was simply that, as a physician, he was in charge of the patient, and any further work would have to be started the next day. Karl leaned his head back and stiffened his neck and back, whereupon his wife explained, "Karl means he won't leave your office." [After a few more remarks about how therapy would proceed, Erickson concludes this initial session in a provocative and emotionally arousing manner by utilizing Karl's impatience and authoritarian attitudes for recovery, as follows.]

The interview was then terminated by the preemptory measure of stating in a most dictatorial fashion, "Now, get up out of that chair. Stagger your way to the office door and get out of here and get to your car and give your wife's tired arms and back a little rest on the way. Get going!"

Karl's startled look was replaced by a flash of anger, followed by an expression of *utterly intense effort* as he proceeded, grabbing a chair, then a bookcase to haul himself to the door already opened by the author. Karl's wife came rushing to Karl's assistance but was firmly cautioned to give him only enough help to keep from falling. Clumsily jerking, twisting, using his wife only to balance himself, Karl made his way to the outside steps. . . . [Karl's *"utterly intense effort"* indicated that his *activity-dependent therapy* had already begun, even though he did not realize it. The next day the wife reported that Karl's movements and behavior had already improved. Erickson continues to utilize Karl's *"self-reliance, as much as possible"* with an apparently authoritarian approach to therapeutic hypnosis and posthypnotic suggestions, as follows.]

. . . as Karl dragged, jerked, and stumbled with a minimum of help to his seat in the office, he was told preemptorily, "Close your eyes. Lower your head toward your chest. Relax as much as you can. Listen to the clock on my desk ticking. Spend the next 15, 20, 30, 40, 50, or 60 minutes going asleep in a hypnotic sleep. Take a whole hour if you want to. I know you can do it in 15 minutes, but you can take the whole hour, and the next hour tomorrow we can spend time doing what could have been done in the 45 minutes left. I'll know when you are in a trance. All you have to do is just go to sleep listening to the clock and waiting for me to talk to you and remain asleep while I talk to you. Get going!"

Within 15 minutes the tension of his facial muscles had altered in the characteristic hypnotic fashion, his swallowing reflex had disappeared, his respiratory rhythm had greatly changed, and he presented an acceptable experience of a deep trance. He was told, "Now listen to me. If you are deep asleep, just nod your head gently up and down." Five minutes later he was still perseveratively nodding his head gently in affirmation. This was taken to signify a deep trance, and the noisy dropping of a heavy paperweight on the floor did not elicit a startle reflex or any alteration in his respiratory rhythm.

From here on in the trance state I told him that I reserved the privilege of using invective whenever I pleased, but *that his cure was in his hands.* He was to walk more and more each day. Within three months he was walking well. On the day he walked 15 miles in the desert around the city, he visited me and told me about it in speech that was very clear. He reversed the anger he had, and used it up in directing his energy into walking and all the other aspects of his rehabilitation. His wife was astonished when she heard him tell me, "I love you as a brother."

[Erickson answered my questions about this case in 1978 well before the current trend in neuroscience research on neurogenesis, as follows.]

Rossi: Do you believe the explanation you gave Karl's unconscious during the trance state was a pseudo-argument, or do you really believe his unconscious was manufacturing new brain patterns? Are new brain patterns being constructed during such rehabilitative efforts?

Erickson: Yes. In my own experience with myself [self-rehabilitation from polio] it seems to be a matter of learning to use muscles in a different way. When I was 60, I went for a physical, and the examining neurologist found that I had divided some muscles into halves, some into thirds. One-third of a muscle was realigned to pull against the outer two-thirds of itself. One-half of a muscle was pulled against the other half.

Rossi: You believe that new brain patterns do develop in physical rehabilitation and that these can be manifest by all sorts of re-adaptations in muscles to recover lost functions. There is a greater plasticity in both the central nervous system and our actual musculature than most of us have dared believe. You would definitely encourage more strenuous rehabilitative efforts and greater expectations for recovery?

Erickson: Yes, Karl was told he was a hopeless case, and so was I when I had polio for the first time at the age of 17. (pp. 321–327)

Controversial as Erickson's therapeutic approach may be, there can be little doubt that he effectively utilized Karl's authoritarian and self-reliant attitudes to activate the novelty-numinosum-neurogenesis effect in facilitating his physical rehabilitation. Considerable research will now be required to assess the role of the novelty-numinosum-neurogenesis effect in facilitating healing in a wide variety of situations, ranging from art and cultural rituals to the placebo effect as outlined in the next two hypotheses.

CREATIVE REPLAY
OF THE NOVELTY-NUMINOSUM-NEUROGENESIS EFFECT
IN THE ARTS, HUMANITIES, AND CULTURAL RITUALS

HYPOTHESIS 5.9. Enriching life experiences that evoke the novelty-numinosum-neurogenesis effect during creative moments of art, music, dance, drama, humor, literature, spirituality, awe, joy, and cultural rituals can optimize the psychosocial genomics of consciousness, personal relationships, and healing.

The entire history of human approaches to healing that evoke the novelty-numinosum-neurogenesis effect—from ancient spiritual rituals of exorcism, shamanism, and fire walking to the still "mysterious" methods of acupuncture and neurofeedback (Hammond, 2001, 2002; Othmer, in press—is the data base for this hypothesis (Achterberg, 1985; Greenfield, 2000; Keeney, 1999–2000). Psychobiological healing that occurs during ecstatic religious experiences of the numinosum involving a combined sense of fascination, the mysterious, and the tremendous (Otto, 1923/1950), has much in common with modern rituals of healing (Dossey, 1993) associated with the self-help groups, twelve-step programs, and the so-called "miracle cures" reported in clinical demonstrations of hypnosis (Barber, 1990). We hypothesize that just as negative states of emotional arousal can evoke the psychosomatic network to initiate gene expression cascades leading to the overproduction of stress proteins and illness, so can positive psychological experiences replay the novelty-numinosum-neurogenesis effect to facilitate gene expression, neurogenesis, problem solving, and healing.

Recent research on modern group healing processes modeled after ancient rituals demonstrates the positive effects of drumming, music, and storytelling on the modulation of neuroendocrine-immune parameters in normal subjects. Bittman et al. (2001), for example, report the enhancement of the immune system as measured by increased natural killer (NK) cell activity and lymphokine-activated killer (LAK) cell activity after 20 minutes of rhythmic drumming, guided imagery, and the telling of two stories (each for approximately 15 minutes). The entire session of rhythmic percussion lasted one hour. The total of 111 age- and sex-matched volunteer subjects, with a mean age of 30.4 years, showed a statistically significant increase in their DHEA-to-cortisol ratios that is considered conducive to healing. A high DHEA-to-cortisol ratio is the reverse of the neuroendocrine/neuroimmune pattern associated with the chronic experience of stress. Bittman et al. (2001) summarized the implications of their research, as follows:

> Based on these preliminary data ... group drumming music therapy—in a man-
> ner similar to that of exercise, laughter, meditation, and other interventions that
> are practiced or enjoyed on a regular basis—has the potential to produce cumu-
> lative or sustaining neuroendocrine or immunological effects that could con-
> tribute to the well-being of an individual facing a long-term condition in which
> elevated NK cell activity is known to be helpful. (p. 46)

The neuroscience research reviewed in Chapter 3, detailing how novel-ty, enriched environments, and exercise initiate gene expression cascades lead-ing to the formation of new proteins and neurogenesis—the novelty-numinosum-neurogenesis effect—is the basis of this hypothesis about the psy-chobiological effects of healing in cultural rituals. In the laboratory, researchers found that activity-dependent gene expression in response to novelty and exercise can double the number of cells generated in the hippocampus (Eriksson et al., 1998; Gould et al., 1998, 1999; Kempermann et al., 1997, 1999; Van Praag et al., 1999). Although it is controversial, it is tempting to see a parallel between highly arousing activities that facilitate activity-dependent gene expression in the laboratory with animals and the high level of senso-ry stimulation and movement found in many colorful cultural rituals of pas-sage with humans (Keeney, 1999–2000). Many reports on cultural rituals of healing (Greenfield, 1994, 2000) note that after a period of high sensory stim-ulation and emotional arousal, there is a compensatory period of rest, relax-ation, and sleep after which healing is experienced. This is reminiscent of the mathematical models of therapeutic hypnosis presented in Boxes 5.2 and 5.3, wherein there is a high-phase of ultradian arousal followed by a low-phase that is characteristic of Kleitman's basic rest–activity cycle and the relaxation response (Benson, 1983; Lazarus & Mayne, 1991).

These parallels suggest that both high and low states of psychobiological arousal have contributions to make to what we could call a continuum of healing associated with behavioral state-related and activity-dependent gene expression. This concept of a continuum of healing is confirmed by many studies that suggest that different belief systems about spiritual and holistic healing emphasize different parts of the same continuum. Glik (1993), for example, compared the belief system and spiritual healing rituals of 93 members of charismatic Christian groups, whose healing rituals emphasized *hyperarousal*, with those of 83 members of New Age healing groups, whose meditative approach to healing lead to quiet states of *hypoarousal*. The two groups were apparently emphasizing and facilitating opposite psychobiological states. Glik found, however, that people in both groups reported more healing experiences than people who did not frame their health problems from a spiritual perspective. The spiritual perspective can creatively replay both the high and low psychobiological states for healing.

This finding is consistent with classical studies (Underhill, 1963) that classify meditative states into two major categories: *via positiva* or active meditation versus *via negativa* or passive meditation. In the *via positiva* approach, attention is focused on an external or internal object, concept, or image with numinous overtones (such as a religious figure or symbol) that generates high psychobiological arousal. The via negativa is passive in the sense that practitioners attempt to clear the mind by eliminating sensory input and activities such as thinking and feeling, so that a low state of psychobiological arousal is experienced. However, current research in the emerging field of neurotheology, utilizing brain imaging of practitioners on either the active or passive path, indicates that both can achieve the desired "absolute unitary being" state (d'Aquili & Newberg, 1999). Such studies point to the possibility of developing a universal psychobiological paradigm that could investigate the therapeutic values of all cultural and spiritual practices in an objective manner. They also imply a more general hypothesis about the role of the novelty-numinosum-neurogenesis effect in mediating the positive expectations and experiences of the placebo response.

THE NOVELTY-NUMINOSUM-NEUROGENESIS EFFECT
MEDIATES THE PLACEBO RESPONSE

Hypothesis 5.10. Placebos entrain pathways of mind–body communication and healing between positive psychosocial expectation and the psychobiological dynamics of surprise in the novelty–numinosum–neurogenesis effect.

Placebo research is consistent in finding about 30% of subjects reporting a therapeutic benefit (Harrington, 1997; Quitkin et al., 1996). In one study of the selective serotonin reuptake inhibitors, there was a relapse rate of 45% after one year; however, 55% of the subjects reported a significant placebo response even after one year (Moller & Volz, 1996). Is the placebo response "merely" the outcome of wishful thinking and expectancy? Or does this research reflect a 30–55% psychobiological healing effect that is measurable on the cellular-genomic-protein level (Ader, 1997, 2000)? These statistics sound impressive to believing observers but marginal or not significant at all to many critics. Controversies about the value of placebos for genuine physiological healing will continue as long as we try to establish the "reality" or validity of placebos by relying on the statistics of outcome research rather than the actual molecular pathways of communication and healing (Berman & Dudai, 2001; Kohara et al., 2001; Moerman & Jonas, 2000; Stefano et al., 2001).

At a recent meeting of more than 20 branches of the National Institutes of Health (NIH), Smaglik (2000) reported on the views of many leaders that the "study of the placebo effect—the role of medically inactive treatments in healing—could help bridge the divide between mainstream and alternative medicine" (p. 349). At this meeting Stephen Strauss, director of the National Center for Complementary and Alternative Medicine, said that study of the placebo effect "is very important for our institute because of all the preconceived biases as to how these traditional therapies must be operating—that they must be operating merely through the placebo effect." The critical scientific attitude, however, cannot accept the validity of mind-body healing without knowing what the actual molecular mechanism of the placebo response is. In the search for such molecular mechanisms, Gerald Fischbach, director of the National Institute of Neurological Disorders and Stroke, noted that the placebo effect is complex and as yet poorly understood, perhaps involving specialized nerve cells that respond to the expectation of treatment. Steven Hyman, director of the National Institutes of Mental Health, reported growing evidence that repeated exposure to a stimulus can reconfigure the brain's circuitry. These views, as reported by Smaglik (2000) and delivered at the highest levels of national authority in health and medicine, reflect the need for an understanding of the missing link between mind and body described here as the novelty-numinosum-neurogenesis effect that mediates the role of expectancy in the placebo response.

We hypothesize that the heightened, numinous sense of fascination, emotional arousal, and positive expectation associated with a novel, brightly colored "sugar pill" can be just as effective as a new and mysterious therapeutic ritual introduced by a healer coming from a faraway country. A recent study by De la Fuente-Fernández et al. (2001) illustrates progress in understanding the psychobiological mechanisms of the placebo response as well as the problems and paradoxes that remain. These researchers summarize their findings, as follows:

> The power of placebos has long been recognized for improving numerous medical conditions such as Parkinson's disease (PD). Little is known, however, about the mechanism underlying the placebo effect. . . . As measured by positron emission tomography, we provide in vivo evidence for substantial release of endogenous dopamine in the striatum of PD patients in response to placebo. Our findings indicate the placebo effect in PD is powerful and is mediated through activation of the damaged nigrostriatal dopamine system. (p. 1164)
>
> Our observations indicate that the placebo effect in PD is mediated by an increase in the synaptic levels of dopamine in the striatum. *Expectation-related dopamine release might be a common phenomenon in any medical condition susceptible to the placebo effect.* PD patients receiving an active drug in the context of a placebo-controlled study benefit from the active drug being tested as well as from the placebo effect. By contrast, in the usual clinical setting, active drugs may be devoid of the placebo effect. We found no evidence that the placebo effect synergistically augments the action of active drugs (in fact, a trend for the opposite was observed), so positive conclusions derived from placebo-controlled studies are not impugned by our findings. (p.1165, italics added)

A problem in interpreting De la Fuente-Fernández et al.'s conclusion that "expectancy-related dopamine release might be a common phenomenon in any medical condition susceptible to the placebo effect" becomes apparent, however, in the context of the research by Waelti et al. (2001) reviewed in Chapter 3. Recall that Waelti et al. found that "during initial learning, when rewards occur *unpredictably*, dopamine neurons are activated by rewards. *They gradually lose the response as the reward becomes increasingly predicted*" (p. 43, italics added). The paradox is that the positive role of "expectancy-related dopamine release" in placebo research by De la Fuente-Fernández et al. (2001) appears to be the opposite of learning research by Waelti et al. (2001), who find that when rewards occur unpredictably, dopamine neurons are activated but gradually lose the response as the reward becomes increasingly predictable—that is, expected. Expectancy has been cast in the apparently paradoxical role of enhancing "expectancy-related dopamine release" in the placebo effect but depleting dopamine associated rewards when they are no longer novel or surprising in learning experiments.

This leads us to conclude that a simple cognitive-behavioral model of expectancy is not sufficient to explain the placebo effect because complex interactions take place between learning and the placebo effect at the molecular level of neurotransmitters such as dopamine. Positive expectation plus the surprise and deeply motivational engagement of the novelty-numinosum-neurogenesis effect are required for synergy between the placebo effect and reward-based learning via dopamine release. Hypothesis 5.10 suggests that the novelty-numinosum-neurogenesis effect is required to bridge the gap between positive psychosocial expectations and the psychogenomic pathways of the placebo effect via dopamine release.

Research summarized in Figure 5.11 suggests how the novelty-numinosum-neurogenesis effect could mediate the rapid action of the placebo response by activating immediate early genes. This could account for the fact that the placebo response appears within minutes but may also disappear when the novelty-numinosum-neurogenesis effect loses its potency through familiarity. Research with the new DNA microarray technology and PET imaging is now needed to explore the time parameters of how psychosocial expectancy, mediated by the novelty-numinosum-neurogenesis effect, could be facilitated with the creative replays of therapeutic hypnosis.

A recent brain imaging study by Petrovic et al. (2002), for example, documents how placebos engage the same brain circuits as pain-killing drugs.

> It has been suggested that placebo analgesia involves both higher-order cognitive networks and endogenous opiate systems. The rostral anterior cingulated cortex (ACC) and brainstem are implicated in opioid analgesia, suggesting a similar role for these structures in placebo. [In this PET study], we confirmed that both opioid and placebo analgesia are associated with increased activity in the rostral ACC. We also observe a covariation between activity in rostral ACC and brainstem during both opioid and placebo analgesia, but not during the pain only condition. These findings indicate a related neural mechanism in placebo and opioid analgesia. (p. 1737)

Since we now know that pain and analgesia engage gene expression (Figure 5.10), this study is a model for future psychogenomic research as implied in Holden's (2002) commentary on the Petrovic study.

> Both the genuine analgesic and the placebo led to increased blood flow in areas of the brain known to be rich in opioid receptors: the brainstem and the rostral anterior cingulated cortex (ACC), which exchanges information with a network of brain regions, including the orbitofrontal cortex, a relatively sophisticated part of the brain known to process emotions. Furthermore, those people who responded most to the placebo—according to their ratings on a scale of 0 to 100 of how much it reduced pain—also showed more rostral ACC activation from the drug. *This . . . provides new fodder for the hypothesis that high placebo responders have a more efficient opioid system.* (p. 947, italics added)

The use of a 0-to-100 scale in the Petrovic study is identical to the use of symptom scaling to fascilitate and assess the analgesic and healing effects of therapeutic hypnosis (illustrated in Part 2). The Petrovic study contradicts the implications of a previous generation of researchers that failed to relate hypnosis to the placebo and opioid analgesia (Hilgard & Hilgard, 1983). More recent research by Rainville et al. (1997, 1999), however, is consistent with Petrovic in documenting how brain areas that mediate pain include the human anterior cingulate, which is implicated in the experience of hypnosis. These studies indicating the same brain localization of the analgesic effects of molecules, placebo, and therapeutic hypnosis are consistent with the emerging science of psychosocial genomics that could unify modern molecular medicine with alternative and complimentary medicine into a unified theory of mind–body healing via therapeutic hypnosis, the placebo, and psychotherapy in general.

SUMMARY

The leading edge of neuroscience is tracing the pathways of mind–body communication on all levels, from the psychosocial to the cellular–genomic, in a manner that makes possible a genuine science and practice of therapeutic hypnosis and the healing arts. Many of the problems and paradoxes of historical hypnosis and the impasse of current hypnosis theory can be resolved by a deeper understanding of the psychobiological parameters of mind–body communication and healing as a complex adaptive system that includes all levels from mind to molecule.

We propose an expansion of the psychological domain of therapeutic hypnosis and the healing arts to include the utilization of the entire cybernetic loop of information transduction that connects the psychosocial environment, the central nervous system, and the psychosomatic networks of the autonomic, neuroendocrine, and immune systems. The dynamics of mind–body communication and healing, via the psychosomatic network of messenger molecules and their receptors, have been integrated into a new field of psychosocial genomics. Psychosocial genomics integrates neuroscience research with the novelty–numinosum–neurogenesis effect in therapeutic hypnosis and the placebo response. It is proposed that the various classes of genes whose expression can be modulated by psychosocial cues—such as immediate early genes, behavioral state-related genes, clock genes, and activity-dependent genes—in the immune, endocrine, autonomic, central, and peripheral nervous systems could serve as a new scientific foundation for therapeutic hypnosis and the healing arts.

Ten hypotheses integrating the molecular model of modern medicine with the psychobiological dynamics of the novelty–numinosum–neurogenesis effect are outlined as a guide for research and clinical practice. The basic idea is that the psychotherapeutic arts engage the creative replay and resynthesis of psychological experience via the state-dependent circular pathways of gene expression, neurogenesis, and healing in the brain as well as the body. Although these hypotheses are visionary in their scope, recent advances in measuring gene expression with DNA microarray technology and PET imaging would make it possible to utilize psychosocial genomics on a practical, clinical level. The clinical possibilities that are already available to us are explored in the innovative approaches to facilitating the creative dynamics of the novelty-numinosum-neurogenesis effect in the psychotherapeutic arts in Part II.

THE

PSYCHODYNAMICS

OF GENE EXPRESSION IN

THE HEALING ARTS

Part Two

THE PSYCHODYNAMICS OF
GENE EXPRESSION IN THE HEALING ARTS

A cornucopia of concepts has evolved in the field of psychotherapy over the past few hundred years, starting with Mesmer and the early pioneers in therapeutic hypnosis and continuing with Freud, Jung, the behaviorists, and now neuroscience. It is high time to take Occam's razor to the whole lot and outline a few simple, positive ways of working with people that are consistent with the newly identified psychobiological processes of gene expression and neurogenesis that lead to healing. The innovative approaches outlined in Part II utilize the fundamental principles of evolution, gene expression, and neurogenesis that are mediated by creative replay and resynthesis, as explored in Part I. These principles are the psychobiological foundation of a new "positive psychology," whose aim is to help people learn to optimize their own natural psychobiology to solve their own problems in their own creative ways.

Chapter 6 outlines the critical transition from the original pathology-based exploration of the psychobiology of stress by Hans Selye to the development of our current approaches to facilitating the creative process in daily life. We introduce the concept of the *breakout heuristic* to illustrate how the creative cycle is replayed on many levels in individual lives as well as in the arts, sciences, and cultural rituals. Although there have been generations of scholarship in these areas, there has been curiously little integration of our knowledge of how the creative process is manifest on all these levels. Chapter 6 brings together objective research on the creative cycle with a practical self-help section outlining how to enjoy creating a great day.

In Chapters 7 and 8 on "The Experiential Theater of Demonstration Therapy," we explore a videotaped example of how to utilize these new, creativity-based approaches in a highly permissive manner. In a sense, this videotaped example of how psychotherapy is taught today, using live demonstrations, contains our entire human heritage: Communities have always come together to replay and recreate their most salient beliefs and practices for self-discovery and healing. This experiential theater of demonstration therapy employs the now-familiar principles of novelty, environmental enrichment, and exercise on all levels, from the deeply private to the public, from serious explorations in self-reflection and co-creation to those spontaneous moments of laughter and simple fun. We speculate about how we may access *immediate early* and *behavioral state-related gene expression,* and then go on to facilitate *activity-dependent gene expression and neurogenesis.* We explore how *implicit processing heuristics* could be used to facilitate the novelty-numinosum-neurogenesis effect in a neuroscience-based approach to psychotherapy.

In Chapters 9 and 10 we explore the new psychodynamics of neuroscience that can be utilized to facilitate creativity, problem-solving, and healing. In Chapter 9 we develop a new class of verbal *implicit processing heuristics* that may facilitate the creative process on the deep psychobiological levels of gene expression, neurogenesis, and healing. In Chapter 10 we integrate everything we have learned to develop a new class of nonverbal *symmetry-breaking heuristics* that engage our numinous sense of wonder in the creative play of imagination to synthesize new possibilities in life. We would like to believe we are integrating the best of the arts, cultural rituals, and spiritual intimations of the past with these new approaches to facilitating gene expression, neurogenesis, and healing in the future.

Chapter 6

POSITIVE PSYCHOLOGY REPLAYING THE FOUR-STAGE CREATIVE CYCLE

How to Enjoy Creating a Great Day and Night

"It is time to put the organism back together again. It is time to visit a much more social gene, a gene whose whole function is to integrate some of the many different functions of the body, and a gene whose existence gives lie to the mind-body dualism that plagues our mental image of the human person. The brain, the body and the genome are locked, all three, in a dance. The genome is as much under the control of the other two as they are controlled by it. That is partly why genetic determinism is a myth. The switching on and off of human genes can be influenced by conscious or unconscious external action. . . . Genes need to be switched on, and external events—or free-willed behavior—can switch on genes. . . . Social influences upon behavior work through the switching on and off of genes. . . . The psychological precedes the physical. The mind drives the body, which drives the genome."

—Matt Ridley
Genome: The Autobiography of a Species in 23 Chapters

"I had first to come to the fundamental realization that analysis, in so far as it is reduction and nothing more, must necessarily be followed by synthesis and that certain kinds of psychic material mean next to nothing if simply broken down, but display a wealth of meaning if, instead of being broken down, that meaning is reinforced and extended by all conscious means at our disposal—by the so-called method of amplification."

—C. G. Jung
The Synthetic or Constructive Method

"What is positive psychology? It is nothing more than the scientific study of ordinary human strengths and virtues. Positive psychology revisits 'the average person,' with an interest in finding out what works, what is right, and what is improving.

It asks, 'What is the nature of the effectively functioning human being, who successfully applies evolved adaptations and learned skills? And how can psychologists explain the fact that, despite all the difficulties, the majority of people manage to live lives of dignity and purpose?' "

—Kennon Sheldon and Laura King, 2001
"Why Positive Psychology Is Necessary,"
American Psychologist

A new view of how positive psychology and the four-stage creative cycle are replays of gene expression, neurogenesis, and healing is introduced in this chapter. We use current research on the pathways of gene expression and neurogenesis to generate new models of psychotherapy and the healing arts. We now know, for example, that brief but manageable periods of acute stress can facilitate optimal performance and healing. Chronic stress, by contrast, leads to replays of gene expression that are described as malfunction, injury, disease, aging, and ultimately death.

In this chapter we begin with a review of the classical conception of stress and the general adaptation syndrome as originally developed by Hans Selye. We then explore emerging views of how the positive psychology of happiness, optimal performance, consciousness, and our perception of free will are associated with the creative replay of gene expression, neurogenesis, problem-solving, and healing. The new orientation toward positive psychology currently developing in the field corrects the psychopathological orientation of most schools of psychotherapy in the last century. In this chapter we explore how the four-stage creative cycle, proposed as the natural psychobiological foundation of psychotherapy in Chapter 2 (see Figure 2.11), is the central dynamic underlying the facilitation of happiness, optimal performance, health, and well-being. We then generalize this four-stage creative cycle as the *breakout heuristic* that provides us with an overview of the essential psychobiological dynamics and *raison d'être* of the positive cultural rituals of transformation in human history. We conclude with a practical plan for optimizing the positive psychology of creative experience during the natural circadian and ultradian replays of gene expression, neurogenesis, and healing in everyday life as well as psychotherapy.

Stress and the General Adaptation Syndrome

The scientific era of studying optimal performance, stress, mind-body communication, and healing began with the early thoughts and lifetime of research by Hans Selye (1956, 1974). We could do no better than to quote Selye's own engaging summary of his discovery of the psychobiological roots of stress and the general adaptation syndrome:

In 1926, as a second year medical student, I first came across this problem of a *stereotyped response* to any exacting demand made upon the body. I began to wonder why patients suffering from the most diverse diseases that threatened homeostasis have so many signs and symptoms in common. Whether a man suffers from a severe loss of blood, an infectious disease, or advanced cancer, he loses his appetite, his muscular strength, and his ambition to accomplish anything; usually, the patient also loses weight, and even his facial expression betrays that he is ill. What is the scientific basis of what at that time I thought of as the "syndrome of just being sick?". . .

How could different stimuli produce the same result? In 1936, this problem presented itself again—under conditions more suited to exact laboratory analysis. It turned out in the course of my experiments in which rats were injected with various impure and toxic gland preparations that, irrespective of the tissue from which they were made or their hormone content, the injections produced a stereotyped syndrome (a set of simultaneously occurring organ changes), characterized by [1] *enlargement and hyperactivity of the adrenal cortex,* [2] *shrinkage (or atrophy) of the thymus gland and lymph nodes, and* [3] *the appearance of gastrointestinal ulcers.* . . .

It soon became evident from animal experiments that the same set of organ changes caused by the glandular extracts were also produced by cold, heat, infection, trauma, hemorrhage, nervous irritation, and many other stimuli. . . . This reaction was first described, in 1936, as a "syndrome produced by various nocuous agents" and subsequently became known as the general adaptation syndrome (GAS), or the *biological stress syndrome.* Its three stages: (1) the alarm reaction; (2) the stage of resistance; and (3) the stage of exhaustion. (1974, pp. 24–27, italics added)

Selye's courageous conceptualization of stress and the general adaptation syndrome was initially received with skepticism but eventually became the foundation of modern psychosomatic medicine. Selye documented, with clinical observations and laboratory research, how chronic mental and/or physical stress could be experienced as physical illness. Although many of Selye's basic observations on stress and the GAS (illustrated at the top of Figure 6.1) have been confirmed, recent research has led to a broader understanding and reinterpretation of his triphasic alarm, resistance, and exhaustion response. Sapolsky (1992), for example, clarified how Selye's third stage of exhaustion, in particular, requires revision:

> In actuality, there is little evidence of such a global Selyean exhaustion of the hormones of the stress-response. Chronic stress is not pathogenic because the body's defenses fail but because, with chronic stress, those defenses themselves become damaging. Basically, most of the features of the stress-response are catabolic and inefficient. During a short-term stressor, their costs can be contained, but with chronic activation, they eventually exact a toll. (p. 5)

An example of the damaging effects of chronic stress on adaptive responsiveness to dangerous life situations is illustrated in Sapolsky's research on psychosocial stress, social dominance, and testosterone levels in male baboons living in the wild. In baboon society subordinate males are under chronic stress and prone to psychosomatic problems, such as a suppression of the immune system and testosterone levels, just as humans are when they are low in the social hierarchy in civil service (Marmont et al., 1991). Figure 6.2 graphs the testosterone levels in dominant and subordinate male baboons that were subjected to the acute stress of being captured and subjected to anesthetization. As can be seen, the adaptive response of dominant males to this acute stressor involved a peak testosterone level within one and a half hours, which then dropped over the eight hours during which measurements were made. The socially subordinate males whose adaptive testosterone response was damaged by chronic stress, however, never managed to respond to the acute stress with a peak in testosterone; their testosterone levels simply dropped immediately in a maladaptive response. That is, the *chronic psychosocial stress* to which the subordinate males had been subjected somehow interfered with their ability to respond immediately and adequately to acute stress in a real-life emergency situation.

FIGURE 6.1 | The four-stage creative psychobiological cycle. Top: Selye's tri-phase stress response beginning with an alarm reaction (A.R.) leading to a stage of resistance (S.R.) and finally the stage of exhaustion (S.E.). Middle: The 90–120-minute basic rest–activity cycle as replayed in everyday life, psychotherapy, and the healing arts. Bottom: The economic cycle as approximately manifest in the four major stages of the stock market.

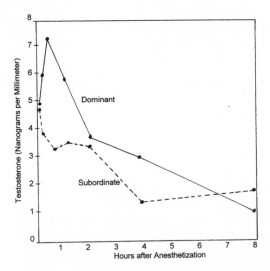

FIGURE 6.2 | The average testosterone levels in dominant versus subordinate male baboons diverge dramatically when the animals are exposed to an identical acute stressor—in this case, being subjected to anesthesia on being captured in the wild. The testosterone levels in the dominant males immediately goes up (solid line) and remains elevated for about one and a half hours in a typical ultradian period of attempted adaptation to the challenge and stress of being captured. The testosterone levels in the subordinate males (dashed line), by contrast, plummets immediately in an apparently maladaptive passive response to acute stress. (With permission from Patricia J. Wynne, artist, Copyright 1990 by Patricia J. Wynne. Redrawn from Sapolsky, R., 1990. *Scientific American* p. 118.)

Chronic stress does not lead to an exhaustion of the arousal (stress) hormones such as adrenaline (epinephrine), the catecholamines, and glucocorticoids (cortisol), as Selye proposed, however. The exact reverse appears to be true. It is the chronic excess of these arousal hormones over time, even when an emergency is no longer present, that leads to the eventual breakdown of mind-body cycle of communication, adaptation, and healing that we typically call "stress" or "the psychosomatic response." In Sapolsky's view, normal psychobiological arousal becomes pathogenic stress when it persists over time because of an apparent loss in the ability to turn off the mind-body's arousal hormones at the cellular-genomic level. The question now becomes: How does the mind-body lose its ability to turn off arousal hormones and maintain optimal homeostasis and adaptation?

From our current perspective we would say that the mind–body has not actually lost its ability to turn off arousal hormones. Rather the mind–body keeps pumping out arousal hormones in a futile effort to solve the acute problem that originally evoked the arousal. This interpretation implies that Selye's third stage of exhaustion is a failure of the creative or adaptive aspect of the general adaptation syndrome on the psychobiological level. That is, there has been a *temporary* failure in generating an adaptive psychobiological response to life's acute challenges, thereby disrupting vital systems of homeostasis.

Life is more than homeostasis, however. Homeostasis means maintaining the same biological state. Positive psychology explores how a well-lived, creative life is a continuous process of psychogenomic adaptation and heroism that is well described in cultural rituals of song and dance as well as the great myths, dramas, and literatures of the world, together with the biographies of accomplished people (Amabile, 2001; Nakamura & Csikszentmihalyi, 2001). The concept of homeostasis, formulated by the great French physiologist Claude Bernard, stated the idea that a fundamental characteristic of living organisms was their ability to maintain the constancy of their internal milieu. The concept of physiological homeostasis was an important step in understanding how the physiology of the body maintained and recovered its equilibrium after experiencing a temporary shock or stress. Physical trauma that leads to a loss of blood pressure, body temperature, or any aspect of internal chemistry, for example, would be compensated for by the responses of many cybernetic physiological systems aimed at restoring homeostasis. Selye's concept of the general adaptation syndrome as a triphasic response to stress was a general description of the biological effort to maintain homeostasis. As Sapolsky suggests, however, something is missing in Selye's conception of the general adaptation syndrome.

Notice how Selye used the terms "stereotyped response" and "stereotyped syndrome" in his description of the general adaptation syndrome in the above quote. What is missing in Selye's description of the general adaptation syndrome is the possibility of breaking out of this futile stereotyped response by initiating a creative response that enables the person to work through the defense and resistance of the second phase, without falling into exhaustion or a psychosomatic syndrome. Selye presents an intimation of the possibility of creative adaptation to stress, however, in the last paragraph of the 1956 edition of his famous book, *The Stress of Life*:

Perhaps the most fascinating aspect of medical *research on stress* is its *fundamentally permanent value to man*. . . . The study of stress differs essentially from research with artificial drugs because it deals with *the defensive mechanisms of our own body*. The immediate results of this budding new science are not yet as dramatic in their practical applications as are those of many drugs, but what we learn about nature's own self-protecting mechanisms can never lose its importance. Such defensive measures as the production of adaptive hormones by glands are built into the very texture of the body; we inherited them from our parents and transmit them to our children, who, in turn, must hand them on to their offspring, as long as the human race shall exist. *The significance of this kind of research is not limited to fighting this or that disease. It has bearing upon all diseases and indeed upon all human activities,* because it furnishes knowledge about the essence of THE STRESS OF LIFE. (pp. 304–305)

Notice Selye's use of the word fascination in the beginning of this quotation. Here we can see how Selye hovered between the relatively rigid and stereotyped traditional biological conception of homeostasis and adaptation of his day and the new concepts of psychogenomic replay that are now available to us with the research on activity-dependent gene expression, neurogenesis, and healing that was summarized as the novelty-numinosum-neurogenesis effect in Part I. Selye's personal experience of fascination is a clue to the numinosity of his prescience about the connection between stress and creative experience that *"has bearing upon all diseases and indeed upon all human activities."*

POSITIVE PSYCHOLOGY
REPLAYS THE FOUR-STAGE CREATIVE CYCLE

We are now in a better position to understand the deeper implications of the natural psychobiological relationships between mind, body, and the four-stage creative cycle in the new dynamic of creative work presented in Chapter 2. A comparison of Selye's 1956 triphasic diagram of the stress response with our new psychobiological conception of the creative cycle presented in Figure 2.11 highlights Selye's prescience about the general significance of the stress response for *"all human activities."* A creative response to life's challenges is the opposite of "stereotyped," highly conditioned or programmed psychological experience and behavior (Erickson & Rossi, 1979, 1981; Sternberg & Davidson, 1995). Selye's phases of (1) the alarm reaction and (2) resistance correspond to stages one and two of our new psychobiological map of the creative process, illustrated in the middle of Figure 6.1. What Selye intuited but left out of his triphasic diagram of the stress response are the creative possibilities inherent in the challenge of stress. Selye focused on the crisis and failure aspect of his third phase as the stage of exhaustion. In our new map of the creative cycle we focus on stage three as a new adaptive response that humans experience as illumination, insight, and a breakout of old patterns of behavior into the new (middle of Figure 6.1). In experimental psychology the classical formulation of the four-stage creative cycle was originally outlined by Wallas (1926) as (1) preparation, (2) incubation, (3) insight, and (4) verification. This highly evolved human skill, which allows creative utilization and transformation of psychobiological arousal and stress, was called *eustress* by Abraham Maslow (1962, 1967) in his theory of meta-motivation. One of the foundational concepts of humanistic and transpersonal psychology, Maslow's theory is one of the original sources of the current emphasis on positive emotions and attitudes in American psychology (Buss, 2000; Fredrickson, 2001; Seligman, 2001; Seligman & Csikszentmihalyi, 2000). There are many other applications of the four-stage creative process in the historical life cycles of social and cultural processes. The image on the bottom of Figure 6.1, for example, illustrates how the four-stage creative cycle is replayed in the bull and bear phases of the stock market (Weinstein, 1988).

One of most interesting areas of research related to Maslow's concept of positive eustress may be the exploration of whether organisms can speed up their own evolution by successfully coping with stress on a molecular level. Current research is exploring whether stress can facilitate adaptive mutations (Chicurel, 2001), as suggested in Figure 1.2, by evoking replay in the complex adaptive system of Gaia, gene, body, and mind along the Ten-Fold Way of evolution. In his *The Origin of Species* Darwin originally suggested that environmental stress resulting from animal domestication by humans could be a selection factor that might affect variability and therefore the evolutionary process itself. Cairns and Foster (1991) shocked the molecular biology community by demonstrating that stress—in particular, the stress of starvation—could increase the rates of compensating mutations in the common intestinal bacteria Escherichia coli. Over the past decade researchers have explored the molecular dynamics of these so-called adaptive mutations in higher organisms with promising though still highly controversial results. In brief, Taddei et al. (1995) found a molecular "SOS" mechanism, activated by the stress of starvation, that increases the number of mutations. Recently McKenzie et al. (2001) demonstrated that, under conditions of stress, an "error-prone polmerase IV" involved in gene transcription is responsible for many adaptive mutations in higher organisms. Susan Lindquist, a cellular and molecular geneticist at the University of Chicago, recently summarized her research exploring whether a stressful environment could facilitate adaptive mutations by saying, "The main point is that, no matter how they arose, [these processes] provide a plausible route to the evolution of new traits" (in Chicurel, 2001a, p. 1825).

There is as yet no research relating adaptive mutations in response to environmental stress on the molecular level with the experience of psychosocial stress and the "evolution of new traits" in humans. We do not yet know whether our psychological approaches to the resolution of the stress generated in stage two of the creative cycle, for example, are related in any way to the adaptive mutations observed in lower organisms. As reviewed in Part I, however, we do know that the *entirely normal everyday expression* (not mutation!) of immediate early genes, behavioral state-related genes, and activity-dependent genes are associated with modulations in behavioral states, emotions, memory, learning, dreaming—and, by implication, the psychological transformations characteristic of problem-solving, identity formation, and new developments in human consciousness. For now, we will utilize this well-documented experimental data base of neuroscience and functional genomics as the psychobiological foundation of the four-stage creative process and the breakout heuristic that we will explore in the following sections.

THE BREAKOUT HEURISTICS
THE CULTURAL FACE OF THE CREATIVE CYCLE

The clinical application of Wallas's classical formulation of the four-stage creative cycle, as it was experienced by young people in psychotherapy, was described originally as the *breakout heuristic* (Rossi, 1968, 1972, 2000a).

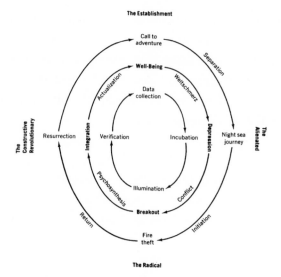

FIGURE 6.3 | The breakout heuristic illustrating how the four-stage creative cycle is replayed in the arts and sciences (inner circle), psychotherapy (middle circle), myth (outer circle), and social/cultural/political processes (outer labels).

Illustrated in Figure 6.3, the breakout heuristic is a generalized paradigm of the four-stage creative cycle as replayed in individual and collective endeavors in the arts, humanities, and culture. From this perspective, creative development in the arts, drama, literature, music, myth, ritual, science, and spiritual practices all involve individual experiences of the breakout heuristic. The inner circle in Figure 6.3 illustrates the classical formulation of the creative cycle as it is experienced in breakthroughs in the arts and sciences (Hadamard, 1954; Rossi, 1972, 1985, 2000d; Sternberg & Davidson, 1995). The middle circle illustrates the four-stage creative cycle as it is experienced in dreams and everyday life as well as in psychotherapy (Rossi, 1967, 1968, 1972, 2000d), as creative individuals "defy the crowd" to find their own way (Amabile, 2001; Sternberg, 2001). The outer circle represents the creative cycle as it is replayed in myth and saga (Campbell, 1956-1968; Neumann, 1959-1968) as well as in the spiritual practices of many cultures (Bucke, 1901; Jung, 1960; Keeney, 1999; Rossi, A., 2002; White, 1972, 1985; Wilber, 1993). The outer labels of Figure 6.3 indicate the same creative cycle as it is replayed on sociopolitical levels.

Let us now take a tour of the creative cycle and the breakout heuristic as they are replayed in the transformations of consciousness, self and society. We will be looking for hints about how human phenomenological experience at each stage of the creative cycle may be associated with the psychobiology of gene expression, neurogenesis, and healing in everyday life and dreams as well as the arts, sciences, and psychotherapy.

STAGE ONE: PREPARATION

Awareness of the Need for the Novelty-Numinosum-Neurogenesis Effect

Even when people think they are content with their lives, there may be evidence of a static quality in their experiences or developmental difficulties in their personality that suggest how gene expression, neurogenesis, and healing—the novelty-numinosum-neurogenesis effect—is not taking place in an optimal manner. Themes such as the following may be evident in their dreams as well as part of their conscious preoccupations throughout the day:

> "I'm lost. . . . I don't know what to do. . . . I'm feeling stale and depressed. . . . I have no energy. . . . I don't understand what's happening. . . . I'm just running around in circles. . . . I don't feel I'm going anywhere."

Boredom and the need for life enrichment seem to pervade our culture. Gaining this awareness is the first step in the creative process. It may be a simple, direct perception of a life-saving need to grow. It would be interesting to have a boredom or lack of optimal growth index that would enable us to compare different cultures and time periods on this initial psychological precondition of the creative process. We would predict that there is an optimal balance between "normal" stability and rapid periods of transition and change in individuals as well as groups (industries, corporations, professions, etc.), tribes, nations and cultures in general. In fact, empirical studies of complex adaptive systems and their evolution (using computer simulations) indicate that (1) complex organisms are more robust than simple ones over time, and (2) development over time is characterized by relatively long periods of "normal" stability that is occasionally punctuated by very rapid phases of change and transition (Elena et al., 1996; Lenski et al., 1994). These findings appear to be consistent with the theory of evolution that maintains that the overall evolutionary process has many periods of relatively little innovation that are periodically punctuated by brief and very rapid experiences of major change (Gould & Eldredge, 1977). We extend this theory of the evolution of life in general to the psychobiological dynamics of gene expression and neurogenesis in the four-stage creative cycle of everyday life.

When we are cut off from an awareness of the new developing within—the novelty-numinosum-neurogenesis effect—our behavior becomes stereotyped and predictable. We overidentify with the self-same attitudes and roles we habitually display within the confines of our daily lives. The most essential dynamic of the being that we call human—the process of growth and transformation—is buried under a host of attitudes and behaviors we call "normal." In the normal dynamics of social life, we learn a few patterns, usually those valued by our parents and society, to the exclusion of the new that feels foreign or too risky once it does manage to emerge into our awareness. Having limited ourselves, without even being aware of it, to a relatively narrow range of life experiences, we unwittingly set the stage for feelings of stasis, inadequacy, inferiority, and ultimately psychogenomic depression, illness, and even death.

A sense of inadequacy can be a useful symptom when it forces us to recognize our need to expand our too-narrowly-defined self. But if we have a *negativistic attitude* toward the novelty-numinosum-neurogenesis effect—if we are immersed in a habitually depreciating attitude toward our own originality—we will not recognize its value when it does emerge. *Lack of awareness of our own uniqueness, together with a negativistic attitude toward the new, are the most typical blocks to development.* With these blocking tendencies we become mired in a sense of inadequacy and fall into depression. From the creative perspective, depression is a classical sign of a need to incubate, to turn inward to the sources of our individuality that are teeming for expression.

STAGE TWO: INCUBATION

Depression and the Dynamics of Natural Variation

To turn inward to respond to the soul and the call of the new seems like a good idea in our society when everything seems to be going too fast on the

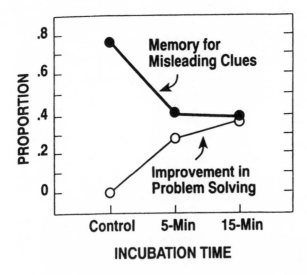

Figure 6.4 | Typical incubation time for stage two of the creative process. Improvement on initially failed problems (because of being stuck on misleading clues) as a function of incubation time in stage two. During incubation, that is, inner work replayed on an implicit level, the memories for misleading cues drop out simultaneously with the improvement in problem solving. (With permission from Smith, 1995.)

outside. But who pays us to do that? Stress, conflicts, depression, and psychosomatic symptoms are experienced in stage two of the creative cycle when it is not hosted willingly, within a positive frame. The creative cycle becomes stuck in stage one when the person is caught in previously established patterns of *state-dependent memory, learning, and behavior* that are no longer adaptive. In popular parlance we say the person has fallen into "a rut."

From this perspective we may say that chronic stress is a condition of being stuck in stage two—the arousal stage—of the creative cycle. The mind-body is continually generating arousal-stress hormones during stage two in futile attempts to cope. The mind-body is struggling to access inner resources for problem-solving and healing but is not succeeding because older, habitual, stereotyped state-dependent response patterns are interfering with the emergence of the new at all levels—from the cognitive and behavioral to the cellular-genomic.

Research on stage two of the creative cycle documents the typical ultradian time parameters of getting into and out of these ruts (Smith, 1996).

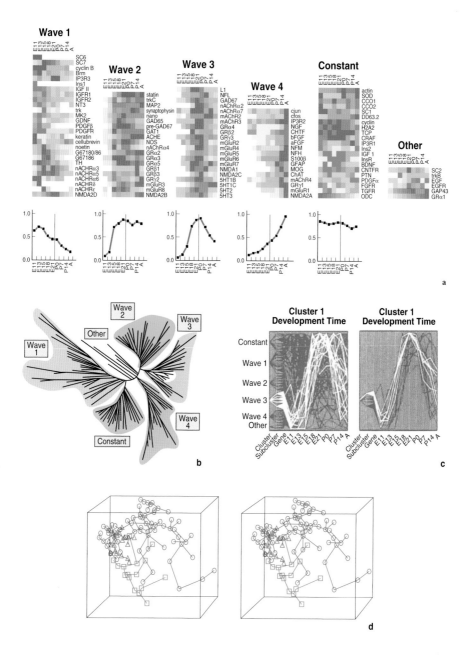

FIGURE 1.3 | The temporal waves of gene expression in the development of the central nervous system (CNS). (With permission from Wen, X., Fuhrman, S., Michaels, G., Carr, D., Smith, S., Barker J., & Somogyi R., 1998. Large-scale temporal gene expression mapping of central nervous system development. *Proceedings of the National Academy of Science*, 95, 334–339. Copyright 1998, National Academy of Sciences, U.S.A.)

Common Elements in the Design of Circadian Oscillators

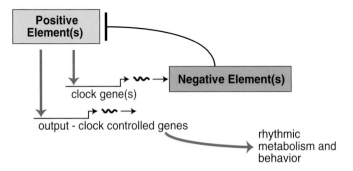

Positive elements in circadian loops:

   kaiA in Synechococcus
   WHITE COLLAR-1 & WC-2 in Neurospora
   CLK & CYC in Drosophila
   CLOCK & BMAL1 (MOP3) in mammals

Negative elements in circadian loops:

   kaiC in Synechococcus
   FREQUENCY in Neurospora
   PERIOD and TIMELESS in Drosophila
   PER1, PER2, PER3, (& TIMELESS?) in mammals

FIGURE 2.2 | A number of the common elements, design, and functions in the creative replay of the circadian oscillator of the gene expression and protein synthesis cycle that generates the rhythms of metabolism, physiology, and behavior. (Reprinted from *Cell*, Volume 96, J. Dunlap, The molecular basis for circadian clocks, pp. 271–290, Copyright 1999, with permission from Elsevier Science.)

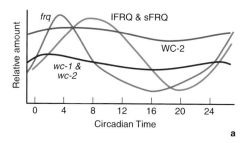

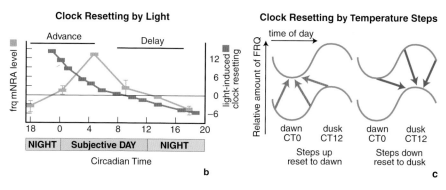

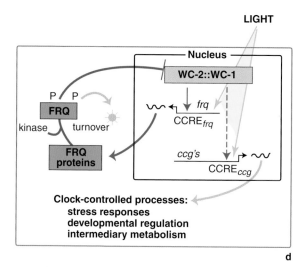

FIGURE 2.3 | The time dynamics of how light and temperature entrain (modulate, regulate) the oscillations of expression of the clock gene Frequency (frq) in the bread mold Neurospora. (A): Temporal regulation of the *frq* gene and the large (lFRQ) and small (sFRQ) proteins. (B): How light resets the *Neurospora* clock. (C): How temperature resets the *Neurospora* clock. (D): An example of how clock genes are involved the stress response, developmental regulation, and intermediary metabolism. (Reprinted from *Cell*, Volume 96, J. Dunlap, The molecular basis for circadian clocks, pp. 271–290, Copyright 1999, with permission from Elsevier Science.)

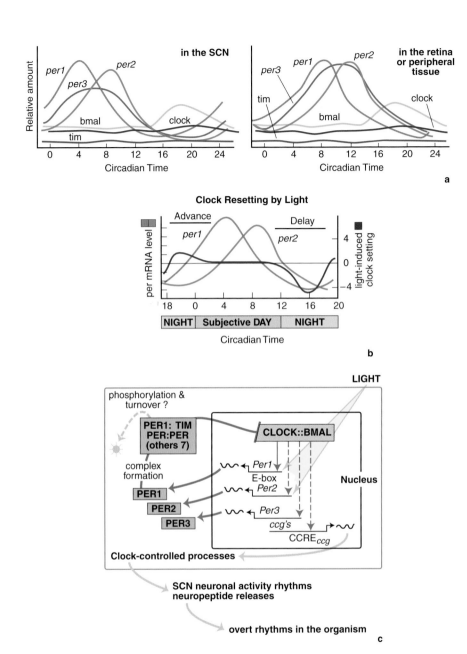

FIGURE 2.4 | Identity and regulation of clock genes in the mammalian oscillator and their roles in entraining the overt rhythms of biology and behavior. (A): Temporal regulation of the per 1, per 2, per 3, tim, clock, and bmal1 (mop 3) genes. (B): How light resets the mammalian clock. (C): Elements of control logic in the circadian oscillatory loop of animals. (Reprinted from *Cell*, Volume 96, J. Dunlap, The molecular basis for circadian clocks, pp. 271–290, Copyright 1999, with permission from Elsevier Science.)

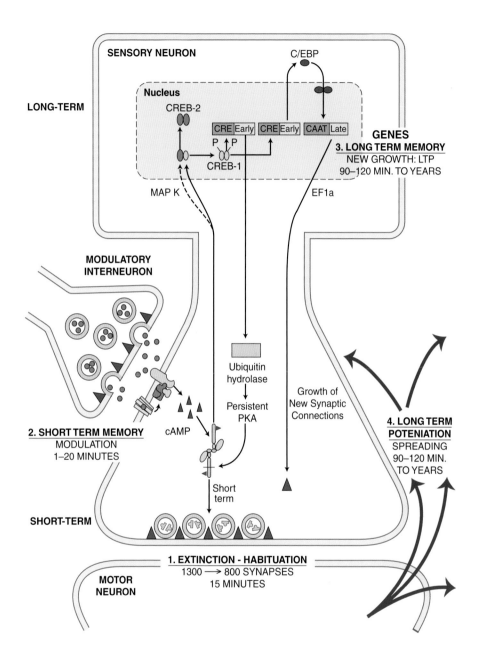

FIGURE 3.2 | Time parameters of short- and long-term memory. (Adapted from *Memory: From Mind to Molecules* by Larry R. Squire and Eric R. Kandel, 1999, p. 143, 2000 by Scientific American Library. Used with permission of W.H. Freeman and Company.)

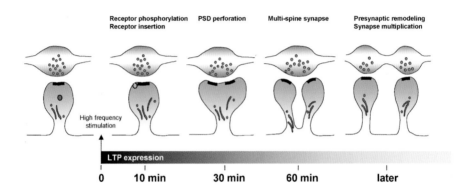

Figure 3.3 | The ultradian dynamics of activity-dependent memory, learning, and behavior by long-term potentiation (LTP) as proposed by Lüscher et al. (2000). Within the first 10 minutes there are measurable changes in gene expression and the activation of receptors that are involved in synaptic communication via neurotransmitters. Within 30 minutes the size of the synaptic spine increases and the stimulated AMPA receptors move to the postsynaptic membrane; this leads to an increase in the size of the post synapse. Within an hour, some synapses divide in two. This leads, in turn, to further growth in presynaptic multiplication and remodeling to create new neural networks encoding memory, learning, and behavior change that is of essence for psychotherapy and many activity-dependent processes of gene expression involved with healing. (With permission from Lüscher, C., Nicoll, R., Malenka, C., & Muller, D., 2000. Synaptic plasticity and dynamic modulation of the postsynaptic membrane. *Nature Neuroscience, 3*, 545–567.)

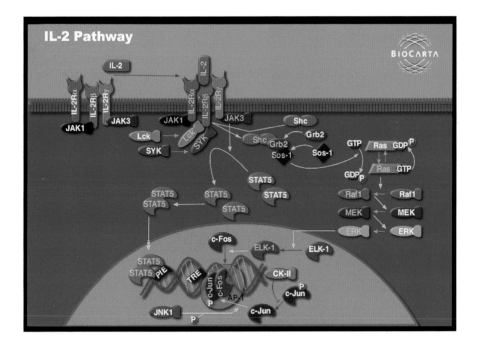

FIGURE 5.2 | The cellular communication pathways of psychosocial genomics. This is a partial illustration of the cellular-genetic dynamics of the interleukin-2 (IL-2) pathway from the surface of the cell (top left) to the nucleus and genes (rectangle at bottom). The IL-2 gene is down-regulated with psychosocial stress and up-regulated with positive psychosocial support and psychotherapeutic interventions. The mitogen activated protein kinases (MAPKs) are a family of transducing (transforming) signals that travel from cell surface receptors, such as IL-2, by multiple pathways toward the nucleus, where gene transcription is turned on and off in an adaptive response to psychosocial factors.

There are three main branches of MAPK signaling known as the ERKs, JNK/SAPK, and p38 pathways (not all shown here). Gene expression usually begins with a primary "molecular messenger of the body," such as a hormone (see figure 2.1) signaling a small G-protein at the surface of the cell. This signal proceeds through a series of "secondary messengers" such as the kinase cascades (e.g. MAPKs), which carry the signal (yellow arrows) toward the nucleus of the cell. These secondary messengers activate (energize) many cellular systems within the cytoplasm by adding phosphate groups and ultimately the transcription factors that turn genes on or off within the nucleus. Notice the signaling role of *immediate early genes* such as c-fos and c-jun in the nucleus of the cell (rectangle at bottom), which can be activated within a minute or two of psychobiological arousal from many sources, ranging from novelty to pain, stress, the basic-activity cycle, and dreaming (see Figure 5.1). (With permission from Biocarta [www.biocarta.com.])

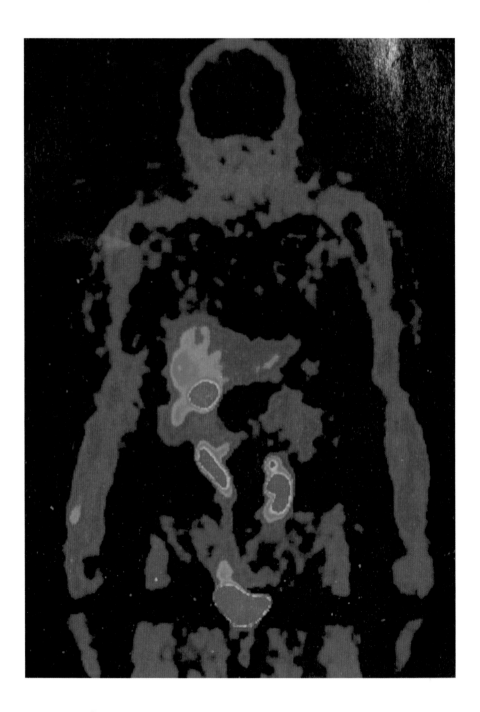

FIGURE 5.9 | A color-coded positron emission tomography (PET) scan of the human body illustrating areas of heightened gene expression in the liver, intestines, and bladder. (With permission from Yaghoubi et al., 2001.)

Figure 6.4 illustrates how discarding memories of misleading clues (sources of stereotyped responses or ruts) is associated with an improvement in problem-solving over the 15 minutes of the typical incubation process—that is, inner work on an implicit level. An apparent failure of Selye's general adaptation response leaves the person treading water but making no progress in stage two of the creative response. For whatever reason, optimal gene expression, neurogenesis, insight, and creative problem-solving are not taking place. This stagnation leads to Selye's description of the stress of life, wherein chronic arousal-stress hormones are released as the mind-body tries to break through the interference (the physiological "resistance" noted by Selye as well as the psychological resistance of classical psychoanalytic theory) experienced in this stage.

Depression—a hallmark of misunderstood incubation—is characterized by a withdrawal of interest from the outside world. When depressed, we become self-absorbed, aloof, or morose. Our emotions seem flat and our mind seems blank. What is happening to a person in this state? Attention has been withdrawn from external concerns so that it can be invested in the inner process of transformation. Depression corresponds to the second (incubation) stage of the creative process outlined in the middle circle of Figure 6.3. In the creative sense, depression is a period of incubation signaling a time for inner work on deep implicit levels.

Depression is not usually understood in this creative context, however. More often than not, it is conceptualized as being symptomatic of sickness. In this stage of the creative cycle, we may very well feel sick as we trudge through unchanging state-bound patterns of thinking, feeling, and behavior. When we become aware of new developments for which we do not feel prepared, we may experience fear and a sense of emotional crisis. This crisis is the conflict between the newly developing patterns on an implicit level and the older, less adequate explicit worldview with which we are identified. Self-division, duality and conflict are apparent in dreams as well as everyday experiences such as the following:

> "I can no longer agree with my parents. . . . I'm not really living my life. . . . something has got to change. . . . I keep going three steps forward and two backward. . . . I cannot live with this shifting back and forth and indecision. . . . Sometimes I really feel I am going crazy. . . . I've got to make up my minds!"

These attitudes of impatience and going crazy are characteristic of people who are unaware of the significance of the unfamiliar feelings, thoughts, and conflict arising out of a creative process on an implicit level that has not yet emerged on the conscious explicit level. Their inability to express the new may become so extreme as to lead to symptoms of stress and mental illness. Many aspects of psychopathology—particularly those of anxiety and stress associated with motivation for high performance—may be understood as symptoms or signals of creative flux and psychological development (Nakao et al., 2001). Crises of consciousness have long been recognized as characterizing the transformational experiences of expanding awareness and spiritual development (Bucke, 1901; Jung, 1960, 1966; White, 1972, 1995).

Other significant signs of creativity at this stage include spontaneous flashbacks and the recall of poignant but troubled memories that are characteristic of the posttraumatic stress disorders. *These flashbacks are often painful replays that often seem to be going nowhere. People wonder why they become preoccupied with these vignettes of an earlier time. But this recall and the possibility of creative replay is the very essence of what the crisis situation is all about!* The early memories are usually loaded with other states of being, with other aspects of the personality that may have been prematurely pushed aside. These state-dependent memories and identities, when sympathetically understood, often point to potentialities within the individual's personality that have not reached their optimal development.

People going through the stress and strain of trying to find themselves are usually in a rebellious mood. It is a rebellion against the old that is no longer adequate in comparison with the new, which holds a promise of a more fulfilling worldview. Of course, many people get lost in their rebellion; they become rebels without a cause. Rebellion can be very destructive when it is used indiscriminately to overthrow what was of genuine worth in the old along with what is no longer of value. The critical factor that makes the difference between constructive and destructive rebellion is whether or not the rebel has any awareness of what he or she is really fighting for and why. Is the young rebel aware, for example, of just what in the old worldview needs to be changed to make life more meaningful for his or her new generation?

Parents locked in old worldviews usually cannot help because they are blind to the new world the young are entering. In our modern world society changes so radically from generation to generation that parents cannot keep up; they cannot foresee what characteristics and attitudes their children will need to create their place in the new generation. Because the elders cannot help effectively, the young have an even more difficult task of finding themselves. Anyone in the crisis phase of the growth process has a triple task: (1) become aware of the inadequacies of the old worldview, (2) seek out new possibilities for the present and future, and (3) test these possibilities to determine which will work.

Passive and negative attitudes toward the new are related to an embarrassing state of *self-consciousness,* which is usually experienced as awkward and painful—but it can also represent an important developmental step in stage two of the creative cycle. Self-consciousness can be a prelude to new patterns of awareness and behavior that set the stage for self-reflection and the possibility of changing in a self-directed way. The degree to which we can become self-aware will influence the entire course of our future development. The issues are crucial and may lead to an identity crisis in this second stage. Some experience this crisis as a state of anxiety and uncertainty about the major questions of life.

Thoughts may replay repetitively, on and on, in waves of anxiety and confusion about current life events and our emotional responses, over which we feel have no control. Current research on the dynamics of transition and change, however, indicates that there is a *natural process of Darwinian variation* that is replayed in behavior when we are no longer reinforced for old patterns that are not adaptive (Carpenter, 2001; Neuringer et al., 2001). Neuringer and his students have documented a number of factors that are significant for an understanding of the psychodynamics of change. Many of these factors reflect an interaction between nature (behavioral state-related gene expression) and nurture (activity-dependent gene expression) that we reviewed earlier in Chapters 2 and 3. For example, there is an inherited aspect of behavioral variability that is well documented in hypertensive (easily stressed) rats (Hunziker et al., 1996) that may be used to model human attention-deficit/hyperactivity disorder (ADHD). These genetically hypertensive animals replayed response variability throughout the experimental conditions, without ever settling into an adaptive response. The normal animals, however, were more focused on the specifics of the reinforcement conditions that led them to adaptive responses. In a later study Miller and Neuringer (2000) found that variability could be reinforced in adolescents with autism to facilitate the learning of new behavior.

A major implication of Neuringer's research is that Darwinian natural variation is a fundamental process in the creative psychobiological dynamics of behavior change. Sigrid Glenn, a behavioral researcher at the University of North Texas comments on Neuringer's research (in Carpenter, 2001):

> Much of behavior analytic work focuses on steady states [i.e., homeostasis and the general adaptation syndrome] and the transitions from one state to another haven't historically been studied as much. Neuringer's work and other work on variability addresses the idea that variability is good—that it's not something to be gotten rid of. This approach allows us to see a parallel between operant, behavioral selection processes and [Darwinian] natural selection processes at the level of species, because without variability there can be no change. (p. 71)

This emphasis on the central role of Darwinian natural variation during phases of arousal and replay in stage two of the creative cycle is illustrated in Figure 2.11. The high state of stress and conflict in the replays of stage two will suddenly be discharged during the next stage of the creative cycle, when we experience a natural variation or breakout of the old stereotyped patterns into a new world of original psychological experience (Gruber & Wallace, 2001; Rossi, 2000d; Stokes, 2001; Ward, 2001).

An interesting account of "An Artist at the Abyss" tells the story of how contemporary artist Jack Lowe confronted himself in an important early crisis in his identity that illustrates a successful shift from stage two to stage three of the creative cycle (Cheverton, 2001).

> . . . Lowe painted. It was a compulsion that had first struck him at Long Beach's Jefferson Junior High School, when he chanced to pass a hallway display case with casts of the eye, nose and mouth of Michelangelo's *David*. "I was fascinated [an experience of the numinosum]. I had never seen anything like that in Long Beach, and I was amazed that someone could reproduce something so skillfully."
>
> But in the early '60s, he hit the wall: "I just painted myself into a corner. I was burnt out. I just wasn't finding my identity. . . . There was no intensity, there

was no essence. . . . They [the paintings] were just shallow."

His solution was brutally direct: "I pulled [my paintings] off their stretcher bars, rolled them up, rented a trailer and took them to the dump and threw them away." Others went to the fire pits on the beach. "I loved doing that—burning 'em." (p. 22)

Jack Lowe went on to become one of the great artists of the San Francisco Bay scene after World War II, where Richard Diebenkorn was both his teacher and contemporary artist along with others. Now, at the age of 78, Lowe is reportedly doing the best art of his career.

When it is successful, a well-focused process of psychological arousal can facilitate the gene expression and neurogenesis that underlies problem-solving and healing. When it is not successful, stress proteins may accumulate (Morimoto et al., 1990; Taché et al., 1989), which do not lead to problem-solving. The critical transition between the maladaptive stereotyped response of Selye's conception of stress versus creative health is experienced by self and society as the shift from stage two to stage three of the creative cycle and breakout heuristic. Under the best of circumstances, stage three is typically experienced as a *creative moment* of insight, new meaning, problem-solving, and healing—all of which are mediated by the novelty-numinosum-neurgenesis-effect.

STAGE THREE: ILLUMINATION

The Novelty-Numinosum-Neurogenesis Effect in the Creative Cycle

The ultimate psychobiological resolution of acute as well as chronic stress requires the private inner work that facilitates the shift from stage two to stage three of the creative cycle. From the perspective of computer simulation research on artificial life cited earlier (Elena et al., 1996; Lenski et al., 1994), this shift is often experienced as a sudden insight that punctuates our normal life so that we experience a mental, emotional, or identity crisis as a very rapid jump in our consciousness, adaptability, and/or fitness. From our neuroscience perspective, a creative experience of novelty, enrichment, and physical exercise can often facilitate this shift from stage two to stage three. A beneficial outcome of the private period of inner work during the peak or crisis stage of the breakout heuristic leads to an experience of creative psychobiological activity that is called by many different names in different cultures: awe, curiosity, wonder, eureka, ecstasy, peak experience, eustress, being motivation, satori, kensho, enlightenment, flow, numinosum, liminality, joy, and love (Maslow, 1962, 1967; Rossi, 1972, 2000a; Wilber, 1993). These creative states are the experiential aspect of the novelty-numinosum-neurogenesis effect.

The novelty-numinosum-neurogenesis effect is a psychogenomic bridge between the experiences of mind (novelty, numinosum) and matter (neurogenesis). It is a key to understanding how the natural selection aspect of Darwinian variation and selection is experienced in the creative cycle. How do we recognize when optimal gene expression and neurogenesis are taking place to punctuate our life experience with a sudden developmental jump? *The sense of the numinosum in the ongoing flow of creative experience is the key selection factor, alerting us to which activity, thought, intuition, etc., merits for further examination and possible development.*

The essence of stage three is the cognitive-emotional recognition and *selection* of a creative psychological experience. More than Darwin's original concept of selection is involved, however (Wesson, 1991). In his theory of neuronal group selection, for example, Edelman (1987; Edelman & Tononi, 2000) has extended Darwin's theory of variation and natural selection to the levels of gene expression, neurogenesis, and brain development from the earliest stages of embryogenesis to psychological experience and consciousness throughout the life cycle. He describes three main tenets of the theory of neuronal group selection: (1) "Developmental selection" on the levels of gene expression and neurogenesis leads to a highly diverse set of brain neurons that make up the brain, from which (2) "experiential selection" leads to changes in the strengths of connections between brain synapses, which eventually favor some pathways over others, and (3) "reentrant neural dynamics" that culminate in the differentiation of states of consciousness. Edelman emphasizes that reentrant neural dynamics are not made up of single feedback loops that are built into the system in a hard-wired fashion. Rather, the reentrant dynamics are made up of multiple parallel pathways that are continually regenerated and resynthesized in the ongoing processes of recall and replay that are so characteristic of the creative cycle.

During the third stage of the creative cycle, people suddenly experience themselves with a new idea, hunch, or intuition. They glimpse new positive possibilities in their lives. There is the excitement of discovery; a new world of understanding is opening up as they *break out* of old patterns. Poets, artists, scientists—all creative workers—prize this experience above all others. It is their *raison d'être*. They are most themselves at just these moments of original psychological experience. They identify these numinous moments as being the best expression of their essential selves (Maslow, 1962, 1967). From the spiritual perspective of Huxley's (1970) *The Perennial Philosophy,* these numinous moments have been described as experiences of self-realization and/or unity with God (White, 1995; Wilber, 1993) and cosmic consciousness (Bucke, 1901).

A sense of *unreality* is highly characteristic of the states of transition and discovery that culminate in an original psychological experience in any domain, from art and science to everyday life, relationships, work, and so on. Dreams seem unreal, fantasies seem unreal, hunches and new ideas and discoveries all seem unreal at first. Everything that is new seems to be unreal on first exposure to it. The new seems unreal precisely because it is still so new that it does not yet have a well-established place in our familiar contexts and ways of understanding. Those odd sensations and qualities of seeming distortion and grotesqueries in dreams are actually new patterns of awareness breaking through to consciousness as they shatter previous contexts and meanings in a seemingly spontaneous manner. That which seems absurd, bizarre, or meaningless in dreams only seems so in relation to the older, more established attitudes and points of view that still dominate the conscious mind. Consider the following poignant description of the initial sense of unreality and disbelief experienced by mathematician Andrew Wiles at the critical moment of correcting an error to finally solve a 300-year-old puzzle by proving Fermat's last theorem (Aczel, 1996):

> Wiles studied the papers in front of him, concentrating very hard for about 20 minutes. And then he saw exactly why he was unable to make the system work. "It was the most important moment in my entire working life," he later described the feeling. "Suddenly, totally unexpectedly, I had this incredible revelation. Nothing I'll ever do again will." At that moment tears welled up and Wiles was

choking with emotion. What Wiles realized at that fateful moment was so inde-
scribably beautiful, it was "so simple and so elegant—and I just stared in disbelief." Wiles walked around the department for several hours. He didn't know
whether he was awake or dreaming. Every once and a while, he would return
to his desk to see if his fantastic finding was still there—and it was. (pp. 132–133)

Many corresponding experiences are reported by creative workers in literature and the arts. Such creative moments are the basis of much art, drama,
and literature of fantasy and fiction. What, for example, are the magical
moments so common in fairytales and myth supposed to mean? Why do children, as well as adults, continue to find delight in reading such fictions? Well,
you might respond, such delights are normal experiences of "enjoyment."
What precisely is this enjoyment? We can understand why nature builds in
the perception of danger and the sensation of pain as an aid to survival, but
what good is the enjoyment of the fictions of drama and tales of fantasy and
wonder? The many experiences of enjoyment and light on explicit conscious
levels are numinous moments that are accompanied by the occurrence of
gene expression, neurogenesis, and healing at an implicate level. This is the
essence of what we proposed as the novelty-numinosum-neurogenesis
hypothesis of Chapter 3.

The "magical fictions" of scientific theory, mathematics, and art are evocative of the experiential sense of the numinosum that humans have when their
deepest motivations are touched and the new is breaking through into their
consciousness. Reading novels of love and adventure, going to the theater and
the movies, enjoying dance, opera and musical performances of all sorts are
training experiences "seeding" us, on an implicate level, with "how-to" manuals of the creative process. The implicit self-organizational phase of the creative cycle is experienced as surprising, unexpected, or magical in stage three
precisely because conscious, explicit processing does not do it. Such self-organizational dynamics on the implicate level replay in a range that extends from
the trivial to the profound, from what seems to be anxiety-provoking and
harmful to the breathtaking peak experiences of creative work, as described
by the mathematician Andrew Wiles.

Developing an attitude of receptivity to the new points of view that
develop in dreams and states of reverie is akin to the hero gaining the boon
in mythology. The new is a gift, a saving grace and sensitive perception that

will enable us to expand our awareness and even recreate our consciousness and identity. The wondrous and mysterious happenings in folk tales and myths correspond to the fascination we feel at the equally mysterious changes in ourselves when the new is recognized. The magical power of foods, gloves, capes, and swords in fairytales corresponds to the magical potency of new understanding to expand our worldview and dramatically change our lives.

Good feelings, love, and a sense of the beautiful frequently accompany new awareness and positive developments in the personality. *Creative moments of transformation from the painful or the ugly to the beautiful can sometimes be observed directly within the dreams.* It is only a short step from such positive and exhilarating dream experiences to the mystical states of consciousness described by cultivated minds throughout the ages. A special sense of harmony, inner light, heightened awareness, and relatedness to the universe is characteristic of both. We may infer that the Zen experience of satori, the yoga experience of samadhi, and the ecstasy of the mystic differ only in degree from the original psychological experiences we are all capable of experiencing in everyday life (Rossi, 2000a; White, 1954, 1995; Wilber, 1993, 1995).

Original psychological experience and new learning are not always accompanied by a numinous creative moment, however. Siegler (1998) has reviewed "microgenetic" research into children's learning that examines the various learning strategies that are explored and adapted in the first decade of life. The microgenetic approach does not actually employ genetics. It is a method of investigating how children learn by imitation, how their reasoning works in mathematics, how they gain an understanding of other people's intentions and beliefs. The microgenetic approach documents how children use a variety of approaches and learn in "overlapping waves" of strategies.

> At any one time, some strategies are cresting, some are waning, some are gaining renewed force, and new ones are forming just below the surface of conscious deliberations. The coexistence of diverse strategies and ways of thinking in the same person is a central characteristic of our cognitive system. This gives people the flexibility to adjust what they do to the demands of the problem and the situation. (In Bower, 2001, p. 172)

These researchers find that gradual change may generate more robust and flexible learning in children rather than the abrupt change that is sometimes dramatically evident in adults experiencing the kind of creative moments described above. New learning is also stabilized when it is accompanied by self-reflection, wherein the child is able to compare the different strategies. Children's learning is often accompanied by physical head, hand, and other body gestures. Skilled teachers match the children's physical gesture to facilitate their learning processes. Research in this word-and-gesture approach documents how children comprehend better when the teacher utilizes children's physical gestures rather than mismatching gestures (Granott & Parziale, in press). This *utilization of gesture or pantomime* (Erickson, 1958/1980, 1959/1980,1964/1980) is entirely consistent with the experience of many clinicians who facilitate the *creative process and insight via inner rehearsal and replay in psychotherapy.* Research into the utilization of hand and body gestures provides a new database for what has been traditionally described as "ideodynamic and state-dependent memory and learning" in mind-body therapy (Rossi & Cheek, 1988). The use of gesture and physical movements to facilitate creative ideodynamic replay, self-reflection, and experiential learning is the essence of the new symmetry-breaking approaches to psychotherapy demonstrated in the next two chapters and outlined with many creative variations in Chapter 10.

STAGE FOUR: VERIFICATION

Reality Testing and Co-Creation of the New

Up to this point in our discussion, the implicate dynamics of the creative process have functioned autonomously: People describe a new idea that came from "out of the blue," an "intuition," "hunch," "inner guidance," or "the experience of grace." We automatically fall into a *depression* when we cannot find fulfillment, and we naturally tend to rebel against any *suppression* of our individuality. But it is not enough to throw off the yoke of the old and fight for a good experience of the new. We usually need to experiment with our new patterns of awareness and learn how to facilitate the ever-changing states of a more rewarding life. We need to make an active, conscious effort to *reality-test* the new to create our unique identity and destiny. This reality-testing process engages a new phase of selection, activity-dependent gene expression, and neurogenesis. In stage four of the creative cycle the natural processes of growth and transformation, occurring on an implicate level, are integrated into our conscious, self-directed efforts to facilitate our own development. The flavor of this positive, active effort to self-facilitate new identity is frequently mirrored in cognitions and dialogue with oneself and others.

The active effort required for the cultivation of our positive nature in this reality-testing stage frequently adds a special dimension of consciousness and self-direction to dream experience. The dreamer sometimes notices that he or she is repeating a part of, or a whole, dream—over and over again. Though this repetition may be experienced as frustrating, it often points to places in the creative cycle where there is uncertainty and a need to replay the tough spot in the process of making changes and integrating the new. The dreamer may be engaged in trying to make the dream work out in a certain way. The dreamer is learning to actively cooperate with the novelty-numinosum-neurogenesis effect. That is, the dreamer is learning how to *co-create* consciousness and his or her own evolving worldview.

Dire events may be set up within one's imagination and the drama of the dream on the implicate level of state-dependent memory and behavioral state-related gene expression, however. The developmental challenge is for the individual to learn to cope with these inner trials by engaging in activities that stimulate activity-dependent gene expression—art, drama, sports— whatever evokes the novelty-numinosum-neurogenesis effect. This creative interaction between *state-dependent memory, learning, and behavior* (SDMLB), *behavioral state-related gene expression* (BSGE), and *activity-dependent gene expression* (ADGE) has not been investigated. No research has been done on it because we have not developed teams of researchers familiar with the new possibilities of integrating the central concepts of functional genomics, neurogenesis, and positive psychology. If this association could be identified experimentally, it would serve as a psychobiological bridge between mind (SDMLB), matter (BSGE), and the role of consciousness and free will in facilitating self-creation (ADGE). Interesting questions could be answered. When the dreamer contrives to creatively replay and rework the dream drama to ensure a constructive outcome (as in lucid dreaming, LaBerge & Rheingold, 1990) on a more explicit level, does this mean we are witnessing an *activity-*

dependent level of gene expression and neurogenesis? Could we witness the molecular footprint of lucid dreaming and creative life experience precisely in this cooperative and co-creative moment that integrates behavioral state-related gene expression with activity-dependent gene expression and neurogenesis?

When the dreamer becomes aware, within the dream, of the different directions the plot can take and the alternative endings that are possible, the threshold to lucid dreaming has been reached. The dreamer finds that he or she now has the surprising ability to intervene in the drama of the dream and direct it in a desirable way. This sense of directing the dream in a constructive psychosynthetic or co-creative manner becomes particularly evident in the twilight period between sleep and consciousness. The so-called "hypnopomic state" between sleeping and waking is a very valuable one to cultivate when one has the luxury of lingering in bed in the morning.

Having survived the breakup of the old world and the precarious task of finding and facilitating the new, we are now truly launched on a path of self-actualization that may be experienced as a spiritual transformation. We do not need to overidentify with others or try to force our worldview on them. Rather, we can use our evolving consciousness to relate easily to others and see things from their point of view. Rapport becomes easy. Misunderstandings are readily resolved. A sense of really being oneself and spiritual well-being prevails.

Can we identify the essence of psychological and spiritual transformation as the experiential manifestations of the novelty-numinosum-neurogenesis effect in the creative cycle? Could this be the existential significance of the transitions from the "dark night of the soul" to the joy and ecstasy of the high peaks of spiritual experience? Much available evidence indicates that those cultures that have a vital, living, spiritual tradition continue to evolve, while those that lose their psychological and spiritual sense of well-being perish. Recent research in positive psychology confirms the connection between psychological health, healing, and a sense of existential or spiritual transformation for many people (Dossey, 1993; Fredrickson, 2001). Rather than speculate about the nature of such transformations, however, science does best with empirical explorations of how they are actually experienced and how they can be best facilitated (Gillett, 2001; Rossi & Nimmons, 1991). This will be our adventure in the next section that explores how to create a great day.

HOW TO ENJOY CREATING A GREAT DAY AND NIGHT

Aldous Huxley, author of *The Perennial Philosophy*, certainly knew how to enjoy a creative day. After an early morning breakfast he would let himself sink into a state he called "Deep Reflection" in his favorite armchair in his study. After about an hour or so, he would awaken spontaneously with words, sentences, and paragraphs teeming through his mind for expression. He would then write easily and steadily throughout the morning, almost as if he were taking down dictation from a boundless source.

What is Deep Reflection and what is its source? Huxley used these words to describe the nature of his Deep Reflection when he explored it with the help of Milton H. Erickson (1965/1980a).

> I use Deep Reflection to summon my memories, to put into order all of my thinking, to explore the range, the extent of my mental existence, but I do it solely to let those realizations, the thinking, the understandings, the memories seep into the work I'm planning to do without my conscious awareness of them. Fascinating . . . [I] never stopped to realize that my Deep Reflection always preceded a period of intensive work wherein I was completely absorbed . . . (p. 89)

After a morning of writing, Huxley would have lunch, rest a bit in the afternoon, and then enjoy some social life and light exercise like a walk. After supper he reentered the creative mode in quite another way. He would relax in his study, browsing a bit here and there amidst his vast home library. He was vaguely aware that he was searching for material for the next morning's writing, but he was not working actively on it with his conscious mind. He was simply enjoying himself, feeling comfortable and relaxed as a prelude to sleep and dreaming, so that he could awaken the next morning primed for another creative day.

Recent research supports the wisdom of Huxley's creative lifestyle. A basic idea emerging from a variety of fields (ranging from ecology and engineering to molecular genomics and neuroscience) is that there is a "complex adaptive system" of psychobiological rhythms of growth, self-reflection, and co-creation underlying creative life. Odum (1988) calls this rhythm the "pulsing growth paradigm" of life.

This complex adaptive pattern of life processes underlies what has been called "the wave nature of consciousness" (Rossi, 1986a, 1990c, 1991, 2001; Rossi & Nimmons, 1991). Huxley was successful in his creative work, at least in part, because he followed this natural complex rhythm in his daily lifestyle. The biographies of many creative workers throughout the ages confirm that there is a highly individualized natural pacing of creativity and rest through-out the day. Let's explore emerging research in this area by looking at a typical day's 90–120-minute basic rest–activity cycle. We will discover many hidden connections between these complex adaptive circadian (daily) and ultradian (many times a day) rhythms and our energy levels, moods, sense of stress, exercise (or lack of), diet, accidents, addictions, and even the natural pacing of our sexuality and relationships.

Midnight to 2:00 A.M.—Deep Sleep and Growth Hormone

The new day really begins during the first two hours of sleep when growth hormone is released to facilitate the growth and repair of tissues throughout the body and mind. As we sleep the pineal gland in the brain releases mela-tonin, a major hormone that also modulates growth as well as sleep and dream rhythms, the immune system, and even the aging process. Meanwhile the liver is synthesizing cholesterol. Although high levels of cholesterol can be dangerous, it is a vital substance that forms the basic building block for the daily production of a variety of hormonal messenger molecules that mediate mind-body communication, sex drive, moods, energy levels, consciousness, and productivity.

We dream every hour and a half, or so, while we sleep. Dreams in this first ultradian period of sleep are usually of the more chaotic and bizarre sort that we find most difficult to understand (Hobson, 1997). Yet it may be this earliest and deepest phase of the sleep-dream rhythm is most intimately associated with the primacy of nature guiding us via behavioral state-related gene expression. We speculate that the implicit processing dynamics of these deepest REM state are furthest away from influence by the explicit experiences of daytime consciousness and cognition. The opportunity for mind-body communication usually comes later in our early morning dreams between 4 and 6 A.M., when our dream ego has more coherence and the strength to interact with the presentations of implicit, unconscious processing in the dream. We speculate that these early morning dreams manifest more evidence of activity-dependent gene expression and neurogenesis, as the dreamer actively copes with waking life problems, as so well illustrated by Davina's dreams in Chapter 4.

The current revolution in sleep and dream research typified by the studies reviewed in Chapter 4 (Ribeirio et al., 1999; Strickgold et al., 2000a, 2000b) emphasizes their creative and prospective significance. Sleep is not simply rest and recovery; it is also an active process preparing us for the forthcoming day. Dreams are not simply a review of the past; dreams involve an active replaying and reintegration of the past with the present and future possibilities, if we are sensitive and astute enough to pay them heed. Jung (1946/1966) describes his original realization of the synthetic nature of sleep, dreaming, and psychotherapy when he said, explaining his break away from Freud's emphasis on psycho-*analysis*, "I had first to come to the fundamental realization that analysis, in so far as it is reduction and nothing more, must necessarily be followed by synthesis" (p. 81).

2:00–4:00 A.M.—Ultradian Cycle and Sleep Problems

That there is a natural ultradian 90–120-rhythm of complex adaptation that continues throughout the nighttime is well illustrated by the tendency for some people to awaken every few hours or so throughout the night. When this tendency is not understood, it can have tragic consequences. Who has not seen the painful pictures of the formerly beautiful Elvis Presley, with swollen face and undue weight gain, in the last years of his life when he was addicted to drugs? The story is told in his biography of how he required a handful of barbiturates to put him to sleep at night. Alas, these drugs would soon wear off and he would awaken with a startle after an hour and a half or so of sleep. No one was there to assure him that he would soon drift back into a more natural sleep. In his alarm at awakening so soon and fearing that he would not get enough sleep to perform well the next day, Elvis would knock on his bed board to signal a member of his retinue to come in and give him another handful of "attack pills" to put him back to sleep. And thus it went throughout the night, as the normal psychobiological systems that regulate appetite, food intake, weight gain, growth, healing, memory, learning, creativity, and many other delicately balanced life processes spiraled out of control, until he finally died well before his time.

Contrast Elvis Presley's early demise with the advice of the 96-year-old Nathaniel Kleitman (1963; Kleitman & Rossi, 1992), who described to me how it was entirely natural for some people to awaken every hour and a half or so throughout the night as a natural manifestation of the basic rest–activity cycle. He claimed that this was true especially as people got older. "So what?" he would exclaim. "When I awaken in the middle of the night, I can enjoy it. I can read! I can think! I can even meditate if I feel like it. I can enjoy myself! I would never take drugs to go to sleep! What for? They cannot be good for someone my age" (personal communication, circa 1990).

4:00–6:00 A.M.—Pain, Meditation, and Lucid Dreaming

Milton H. Erickson also had experiences of awakening out of sleep in the middle of the night, most often to deal with pain that arose as an aftereffect of the two bouts of polio he had experienced, first when he was 18 and then again in his early 50s. Erickson (1966/1980, 1980b) was an acknowledged master in dealing with pain problems through the use of his many innovative approaches to therapeutic hypnosis. He was very successful in using self-hypnosis for pain control in his own life—with one exception. When I spoke with him about it when he was in his 70s, he assured me that he was always successful in his personal pain control when he was awake, and he was always comfortable when he fell asleep at night. He also assured me that through the use of posthypnotic suggestion, most of his patients remained pain-free throughout the night. He did not know why, however, but for some reason or other, he would occasionally awaken with pain after the first few hours of sleep or sometime toward morning. Erickson had not heard of ultradian rhythms or the basic rest–activity cycle of an hour and a half, or so. To reassert his pain control, he would have to awaken his wife, who then had to help him sit up in bed so that he could fully awaken, and then listen to him as he described to her his process of self-hypnosis. Erickson told me it was necessary for him sit up, and even use his wife as an audience, in order to awaken fully before he could proceed with pain control in the middle of the night.

Erickson's experience provides anecdotal evidence for the idea that arousal is important in many aspects of therapeutic hypnosis described in the previous chapter. How did Erickson do it? Following is the story of how he used self-hypnosis in the middle of the night, as he told it to me. I can only paraphrase his words from memory. I use ellipses (...) to indicate Erickson's dramatic pauses throughout his description.

Can you imagine how frightening it is for me in my condition to feel an intense, sharp, biting pain going right through my shoulder? I have to get fully awake even in the middle of the night, sit up, and fully experience that pain before I can do anything with it. . . . After a few moments or so, I realize it feels like a hot wire going right through my shoulder . . . then it seems I forget myself for a while . . . and after a while I suddenly notice, what's this? . . . It seems as if the hot wire is now resting on top of my shoulder!

[At this point Erickson glances at me sharply, as if to check whether I understand the significance of this sensory-perceptual shift as the first indication of the spontaneous healing characteristic of therapeutic hypnosis. At no time does Erickson suggest or tell himself what changes will take place during his self-hypnosis. He has no conscious expectancy of how the pain relief will take place. He appears to be simply observing and reporting how an implicate process evolved entirely on its own. The process continues with a life of its own, somewhat as follows.]

I wonder about that hot wire that is now resting on the top of my shoulder. Can you imagine what a great relief it is to me to realize it's only on top of my shoulder, not inside, where it was really frightening to me? . . . And I seem to forget myself for a little while again. . . . Then, what is this!? . . . Now it seems as if the whole top of my shoulder is hot . . . so hot that I can hardly stand it! But it is much better this way because I realize it is starting to spread. . . . After a while I feel the heat spreading down my arm to my elbow. . . . What a relief that is to me, for I now realize that as it spreads, it's less hot. I sense it slowly moving down to my wrist and then my hand . . . by this time its only warm . . . pleasantly warm, in fact.

Pleasant warmth . . . umm . . . with the warmth I begin to feel a pleasant, comfortable fatigue!

[At this point Erickson again glances sharply at me with quizzically arched eyebrows. He wants to know if I am aware of the significance of this feeling of "comfortable fatigue." Of course, I am not, so I silently shake my head slowly, no. Erickson proceeds to explain pedagogically.]

From frightening pain in the center of my shoulder, where the bones may be falling apart, to a hot wire feeling on top of my shoulder. From a hot wire to intense heat all over the shoulder. The heat gradually becomes warmth as it spreads down my arm. The warmth becomes comfortable and I feel fatigue! Warmth, comfort, and fatigue . . . fatigue as in tired . . . tired as in *sleep! I'm beginning to feel comfortable and sleepy!* I tell my wife I am sleepy and to please shift me to a sleeping position . . . thank you, and good night!

For Erickson, self-hypnosis for pain relief in the middle of the night required a sense of arousal and a trust that therapeutic transformations would happen, all by themselves, if he allowed himself to be in a receptive state. Exactly what the therapeutic transformations of his sensory-perceptual and psychobiological sifts would be, however, were left to an implicate or unconscious level of processing. This is very difficult to explain from a conventional theory of hypnosis as a response to suggestion, but it could be explained as a natural result of expectancy. But expectancy of *what*? I would propose that it is an expectancy, learned through long, if unconscious, experience, that the natural ultradian shifts in our basic rest–activity cycle (not that we would use these terms) that will allow us, sooner or later, to slip into the comfort of rest and healing that takes place every hour and a half or two. In the next chapter we will see another example of the heat of transition in therapeutic hypnosis that arises spontaneously and appears to be psychobiological reality rather than a metaphor.

Research has documented that the early morning period around 4–6 A.M. is the lowest point in our daily temperature (that normally varies between 97.88 Fahrenheit [36.3 degrees centigrade] around 4 A.M. to a peak of 98.42 Fahrenheit [36.9 degrees centigrade] at about 11 A.M.). It is no coincidence that the early morning low is the period when there are more traffic accidents due to a lack of attention, while the late-morning peak period in temperature coincides with a peak in mental attention as well as physical strength. It may seem to be a contradiction that most spiritual traditions regard this low period of 4–6 A.M. as the clearest and most beneficial meditative time of the day. This apparent contradiction can be resolved when we realize that their circadian cycle may be shifted because these meditators tend to go to sleep at sundown, so that by 4 A.M. they have already had a full night's sleep. Lucid dreamers, who are able to recognize that they are dreaming while dreaming, also report this early morning period as providing one of the best opportunities for communicating with their inner selves. If this "inner self" is interpreted as a creative source, then this is evidently one of the best times of the day for accessing it.

6:00–8:00 A.M.—Creative Awakening in the Morning

Awakening in the morning is a process of psychobiological arousal generated by behavioral state-related gene expression. We are able to cross the threshold between sleep and waking, in part, because of the activity of many genes associated with informational and energy processes at the cellular levels. The CYP17 gene, for example, codes for a protein that functions as an enzyme for converting cholesterol into cortisol, testosterone, and estradiol. As we noted in Chapter 2 (Figure 2.11), the threshold between sleep and waking is associated with circadian peak levels of cortisol and testosterone at this time. Ridley (1999) regards CYP17 as an example of what he calls a "social gene" because the hormones it codes for are associated with social interactions as well as awakening, performance, stress, and emotional arousal. From our perspective we would regard the activity of CYP17 as an example of behavioral state-related gene expression when it modulates the behavioral transition between the states of sleep and waking. Along with the families of immediate early genes (such as per, jun, and the clock genes), CYP17 modulates the circadian cycle of consciousness itself.

Awakening in the morning is "a call to adventure" from the mythological perspective. A major creative issue in the morning for most people, however, is their lack of awareness of how they changed during their night of sleep. They are usually not aware of how their consciousness, memory, learning, identity, and behavior are continually changing on an implicit level during sleep and dreaming, as we reviewed in Chapters 2, 3 and 4. Many creative workers, however, report that the seed of a new idea is often found in their first few minutes after awakening out of a dream.

The philosopher René Descartes, for example, is said to have enjoyed the luxury of lingering in bed to dwell on these pristine early morning thoughts. His habit of accessing the inner mind-body source led to one of the most important discoveries in all of mathematics. One morning, as he watched a fly crawl across his ceiling, he suddenly realized that it would require only two numbers to locate the fly precisely and thus discovered the famous Cartesian coordinates that became the basis of analytical geometry and higher mathematics.

As our energy and sexual hormones approach their maximum, we awaken. Even the consistency of our blood is a bit thicker as we get ready for daily activity. Yes, theoretically this would be the best time to have sex, even though few of us do. Apparently we are too distracted by what we imagine are all the other important things we are supposed to do this morning. But during honeymoons and vacations . . .

Anyway, within an hour and a half or so, after we awaken, we are already receiving mind-body cues (yawning, stretching, and wanting another cup of coffee) for our first ultradian rest and recovery period. I think this is the time when Huxley sank into his easy chair in his study for his Deep Reflection and would pull together the new that was synthesized within his dreams so that he could express it in his creative writing.

8:00–10:00 A.M.—Prime Creative Working Time for Larks

For larks this is usually the most productive time period of the day for doing creative work. They are still close to the dynamics of gene expression, neurogenesis, and problem-solving that have been taking place in the final stages of sleep and dreaming. In fact, if the creative work that has been taking place on an implicit level during sleeping and dreaming are to be realized, they need to be reinforced as soon as possible after awakening. That is why it is a good idea to write down dreams upon awakening—even before getting out of bed. If one is in the excellent habit of asking the unconscious for help on important questions or issues during sleep, then the first few moments of awakening is the ideal time to wonder if an answer is now more apparent than it was prior to sleep. It is also fine if no specific dream can be recalled. Often, the first few thoughts of the day, whether dream-related or not, are pregnant with the new that needs to be developed further.

Children can be helped to utilize this special waking time as well as adults. One of the first bits of practical wisdom children need to learn is that *"the morning is wiser than the evening."* When first approaching new topics in reading, writing, and arithmetic, so much of it is strange that initially it may seem impossible to grasp. Things that seem so difficult in the evening when they are doing their homework will become clearer in the morning. There is something like a magical genie that helps us learn while we sleep and dream. (Of course, neuroscience now calls the magical genie "gene expression and neurogenesis.") Children can easily be taught to look forward to the new day so they can notice what the genie has helped them learn while sleeping. They are usually delighted to realize how easy it is now to learn what seemed so difficult the previous night.

10:00 A.M. to Noon—Creative Work and Choice

This is the time of the day when consciousness, creativity, and energy are at their daily peak for many people. This is the period when it may be best for some people to exercise, which can turn on and coordinate many of the hormonal messenger molecules and energy systems of mind and body to optimize performance via gene expression in neurogenesis and healing.

Table 6.1 summarizes how many people experience the creative moment of choice between the ultradian healing response and the utradian stress syndrome in everyday life—with little realization of its significance. Most people are simply irritated when they feel tired every hour and a half or two throughout the day. Table 6.1 outlines the two basic alternatives we face during those critical transitions every few hours, when we have a choice of engaging either the *"ultradian stress syndrome"* or the *"ultradian healing response."*

The poem "Vacillation," written by William Butler Yeats in 1934, describes a breakthrough experience of choice and transition.

My fiftieth year has come and gone,
I sat, a solitary man,
In a crowded London shop,
An open book and empty cup
On the marble table-top.

While on the shop and street I gazed
My body of a sudden blazed;
And twenty minutes more or less
It seemed, so great my happiness,
That I was blessed and could bless.

TABLE 6.1 | THE CREATIVE CHOICE BETWEEN THE ULTRADIAN HEALING RESPONSE AND THE ULTRADIAN STRESS SYNDROME IN EVERYDAY LIFE.

The ultradian healing response	The ultradian stress syndrome
1. *Preparation—recognition signals:* An acceptance of nature's call to rest and recover strength and well-being that leads to an experience of comfort and thankfulness.	1. *Preparation—take-a-break signals:* A rejection of nature's call to rest and recover strength and well-being that leads into an experience of stress and fatigue.
2. *Incubation—accessing a deeper breath:* A spontaneous deeper breath comes all by itself after a few moments of rest as a signal of slipping into a state of relaxation and healing. Now is the time to explore the deepening feeling of comfort that comes spontaneously and to wonder about the possibilities of mind-gene communication and healing with an attitude of "dispassionate compassion."	2. *Incubation—high on hormones:* Continuing effort in the face of fatigue leads to the release of stress hormones that short-circuit the need for ultradian rest. Performance goes up briefly but at the expense of hidden wear and tear and further stress. A need for artificial stimulants (caffeine, nicotine, alcohol, cocaine, etc.) becomes urgent, and a process of addiction, involving uncontrolled highs and lows, is established.
3. *Illumination—mindbody healing:* Spontaneous fantasy, memory, feeling-toned complexes, active imagination, and numinous states of being are orchestrated for healing and life reframing.	3. *Stress—malfunction junction:* Many mistakes creep into performance, memory, and learning; emotional problems become manifest, including depression or irritability and possibly abusive behavior toward self and others.
4. *Verification—rejuvenation and awakening:* A natural awakening with feelings of serenity, clarity, and healing, together with a sense of how to enhance creativity and performance.	4. *Verification—the rebellious body:* Classical psychosomatic symptoms now intrude and rest becomes essential. There is a nagging sense of failure, depression, and illness.

It is interesting that there is a cultural history of 20 minutes as a typical time period of peaks, as well as troughs, in our moods in everyday life (Deikman, 1980; Freed, 1991; Pennebaker, 1997; Smith & Tart, 1998). We hypothesize that human clock genes are entrainable or modifiable in their expression to some degree by psychosocial cues but not by much—perhaps about 20 minutes, or so. It would be an interesting research project to determine if such entrainment windows occur during the 20-minute peaks and troughs of the basic rest–activity cycle. Stress begins to build when we try to stretch the highs and lows too far beyond their natural limits.

Noon to 2:00 P.M.—Lunch and the Ultradian Diet

An ultradian approach to diet consists of six small meals of 200 to 300 well-balanced calories spaced throughout the day to coincide with our natural 90–120-minute hunger rhythm. That such an ultradian eating pattern lowers cholesterol, reduces the need for insulin, and is generally health-promoting has been well documented by the Jenkins research group in *The New England Journal of Medicine* (1989). People can enjoy a light meal and then reward themselves with a 20-minute ultradian healing period, during which the body will make an efficient beginning in digesting the food to the point where they feel completely satisfied.

After lunch we tend to experience the post-lunch dip in our consciousness and alertness. Most of us would like to take a comfortable nap about this time—which brings up the very interesting research that has been done recently in this area of napping.

2:00–4:00 P.M.—Creative Naps and the Ultradian Healing Response

Research on sleep and alertness by David Dinges and Roger Broughton (1989) uncovered fascinating findings on the ultradian dynamics of naps. If we take a nap in the morning, we are more likely to experience a large proportion of the kind of REM sleep associated with dreaming. If we take an afternoon nap, it will contain a larger proportion of the kind of deeper slow-wave sleep associated with the release of growth hormone and its contributions to optimal functioning of the immune system as well as repair and rejuvenation. While many seniors complain that they cannot get a full unbroken eight hours of sleep at night, current research documents how, contrary to popular belief, those who sleep eight hours or more at night actually die sooner than those who sleep six hours or less (Kripke et al., 2002). Further research is required to clarify the controversial issues involved in this surprising finding.

Seniors really have to do more than nap their way to well-being, however. As we age, we tend to fall into a newly recognized condition called "low amplitude dysrhythmia" (Lloyd & Rossi, 1992). If we do not challenge ourselves a bit with exercise and effort every day, we won't push ourselves to peaks of top performance, as illustrated in the ultradian performance rhythm in Figure 2.11. As we know, this view is entirely consistent with neuroscience research that documents how novelty, enrichment, and physical exercise are needed for optimal gene expression and neurogenesis throughout the day.

Insufficient exercise throughout the day has hidden but insidious consequences because there is less need to descend into deep slow-wave sleep at night, which is required for the growth hormone production that is used for tissue repair and overall rejuvenation. A "good night's sleep" is not experienced, leading to less or no exercise the following day and even less need to go into deep sleep that night. The so-called "low amplitude dysrhythmia" sets in, perhaps leading to depression, another condition in which low amplitude dysrhythmia is apparent.

Exercise is currently regarded as an important component for dealing with depression and some aspects of the aging process. For young people, aerobic exercise for about an hour a day is optimal; but research is not yet consistent about what is best for older people. A full appreciation of the implications of our ultradian 90-120-minute basic rest–activity cycle, however, suggests that it may be better for older adults to enjoy 15–20-minute exercise periods three or four times a day rather than an exhausting effort for an hour or more once a day that might overtax the body.

4:00–6:00 P.M.—Breaking Point, Nurture, and Healing

This the period that some biologists call the "breaking point," because it is a time of important shifts in gene expression, neurogenesis, and psychobiological experience on many levels. This is not the time to engage the outer world in challenging and difficult tasks. In India there is an old saying, "If you want to make an enemy of a friend, plan to meet at four in the afternoon."

Many people report that this late-afternoon period of their day as offering the deepest ultradian healing-restoration period of their day. Many of the psychotherapeutic methods of free association, inner dialogue, active imagination, and self-hypnosis seem to be particularly profound at this time period (Tsuji & Kobayshi, 1998). If one has been skipping one's rest-rejuvenation periods throughout the day, however, this is typically the time when energy and emotions crash. The misinformed interpret this emergency state of physical fatigue as a psychological depression that means they are guilty of having done something wrong—they feel they are just not good enough. Self-esteem sinks to zero in spite of all efforts to bolster it. The glass always appears half empty instead of half full. This is not the time to make decisions and plans but a time for self-nurture and mutual support from family and friends.

In the precarious state of being stressed, overtired, and worried toward the end of the workday, some people try to bolster themselves artificially with drugs or too much food. From the psychobiological perspective, this pattern of overriding our ultradian need for rest and rejuvenation is the basis of our addictive society. The so-called "happy hour" in the cocktail lounge is anything but that. People need gentle social support and nurture, not artificial chemical stimulation at this time. Twelve-step programs, such as Alcoholics Anonymous, provide social support precisely at the time, when addictive behavior is most tempting. This is the time to turn to family and friends for support and healing. It may also be a good time to practice the final ultradian healing response of the day.

6:00–8:00 P.M.—Family, Social Support, and Healing

This period in the late afternoon, dinner, and early evening is the time of day when most families reassemble, for better or worse. Returning to a happy, supportive family can be a welcome and healing period. It is an optimal time for nurturing others and ourselves. However, if we have ignored our ultradian calls for restoration and healing throughout the day and return home overwhelmed by stress and fatigue, this period can be a time of misunderstanding, arguments, and mental and physical abuse.

8:00–10:00 P.M.—Social Engagement and Self-Nurture

This evening period may be a quiet time of family togetherness, solitary journaling or meditation, hobbies, or reading; it may, on occasion, involve more extraverted activities such as attending the theatre, movies, symphony, or social affairs. From the creative perspective all these activities, from the most solitary to the most social, can be best conceptualized as *implicit processing heuristics*. They are all sending messages and cues as well as providing information that, hopefully, will evoke creative replays of gene expression and neurogenesis in our sleep and dreams. For larks, activities at this time can be best understood as preparation for the the rest and rejuvination that will hopefully take place during sleep. For owls, however, this is the best period for creative work and/or play. Owls can be the life of the parties in the evening and night. Larks need the courage to recognize they need not try to compete with owls by overstimulating themselves at this time. This period offers important clues to couples contemplating marriage. If one is an owl and the other is a lark, problems are likely to pop up at this time of day. One wants to party while the other wants to rest. Wisdom suggests they work this out in the courtship period, if they hope to enjoy a happy marriage.

10:00 P.M. to Midnight—Engaging the Creative Dream Cycle

During this time the mind-body prepares for sleep and rejuvenation at the deepest levels. There is a curious peak in cell division, for example, around 10 P.M. that seems to set up natural periods of growth and healing as we approach sleep. Recall that this is the period during which Huxley enjoyed his relaxed browsing in his home library. There can be something profound about comfort and quiet enjoyment during this final ultradian period of the day. Nature is telling us that it is okay to relax our grip on the outer world. She needs the time to do her inner work of healing and re-creation at the deepest levels of sleep and dream that we will soon enter. Clock genes and the natural process of behavioral state-related gene expression (Chapter 2) are initiating the dynamics of synthesizing the new proteins and hormones that will lead to rest, rejuvenation, and healing during sleep.

What Is Art?

What could art be other than the facilitation of the novelty-numinosum-neurogenesis effect in our positive experiences of creating a better brain? An anecdote from the life of Miró illustrates how the four-stage creative cycle is played out with infinite variety in experiencing novel and numinous art. Schneider (1959) describes how in the 1920s, Jacques Doucet, a leading *couturier* of Paris, was brought to the studio of the young Miró by two friends—artist André Masson and poet Louis Aragon.

> The elegant old couturier sat down, flanked by Masson and Aragon, while Jacques Viot, Miró's dealer, stood in a corner. One after the other, Miró placed his canvases on the easel. The first were still lifes painted in a sharp, crystalline manner derived from cubism. "Quite incisive," M. Doucet commented curtly. But then, as Miró began bringing out painting upon painting from which ordinary reality—even in the fragmented, prismatic version in which the cubists had represented it—had vanished, Doucet lapsed into silence [*stage one: Doucet preoccupied with the new art*]. It lasted for an excruciating hour or two [*90–120-minute ultradian basic rest–activity cycle*]. Not until he was in the street again with his two guides did Doucet speak up: "Well, my young friends, you have fooled me often, but you won't fool me this time. Your friend is mad, stark mad, and that man in the corner was his keeper. While in there, I said nothing, for fear that he might grow excited. Why, he might have taken his brush and splashed it on my face!" [*stage 2: negativity, conflict, stress and emotions*]. Three weeks later, Masson met the art patron again. "You know, I have bought two paintings by that friend of yours," said Doucet. And he added, " I have hung them at the foot of my bed. When I wake up in the morning, I see them, and I am happy for the rest of the day" [*stage 3: valuing the new; stage four: the new facilitates a better life*]. Very simply, the aged dress designer had put his finger on the secret of Miró's art and of its appeal: a spontaneous, vivid, fresh elation, instinctively felt and immediatly contagious. It is a feeling experienced unconsciously—sometimes even unwillingly—by every spectator [*excellent description of numinous experience*]. (p.70, italics added)

SUMMARY

This chapter began with a review of Hans Selye's classical conception of stress and then explored emerging views of how the positive psychology of optimal performance and the basic rest–activity cycle are associated with the four-stage creative cycle. Brief periods of manageable stress can facilitate optimal performance and healing, whereas chronic stress leads to malfunction, injury, aging, and ultimately death. We generalized the four-stage creative cycle as the *breakout heuristic* that provides an overview of the essential psychobiological dynamics and *raison d'être* of cultural rituals of transformation throughout human history. Finally, we created an ultradian outline for optimizing creative psychogenomic experience to give a new answer to the eternal question: What is art?

In the next two chapters we explore how to facilitate the positive psychology of healing and problem-solving with a new cultural ritual that is described as *the experiential theater of demonstration therapy*. We will document how the novelty-numinosum-neurogenesis effect can be facilitated by the new neuroscience of positive psychotherapy, therapeutic hypnosis, and the healing arts.

Chapter 7

THE EXPERIENTIAL THEATER OF DEMONSTRATION THERAPY: PART A

The Preparation and Incubation Stages of Creative Work

"The induction and maintenance of a trance serve to provide a special psychological state in which patients can re-associate and reorganize their inner psychological complexities and utilize their own capacities in a manner in accord with their own experiential life. Hypnosis does not change people nor does it alter their past experiential life. It serves to permit them to learn more about themselves and to express themselves more adequately."

—Milton H. Erickson
"Hypnotic Psychotherapy"

Introductory college courses in the humanities remind us that theater, drama, art, and literature over the millennia had their origins in bringing the community together for mutual experiences of sharing, bonding and healing. Throughout human history, indigenous societies around the globe have had hubs of social and spiritual communion. Current research in the psychobiology of gene expression annotates these deep cultural traditions with a profound view of how mind and molecule modulate each other on multiple levels during group experience.

Schools of psychotherapy and the theater have tiptoed around each other, with mutual wariness, over the past hundred years. The original theater school of Stanaslavski in Russia and its offshoots in the teaching of Strasberg and the Group Theater in America were adopted by the new psychodrama of Moreno, the Gestalt approach to psychotherapy of Fritz Pearls and many of its stepchildren in Transactional Analysis, redecision therapy, and the concept of demonstration therapy, which I originally discussed with Carl Rogers two decades ago.

Therapeutic hypnosis, meanwhile, had its origin in the *baquet* of Mesmer, a kind of dramatic group therapy of altered experiential states that was believed to facilitate healing. Today, deeply engaging experiential states that have much in common with drama are used in training clinicians to facilitate the process of psychotherapy (Rossi, 1972, 1985, 2000a; Zeig, 1997). After two centuries of pursuing every-which-way theory, we are now in a position to investigate how positively oriented therapeutic hypnosis evokes the novelty-numinosum-neurogenesis effect underlying healing. The experiential demonstration of a positive, activity-dependent approach to therapeutic hypnosis explored in this chapter is another step in that direction.

THE PARADOXES OF DEMONSTRATION THERAPY IN THE HEALING ARTS

In a huge auditorium about a thousand therapists gathered to observe a demonstration of therapeutic hypnosis at a professional teaching workshop sponsored by the Erickson Foundation. The late Milton H. Erickson, M.D., who was the founder of the American Society of Clinical Hypnosis, was the prototypical teacher who reactivated an interest in the medical applications of therapeutic hypnosis two generations ago with teaching seminars of this sort. These educational workshops can be attended only by professional therapists who have a license to practice their clinical specialty (medicine, dentistry, psychology, psychiatry, social work, etc.), and students training in these areas.

There are many paradoxes about these teaching demonstrations or "demonstration psychotherapy." My first awareness of these paradoxes arose during conversations with the late Carl Rogers, who emphasized that these demonstrations were actually a special modality of the healing arts that should be carefully studied and perhaps licensed in their own right. Rogers recalled that in his more than 30 years of experience conducting such demonstrations, he found that many subjects (clients, patients), seen only once, responded very quickly and had profound therapeutic experiences which they vividly recalled for the rest of their lives. Why should this be true in a profession where a common practice is to have a (1) lengthy number of leisurely therapeutic sessions, (2) spread sometimes over years, and (3) conducted in strict privacy that is tightly bounded by laws of "professional confidentiality"?

Our interpretation of current neuroscience research suggests that demonstration therapy may be effective precisely because it provides a positive psychosocial experience of novelty, environmental enrichment, and exercise that are prime conditions for generating new memory, learning, and behavior. The psychobiological arousal experienced by revealing oneself in front of a large audience can evoke natural processes associated with immediate early genes, behavioral state-related gene expression, as well as activity–dependent gene expression and neurogenesis. A once-in-a-lifetime experience with demonstration therapy is a special psychosocial event that has much in common with the healing rituals of many cultures. These positive cultural rituals range from the challenges of adolescent transition to adulthood, such as the walkabout of Australian Aborigines, to the ceremonies of shamanism and spiritual practices as well as prayer, psychotherapy, and meditation. The positive psychosocial genomics evoked by these cultural practices vary widely in relation to *levels of arousal evoked:* from the heightened alertness commanded by loud drums, dancing, and music to the apparent opposite in Quaker quietude, prayer, and the relaxation response of the classical hypnosis initiated with suggestions for sleep and comfort. It initially may seem curious to express it in this way, but from our current psychobiological perspective, prayer and spiritually oriented cultural practices may be regarded as creative mind-body replays of psychosocial genomics that, in the best of circumstances, could generate real healing on a molecular level.

The common denominator, the bottom line, that these varying cultural healing practices share are the phenomenological experiences of what current neuroscience identifies as novelty, experiential enrichment, and exercise. Until now it was not understood how the cornucopia of therapeutic practices, invented and reinvented by different cultures with different philosophies and belief systems, could propose apparently opposite approaches that seemed to facilitate a common core of healing. We now recognize that the apparently opposite approaches (e.g., arousal versus relaxation) engage different phases of the same psychogenomic process. The resolution of many paradoxes of healing can be found in the common core of gene expression, neurogenesis, and healing at the cellular level that can be evoked by the dynamic human experience. *Replaying with creative variations the psychosocial genomics between consciousness, behavior, and gene expression, generates the structural, energetic, and informational dynamics of life* as a complex adaptive system (Rossi, 1996a).

While the demonstration of therapeutic hypnosis that follows takes an exploratory step toward the integration of traditional practices of healing with modern research in neuroscience, we must also emphasize its limitations.* There are, as yet, no well-controlled, peer-reviewed, published clinical studies documenting the sometimes irreverent and idiosyncratic approaches used in this demonstration. It is presented here for its possible heuristic value in inspiring neuroscience research into the conditions needed for facilitating psychosocial genomics in our evolving Western approaches to healing. While this demonstration may have some value in teaching and exploring the issues of "demonstration psychotherapy," it cannot be recommended as a general model for the therapeutic arts until the requisite clinical trials of its efficacy are well replicated.

A POSITIVE ORIENTATION
VIA CREATIVE ACCESSING QUESTIONS

**Facilitating Immediate Early Genes and Behavioral
State–Related Gene Expression [time on videotape: 00:45]**

ROSSI: For most of us, hypnosis is really about healing . . . is it not? So I would like to ask if there is anyone in the audience this afternoon who is really in *an acute state of distress? Chronic distress, physical, mental, pain?* Someone who has really got *an issue that they feel they can do some effective work with this afternoon?* So, this is not merely a demonstration, this is the real thing!

Notice that in these seemingly casual introductory remarks to the audience, before I've even met a volunteer, I ask a series of positively oriented questions with the intent of focusing everyone's attention on what we are doing and who should volunteer. I began with the positive statement/question, (1) *". . . hypnosis is really about healing . . . is it not?"* Erickson regarded the peculiar form of this type of question that ends with *"is it not?"* as a mild

---

* This demonstration is available on videotape, "A Sensitive Fail-safe Approach to Hypnosis" (IC-92-D-V9), and is availble from the Milton Erikson Foundation: office@erickson-foundation.org; www.erickson-foundation.org

therapeutic double-bind that disarms potential resistance by stating the negative as well as the positive, so that the subject does not have to waste any effort in negation and denial (Erickson & Rossi, 1975).

I do not ask for just anyone to volunteer. My apparently open-ended request for a volunteer was actually very specific in asking for someone who was (2) in an "*acute* state of distress . . . *chronic distress, physical, mental, pain?*" I use the word *acute* because I would like to demonstrate with a volunteer who is already disposed toward a psychobiological state of arousal characteristic of stage two of the creative process (see Figure 2.11). I then become even more specific and focus sharply on the positive point of asking for (3) "someone who has really got *an issue that they feel they can do some effective work with this afternoon.*" For a brief teaching demonstration, I definitely want to work with someone who is ready to "*do some effective work . . . this afternoon.*" I then raise the ante of (4) positive psychobiological arousal and heightened expectation even further with my final strong injunction, "So, this not merely a demonstration, this is the real thing!"

The apparently casual and permissive attitude of the first three questions facilitates a *positive, receptive mental attitude,* a kind of *yes set* (Erickson, Rossi, & Rossi, 1976), that is sharply punctuated by a statement, delivered as an exclamation point, that in itself tends to facilitate a positive state of highly focused psychobiological arousal—on "the real thing!" From the neuroscience perspective, it will be important to determine the degree to which immediate early genes and behavioral state-related gene expression are actually evoked with such introductory remarks in this initial *preparation stage of the creative process.* From a clinical perspective it will be interesting for readers to determine for themselves the degree to which the purpose of these four introductory questions/remarks was realized in this therapeutic session.

A RELUCTANT VOLUNTEER
EVOKES AN ACTION-ORIENTED THERAPIST

Minimal, Nonverbal Cues and Positive Expectation

As I gaze into the audience looking for a volunteer, I notice a sudden flurry of activity in the front rows off to one side. I immediately recognize that a small group of people seems to be gently pulling and pushing a young woman, apparently in an effort to encourage her to volunteer. I quickly look away from this small group to the general audience, since I prefer not to encourage or influence the commotion. When I glance back a moment later, however, I see that the young woman is now hesitantly walking down the aisle and coming up the steps of the stage. She seats herself next to me and signs the professional release document that is required. Without a word I catch her eye and then gaze pointedly at the small lapel microphone that is on the table between us. She picks up the microphone and attaches it to herself. She sits back and looks expectantly at me.

This was our first nonverbal, action-oriented transaction. Without quite recognizing its implications, she has shifted from reluctant to willing volunteer in the course of at least five overt acts of cooperation: She walked down the aisle and up the steps of the stage, she sat down next to me, she signed the release document, she picked up my nonverbal cues and attached the microphone to herself, and perhaps most importantly, she gazed at me with that apparently *positive expectation* that Erickson called "response attentiveness." She responds clearly and concisely to my initial verbal queries. If therapeutic hypnosis is defined as a state of highly focused attention, then we could say she is already in a light therapeutic trance (Erickson, 1964a/ 1980).

STAGE ONE:
PREPARATION: FACILIATING THE NOVELTY-NUMINOSUM-NEUROGENESIS EFFECT

Basic Accessing Questions and Implicit Processing Heuristics [3:12]

> ROSSI: [glancing at her nametag] I see your name is Celeste?
> CELESTE: That's right.
> ROSSI: How can you be helped today?

The first introductory question about her name confirms her yes set, and the second immediately focuses her attention and expectancy of a therapeutic process in this highly *novel and unusual (i.e., arousing) setting* of being on stage, in the spotlight of a video camera, in front of a large crowd of people. This stage setting tends to evoke the novelty-numinosum-neurogenesis effect. While they may seem to be very *permissive* and general, these basic accessing questions evoke the problematic *state-dependent memory, learning, and behavior systems* she needs to *replay and creatively reframe* in this therapeutic encounter. These questions tend to evoke and set in motion whatever specific motivations and inner resources she has for healing. We all have an acquired expectancy that important events transpire in front of documentary video cameras. Utilizing this *expectancy* is a very powerful stimulus, an *implicit processing heuristic,* for *facilitating psychobiological arousal and meaningful transformations.* This is the essence of the therapist's task in this preparation stage of the creative cycle: Asking basic accessing questions that will launch patients into their own inner journey of problem-solving and healing.

> CELESTE: I have rheumatoid arthritis, and today I am especially flared up from it. I'm not sure exactly why, it might be the rain or it might be a cold I've had, but in general I'm just concerned about the rheumatoid.
>
> ROSSI: How long have you had this?

This is another apparently simple, obvious, explicit question. On an implicit level, however, it initiates an inner process of *self-reflection, recall, replay, and the possibility of new learning and therapeutic resynthesis.* Such initial questions, in the process of history-taking in a culturally sanctioned therapeutic context, are aimed at more than fact-finding. They are basic accessing questions, a class of *implicit processing heuristics,* which initiate the first stages of the psychobiology of the creative cycle. The therapeutic intent of these questions is to help people find their own creative edge via an experience of the novelty-numinosum-neurgenesis effect.

> CELESTE: About 10 years?
>
> ROSSI: What kind of therapy have you had?

This is another basic accessing question searching for attitudes, experiences, and inner resources that could be utilized to facilitate the therapeutic encounter.

> CELESTE: Western medicine. Lots of medicine. Acupuncture.
>
> ROSSI: You've had thorough Western traditional medical work-
> ups and you are really diagnosed with rheumatoid arthritis?

Celeste responds immediately and concisely by identifying her broad worldview that embraces Western medicine as well as an alternative approach, such as acupuncture. Being a psychologist, I am always concerned that people have had a Western medical examination and diagnosis when significant medical symptoms are involved. I prefer to have some form of medical assurance that a psychobiological approach is warranted as part of a team approach.

> CELESTE: Right. Yeah, and I've learned over 10 years that
> Western medicine only works with symptoms, and it hasn't really
> done anything for any kind of *healing*. Of that, I'm certain. So,
> I've tried acupuncture and I've tried a lot of different things, but
> I've found it very difficult to have it penetrate.

She now definitely identifies her frustration with what she perceives to be the limitations of traditional Western medical practice and perhaps part of her motivation for attending this hypnosis training workshop. I realize I will have to take special care to accept and utilize her attitudes, while mediating between them and my own more generous conception of the current and future prospects of healing within the neuroscience worldview of Western medicine. I notice, however, that she uses the word *healing*, which I used in my first introductory sentence to the audience. This may suggest that we already share much in our therapeutic worldview and augurs well for our therapeutic alliance, but I must continue with further basic accessing questions to deepen our developing rapport.

> ROSSI: You've tried Western medicine. You've tried acupuncture.
> What else?

Here I'm searching for more information as well as for mutual accord so that we can achieve more of a mind-meld (to borrow a term from *Star Trek*). I continue the permissive, nondirective style of replaying and repeating her words. In this type of demonstration therapy, my repetition of her soft-spoken words is as much to reassure myself and the audience that I have heard her correctly as it is a use of the reflective technique of Carl Rogers' client-centered therapy.

From the perspective of neuroscience research, however, this repetition is a kind of *replay*—a type of *implicit processing heuristic*—that facilitates the recall, reassociation, reorganization, and resynthesis that may optimize activity-dependent gene expression and neurogenesis on a molecular level.

> CELESTE: Well, I was in Brazil one year and I saw a number of healers. I did some energy work there and there was some effect from that.

She confirms her sophistication as a seeker of alternative healing approaches and provides an important indication that she can receive some therapeutic "effect." This augurs well for our work, but it may also foreshadow difficulties if she has learned therapeutic techniques that are not optimal for her progress. Previous therapeutic approaches that she may be applying to herself by rote could interfere with the creative dynamics of the novelty-numinosum-neurogenesis effect we hope to facilitate in this session. Nonetheless, I feel comfortable enough to take an exploratory step by introducing *symptom-scaling* as an *implicit processing heuristic* that may facilitate self-reflection and inner work.

STAGE TWO:
SYMPTOM-SCALING AND INCUBATION

The Induction of Therapeutic Hypnosis? [4:44]

> ROSSI: On a scale from 1 to 100, where 100 is the worst you have ever experienced your symptoms, Celeste, what are you experiencing right now?
> CELESTE: About 75.
> ROSSI: Very good. Okay, we've really got something to work with!
> CELESTE: Oh, yeah!

Celeste smiles broadly and laughs at this point. An appreciative murmur goes through the audience. It may seem paradoxical that everyone seems pleased with a rating of 75 subjective units of discomfort (SUD) out of a possible 100. If there is any rational reason to be pleased, it is that the lore of therapeutic hypnosis and current neuroscience research (Dudai, 2000; Shimizu et al., 2000) maintains that a person needs to experience at least a little bit of the state-dependent encoding of a symptom in order to access and replay it in a therapeutic manner. The problem has to be experienced in the here-and-now before patient and therapist can work with it. That is, problems are *state-dependent*; the patient must be experiencing the psychobiological state that encoded the symptom or problem before it can be accessed, replayed, and reframed in a way that is experienced as an insight, problem resolution, cure or healing (Erickson et al., 1986; Rossi, 1986/1993; Rossi & Cheek, 1988). Every recall accesses and "opens" the state-dependent, psychobiological encoding of memory and learning so that it may be modified, restructured, reframed, and re-encoded in a new way (Nader et al., 2000a, 2000b). In this sense, symptom-scaling is an implicit processing heuristic that initiates a psychodynamic process of *self-reflection, inner work, and incubation* characteristic of stage two of the creative cycle.

Does symptom-scaling initiate a process of focusing attention that is characteristic of hypnotic induction? An answer to this question depends on how we define hypnosis. If therapeutic hypnosis is regarded as a process of focused attention wherein the state-dependent encoding of a problem may be accessed and replayed for reframing and resolution, then, certainly, symptom-scaling is an excellent means of initiating it. Symptom-scaling is particularly interesting in this regard because it may be an approach to integrating left and right cerebral hemispheres and bridging the problematic dissociations between them. Research on the use of brain imaging supports the idea that hypnosis may, in fact, integrate various areas of brain processing (Rainville et al., 1997, 1999). This is a research area that clearly needs to be developed further in the future.

ROSSI: All right! *Whew.*

I exhale a slow, barely audible sigh of relaxation that is typical of my personal way of being. It may serve as an indirect cue for Celeste to relax, but that was not my intention. Celeste's successful experience of symptom-scaling is an initial step on the path of accessing the state-dependent encoding of her symptom. Sympton-scaling tends to initiate a mind-body connection that will make further progress along the symptom path possible. From the perspective of traditional therapeutic hypnosis, her hypnotic induction has been successful—symptom-scaling requires an inner focus of attention that is now evident as she successfully engages this novel experience of assessing the intensity of her symptom at this moment. From the traditional perspective of therapeutic hypnosis, the next step is to deepen her inner self-engagement to facilitate the novelty-numinosum-neurogenesis effect. How could that be done here?

This is the essence of the therapist's task in stage two of the creative cycle: recognizing and supporting the patient's state-dependent encoding of symptoms and problems as well as the inner resources for dealing with them. On the videotape I now appear momentarily distracted as I look off in reflective thought, searching for what to do next to further engage her in the inner work on her symptom. I tentatively decide to explore what I call "the symptom path to enlightenment" (Rossi, 1996a), which typically begins with a direct focus on the symptom that the patient is experiencing in the here-and-now.

AN ACTIVITY-DEPENDENT, SYMPTOM PATH APPROACH TO PSYCHOBIOLOGICAL AROUSAL

Minimal Cues and Symmetry-Breaking to Facilitate the Novelty-Numinosum-Neurogenesis Effect [5:19]

ROSSI: I notice your hands.

This is a simple observation that shifts her mind-body focus to the symptom area in order to assess whether she is receptive and ready to explore her manifest symptom directly.

CELESTE: Uh-huh. That's where it shows the most.

Indeed, the video camera records the severe swelling and distortions in the rheumatoid structure of the joints of her fingers and hands. With Celeste's receptive remark, *"That's where it shows the most,"* I immediately recognize that she probably is ready to work on her arthritis and that her hands could be the ideal locus for an *activity-based symptom path approach* to focusing her attention, using the naturalistic, utilization modality of therapeutic hypnosis (Erickson, 1958/1980, 1959/1980; Rossi, 1996a).

I swivel in my chair to face her directly and silently model a symmetrical position of resting my hands very lightly on my lap. With a glance or two back and forth from my hands to Celeste's hands, I nonverbally signal Celeste to do the same—and she does (Figure 7.1). I usually begin with a symmetrical hand position in which the hands are held a few inches above the lap, but in this case, not yet sure of her strength, I indicate a position barely touching the lap. A full resting position on the lap is not a desirable way to begin, since it implies that the client will rest in quiet repose while the therapist does all the active verbal therapeutic work, as is typical in traditional hypnosis. Here, however, I am demonstrating an *activity-dependent approach to psychobiological arousal* with the goal of evoking the activity-dependent gene expression, neurogenesis, and the possibility of healing.

Detecting and offering minimal nonverbal cues have been described as highly efficacious for facilitating therapeutic hypnosis (Erickson, 1964a/1980). What is so special about minimal nonverbal cues? They could function as *implicit processing heuristics* insofar as they may operate on a more implicit and/or right-hemispheric level, below the explicit level of consciousness.

> ROSSI: Yes. Can you turn toward the audience so that you can
> . . . it is okay to share some of that? There are really some obvi-
> ous distortions you can see in the joints of your fingers.

Up to now Celeste has been very circumspect in showing her hands. It appears that she habitually tries to hide them from view. My question asking her to expose her hands to everyone's scrutiny, followed by my bold statement of their condition, certainly activates further psychobiological arousal. I am "upping the ante" by seeking to confirm her willingness to become more engaged in the *activity-dependent therapeutic process*. She easily complies

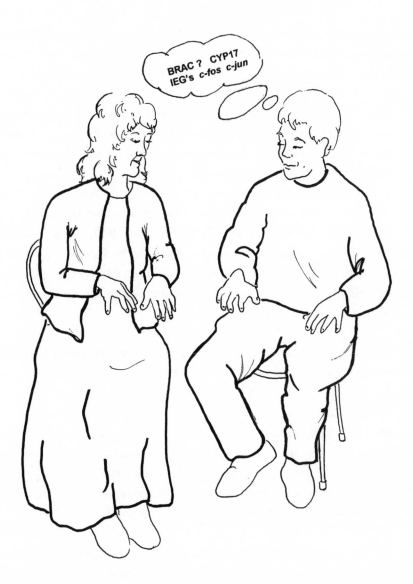

FIGURE 7.1 | Stage one: The therapist models a delicately balanced and symmetrical hand position a few inches above the lap to engage an activity-dependent approach to therapeutic hypnosis and the creative cycle. The therapist explores what stage of the basic rest–activity cycle (BRAC) the patient may be experiencing initially in order to facilitate further progress. He wonders whether CYP17—the social gene—is becoming engaged as a natural manifestation of the psychotherapeutic transference. Immediate early genes (IEG's), such as c-fos and c-jun— associated with a creative state of psychogenomic arousal, problem-solving, and healing—will become engaged to the extent that meaningful novelty, enrichment, and exercise are facilitated.

with this request and even elaborates on a physically revealing description of her arthritic hands. We can now say that a therapeutic temenos has developed, containing us as partners in our quest for her healing.

> CELESTE: Yes. I've got a couple of tendons that are ruptured and a couple of joints that are swollen. And my hands are . . . I can get it all over my body, but my hands are the place that it is usually at most of the time, or all the time.
> ROSSI: I see, most of the time or all the time? And this is where it is about 75% level of intensity right now . . . or is it . . . ?
> CELESTE: Um. I also feel it throughout my body. So it's kind of an achiness. I would say it is about 75% in both tendons. . . . It is an achy, fluish feeling in my hands and feet. My feet get it, too.
> ROSSI: Okay. I like that you began with a description of an aching, fluidish feeling?

Here I make a mistake in hearing what Celeste is saying—I am confused between *fluish* and *fluidish*. She now carefully corrects me.

> CELESTE: Fluish.
> ROSSI: Fluish! Okay.
> CELESTE: Yes, like I have a virus or something.
> ROSSI: Oh, I see, like a flu. Uh-huh. *Stay with it now*, Celeste, tell me just where you are experiencing it and give me a little commentary on what exactly you are experiencing.

My words are intended to launch Celeste on the "symptom path to enlightenment" (Rossi, 1996a). Focusing in a receptive manner on the sensations of the actual sensations of symptoms being experienced in the here-and-now tends to evoke the state-dependent, psychobiological encoding and psychodynamic associations to the symptom. The symptom path can be regarded as an implicit processing heuristic that accesses a problem area for resynthesis and healing.

Unfortunately, I make a verbal error in using the phrase "stay with it now." Of course, I don't really want her to *stay with* any sensation or experience, even if it were possible. It would be better to facilitate her ongoing process with something like "going with that" or "continue exploring whatever comes all by itself until . . ."implying that her "fluish" sensations would change, as they must. It is precisely in this Darwinian process of natural variation and selection on all levels—from gene expression and neurogenesis to shifts in sensations, emotions, and thinking—that give rise to the possibilities of therapeutic transformations.

CELESTE: Right now?

ROSSI: Yes!

CELESTE: [pause] My shoulders ache.

RECALL, REPLAY, AND THERAPEUTIC TRANSFORMATIONS

Time-Binding and Incomplete Sentences as Implicit Processing Heuristics [7:03]

ROSSI: Stay with that ache for a moment Celeste and let's see what happens with it when you focus on it now.

Long pause as she apparently reflects privately. Then she swallows. I would now prefer to say, "Go with that ache just for a second or two until . . . ?" rather than, "stay with . . ." which, as noted, has possibly unfortunate connotations of unchanging conditions.

"Going with . . . until . . . ?" is a time-binding implicit processing heuristic that sets in motion an inner process of sensory-perceptual-motor search that usually evokes a change in the experience and state-dependent encoding of symptoms and problematic states. Tacking on the word until, said in a questioning tone and in an incomplete sentence, tends to facilitate a nondirective, reassuring, time-binding, safety factor that implies the symptom will continue but only for a "second or two until . . ." implying the inevitability that something interesting and unexpected will happen that may be of therapeutic value.

ROSSI: That's right. And what are you experiencing now?

RESISTANCE AND DISSOCIATION
VERSUS CREATIVE REPLAY

Open-ended Questions and Therapeutic Double Binds [7:42]

CELESTE: Uh, at first *I resisted feeling* it because I think, lots of times, I kind of separate, I kind of separate from my body so I don't feel it as much, and then, for some reason, I just started relaxing.

Resisting a sensation, perception, or feeling is a type of dissociation that many chronic pain patients learn by themselves. Many therapists try to teach patients how to dissociate as a therapeutic process of pain control. Valuable as dissociation may be, however, it is the opposite of the accessing and replaying approach I am attempting to facilitate here. This is another example of the paradoxes of therapeutic hypnosis. Dissociation can be a process of resistance or an aspect of a therapeutic responsiveness, depending on how it is used. Current neuroscience research, however, implies that the transformations that normally take place during recall, replay, and resynthesis may be a more efficacious approach to facilitating therapeutic transformations (see Chapter 3; Nader et al., 2000a, 2000b; Petrovic et al., 2002).

ROSSI: Uh-huh, and how about that discomfort in your shoulders? Is it getting better, or worse, or changing in any way, or what?

Notice how this open-ended question is not easily answered by the conscious mind when symptoms or vague sensations and feelings are involved; it requires an inner search to be answered. It is this inner search that has potential therapeutic value. *Open-ended questions that require a fresh, activity-dependent, sensory-perceptual inner search* to determine whether discomfort, symptoms, or problems are getting "better, or worse" is a *fail-safe, double-binding,* implicit processing heuristic facilitating *symptom-scaling* that monitors and further impels the patient along the experiential symptom path. It is fail-safe in the sense that whatever the answer—better or worse—the patient will receive support from the therapist. It is a therapeutic double-bind because it may move the patient's process along a therapeutic path, whatever the answer (Erickson & Rossi, 1975; Erickson et al., 1976; Rossi & Jichaku, 1992). The symptom getting better is, of course, the ultimate desideratum. The symptom getting worse is a temporary desideratum in the sense that it means the patient has the courage to allow a greater intensity of engagement with the state-dependent encoding of the symptom, which may provide an even more potent possibility of resynthesizing it in a therapeutic manner.

THE FIRST THERAPEUTIC RESPONSE

Compliance or the Novelty-Numinosum-Neurogenesis Effect? [8:14]

> CELESTE: [long pause, then she wets her lips.] It feels like it kind of spread out and got lighter.
> ROSSI: Very good! Go with it now, Celeste. Go with the spreading out, getting lighter, and let's see what happens next.

When the sensations of a symptom spread out and get lighter, it is a very reliable indication that the patient is well engaged in the accessing, defusing, and resynthesizing of the state-dependent encoding underlying it. Note that Celeste experiences this spreading and lightening of her symptoms spontaneously, without any direct suggestion for this type of response by me. That is, I did not suggest anything like, "Those symptoms will spread out and get lighter." The fact that this therapeutic response was not directly suggested gives us some reassurance that her experience is something more than expectancy or mere compliance with suggestions. We take this to be an example of Erickson's words, cited in Chapter 3 (1948/1980):

Direct suggestion is based primarily, if unwittingly, upon the assumption that whatever develops in hypnosis derives from the suggestions given. It implies that the therapist has the miraculous power of effecting therapeutic changes in the patient, and disregards the fact that *therapy results from an inner re-synthesis of the patient's behavior achieved by the patient himself.* It is true that direct suggestion can effect an alteration in the patient's behavior and result in a symptomatic cure, at least temporarily. However, such a "cure" is simply a response to the suggestion and does not entail that re-association and reorganization of ideas, under-standings, and memories so essential for an actual cure. *It is this experience of re-associating and reorganizing his own experiential life that eventuates in a cure,* not the manifestation of responsive behavior, which can, at best, satisfy only the observer. (pp. 38–39, italics added)

Our open-ended, activity-dependent symptom path approach is the opposite of the view that defines therapeutic hypnosis as a process of programming or conditioning the patient, with the therapist operating as the master programmer. Rather, we view this apparently *spontaneous experience of healing that is discovered by the patient with a sense of surprise as a manifestation of the novelty-numinosum-neurogenesis effect.*

New research approaches combining brain imaging and gene expression assessment with DNA microarrays will be important in assessing this psychobiological view of the healing process.

THE LOCUS OF CONTROL REMAINS IN THE PATIENT

Courage and Permission for Private Experiencing as Implicit Processing Heuristics [8:58]

ROSSI: [long pause] That's right, the *courage to really go with it,* Celeste, and *occasionally saying a few words or a sentence . . . only what I need to hear* [pause] *to help you further.* [long pause]

Notice what is implied in this single sentence offering support that is often needed in the typically difficult inner work of stage two of the creative cycle. The phrase *"the courage to really go with it"* is a powerful encouragement that most people are happy to receive. The illusion that hypnotherapy is an easy process to experience is held by many. Supposedly the therapist puts the patient into a comfortable and deeply relaxing state of something like sleep and utters some magical metaphors and suggestions that automatically accomplish all the healing. Heck, some therapists can just snap their fingers or make a hypnotic gesture, like the comic strip character, Mandrake the Magician, and patients fall into profoundly healing trances. Nothing to it! No doubt about it! As already noted, standard scales of hypnotic suggestibility suggest that this type of near-instant induction may happen with about 5-10% of the general population. For the other 90-95% of us, however, something vastly more is involved. Patients need courage to continue with the often-difficult experiences of emotional arousal and inner work so characteristic of stage two of the creative cycle.

The phrase *"occasionally saying a few words or a sentence"* sounds permissive, nondirective, respectful, and open-ended, so it is easy for most patients to accept. Notice, however, how this phrase functions as a highly focusing implicit processing heuristic. Saying a few words or a sentence *only occasionally* makes the task of communicating with the therapist seem very easy. It also implies, however, that patients will spend most of the time doing important, private, inner work.

The phrase *"only what I need to hear* [pause] *to help you further"* is another supportive and permissive implicit processing heuristic that empowers patients to receive, recall, and replay their most important experiences in a private manner. Patients are also empowered by the act of determining what to report to the therapist. *The locus of control over the therapeutic process remains within the patient, where all the really important inner work is taking place,* rather than in the therapist, who can have only marginal intuitions of what the patient really needs to do.

Another considerable benefit of this private and permissive approach is that it tends to lower the so-called "resistance." Patients do not have to waste precious time and mental energy in being constantly concerned with the task of translating their often vague, implicate (right-hemispheric) experiences into an explicit, verbal (left-hemispheric) form just to please the therapist.

ROSSI: Mm-humm . . . [long pause]

The hum of *"um-humm"* is an all-purpose implicit processing heuristic that can be intoned in many different ways to communicate many different meanings, ranging from the critical and demanding to simple, nondemanding permissiveness. "Um-humm" can serve as an indirect question; an expression of warm support and understanding; encouragement to continue; a solicitation for communication; an expression of curiosity, wonderment, and numinosity; a murmur of eroticism, boredom, disdain, disinterest, or impatience; as a cue to hurry on or slow down; an insinuation to be more truthful; even as a slap in the face. It is a wonderful practice for students, walking along their way somewhere, to intone *um-humm* in as many different ways as possible. Partners can enjoy exploring their relationship by using nothing more than their skills at varying their um-humms. Erickson noted an actor who claimed to be able to say no in at least 40 ways. But that is another story for another time.

ROSSI: And is that going well?

A simple question after a long pause, particularly if the patient is frowning or manifesting facial cues of negative inner experiences (such as anxiety or fear), reminds him or her that, after all, the therapist really is attentive and there to help things *go well*. Different patients and therapists, of course, have different levels of tolerance for ambiguous periods of silence in psychotherapeutic work. I have found the range between a few minutes to 15 or 20 minutes to be most productive. A glance at Table 2.1, listing ultradian rhythms in human experiencing on all levels—from the molecular-genomic to the cognitive-behavioral, will reassure most readers of the many important psychobiological processes of healing that can take place in this time range.

CELESTE: The words aren't really coming out, I need to go deeper inside.

Here she gives us a wonderful confirmation that it does require time for significant inner work to take place on an implicit level before words are available on an explicit level. Why are the words not coming out? Could this be an indication that she has shifted from the left-hemispheric verbal modality to a right-hemispheric mode of nonverbal processing that is more characteristic of hypnotic experience (Spiegel & Barabasz, 1988). Research is now needed to determine to what extent the nonverbal approach described above contributed to this right-hemispheric mode of implicit experiencing.

Supporting Novel,
Numinous, and Enriching Experiences

"I Can't Believe I'm Doing This" [10:20]

ROSSI: Okay. Let's let that happen.

With these mundane words I utilize and support her expressed need to "go deeper inside." There is then a long pause, punctuated by many minimal facial indications and slight finger movements indicating that Celeste is indeed focusing inward. I would normally allow it to continue longer but, given the time limitations of this demonstration, I prompt her on with an unnecessary and possibly annoying question that I would not recommend anyone to use.

ROSSI: Is it easy for you to go deep inside right now?
CELESTE: Pretty much. For one second I thought, *I can't believe I'm doing this!*

Fortunately, my useless and potentially annoying question is received with remarkably good grace and even interest by Celeste. Her exclamation of not believing she is doig this suggests she is experiencing the novelty-numinosum-neurgenesis effect.

ROSSI: That's right. Stay with that. You can't believe you are doing this.

I utilize her exact words here to support her possibly novel, numinous, inner experiencing that could be initiating gene expression and neurogenesis. But I really have to get this misleading phrase, "stay with that," out my verbal repertory!

CELESTE: But I kept doing it.
ROSSI: Yes.
CELESTE: I keep saying that.

TO COMMENT OR NOT TO COMMENT

Dissociation versus Activity-Dependent Facilitation Work [11:22]

ROSSI: Yeah. I notice from the pulsations in your neck that your heart is really beating quite rapidly. Can you feel that?

Of course, I am really showing off to the audience how sensitive and clever I am in picking up minimal cues! Recognizing such minimal cues of psychobiological arousal was one of Erickson's favorite indicators that the patient was becoming engaged in finding and experiencing something of interest (novel, numinous, and motivating) in relation to the personal problem or symptom. He usually did not comment on them to the patient, however, because he respected the patient's unconscious need for privacy so much that he preferred that the signals generated by the unconscious be allowed to occur without the conscious mind's awareness. To comment verbally on such minimal behavioral cues of arousal just as they are taking place, particularly with a shy subject, could interfere with the arousal process. I break this rule here because I am trying to draw the attention of the audience to another sign that Celeste is entering stage two of the creative cycle, which is usually characterized by arousal rather than relaxation.

In asking Celeste, "Can you feel that," I am also exploring the possibility that her conscious awareness might facilitate her process of arousal. I am beginning to suspect that her previous experiences with alternative methods of healing may have taught her to use dissociation as a means of escaping the painful symptoms rather than accepting and utilizing them to recall, replay, reframe, and resynthesize, as I am exploring with her here. I don't want her to dissociate and lose touch with her natural psychobiological arousal at this early stage of the inner work. This is another open question that requires further research. When does commenting on the patient's minimal cues of psychobiological arousal function as an implicit processing heuristic that can facilitate the inner work, and when does it do the reverse by inhibiting it?

CELESTE: Mmm. [pause]

She appears to be trying to dissociate herself by acting very quiet and not using words.

ROSSI: You are trying to go inside, but that *poor old ticker* is . . . [pause]. What would you say you are experiencing exactly now?

I am continuing to notice the rapid pulse in her throat when I hesitantly say "but that *poor old ticker* is . . ." I do not complete this phrase because I would rather not take the chance of misinterpreting its significance. By asking her this direct question about what "you are experiencing exactly now," however, I am attempting to change the course of what she thinks she is supposed to be doing. I don't want her to dissociate to avoid the arousal and activity-dependent work of stage two of the creative cycle. I want her to witness her own state and go with her naturally ongoing process of psychobiological arousal and inner work rather than escaping prematurely into relaxation and quietude. I am attempting to help her recognize that she is in stage two of the creative cycle wherein self-awareness and self-acceptance are very important to optimize the inner work.

ROSSI: You are trying to go inside [pause] yet? [pause]
CELESTE: I feel like I'm inside about down to the waist.
ROSSI: Say that again?

I guess I really don't want to hear what she is saying! How can I be facilitating the activity-dependent arousal and psychobiological inner work needed for gene expression, neurogenesis, and other fancy stuff like that, if she is dissociating? Nonetheless, I am mindful of my responsibilities to allow her to experience what she needs as well as what the audience needs to witness (such as how dissociative experiences can be utilized in therapeutic hypnosis), so I realize I need to adjust my own attitudes and ambitions and make the best of where she actually is.

CELESTE: I feel like I'm inside about down to my waist.
ROSSI: All right, you feel like you are inside about down to your waist. It's like about halfway?
CELESTE: Yeah.

I am now beginning to marvel at the specificity of her skills at sensory dissociation, given her description "inside down to my waist." So I decide to give up my own orientation and shut up at this point in order to let her do it her own way after all. Maybe I am wrong about trying to facilitate her psychobiological arousal and awareness rather than her dissociation. Maybe she was frightened about my earlier observations about her "poor old ticker."

CREATIVE CONFLICT RESOLUTION
VIA TWO LEVELS OF FUNCTIONS

Dissociation and Multiple Levels of Experience as an Implicit Processing Heuristic [12:46]

ROSSI: [long pause] And now?
CELESTE: Finally. [pause] I feel that [pause] I could almost finally lose track of where I am.
ROSSI: You could almost lose track of where you are. Uh-huh. I tell you what, Celeste, continue with what you are doing, but instead of trying to lose track, suppose you simply really tune into what is happening within you. Really, instead of trying to do something to yourself, dissociating, or whatever, I'm asking you to be an accurate reporter of what is really going on. [long pause]

I guess I really can't let go of it. This may be one of the dangers of demonstration therapy. The therapist is tempted to strut his stuff and make a procrustean bed of his orientation. I am back into the tempest in a teapot of trying to facilitate her arousal and awareness while she seems intent on forgetting herself. She believes that her job of going into hypnosis is to dissociate to the point of losing track of "where I am." This certainly can be of value and is prescribed in the traditional approach to hypnosis. I therefore change my strategy. I suddenly recognize that this is an ideal opportunity to facilitate *functioning on two or more levels* as an implicit processing heuristic. I therefore tell her to *"continue with what you are doing"* and then add to it by hitchhiking the new suggestion to be an accurate reporter to this ongoing behavior at the same time (Erickson, Rossi, & Rossi, 1976; Erickson & Rossi, 1979).

This, by the way, is an interesting example of a conflict resolution tactic between patient and therapist—and perhaps life in general. Why should it be either/or? Why not explore the possibility of doing both at the same time— this *and* that—reveling in whatever unexpected creative response emerges from such *enriched* inner processing on more than one level at the same time? Is this not the true point of mediation and meditation?

SYMPTOM-SCALING TO REPLAY AND CREATIVELY REFRAME STATE-DEPENDENT EXPERIENCE

Casual, Humorous, and Playful Games of Yo-Yoing Consciousness to Facilitate Behavioral State-Related Gene Expression? [13:53]

> ROSSI: For example, what number are you experiencing on the pain or discomfort or arthritis involvement right now?
> CELESTE: I'm . . . [pause] I'm about 65.
> ROSSI: [with a self-satisfied smile on his face] 65%? Do you think you can bring it back to about 70 or 75%?
> CELESTE: Yes.
> ROSSI: Okay, let's see if you can try that, Celeste. Let's see if you can let it get . . . worse, as a matter of fact, quite worse . . . before it gets better.

Notice how the words *"worse before it gets better"* achieve a close *apposition of opposites* that may function as *a double-binding implicit processing heuristic* (Erickson, Rossi, & Rossi, 1976). When a person is in a right-hemispheric mode of experiencing, it is not easy to sort out logically what is being said when words with opposite meaning are closely juxtaposed in this manner. The subject therefore tends to fall back on easier implicit (unconscious) levels of processing.

Erickson used the phrase *"the yo-yoing of consciousness"* to describe this casual and playful process of encouraging patients to let their symptoms get worse, then better, over and over again. He insisted that it should be casual as well as playful. This lighthearted attitude is a fail-safe way of exploring suggestions, since any so-called failure can be dismissed easily with humor. When I began my studies with Erickson, I was very dubious about the value of such "casual play." From our neuroscience perspective, however, we can now understand how such novel play is actually an experience of natural variation and conscious selection that may promote activity-dependent gene expression, neurogenesis, and healing. Who will be the first to use brain imaging devices and DNA microarrays to assess this hypothesis?

ROSSI: [pause] How about now?
CELESTE: Probably about 80. [She smiles.]

My tactic of encouraging multiple levels of experiencing seems to be succeeding. Her smile indicates that she is proud of her abilities. She is catching on to the casual and playful game of gaining skill over her experiential states. Letting a symptom get "worse before it gets better," only for a second or so, is a safe, time-binding approach to accessing and replaying a state-dependent process as a prelude to resynthesizing and reframing it therapeutically.

Does this accessing and replaying of state-dependent memory, learning, and behavior facilitate *behavioral state-related gene expression* as well? If so, would it mean that we are actually facilitating mind-gene communication? State-dependent memory, learning, and behavior are replays of *mind* that may facilitate creative replays of behavioral state-related and activity-dependent gene expression to optimize neurogenesis and healing. This is the essence of our original proposal in Chapter 3 that *creative replay, which involves natural variation and conscious selection, is the basic mechanism of mind, memory, psychotherapy, and healing.* The ultimate desideratum for the future of therapeutic hypnosis and the healing arts would be the experimental demonstration of the possibilities and parameters of the creative replay of mind-gene communication to facilitate healing at the molecular level (Rossi, 1994c, 1996b, 1999a, 1999b, 1999c, 1999d, 1999e, 2000a).

ROSSI: Very good, Celeste, go for that now. . . . Let's see if you can get it even worse than that. Just really go for it now. The courage just to really let yourself get into it. [pause]

A SURPRISING BREAKTHROUGH

Getting Hot and Sweating as Mind-body Signals of Intense Inner Work [15:25]

ROSSI: And tell me a little bit about the worst that you are experiencing as you are doing it.

CELESTE: My hands are really hot.

ROSSI: The hands are really hot! Fantastic, Celeste. Do they often get hot like this?

CELESTE: Yes.

At least two important things are happening here. Getting hot is a reliable indicator that the subject is engaged in intense inner work that is accompanied by sympathetic system arousal. Getting hot and even sweating in a body area experiencing a symptom is a signal that the symptom may be in a creative process of transformation from pain to warmth, comfort, and healing, as illustrated by Erickson's personal experience of nighttime pain from his polio reported in the previous chapter.

ROSSI: Very good. Let them get even hotter, let's see what that leads to.

I am happy with the heat that indicates she is getting more involved in stage two of the creative cycle, but the heat is still only an indication that she is on the symptom path. She is not yet at the goal of experiencing a creative healing response typical of stage three of the creative cycle. With the words, *"let's see what that leads to,"* I am offering another implicit processing heuristic to further her progress, hopefully, toward stage three.

[pause]

ROSSI: Are they getting hotter?

CELESTE: They are starting to sweat.

ROSSI: Huh?

CELESTE: They are starting to sweat, I think.

Accessing and Replaying State-Dependent Discomfort

Humorous Hyperbole to Facilitate Therapeutic Replay [16:08]

ROSSI: Oh, they are starting to sweat! That's excellent. Really, let them sweat. Let's see *what kind of . . . incredible discomfort can continue to develop* there?

By asking *"what kind of incredible discomfort can continue to develop,"* I am, of course, indulging in humorous hyperbole. Notice, however, I am also encouraging her to further amplify a heightened sensory quality of skillful experiencing that may help her access and replay the state–dependent aspects of her discomfort. She may eventually be able to utilize this skill to reframe her discomfort successfully into another more agreeable experience in stage three. *Humor* is another way of facilitating the simultaneous experiencing of two or more levels of being to optimize self-reflection and creative inner work.

ROSSI: [She swallows; pause] Is it still getting worse, Celeste?

Psychobiological Arousal and the Novelty-Numinosum-Neurogenesis Effect

A Homology Between Dissociation in Hypnosis and Stage Two of the Creative Cycle? [17:02]

ROSSI: [Celeste smiles as somewhat strange and apparently involuntary, minimal movements begin to take place in the fingers of her right hand, as illustrated in Figure 7.2] You are smiling a little bit, so I can't tell if it is getting worse or your having fun or what?
CELESTE: *My right hand is doing some stuff.*
ROSSI: Your right hand is doing some stuff?
CELESTE: Yeah. My left one feels like lead, but my right one is doing some stuff, but my left hand. . . .

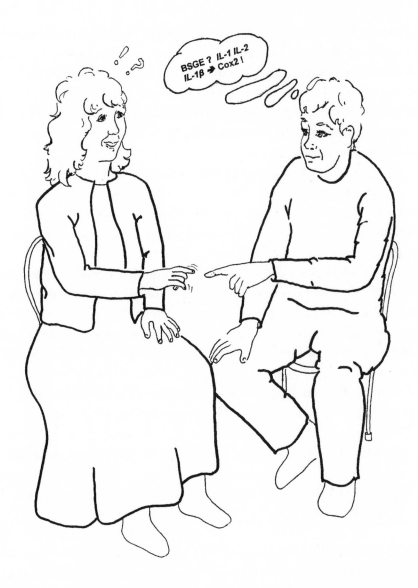

FIGURE 7.2 | Stage Two: Evidence of psychobiological arousal and behavioral state-related gene expression (BSGE). Celeste evidences surprise when her "right hand is doing some stuff" (unusual sensations and involuntary movements) that was not suggested by the therapist. Unexpectedly her hands also become hot. These are typical manifestations of pscychobiological arousal in stage two of the creative process that we hope to channel into a therapeutic process of salient self-engagement and healing. The therapist wonders about how to engage the psychogenomic dynamics of immunological variables such as interleukin-1, 2, and 1ß associated with Cox2 that has been implicated in rheumatoid arthritis which is Celeste's presenting symptom.

Notice the implications of what she is saying. She is not saying that *she* is doing some stuff with her right hand or that she is making her left hand heavy. She is not making any claim of voluntary behavior in her sensory and motor experiences when she says, *"My right hand is doing some stuff."* Her words imply that the finger movements and sensations are being experienced as autonomous, dissociated, and involuntary. I have not suggested these movements and sensations in any way.

Involuntary movements and peculiar sensations such as "heavy as lead" are commonly suggested as "challenges" in standardized hypnotic susceptibility scales. The experimentalist hopes to convince the subject that such suggested, unusual experiences are experiential proof that the subject is in a trance and under the power and control of the hypnotist.

Here we interpret Celeste's spontaneous experiencing of these peculiar movements and sensations as an example of the homology between the induction process of classical hypnosis and stage two of the creative cycle. This homology can be used as a microscope helping us generate more fruitful explorations of the natural psychobiological dynamics of healing in hypnosis (Erickson et al., 1983). When patients appear to be fascinated with unusual sensations and trembling movements—as Celeste is here—they may be experiencing the novelty-numinosum-neurogenesis effect.

> ROSSI: Oh, I see! Your left hand is like lead, and your right one is like . . . ?
> CELESTE: *I don't know, I think it started shaking a little bit, or something.*

Notice again the natural language of psychological dissociation when she acknowledges that *she does not know* and "I think *it* started shaking" rather than *"I started shaking it."*

> ROSSI: Oh, yes, I see it now! It is showing some micromovements. Thank you for pointing that out to me. Okay. *Let's see what continues to happen in both those hands.*

Notice the subtle language of dissociation that I am reflecting back to her when I say, *"Let's see what continues to happen in both those hands."* I do not use the typical everyday language of voluntary behavior such as, *"Let's see what you do with your hands."*

ROSSI: [pause] Even while you continue with the discomfort, huh?

The video camera documents a rather remarkable series of unusual finger and hand movements throughout this playful exploratory period.

ROSSI: There you go, very fine, yes! [pause] Do those fingers typically move that way, Celeste?

I'm using the language of dissociation again by asking her about *"those fingers"* rather than *her* fingers. This question may assess whether her state of dissociative movements continues, or whether she has returned to normal voluntary movement. The use of dissociative language, of course, tends to reinforce a continuing dissociation of her finger movements. The video shows me pointing closely to the dissociated micromovements in her right hand to draw the attention of the audience to them (Figure 7.2).

CELESTE: I don't think so.

This response implies she is still experiencing dissociated movements, since *"those fingers"* do not typically move that way in everyday life.

ROSSI: Very fine. As if they have a life of their own. Let's see how they continue to express themselves.

Since dissociation and involuntary movements seem to be one of her natural hypnotic talents, I support them further with the language of therapeutic dissociation when I say, *"as if they have a life of their own."* What is *therapeutic dissociation?* Haphazard dissociation is considered to be "psychopathological" when it is out of control of both the patient and the therapist. *Therapeutic* dissociation, however, is when the dissociation can function as an implicit processing heuristic facilitated by the therapist.

ROSSI: Yes, my goodness. That's right, really allowing that to continue. That's it, very fine, those fingers lifting and kind of vibrating in a hesitant way.

Celeste and I both contemplate the involuntary movements with evident satisfaction. Unfortunately, these micromovements are not easily evident on the videotape.

PERMISSIVE QUESTIONS AS IMPLICIT PROCESSING HEURISTICS

Hand Levitation to Facilitate Activity-Dependent Gene Expression, Neurogenesis, and Healing? [18:50]

ROSSI: That's it, very fine. Allowing that sensitivity . . . very good, now it's like, it's like three or four of *those fingers are lifting completely up?*

As I stare intently at Celeste's right hand, it seems to me that some of her fingers are lifting up and staying up rather than simply bobbing up and down or vibrating in an uncertain manner, as they were previously. Such subtle shifts in semiautonomous behavior may be important indicators of therapeutic shifts along the symptom path to stage three so I study and support them very carefully.

ROSSI: That's it!

A moment after I asked the question, *"those fingers are lifting up?,"* her fingers remain up. She now appears to be responding in a very positive and sensitive manner to my suggestions, even when they are made as permissive questions. My *permissive question is an implicit processing heuristic* that functioned as a rather *fail-safe way of assessing her suggestibility at this point.* Since she responds so well to my suggestion, I interpret this to be a moment when I could further facilitate her inner involvement and creative work by permissively questioning whether her entire hand will levitate, as follows.

ROSSI: Yes, my goodness! I wonder if the whole hand is going to continue lifting? It's an extraordinary sensitivity, very fine. Very fine, Celeste. Even while that hand continues, I can continue talking with you, Celeste. My goodness. Do you often have an experience of hand levitation this way?

This is a classical approach to heightening and utilizing hypnotherapeutic suggestion. Hitchhiking one suggestion onto an ongoing pattern of behavior is another implicit processing heuristic that tends to reinforce the dissociated nature of her hand lifting.

CELESTE: I've never done it before.

Can this subjective report be taken as evidence that Celeste is experiencing something (1) *novel, surprising,* and (2) *enriching* that may be evoking *activity-dependent gene expression, neurogenesis, and healing?* If we add the observation that her hand levitation could be taken as (3) a *novel form of exercise,* then we have all three of the conditions that current neuroscience has found associated with gene expression, neurogenesis, and healing (see Chapter 3). Research utilizing brain imaging and DNA microarrays may be an approach to this question about the psychobiological dynamics of healing in hypnosis and the therapeutic arts.

ROSSI: Never done it before. Have you witnessed other people?
CELESTE: That's why I was smiling.
ROSSI: Oh, that's why you were smiling. Have you seen other demonstrations with hand levitation?

Her hand visibly jumps up a bit, as is highly characteristic of involuntary hand levitation movements. Apparently even my casual and permissive question that simply used the words *hand levitation* is received in a sensitive manner that Erickson et al. (1976) called "response attentiveness." This response attentiveness, which is so characteristic of hypnotic behavior, now appears to be a primed or heightened readiness to respond that could be optimizing activity-dependent gene expression and neurogenesis.

ROSSI: That is it! Continuing that wonderful levitation.

SURPRISING IMPLICIT INVOLUNTARY MOVEMENTS

Ambiguity as an Implicit Processing Heuristic to Facilitate the Novelty-Numinosum-Neurogenesis Effect Associated with Dopamine Release? [19:48]

Celeste's right ring finger now vibrates sharply, very noticeably, apparently in immediate response to my suggestion, *"continuing that wonderful levitation."* But my suggestion was for her hand to lift, not her finger. Why does her *ring finger suddenly lift* here, as if intruding in a rather surprising way?

CELESTE: Umm . . . [quietly, uncertainly].
ROSSI: That's it, continuing *that wonderful lifting.*

I just continue with this ambiguous *"that wonderful lifting"* that could apply to either her hand or her finger. I do not mention anything about her ring finger lifting at this point, because I am not sure she noticed it. If she did not notice it, I would not want to interfere with the privacy of her implicit level response by talking about it.

ROSSI: [More immediate sharp finger movements occur, apparently in response to my words.] Yes, that's it!

I am so surprised at her ring finger sharply lifting that I feel myself in a mild state of shock (Rossi, 1973a), and I cannot respond to it in any way other than a stereotyped reinforcement, *"Yes, that's it!"*

CELESTE: I just didn't expect it, because I don't think you said anything.

Celeste may be referring to her sudden ring finger uplift, but I still cannot be sure. Although I have been suggesting hand levitation, I certainly did not suggest that her right ring finger suddenly lift and vibrate. It is precisely such *unexpected* ideodynamic finger movements, however, that may be signals of something important being experienced on an implicit level, which may, or may not, eventually reach the explicit level of consciousness (Rossi &

Cheek, 1988). This lack of expectancy or predictability is associated with the release of dopamine in novel learning experiences (Waelti et al., 2001, reviewed in Chapter 3) as well as the placebo response (De la Fuente-Fernández et al., 2001, reviewed in Chapter 5) that we have summarized as the novelty-numinosum-neurogenesis effect.

I allow the *ambiguity* of what we are talking about to remain as a further implicit processing heuristic in the following remarks that could apply to either her finger or her hand lifting. The finger signaling approach developed by Cheek (Cheek, 1994; Rossi & Cheek, 1988) could be employed at this point to utilize Celeste's finger movements to identify the sources of her problems and symptoms and then initiate a highly focused way of working with them therapeutically. In this demonstration session with so large an audience, however, I decide not to use that approach to psychodynamic insight because I respect her privacy. It is remarkable to recognize just how much therapeutic progress patients can make without the therapist trying to make them explicitly aware of their personal psychodynamics.

IMPLICIT PROCESSING HEURISTICS
VERSUS CLASSICAL HYPNOTIC SUGGESTION

Replaying Variability on the Creative Edge [19:56]

ROSSI: That's right, you didn't expect it, because I did not say anything. And yet it's happening all by itself, and now the hand lifts even more, and there is an extraordinary self-sensitivity, while I am sure some people up close can see it now, right? Very good, Celeste, just allowing that to happen all by itself, because we know, of course, that *healing takes place all by itself, too, very often, does it not?*

This is a use of one of Erickson's favorite therapeutic double-binding questions, "It is, is it not?" (Erickson & Rossi, 1975), that I would now call an implicit processing heuristic. This double bind may be *therapeutic* in the sense that it tends to fix or hold the patient's attention in ambiguous suspension for a meditative moment, while a lot of *inner processing* takes place on an *implicit* or unconscious level. It is a heuristic in the sense that it may help facilitate inner discovery and problem-solving, but it does not determine what the exact outcome of that discovery will be. It is not a method of manipulating or controlling a person's mind because it is not a reliable way of making a person do anything in particular.

The therapeutic double bind is like a "wild card" that may facilitate the intensity of a person's inner processing, but it does not determine what the outcome of that inner work will be. An implicit processing heuristic is like stepping on the gas for a brief moment, but it does not determine the direction of the car's movement. Herein is precisely the difference between implicit processing heuristics and the traditional concept of suggestion in hypnosis theory: Implicit processing heuristics may focus attention and an optimal level of inner work, but they do not presume to be a method of facilitating compliance or programming with any particular *suggestion about what explicit or manifest form of behavior should take place.*

Understanding the difference between the intent of implicit processing heuristics versus traditional direct hypnotic suggestion is crucial. Implicit processing heuristics are just that: they only presume to facilitate a momentary suspension of explicit, consciously directed thinking to permit more intense activity on a more unconscious, implicit level. Hypnotic suggestions, on the other hand, presume to do much more. Indeed, the intent of classical hypnotic suggestions is to modify behavior by inducing compliance. A great deal of recent research has documented how Erickson's indirect suggestions, for example, may be more acceptable by many people in a clinical context, but they are not any more effective than direct suggestion in helping subjects get a higher score on the classical hypnotic susceptibility scales (Barber, 2000; Matthews, 2000; Peter & Revenstorf, 2000). The implication of such research is that Erickson's indirect approach to hypnotic suggestion, used by countless hypnotherapists, is not effective. The controversial question, however, is, "effective for what?" While Erickson did important experimental work in his early career on using hypnotic suggestion to manipulate subjects' sensations, perceptions, emotions, thinking, and behavior in a predictable manner (Erickson, 1938/1980, 1954/1980, 1980c), in his later career as an innovative therapist, his approach could be more aptly described as facilitating patients' private creative processing rather than compliance with the therapist's directives.

Erickson's approach of facilitating creativity in the process of hypnotic psychotherapy is well indicated in the quote at the beginning of this chapter. Erickson (1948/1980) describes hypnotic psychotherapy, as he employs it, as "a special psychological state in which patients can re-associate and reorganize their inner psychological complexities and utilize their own capacities in a manner in accord with their own experiential life" (pp. 38–39). This perspective is very different from the classical notion of hypnotic suggestion as a method of conditioning, programming, or forcing compliance with the therapist's directives. Notice how Erickson makes liberal use of the "re" prefix to bring patients back to *re-calling, re-playing, re-framing, and/or re-synthesizing inner dynamics* within the privacy of their own minds, utilizing their own resources. In this sense Erickson's psychotherapeutic approaches were more akin to the current neuroscience idea of therapy as a creative process where every recall is an opportunity for a reframe and resynthesis of memory, learning, and behavior (as discussed in Chapter 3, Nader et al., 2000a, b) and what we call *implicit processing heuristics to facilitate psychogenomic replay.*

CELESTE: Yes.

ROSSI: Yes . . . very fine. Really appreciating a little bit of a miracle happening . . . that hand going up, even though I didn't say a word about it. That's it, very fine, yes, very fine Celeste. Very, yes. Yeah. [Said in a very soft, conspiratorial whisper, with a somewhat humorous tone. Celeste's hand again lifts noticeably.] Mmm-mm, that's right. *And does it really seem to have a life of its own?*

CELESTE: Kind of.

Celeste smiles as she says this. Let's appreciate her language. She does not absolutely agree by saying, "Yes, it has a life of its own." She says, "Kind of." Her hand movements are semiautonomous, not completely dissociated. She may be experiencing that *creative edge between implicit and explicit, voluntary and involuntary, behavior.* In a sense, her behavior is on the border between free will and determinism. In the language of complex adaptive systems theory, her hand movements—and the more or less autonomous (dissociated) behaviors of shakiness, tremors, and/or very fine hand, finger, arm, or head vibrations experienced by some hypnotic subjects and patients—could be a *replay or iteration of "computing on the edge of deterministic chaos,"* where life processes are most innovative and interesting (Morowitz & Singer, 1995). This perspective may provide a more useful interpretation of hypnotic behavior than the traditional view that such dissociated movement is a sign of psychopathology.

Whatever language we use, we are trying to describe the therapeutic state of heightened sensitivity, variability, and selectivity that is experienced during creative therapeutic transformations. Recall the interesting fact about all living systems from complex adaptive theory: *Heightened variability in behavior is an important characteristic of health.* Recall the heart interbeat variability example from Chapter 2, where greater variability was associated with health while uniformity was associated with dysfunction. Heightened variability in behavior is associated with a response readiness to quickly adapt to the natural, ongoing changes typical of life at all levels, from mind to gene.

ROSSI: Kind of . . . [laughing] yes.

Chapter 8 continues our exploration of Celeste's experience with demonstration therapy.

SUMMARY

In this chapter we undertook a bold adventure in integrating many historical and cultural rites and rituals of healing into what we called "the experiential theater of demonstration therapy." This new/old healing format emphasizes dramatic, novel, and numinous experiences as an important modality in psychotherapy and the healing arts. Insofar as these experiences facilitate changes that favor dreaming, fantasy, comfort, and relaxation characteristic of low–phase hypnosis, they may facilitate *behavioral state-related gene expression and associated pathways of mind-body healing on a molecular level.* Insofar as these experiences facilitate states of arousal that are characteristic of high-phase hypnosis, they may facilitate the *activity-dependent gene expression associated with the novelty-numinosum-neurogenesis effect.*

The first two stages of the creative cycle—preparation and incubation—were facilitated in this demonstration by a series of implicit processing heuris-

Table 7.1 | Implicit processing heuristics that may facilitate gene expression to optimize neurogenesis and healing during the preparation and incubation stages of creatively oriented psychotherapy and the healing arts.

- Ambiguity

- Apposition of opposites

- Casual, happy, playful attitude

- Humor

- Fail-safe alternatives covering all possibilities of response

- Fascination

- Hitchhiking suggestions onto ongoing behavior

- Ideodynamic hand-, finger-, and head-signaling

- Metaphor and stories

- Novelty-numinosum-neurogenesis effect

- Open-ended questions not easily answered by the conscious mind

- Private inner work

- Recalling, replaying, reframing, and resynthesizing

- Surprise

- Therapeutic double binds

- Therapeutic dissociations

- Time-binding, incomplete sentences, and dangling phrases

- Two or more levels of functioning

- "Um-humm" used in many ways

- Use of "until . . ." to build time-binding expectation

tics (listed in Table 7.1) that are fail-safe approaches to inner work and healing. These implicit processing heuristics tend to evoke free-form and open-ended experiences characteristic of Darwinian natural variation and selection on all levels from mind to gene. Responses to implicit processing heuristics are unpredictable, however, as is characteristic of the creative behavior of *complex adaptive systems* on all levels in nature.

Implicit processing heuristics tend to facilitate open-ended, creative replay, whereas *classical hypnotic suggestions* purport to direct behavior in a deterministic and predictable manner. We usually cannot predict what response we will get to implicit processing heuristics; with hypnotic suggestions, on the other hand, we hope to get the behavior suggested. *Implicit processing heuristics* and *classical hypnotic suggestions* have advantages as well as limitations. In the next chapter we will continue our experiential theater of demonstration therapy as Celeste moves into stages three and four of the creative process. We will be treated to the spectacle of the dueling banjoes of implicit processing heuristics and classical hypnotic suggestion in facilitating *behavioral state-related gene expression and/or activity-dependent gene expression* in neurogenesis and mind-body healing.

THE EXPERIENTIAL THEATER OF DEMONSTRATION THERAPY: PART B

The Insight and Verification Stages of Creative Work

"The three persons reported upon are examples of dozens of others that this author has seen over the years, and the results obtained have been remarkably good despite the fact that the patients were seen only on one occasion for an hour or two.

"In each instance hypnosis was used for the specific purpose of placing the burden of responsibility for therapeutic results upon the patient *himself after he himself had reached a definite conclusion that therapy would not help and that a last resort would be a hypnotic "miracle." In this author's understanding of psychotherapy, if a patient wants to believe in a "hypnotic miracle" so strongly that he will undertake the responsibility of making a recovery by virtue of his own actual behavior and continue his recovery, he is at liberty to do so under whatever guise he chooses,* but neither the author not the reader is obligated to regard the success of the therapy as a hypnotic miracle. *The hypnosis was used solely as a modality by means of which to secure their cooperation in accepting what they wanted. In other words,* they were induced by hypnosis to acknowledge and act upon their own personal responsibility for successfully accepting the previously futilely sought and offered but actually rejected therapy. . . . the use of hypnosis as a technique of deliberately shifting from the therapist to the patient the entire burden of both defining the psychotherapy desired and the responsibility for accepting it.*"

—Milton H. Erickson
"The Burden of Effective Psychotherapy"
(italics added)

We are almost halfway through our session of the experiential theater of demonstration therapy. Our initially reluctant subject, Celeste, is a young woman who has very evident rheumatoid arthritis, particularly in her hands.

An activity-dependent approach to therapeutic hypnosis was facilitated, using symptom-scaling to focus her attention and initiate her own private experiential path through stages one and two of the creative cycle. A series of implicit processing heuristics were used to encourage open-ended creative play, wherein she experienced many surprising and mildly dissociated phenomena that are highly characteristic of therapeutic hypnosis. There was no attempt to analyze or interpret her behavior in any way. The entire focus was on providing her with a safe theater in which she could explore her own inner experiences of creative work and healing in front of her peers.

This chapter continues with a surprising and humorous exploration of her individuality, as she stumbles upon the psychodynamic sources of her rheumatoid syndrome. We revel in the fresh novelty and numinous aspects of her original creative experiences and delight in their apparent unpredictability. A variety of implicit processing heuristics are offered to help her find the precarious balance on the delicate edge between stability and chaos, where complex adaptive theory indicates that life is computing at its creative best. It is fascinating to witness the *Darwinian natural variation and selection* that takes place in her consciousness and in her body language as she explores this creative edge. *When the therapeutic process is going well, it is the patient who guides the therapist, not the reverse, as practiced in conventional therapy.* Throughout I speculate freely about how this dramatic therapeutic theater may facilitate activity-dependent gene expression, neurogenesis, and healing, via the novelty-numinosum-neurogenesis effect.

RECOGNIZING AND REINFORCING NOVELTY AND NEUROGENESIS

A New Hypothesis about the Nature of Hypnotic Dissociation [21:13]

CELESTE: It's going over to the side [laughing].
ROSSI: What's that!?

Celeste notices, with evident amusement, that one of her hands is slowly drifting to the side in that typically autonomous, dissociated manner of hypnotic experiencing—as if it had a life and will of its own. With my evocative question, "What's that!?" I am commenting on a peculiar and novel finger and hand movement that suddenly takes place. With this question I hope to help Celeste explore anything *novel, numinous, and/or dissociative* in her experiencing and behavior. In the altered-state theory of hypnosis, dissociation has been interpreted as a cardinal and defining characteristic. When it is out of control (as in hysterical states), dissociation is regarded as a form of psychopathology. In therapeutic hypnosis, however, dissociation can be utilized as a tool to facilitate the reorganization and reexperiencing of the patient's inner life (Rossi, 1996a).

From a neuroscientific perspective, dissociative behavior and experiencing that are novel, numinous, and surprising may be a sign of gene expression, neurogenesis, and healing—what we call the novelty-numinosum-neurogenesis effect. Novel behavior may be initially experienced as a bit "dissociated" precisely because it is so new that people do not yet recognize and acknowledge that it belongs to them. What is experienced as novel is not yet recognized as *normal*. From this subjective perspective, then, novel (rather than typical or stereotyped) dissociative behavior can be interpreted as the initial and often awkward emergence of the new in stage three of the creative cycle.

STAGE THREE:
FACILITATING ILLUMINATION IN THE CREATIVE CYCLE

Ambiguity, Humor, Not Knowing, Whispers, and
Therapeutic Double Binds as Implicit Processing Heuristics [21:16]

CELESTE: I think it is going over to the side.
ROSSI: Yes, it is going over to the side—that's why I mentioned it. A life of its own. I didn't say anything about its going off to the side, did I? *I haven't* been whispering any secrets to you, *have I,* that the audience hasn't heard? No, this hand really has a life of its own [humorously].

Notice the language of dissociation she uses—*"it is going over to the side"*—rather than "I moved it to the side." My use of *"I haven't . . . have I . . ."* is another example of Erickson's therapeutic double bind that may evoke ambiguity and possibly facilitate her inner search and processing. In the previous section I did, in fact, whisper to her in a *very soft, conspiratorial,* but humorous way. The humor makes it fail-safe, but the soft, conspiratorial whisper is itself an implicit processing heuristic that may evoke hidden secrets that have important psychodynamic significance within the subject on an unconscious level. I am not making any effort to get her to talk about hidden secrets. I am simply evoking the possibility of activating highly motivating implicit dynamics to further fuel her inner work, as she progresses toward stage three of the creative cycle.

More exploratory and novel movements now take place in Celeste's fingers, hands, and arms for an extended period of time. I attempt to utilize this body language by suggesting the numinous possibility that it may have stories to tell that go beyond her typical range of consciousness as follows.

> ROSSI: *I don't know* if it has stories to tell us today? . . . *I don't know* if it harkens to something beyond itself?

These *not-knowing questions* are implicit processing heuristics that may facilitate that peak period of private inner work between stages two and three of the creative cycle. Stories can be metaphors for facilitating a recall, review, and possibly a creative replay and resynthesis of the major motivations and archetypal themes of a person's life.

THE THERAPIST'S CREATIVE UNCERTAINTY

Truisms Co-Creating Rapport and a Mutual Quest [21:50]

> ROSSI: That's it! I really am not sure just how it is connected with healing.

This is an example of what I like to call "utilizing the therapist's creative uncertainty." Of course, I am *not* sure how her private experiencing is connected with healing. *This is a truism that is characteristic of all implicit processing heuristics.* It is a truth that no one could deny and, as such, it tends to facilitate a co-creative attitude of rapport, trust, and mutual quest between therapist and patient. Some student therapists initially experience distress with this creative approach of using implicit processing heuristics, because the therapist is never quite sure where the patient is in this process. Yet precisely in this uncertainty is the potential for creativity.

Given that Celeste is experiencing positive rapport in a context of creative play and mutual therapeutic quest, I personally experience great relief in admitting, "I don't know." The most common professional stress psychotherapists experience, no matter what their theoretical orientation, is in navigating between the patient's desperate hope for "expert" help and the therapist's realizing that one cannot possibly know everything needed to facilitate each patient's highly individualized needs for healing.

The more we learn about the many levels of the therapeutic process—from mind to gene as a *complex adaptive system wherein modulating one variable inevitably influences the expression of many other variables in the essentially nonpredictable manner of the nonlinear dynamics of deterministic chaos theory*—the more we realize that every psychotherapeutic encounter is an open-ended adventure whose outcome cannot be predicted with certainty. Each person presents one *complex adaptive system,* wherein modulating one variable inevitably influences the expression of many other variables in the essentially nonpredictable manner of nonlinear dynamics. *The mission of psychology is not the prediction and control of behavior, as elementary textbooks would define it. The best we can promise our patients is an informed respect for understanding their needs to grow in their own ways, along with some professional skill in facilitating their creative paths.* The current revolution in the neuroscience of understanding the relationships between gene expression, neurogenesis, and healing certainly will enhance our therapeutic skills, but we will all do well to recognize that *creative uncertainty will always play an important role in our work.* How to enjoy and thrive on this creative uncertainty, rather than buckle under the stress of trying to promise the impossible, is a delicate art each therapist must develop for his or her own well-being. This is another way of understanding how to "manage the transference and countertransference" that Freud and Jung agreed was the essence of the therapeutic encounter.

———————

PRIVACY AND CHOICE
IN THE SUDDEN EMERGENCE OF THE NEW

A Time-Binding Implicit Processing Heuristic [22:01]

CELESTE: Actually, *I just thought of something.*

This kind of remark by a patient is usually good news at this stage of creative inner work. Here it indicates that Celeste may be having an insight that is a goal of stage three of the creative process. Her privacy in front of such a large audience must still be protected, however. I therefore attempt to help intensify her inner work while maintaining her privacy by continuing to facilitate psychobiological arousal and creative tension via hand levitation with a *time-binding* implicit processing heuristic, as follows.

> ROSSI: You just thought of something. Um-humm. *Keep it private for a moment, Celeste. Only after a moment or two,* you can carefully consider whether you want to share that with the audience or continue to keep it private.

Notice the empowering utilization of her *private choice as an implicit processing heuristic.* Acknowledging that there is something important to incubate accords plenty of time to the task so that it can be well done.

> ROSSI: That's it, lifting.

Her hand immediately responds with a little jump up and continues lifting steadily. She is obviously functioning on multiple levels at this point. On the inside she is probably still processing what she "just thought of," while on the outside she is experiencing a mildly dissociated state of hand levitation—which is itself an involuntary inner process. This is another example of how the complex adaptive system of psychobiological dynamics can become so interdependent that one's head spins a bit trying to make rational sense of it all.

ROSSI: That's right, lifting and bumping up against the edge of the chair, but that's not stopping it. It's the sensitivity, isn't it? Incredible!

An unfortunate knocking noise from the microphone occurs. Her leg was apparently getting entangled with the microphone wire. I reach over to free the wire from her leg, which does not seem to disturb Celeste's inner focus at all. This apparent imperviousness to outside distractions is presumed to be one of the values of hypnotic focusing.

ROSSI: That's right, and I'm going to adjust this wire. I'm going to take it off your leg so we can release the . . . that's it. There you go. Very fine, Celeste. Is that hand really going up or are you conning us all [humorously]?

I humorously play the devil's advocate here to distract her further from the ticklish task of freeing her leg from the wire. This humorous distraction was probably not necessary; I really only did it to demonstrate Erickson's typical use of humor in such situations.

CELESTE: No [smiling broadly, then laughs].

THE SYMPTOM PATH LEADS TO BETTER FEELING

Darwinian Natural Variation and Selection of Psychological States: A Psychobiological Conception of Free Will? [23:10]

CELESTE: It just feels better going up.

Herein is a mystery of hypnotic healing. Why should her hand feel "better going up," in an apparently involuntary experience of levitation? Conventional theory would say this is an example of the spontaneous dissociation between movement and sensation so characteristic of therapeutic hypnosis. As we just speculated above, however, psychological dissociation may be conceptualized as the first stage of experiencing something new being synthesized on an unconscious, implicit level—which gradually, or suddenly, becomes manifest on an explicit, conscious level during stage three of the creative process.

From this perspective, we could say that Celeste is using this demonstration as an opportunity to play or experiment with the phenomenology of her own psychological experience. We can only wonder if this is an example of what Carl Jung meant when he said that we are all experiments of nature. From a Darwinian perspective we could say that her so-called dissociation is an experience of the *natural variations* in her phenomenological experience that her consciousness may then *select for healing*. Following the symptom path provides patients with an opportunity to self-reflect and become aware of the *natural variations* in their psychological experience. This novel awareness provides their consciousness with an opportunity to *select* what they want to facilitate in these natural variations.

Nature provides the *natural variations* (e.g., via deterministic chaos or the essential uncertainty of quantum theory) of psychological experience. Human consciousness then has an opportunity to function with apparent free will to *select* and co-create psychological states. The most numinous aspect of human maturation involves the emergence of this ability to self-reflect and co-create psychological states (Rossi, 1972, 1996a, 2000a). From a practical psychotherapeutic perspective we could conceptualize *free will* as the psychobiological dynamics of self-reflection and co-creation. We could say that this is what creative consciousness is all about. Obviously people do not always recognize and use their ability to select their own optimal psychological states. It is therefore an important therapeutic task of stage three to help people recognize, select, and co-create the best of their ongoing psychological experience. I now try to do this, as follows.

> ROSSI: It just feels better going up. Very fine! Yes, *it is finding its own way of feeling better, isn't it?* My goodness, it does seem to be taking an odd turn doesn't it? . . . And what are you exactly experiencing in that hand, as it continues all by itself that way?

I really am experiencing a lot of creative uncertainty with this long-winded speech. I know that I am trying like crazy to support what I perceive to be an important stage-three development when I rhetorically ask, "Yes, *it is finding its own way of feeling better, isn't it?"* Notice my use of Erickson's mild therapeutic double bind by tacking on *"isn't it?"* at the end of this implicit processing heuristic.

ACNOWLEDGING CREATIVE UNCERTAINTY

The Novelty-Numinosum-Neurogenesis Path of Healing [23:51]

CELESTE: Just now it's started shaking a little bit.
ROSSI: Yeah, I noticed that. [Celeste's hand and arm continue to oscillate up and down in an exploratory manner throughout this period.] Yes. Hm-mmm. That's right. That's it. Are you doing that voluntarily, or is that merely going up and down in the air for the moment by itself?
CELESTE: I wasn't thinking about it.
ROSSI: What's that?
CELESTE: I wasn't thinking about it.
ROSSI: Okay. Yes. My goodness! It really seems to be . . . my goodness.

As Celeste's hand continues to bob uncertainly at the wrist, I continue to exclaim "my goodness," playing the incredulous innocent in this numinous wonderland of creative uncertainty and high hopes for healing. Expressing my own genuine sense of numinous wonderment is a way of modeling, dramatizing, and possibly facilitating her own experience along the novelty-numinosum-neurogenesis path of healing.

EMERGING STAGE THREE INSIGHT WITH TEARS

Accessing State-Dependent Memory with Metaphor and Body
Language [25:00]

CELESTE: What I remembered was that this is the first hand, the first place, where I got rheumatoid.
ROSSI: I see! This is what Celeste remembered [speaking to the

audience]. This is the first hand, the first place that she got rheumatoid. So that is a very good place to be in. That is where the body signals were coming in—the first place that led to the arthritis. So let's see if we can get the message even clearer.

[addressing Celeste] I don't know if that is going to come in the form of words, or memories, I don't know if this is going to be in the form of emotions, what you were actually experiencing at the time the arthritis began. But I do know that something is really . . . [pause] just continue receiving that privately within yourself, Celeste. That's it. Very fine. It's like there is a *symphony of little movements in those fingers. Little pantomimes of talk, of messages* . . . that's it. Just receiving that privately within yourself. That's . . . my goodness. And you know you can say anything to me that I need to hear. That's it! *However briefly, just enough to help you further.* My goodness.

I signal to the audience by mimicking Celeste's up and down forefinger movements. I am hoping to facilitate her insight about *"what I remembered"* of the experiential source of her arthritis by using metaphors of *"a symphony of little movements in those fingers* (Cheek, 1994; Rossi & Cheek, 1998). *Little pantomimes of talk, of messages."* While giving her a permissive suggestion for communicating verbally, I also continue to carefully guard her privacy by adding, *"However briefly, just enough to help you further."*

ROSSI: Yes. It's like each finger is getting into an act with its own . . . yes. Hmmm. That's right. [Her fingers continue their apparently autonomous, involuntary, unusual movements.] That's it. And really continuing with the courage to receive that, Celeste. That's right. Yes.

I signal to the audience with a finger movement on my cheek, indicating that a tear is beginning to trickle down Celeste's face. Such quiet tears are usually an indicator of important inner experiencing. In the current context it may be signaling the significance of what she is remembering—possibly a stage-three insight.

ROSSI: Uh-huh. Yes. That's right. My goodness. If fingers could speak? Yes [laughs]!

Here I offer another metaphorical question hinting that Celeste may want to speak but keep the main focus on the fingers, wherein ideodynamic movements and sensations may be accessing state-dependent encoding of the meaningful memories.

ROSSI: That's right, my goodness. Now the others are getting into the act, aren't they [said humorously]?

Several of Celeste's fingers are moving erratically and rapidly, as is highly characteristic of critical points in the psychobiological ideodynamics of finger signaling.

CELESTE: [smiling in acknowledgment] Yeah.

RECOGNIZING AND SUPPORTING THE NEW

"Weird" Feelings May Imply Novelty, Numinosum, and Neurogenesis [28:25]

ROSSI: So what are you feeling about that?
CELESTE: Kind of weird [laughs].
ROSSI: Kind of weird, yes it is weird. Weirdness is the best, Celeste! Because if something's weird it means . . . it's something a little bit different, huh?

Notice the casual but very important therapeutic reframe here. *Weird* has ambiguous connotations in our generation: It could be unusual, odd, strange, bizarre, scary, possibly harmful but interesting and perhaps worth investigating further. I reframe the possibly negative connotations of the word *weird* with the permissive, open-ended question, *"it's something a little bit different, huh?"* Hopefully this question about her "feeling" in the context of her tears will function as a *motivating implicit processing heuristic* empowering her to explore what is new and of value at this moment.

This is the essence of the therapist's task in stage three of the creative cycle: recognizing and supporting the new and numinous even when the patient fails to recognize its significance. The numinous is often criticized with a derisive attitude by patients who have experienced much negativity in the past.

The therapist needs to reframe this negative bias with a positive spin. The weird feeling may be her way of experiencing the psychobiological dynamics of the novelty-numinosum-neurogenesis effect that becomes manifest at that high point of private inner work just before stage three, the experience of illumination, insight, or healing.

> CELESTE: Mm-mm.
> ROSSI: Something you're not really used to . . . ?

This is an important phrase, a typical implicit processing heuristic helping her focus on the novel and enriching experiences in her stage-three processing.

> CELESTE: Western medicine never did this.
> ROSSI: What's that?
> CELESTE: Western medicine never did this [smiling].

Since *"Western medicine never did this,"* we may assume this is a *novel and numinous experience* for her. Since she is smiling, we are assured that she approves and enjoys her experience. This combination confirms that she is now successfully engaged in a numinous healing experience.

> ROSSI: No, [laughing] Western medicine never did this. But we are making connections. Very fine. Let's see where it leads to next. My goodness. My goodness, those fingers seem to be . . . that's right.

NOVEL, DRAMATIC, AND
UNEXPECTED THERAPEUTIC MOVEMENTS

Creative Sensory-Motor Replay Facilitating Healing [29:17]

Celeste's fingers spontaneously open widely for the first time in this session, as if she were now holding a ball. I am amazed by this sudden, novel, apparently normal and free coordinated movement that is very different from any other movement she has made thus far.

> ROSSI: Very fine. Really experiencing exquisitely, just what's happening in that hand. Hm-mmm.

I am strongly focusing her attention on the experiential sensory-motor components of her apparently healing experience. Could this be an example of how a replay of symptomatic movements and sensations may lead to gene expression, neurogenesis, and healing, as implied by the neuroscience research reviewed in Part I? Only further research can tell.

> ROSSI: My goodness. Huh?

Two of Celeste's fingers are going into a very fine vibratory movement that may or may not be evident on the videotape (time, 29:40). It is interesting to notice, by the way, that this high period in her therapeutic experience is taking place almost exactly in the middle of this hour-long session. This is very desirable since it will give us enough time to discuss, verify, and support the therapeutic work later in the session.

> ROSSI: That's right . . . mm-hum. Yes, Celeste. Yes. Uh-huh. That's right. Yes! Really. Can you experience what's happening right now? [Celeste's hand rotates at the wrist so that the palm is now facing her.] Are you aware of how that hand is turning in such an interesting way?

I continue to reinforce her focus on the curious, novel, and possibly healing psychobiological dynamics of her apparently autonomous sensory-motor movements and experiences.

A PLAYFUL GAME OF NOT-KNOWING

Replays Generating New, Humorous, and Creative Exploration [30:31]

ROSSI: [Her hand continues rotating very slowly.] My goodness, I think there is something being said there. [Three or four fingers of Celeste's right hand are making autonomous micromovements.] I don't know what [laughs]. But . . . yes, that's right. Yes. Incredible! Mm-hmm. Mm-hmm. My goodness, it is really doing something, and . . . ? And are you receiving something about what's happening there with those fingers—especially one of them, the middle one, that's one of my favorites for some reason. I find myself focusing on it. *What's it trying to tell us?*

CELESTE: [Smiling broadly as her hand turns at the wrist so that it appears as if her finger may be pointing at me.]

ROSSI: Don't point at me, *I don't know* [laughing]!

CELESTE: No. It feels like it's pointing?

ROSSI: It is pointing! What does it seem like it's pointing to?

CELESTE: I don't know [laughing].

ROSSI: Well, I don't know either!

CELESTE: I guess it is [pointing] at you [laughing].

This humorous, playful game of *not knowing* is in striking contrast to the traditional therapist stance as a powerful, all-knowing expert. My jesting could be justified as a way of utilizing Celeste's nontraditional behavior of exploring alternative therapies. However, I really don't know what is happening at this moment. The simple truth is that most therapists and patients are blind to what is really important most of the time on an implicit level. We are certainly blind to what is happening at the level of *behavioral state-related and activity-dependent gene expression and neurogenesis* during our attempts at therapeutic facilitation. *Humor and playful dramas have been enacted by court jesters throughout human history as a way of trying to tell the king the truth. Here we recognize that the so-called truth is always evolving in creative ways we do not expect. Wisdom is in knowing how to facilitate and recognize such emerging creative truths in the playground of subjective human experience.*

Facilitating the Creative
Edge between Stability and Chaos

Spontaneous Catalepsy in High-Phase Hypnosis [32:00]

ROSSI: Yes, wow! Now it's doing a little dance all by itself, pointing up, isn't it? Yes. But you know, I'm so curious, what's that feeling like? Is that, are you . . . ?

Here again I am focusing and facilitating a positive appreciation of her exploration of whatever may be novel, numinous, and motivating in her curious ideodynamic hand and finger movements.

CELESTE: I feel like I can't move my hand, except for exactly where it is.
ROSSI: Uh-huh? You can't move your hand, except for where it is. How interesting, 'cause it's really . . .
CELESTE: It's kind of locked there.
ROSSI: It's kind of locked there, uh-huh?
CELESTE: Like I can't open it, but I can't close it.

She is apparently experiencing a quiet, spontaneous, and involuntary catalepsy that is characteristic of high-phase hypnotherapeutic work. Inexperienced observers sometimes wonder if the person experiencing this kind of permissive hypnosis is really in an altered state when he or she talks in an apparently normal manner. It is just such peculiar verbal reports as these, when Celeste says *"I can't move my hand"* and *"It's kind of locked,"* that support the notion that she is in a special state called "trance." The fact that she seems to stumble into this cataleptic experience accidentally, without conscious intent or any suggestion from me, leads us to hypothesize that this momentary suspension of movement may have ideodynamic significance on an implicit level. It would be interesting indeed if the neural location of activity during such states could be identified with brain imaging techniques (Petrovic et al., 2002).

> ROSSI: You can't open it, you can't close it. But it really is doing both things. It's opening and closing. Have you ever experienced anything like this before?
> CELESTE: Hum-mm [apparently indicating, no.]
> ROSSI: I wonder if you are curious to look at that hand and see what it's doing, or would you rather just allow it to continue . . . ?

Although I never suggested that she close or open her eyes, she has repeatedly done so throughout this demonstration. I usually do not comment on eye behavior, but I do take it as some indication of whether a subject needs outer reality orientation (eyes open) or is comfortable exploring and working within, on his or her own (eyes closed is most characteristic during stages two and three of the creative cycle). I tend to remain silent when the patient's eyes are closed and he or she is raptly focused on inner work.

At this point I am interested in evoking curiosity by focusing her attention on her experiencing to further facilitate the symptom path approach. Looking at her hand's novel and odd behavior could be an implicit processing heuristic that motivates new awareness, self-reflection, and the co-creation of healing.

> CELESTE: Mmm . . . [Celeste opens her eyes and looks at her hand in fascination for a moment.]
> ROSSI: Weird. Huh?

I am being very careful here by using the word *weird* she introduced earlier to describe her surprising and perhaps numinous experience. I thereby hope to utilize her own worldview to further facilitate her numinous experience.

> CELESTE: Yeah, weird [laughs]!
> ROSSI: Have you ever seen it doing anything like that?
> CELESTE: No.
> ROSSI: Oh my goodness. Those . . . I've never seen . . . whew. . . . It really seems as if it is trying to stretch and open up more and more. Does it feel like that to you?
> CELESTE: Yeah. Feels like it got locked. It does still feel locked. But it still feels a bit like it's stretching.

ROSSI: Yes. Feels like it got locked but it still feels like it's stretching.
CELESTE: Yeah, yeah.

Until now, she has been exploring these ideodynamic movements in her right hand, where she first experienced her rheumatoid arthritis. Now I notice more activity in her left hand and so begin to humorously comment on it to facilitate it further.

ROSSI: Yes, now this other hand seems to be wanting to get into the act. [Audience laughs.]

I presume audience members are laughing because they recognize my not-so-subtle effort to facilitate and further extend the dissociative aspects of her experience on whatever therapeutic path they may explore.

ROSSI: Is this more on a voluntary level or are these hands . . . this other hand moving by itself, too?

I am still playing around with questions that may help her stay on creative edge between voluntary and involuntary (dissociated) behavior, where complex adaptive systems theory suggests the most intense "computing" takes place—on the edge between stability and chaos.

Voluntary Behavior
and Free Choice in Therapeutic Hypnosis

Symptom-Scaling and Paradox to Facilitate Therapeutic Progress
[33:50]

CELESTE: I wanted to see if this other hand was locked, too.
ROSSI: Oh! It's not locked, is it? It's not locked. Okay. By the way what number would you say your state is in. It got up to 80, 80, 70. Where are you now?
CELESTE: Well, the rest of my body feels good. *My hands feel worse.*

This is a curious and mildly alarming report: *"My hands feel worse."* I cope with this seeming setback by requesting further confirmation via symptom-scaling. Hopefully, sympton-scaling will function as an implicit processing heuristic to get us back on a healing path.

ROSSI: So the rest of your body feels better, but your hands are worse. That is wonderful.

I use *wonderful* in a way that is clearly paradoxical because I don't know what else to do in my mild sense of shock at her hands feeling worse. Ever optimistic, I now try to utilize the worse feeling in the hands by channeling and reframing it as a part of the normal work along the symptom path. Actually, this is a *truism*. It is entirely typical for patients to go back and forth in experiencing symptoms worsening and then getting better while they are shifting uncertainly between stages two and three of the creative cycle.

UTILIZING A SEGMENTED
TRANCE TO FACILITATE HEALING

Focusing Activity-Dependent Gene Expression and Healing in High-Phase Hypnosis? [34:15]

ROSSI: Let's let it . . . all that misery flow into that hand. Let all the worst be there. Because it seems to . . . I mean, I know you're . . . to me, it seems to be having fun, I'm sorry [laughs].

Suggesting that *"all that misery flow into that hand"* is a way of facilitating what Erickson (Erikson & Rossi, 1985) called "the segmentalized trance" that could function as a therapeutic crucible wherein the symptom is dealt with safely while the rest of the patient's body remains comfortable. From a current neuroscientific perspective, we could hypothesize that a segmentalized trance functions as an implicit processing heuristic that could facilitate highly focused, activity-dependent gene expression and healing exactly where it is needed. Black (2001) has described how the target organ (in our case, arthritis in Celeste's hands) can generate neurotropic factors—growth-enhancing hormones such as nerve growth factor (NGF)—that signal and guide neurons in the central and peripheral nervous systems to the target organ to facilitate healing at the cellular-genomic level. This is only one of many possible molecular-genomic mechanisms that now needs to be investigated as a pathway of mind-body healing.

CELESTE: Probably.

ROSSI: But you are saying it is not feeling comfortable in that hand as it's moving?

CELESTE: Well, the movement feels good—the little movements.

ROSSI: Uh-huh.

CELESTE: In fact, it feels that it could be fun or it just feels like that feels good.

ROSSI: So the little movements feel good. That could be fun.

CELESTE: Yes.

ROSSI: So what's that . . . ?

CELESTE: The part that hurts is the locked part. It hurts on the wrist and it hurts on . . . [Celeste tests her wrist by moving it gently back and forth].

ROSSI: Oh, the wrist hurts.

CELESTE: Yeah.

ROSSI: Oh, I see.

CELESTE: Or if I try to open it or close it, it hurts.

ROSSI: I see, trying too hard to open it makes it hurt, but so long as it has little movements by itself, it is okay, it's like fun?

CELESTE: Well, the wrist still kind of hurts.

ROSSI: Huh?

CELESTE: The wrist still hurts. It kind of . . . ?

ROSSI: The wrist still hurts.

CELESTE: Holding like that.

ROSSI: All right! Okay, let's see what happens *next*.

The attitude of creative play allows us to explore and discuss the pain frankly without too much despair. Her pain at this point may be the price we are paying to explore her symptom via psychobiological arousal rather than the traditional hypnotherapeutic route of relaxation, comfort, and near sleep.

In the language of Chapter 5, we have been facilitating "high-phase hypnosis" rather than the dissociative and passive but comfortable "low-phase hypnosis." When in the middle of the high-phase of experiencing the state-dependent sources and psychodynamics of a symptom, it may seem to get worse before it gets better. This is the rationale of cautiously encouraging Celeste to continue along the symptom path to experience whatever comes next. She is engaged in creatively replaying the state-dependent dynamics of her symptoms to facilitate the possibility of a therapeutic re-synthesis and healing. It is at such delicate moments as these that makes us realize how urgently we need the new tools of brain imaging, DNA, immune system, and protein microarrays to guide our work (Chin & Moldin, 2001).

CREATIVE REPLAY TO DEEPEN INSIGHT

Utilizing "Primitive" Psychodynamics of the First Arthritic Symptom? [35:24]

CELESTE: It's actually my wrist that first got it.

ROSSI: Oh, it was your wrist that first got it!

CELESTE: Yes.

ROSSI: So it's the first to complain and it's the last that's going to give up this complaint [laughing].

CELESTE: Mm-mm.

ROSSI: Again, I really wonder . . . what kind of . . . ?

CELESTE: When I looked at it, it looked kind of *primitive or something.*

ROSSI: Yes, it did. That's exactly a wonderful word, *primitive.* I've never thought of it, but it does seem to fit, *primitive.* So go with that, Celeste, really primitive. Let's . . . that's it . . . really primitive . . . great . . . really primitive. Are you curious to take another peek to see if it's more primitive now?

CELESTE: Yeah. Yeah [laughs].

ROSSI: It is more primitive!

CELESTE: Look at it.

ROSSI: Yes. Good. Go with that primitiveness now. Yes, go with

the primitiveness and let's see what primitive . . . really the courage now, Celeste, simply to receive the primitiveness.

I sit back in my chair, since Celeste now seems poised to revert to a stage-two incubation process to work with her new insight about her perception of the primitive appearance of her arthritic hand. Notice that this is her perception and interpretation, not mine. I immediately seize on her word, though, and *utilize it in a naturalistic manner* (Erickson, 1958/1980, 1959/1980) to make contact with her ongoing numinous replay of the experience of her own symptom and personal identity. It is not the therapist's job to interpret behavior at this stage. The therapist's task, at this point, is to simply recognize the patient's symptoms and insights and encourage further exploration with natural variation and selection on a more or less implicit or unconscious level. There will be time in stage four of the creative process for more conscious reflection, analysis, interpretation, and selection of what is really important for reframing and recreating her life.

There is now a long period wherein her fingers make many small, exploratory movements very gingerly. They are silently signaling the implicit levels of creative replay that are taking place on an unconscious level. This is highly characteristic of the creative replays with natural variation that take place during those periods of private creative inner work.

ROSSI: And where is that taking you now, Celeste?

I am emphasizing the symptom path with the words *"where is that taking you now."* Notice how this phrase may function as an implicit processing heuristic facilitating her unconscious replays while she remains a mildly dissociated observer and reporter of her involuntary experience, rather than attempting to construct her experience on a conscious level. This is a good example of how she is exploring multiple levels of inner experience in her search for the healing path.

NOT-KNOWING INITIATING CREATIVE REPLAY WITH NATURAL VARIATION AND CONSCIOUS SELECTION

Optimizing the Natural Flux of State-Dependent Symptomatology
[37:13]

CELESTE: I don't know.

ROSSI: You really don't know?

CELESTE: *Well, for a minute it brought me back to where I was when I first got it. But . . .*

This is an important breakthrough moment wherein Celeste experiences a state-dependent flashback to the origin of her arthritic symptom! (Erickson et al., 1986) This could engage creative replay to resynthesize the psychogenomic encoding of her symptoms to initiate neurogenesis and healing as proposed previously in Chapter 3. Unfortunately her consciousness falters as she hesitates with "But . . ." so I immediatly jump in with support as follows.

ROSSI: Yes, for a minute it brought you back to where you were when you first got it, so let's see if you can really go back there when you first got it. And let's see if there is anything . . . primitive? [Long pause, as Celeste is obviously engaged in inner work, her fingers gently swaying in an apparently autonomous manner.] That's right, Celeste, staying with that when you first got it. That's it! Staying with that! All about that, and most of that can remain private within you, of course. That's it. Really, privately within yourself. [Celeste's fingers showed a renewed burst of activity, seemingly in response to the word *private*.] The whole complete experience. Everything that's going on.

I am again trying to support *the flux of creative variability* and dynamic movement along the symptom path. Celeste responds by replaying another burst of finger movement.

ROSSI: Yes, that's right, very good. Yes, mm-hmm. That's right. [Celeste swallows.] And now . . . ?

I am inquiring about her swallowing behavior without necessarily bringing her attention to it by naming it. The ambiguous *"And now . . . ?"* gives her optimal freedom to talk about whatever may be necessary.

Taking a Break from Therapeutic Work

Body Language Signaling an Ultradian Rest Period? [39:20]

ROSSI: [Celeste continues gingerly making finger movements.] Mm-hmm . . . ? [Celeste now smiles and seems to be exploring, on an apparently voluntary level, slow, delicate micromovements with her fingers.] That's right [laughs]. It's enough to bring a smile to your face.
CELESTE: Yes.
ROSSI: I don't know if you are having a good time, though, or a tough time.

Here I make a permissive, nonintrusive statement of *uncertainty and not-knowing* that lets her know she can speak—if she wants to.

CELESTE: *It's just kind of a relief to pull it back.* It's just . . . I can hardly move it.

I am very intent on noticing the slightest nuances of her behavior as I watch for any opportunity to move the healing process along in an active manner. I now suddenly notice what was right in front of my nose but to which I had paid no attention until Celeste makes this observation. Her hand has been moving slightly backwards, as if to disengage from the process. Amazingly, I had not noticed it. This is certainly a good example of the limitations imposed by the therapist's own preoccupations. Heck, even Freud and Jung cautioned their followers not to pursue healing too zealously.

ROSSI: I see. *It's a relief to pull it back?*
CELESTE: Yeah.

Why is it a relief for her to pull her hand back? Is this new behavior an expression of resistance? Or is it a reflection of her need to take a little healing rest-break? When she says, "I can hardly move it" in the context of having been hard at work for almost 40 minutes, it may mean that she is tired and is naturally slipping into the ultradian rest phase. Whatever. Since we really don't know, I simply encourage her further along the symptom path of going with whatever is happening. Moving back could be a spontaneous therapeutic movement into low-phase hypnosis and symptomatic relief from her hands feeling worse. In fact, as we look again at the diagram of the four-stage creative process in Figure 2.11, we notice that spontaneous relaxation is characteristic of the shift from stage three to stage four.

METAPHORICAL WISDOM OF BODY LANGUAGE

Shock and Surprise at Novel, Involuntary Catalepsy
[39:54]

ROSSI: Yeah! Go for that! Go with that pulling back. Very good. Very good.

Celeste's hand visibly moves back in response to my permissive suggestion. This seemingly eager backwards movement confirms the wisdom of letting her complex adaptive system do what it needs to do when it needs to do it. Thank heavens for the metaphorical wisdom of body language! Body language is no mere metaphor; it is real and we need to learn to be sensitive and responsive to its messages. One of the easiest entry points for developing this sensitivity to the body's messages is through the ultradian alternations between activity and rest that occur throughout the 24-hour circadian cycle.

CELESTE: *I mean, it's shocking.*
ROSSI: It's shocking?
CELESTE: *It's shocking that I can't move it.*
ROSSI: Now you can't move it?
CELESTE: *I can, but it's slow.*
ROSSI: Yeah. Yeah. That *wonderful contrast between what it's doing*

and what you can do and can't do. Just go with all those . . . yeah, that's it.

Waking up to the ultradian wisdom of her body language, I now strongly support this *"wonderful contrast between what it's doing and what you can do and can't do."* Why is the contrast so "wonderful"? From a neuroscientific perspective it may be wonderful because it is precisely here—at the threshold between involuntary, implicit processing versus voluntary, explicit consciousness—that new insights and behaviors become manifest. To help people become aware of such "creative moments" (Rossi, 1972, 2000a) is to encourage them to value, support, and receive insights in their natural everyday experiencing as well as in their therapeutic work.

It is rather humbling that even as I play the role of the so-called therapist in this demonstration, I must rely on *her* mind-body wisdom to correct my preoccupations and concerns about the apparent worsening of her pain between the times 33:50 and 35:24. Recall that the peak of her psychobiological arousal in this therapeutic session occurred within these few minutes. A tangential thought: This emotional peak was called the crisis in the historical literature of therapeutic hypnosis and the healing arts more than a century ago. This period in the middle 1800s was immortalized by the image of the famous Charcot holding a mental patient, who was draped backwards over his arm in a full-blown hysterical crisis, as he demonstrated his technique before a group of famous physicians in the Salpêtrière hospital (Tinterow, 1970).

THE THERAPIST FOLLOWS THE PATIENT'S LEAD

Surprising Fist Formation and the Psychodynamics of Self-Empowerment [40:20]

ROSSI: [Celeste's hand suddenly makes another slight, spasmodic movement and then surprisingly forms a fist.] Oh, my goodness? Something new seems to be happening?
CELESTE: Yeah! [Her fist tightens further.]
ROSSI: Wow! Yes, something new is beginning to happen!

My tone is obviously enthusiastic in response to this new and unexpected behavior by Celeste's hand. Celeste's clenched fist turns toward the audience momentarily. I now strongly support this creative moment.

ROSSI: That's it! That's it. Really!

I now seem to be in therapist heaven as I suddenly realize how the classical psychodynamics of withheld anger and reactive symptoms could be involved in Celeste's rheumatoid symptoms. If this interpretation were correct, perhaps her therapeutic progress could be facilitated by anything that would help her to experience self-empowerment. From this point on, I do everything I can in this session to further Celeste's activity-dependent experiences of self-empowerment by encouraging exploratory finger and hand movements, playful shadow-boxing (as in Figure 8.1), using strong language and humor to support her emotions and evolving insights—and, finally, the standing ovation she receives from the audience at the end (as in Figure 8.2, p. 389).

ROSSI: That's fantastic! [Celeste's hand again makes a fast clinched movement, to which I respond by imitating her movement with my fist and verbalizing enthusiastic, emotional support.] Mm-mm, that's it. Wow! How does it feel to let it really do that? [Celeste now extends her fingers up in the air, as widely as possible, as if for all the audience to see. She seems to be testing how widely apart she can stretch those fingers.] Yeah, like that? Yeah!

I previously imitated her clenching fist, and now I imitate her surprising behavior of stretching her fingers as wide as possible, verbalizing my utter surprise and delight. The important thing to notice is that I am now following Celeste's creative behavior rather than the other way around, as is supposed to happen in misguided views about programming people.

ROSSI: My goodness!

FIGURE 8.1 | Stage Three: Illumination and activity-dependent gene expression. Celeste experiences playful activity-dependent exercise as a creative breakout of her typically restrained hand and finger movements associated with her rheumatoid arthritis. Future research will be needed to determine if activity-dependent gene expression (ADGE)—such as the CREB genes associated with new memory and learning—as well as the ODC and BDNF genes associated with neurogenesis and physical growth are actually being engaged.

PATIENT'S SURPRISE AT HER
OWN NOVEL THERAPEUTIC BEHAVIOR

Group Validation of a Breakthrough Experience [41:14]

CELESTE: *I sure don't know what this is* [laughing].
ROSSI: [Celeste continues to laugh as her hand opens and closes several times slightly.] You sure you don't know what this is [laughs]?
CELESTE: No.
ROSSI: Well, I'm starting to get some hints. But those are just my private fantasies, I suppose. All I know is that I sure am getting a charge out of this. [Celeste, Rossi, and the audience laugh together—everybody is getting a charge out of it.] Very good that's it. Yeah. Very good!

A numinous, co-creative experience that is highly characteristic of all cultural rituals of healing.

CREATIVE REPLAY
EXPLORING SELF-EMPOWERING MOVEMENTS

Accessing Therapeutic "Energy" [41:43]

CELESTE: It feels like, um, when you hyperventilate and your hands get stiff from it. If anyone's done that? It is sort of what it feels like. It feels like all this energy is locked up in my hand.
ROSSI: Yes. It's like when you hyperventilate and like your . . . all your energy is locked up in that hand. And that's what we are getting to! That energy that's locked up there.

I actually dislike the metaphor of *energy* to describe therapeutic breakthroughs. I vastly prefer the more current metaphors of *information, mindbody communication,* and *creative processing on the edge of chaos* to describe what is going on here. But I follow my professional training and dutifully adopt Celeste's words, as I clench my fist imitating her movements.

CELESTE: Yes.

ROSSI: Incredible. *That's it!* Really go ahead. Yes. Yes!

With an enthusiastic *"That's it!"* I am attempting to further empower Celeste's exploration of her ability to clench her fist, stretch her fingers, and so on. This could be a creative moment of activity-dependent gene expression and healing.

ROSSI: That's it! Wonderful.

Even though Celeste apparently is not observing how I am clenching my fist, her hand makes a fist again, seemingly in response to my urging tone.

ROSSI: Wow. By golly!

CELESTE: [Celeste now smiles as her clenched fist points toward herself.] Wow, my wrist is stiff.

ROSSI: Huh?

CELESTE: That wrist is really stiff.

Celeste is moving her hand back and forth at the wrist, testing its range of movement. She is evidently picking up my enthusiasm, using my "Wow!" word. A therapeutic mind-meld—a shared phenomenological therapeutic experience—is obviously taking place (Rossi, 1972, 1985, 2000a).

PSYCHOLOGICAL SHOCK FACILITATING CREATIVE REPLAY

The Strange Intimacy of Swearing for Psychobiological Arousal
[42:27]

ROSSI: Yes. The wrist is really stiff. *Hell,* it hasn't been moving for a while.

I now imitate how Celeste is circling her wrist as a murmur of laughter ripples through the audience. Why is the audience laughing at this point? Is it because I used the swear word "Hell"? I usually do not use *hell* in my everyday speech and certainly not when I am doing professional therapy demonstrations. I rationalize to myself at this moment that I used the swear word in a mildly humorous manner to embolden Celeste to shake herself free from what I imagine may be her inhibited behavior (Rossi, 1973a).

Using the word *hell* in this context has many layers of complex, adaptive meaning. Most importantly, it may evoke a mild sense of shock and psychobiological arousal (sympathetic system arousal) and the attendant possibility of *behavioral state-related gene expression* and the accessing of *state-dependent memory, learning, and behavior* that could lead to the replaying of some unfinished business regarding significant life events. Replaying, particularly in the context of a therapeutic session, may generate a process of Darwinian natural variation and selection that would facilitate the possibility of healing psychogenomic responses.

If this replay accesses a deep vein of numinous motivation, Celeste may become engaged, on explicit levels of experience, in a more extended psychobiological process of (1) *activity-dependent gene expression,* (2) *protein synthesis and neurogenesis in her brain,* as well as (3) *the recruitment of stem cells for healing in the rheumatoid or otherwise dysfunctional tissues throughout her body.* This may be a lot to ask of using *a single swear word,* but it would be a wonderful economy of effort if future research could validate that it actually happens. Notice that I speak of swearing *humorously with* a person, not *at* a person. Swearing *with* a person in a humorous and mock conspiratorial tone is a way of creating a *strangely binding intimacy* (rapport, transference, and countertransference) for a moment that could evoke a person's psychobiological dynamics, from mind to gene. Swearing *at* a person *seriously,* of course, is something else and is unethical in psychotherapy. It would be interesting to explore the creative possibilities of swearing seriously *at* people in everyday life: Could serious swearing *at* someone actually engage the psychobiology of gene expression and creative replay? Or would it invariably result in inhibition, a dampening of gene expression, and depression?

In the context of demonstration therapy, the professional audience plays an important but little investigated role in relation to controversial behavior. The situation may be confusing for the demonstrating therapist, whose attention is now divided between teaching and pleasing the audience versus the needs of the patient. The audience might lend its humorous support to Celeste's empowerment, whether or not it is really good for her at this point. This could be dubious behavior on the part of both the audience and the therapist. Use of swear words, therefore, cannot be recommended for general clinical practice until its value is assessed with further research.

Perhaps a limitation of the demonstration therapy modality is the possibility that the therapist responds with hubris and a loss of sound professional perspective and judgment, even when the therapy appears to be going well. We may be witnessing this here and in a few of the following sections. Let's carefully monitor the ethical and psychodynamic issues embedded in this evolving, action-oriented therapeutic situation. Just how far can the therapist go in applying current neuroscientific research that reports how novelty, environmental enrichment, and physical exercise lead to gene expression, neurogenesis, and possibly healing?

> **ROSSI: My goodness! It's about time we had a little action here [humorously]!**

Notice my spontaneous and unpremeditated *apposition of opposites* when I juxtapose hell and goodness in these adjacent sentences. This may or may not have been a useful implicit processing heuristic at this point; the apposition of opposites may function as a therapeutic double bind (Erickson et al., 1976; Erickson & Rossi, 1975; Rossi, 1996a) and is also reminiscent of Freud's (1910) concept of the antithetical meaning of primal words that we will touch upon again in Chapter 10.

PAIN AND GENE EXPRESSION:
ADRENALINE HIGH OR GENUINE HEALING?

The Limitations of Current Psychotherapy on the Phenotypic Level
[43:03]

> **ROSSI: Stiff a bit too long, I would say. My goodness. Yeah.**

As I evaluate my remark here, a few years later, I am really embarrassed and chagrined. Who am I to make such a negative pronouncement as *"stiff a bit too long, I would say"*? How could any therapist make such a remark without really knowing what is happening at the molecular-genomic level? This is an unfortunate limitation of our current psychotherapeutic modalities: Judgments can only be made on the phenotypic or behavioral level of observation. Psychotherapy, in all its current manifestations, does not have access to the genotypic level, where much of the significant therapeutic action needs to take place. The general lore of current psychotherapy supports the value of some phenotypic levels of clinical judgment, as was attempted here. We clearly have a great need for the new methods of genomic neuroscience that can help us make judgments based on the genotypic level in the future, however (Chin & Moldin, 2001).

> CELESTE: Mm-mm. [She makes more exploratory movements as she glances at my imitation of her previous movements.]
> ROSSI: Yes, how does that feel when you do that? Yeah. How does that feel?

I ask these questions with a growing sense of concern as I observe Celeste's bolder swiveling of her wrist. Is it really good for her? The videotape illustrates how we are engaged in a co-creative activity wherein each is imitating the other's hand and finger movements while simultaneously interweaving our own natural variations, as if we were seeking something but didn't know quite what.

> ROSSI: Yeah. How does that feel?

I am really concerned now: Are her movements helping or hurting her?

> CELESTE: It's a little more normal.
> ROSSI: Huh?
> CELESTE: It feels a little more normal.

I am relieved to hear the good news that Celeste's movements are helping her wrist feel *"a little more normal."* But what does *normal* really mean in this context? All good athletes know, for example, that in the heat of their best performance they are on an "adrenaline high" wherein their great performance may be pushing them "over the top" and actually risking injury (since the same adrenaline that facilitates performance can also mask temporarily the normal warning signals of pain). Does Celeste's experience of *"a little more normal"* really signify a genuine and lasting therapeutic response? Or is she merely on a brief adrenaline high? We have no way of knowing at this time. We do not yet have the real-time diagnostic tools for assessing this critical third stage of the creative cycle in the psychotherapeutic situation. Research using functional brain and body imaging techniques (such as fMRI and PET) and DNA microarray assessment of gene expression will be useful here.

> ROSSI: It feels a little more normal, right! It *almost* looks normal, too!

I can't believe I really made such an awkward and condescending remark— "It *almost* looks normal too!" I guess I'm getting tired without realizing it. Celeste restores a proper sense of perspective with her next observation.

> CELESTE: My wrist still hurts, though.
> ROSSI: It hurts still. But, yeah. Yes. Great. My goodness! Whoa! All right! Fantastico! Yes. Yes. Gosh. Mmm-mmm. Yeah. Mmm.

If I seem confused in my comments, it is because of my *creative uncertainty* between concern for her hurt, *the fact that some pain may be associated with activity-dependent gene expression and healing* (see Figure 5.11), and wanting to support Celeste as she now tightens her fingers in a clenched fist as if testing her movements and sensations. I hypothesize that such activity-dependent testing could facilitate creative replay and psychogenomic healing.

> ROSSI: Mm-mmm. Yes. My gosh. Do you really still want to keep your eyes closed, or do you want to watch some of this action? Look at that!

I am now suggesting that Celeste open her eyes for some reality-testing and perspective on whether she really wants to clench her fist so tightly. She peeks at her hand and laughs as her fist clinches tighter and she closes her eyes again. She is apparently okay with her right hand boldly clenching with force.

REPLAYING THE POSSIBILITIES
OF NATURAL VARIATION AND POSITIVE SELECTION

Self-Reflection and the Novelty-Numinosum-Neurogenesis Effect [43:57]

CELESTE: I'm embarrassed to watch it [laughing].
ROSSI: What's embarrassing about that? Yeah. What's embarrassing about that? Huh? Tell me.

I now seek to facilitate her self-observation skills particularly about anything positive, such as laughing embarrassment. There is a humorous edge to this sudden reversal of my previously protective attitude toward her privacy. My now seemingly outrageous demand for a public sharing of her private experience is a probe seeking to facilitate her self-awareness along with the nascent novelty-numinosum-neurogenesis effect that may be associated with it.

CELESTE: I'm not sure.
ROSSI: I really, really want to know [humorously]. What's embarrassing about it?
CELESTE: It makes me more aware of what's going on here.
ROSSI: Yeah. What's going on here? [Celeste glances at the audience and smiles in apparent self-consciousness.] A lot of people.
ROSSI: My gosh! Mmm. Yeah! How does it feel when the hand actually closes like that, Celeste?
CELESTE: *Well, how does it feel? It feels kind of good. It feels strong, actually.*
ROSSI: Yes, it feels kind of good, it feels kind of strong, actually. That's what I want you to continue exploring, Celeste. That good feeling. That strong feeling.

Thank heavens! Even the worst of therapists sometimes get a break as they muddle through their creative uncertainty toward the end of a session. I am now falling all over myself encouraging her to replay the good, strong feelings she is experiencing in contrast to the previous painful and hurtful feelings. This is another aspect of the therapist's most important work in stage three of the creative cycle: Replaying the positive possibilities of *natural variation* and conscious creative *selection* of optimal psychobiological states.

MULTIPLE-LEVEL HUMOR OF
THE OBSERVED OBSERVING THE OBSERVER

Facilitating Activity-Dependent Gene Expression, Neurogenesis, and Healing? [44:45]

ROSSI: That's it. By golly. Whew. [With a contemplative attitude, Celeste twirls her clenched fist at the wrist with apparent ease and comfort.]
ROSSI: All right. Mmm. Mmm . . . yes. Incredible. Does it still feel? . . . mmm. All right!

Celeste glances at me and notices that I am twirling my fist, just as she is. I notice her noticing and we both break out in laughter. This is a fun manifestation of the observed observing the observer and the observer noticing that he is observed! This could only happen in a complex adaptive system where all parts play off each other in sometimes humorously absurd but possibly facilitative ways.

ROSSI: To hell with all that verbal therapy business [laughing heartily], right?
CELESTE: Right!
[There is a roar of approving laughter from the audience].

Are we facilitating activity-dependent gene expression, neurogenesis, and healing here—or what? Are we transcending a few limitations of the past centuries of psychotherapy as a lengthy, serious procedure of analysis and interpretation limited to the verbal level as we move on to a more action-oriented facilitation of the patient's own creative process—or what?

ROSSI: Let's have a little real action here! My gosh!

Insights into the
Psychodynamics of Rheumatoid Arthritis

Facilitating Self-Empowerment with Humor [45:44]

CELESTE: When I first got this [rheumatoid arthritis] I put a plastic stretch thing around my wrist. There was one person that was, um, kind of in a power position over me that had given me a lot of grief.

ROSSI: This guy in a power position over you who had given you a lot of grief?

CELESTE: Yeah, and asked to see it. And I remember I clenched up my fist like that, and I said, "You can see it."

ROSSI: Come a little closer [laughing].

[Further audience laughter]

I empower her further with humor by interpreting her fist clenching to mean, *"Come a little closer and I'll sock you one."*

ROSSI: You can see it [laughing]!

CELESTE: Yeah [laughing].

ROSSI: Oh, boy. So this is the beginning of this so-called grief, this character, huh?

CELESTE: It was a woman.

ROSSI: Huh?

CELESTE: It was a woman.

ROSSI: It was a woman.

CELESTE: Yeah. She was a character, too.

ROSSI: She was a character, too. And to what degree did you **knuckle under her**?

Poor pun intended! This is an effort to utilize her metaphoric arthritic body language, as in the double meaning of "knuckle under."

CELESTE: Unfortunately, I was in a position where I couldn't say anything and was told to weather it for a while.

ROSSI: You couldn't say anything and you were told to just weather it.

CELESTE: Yeah. And I weathered it for, about, I think for about a month before this happened.

Is Celeste actually recalling the genesis of her arthritic syndrome, even as she replays and reorganizes it with therapeutic insight at this point? If so, this would indeed be a stage-three insight in the creative process. It is precisely at such delicate moments of psychological transformation, when important memories are recalled, replayed, and reconstructed in a therapeutic manner, that we presume that the gene expression and protein synthesis cycle is reactivated, as described in Chapter 3 (Dudai, 2000; Nader et al., 2000a, 2000b; Shimizu et al., 2000).

ROSSI: You weathered it for a month—kept your mouth shut—before this happened.

CELESTE: Yeah.

ROSSI: That's how long it took to develop this [arthritic syndrome]. A month of holding that stuff in and just weathering it.

UPDATING THE PSYCHODYNAMICS OF FREUD AND JUNG

Fragile Insights in Short-Term Memory during Creative Moments
[47:18]

CELESTE: Yeah. Um. Yeah. I just thought of something, but I can't . . . but I forgot it right away.

ROSSI: What's that?

CELESTE: I just thought of something but forgot it right away.

This is an interesting example of short-term memory failure to "hold on to" an intrusive thought that could be another important insight. It reminds us again of current neuroscience research on the shaky memory trace (Dudai, 2000) in the moment when memories are being recalled and transformed. Freud (1901/1938) regarded such lapses of memory as aspects of psychopathology. Jung (1911/1970) regarded them as indicators of the activation of emotional complexes. Genomic neuroscience now adds that such memory lapses can be indicators of creative moments of self-transformation (Rossi, 1999b, 1999c, 2000a, 2000d). Now Celeste sits back and contemplatively looks at her fully outstretched arm and hand. I imitate her movement in an effort to stay in touch with exactly where she is, experientially and behaviorally.

> ROSSI: Yeah. Well, to hell with it, it can't be that important [with humorous sarcasm].

This is the third time I have used the word *hell*. I really am on some kind of a roll. This is not a normal pattern of language for me. Am I using the swear word in an appropriate manner to further facilitate the spontaneity of Celeste's natural behavior? I even reinforce her immediate forgetting here, in defiance of the normal prescription in psychotherapy to recall as much as possible. The audience roars in laughter at this apparent flouting of the normal rules of therapeutic inquiry.

THE PSYCHODYNAMICS OF INSIGHT

A Tongue Slip Revealing the Genesis of a Symptom [47:44]

ROSSI: My goodness.
CELESTE: Oh, I know, it stopped the grief.
ROSSI: It stopped the grief?
CELESTE: This person was giving me. That's the unfortunate part.
ROSSI: Really, you are saying? Wait a minute. Let's hear what you are saying. When you got the symptom, it stopped the grief?
CELESTE: This person started—stopped—giving me the trouble.

It is wonderful to have a good video record of this tongue slip that really was no tongue slip. It is actually a condensation of two thoughts about the psychodynamics of her symptom. The supervisor's critical attitude was the source of the stress that "started" Celeste's arthritic syndrome. This same supervisor "stopped" her critical behavior when Celeste's arthritis became manifest and thus inadvertently reinforced it. When Celeste says "This person started—stopped—giving me trouble," it is therefore not really a tongue slip in the psychopathological sense. It is a mental condensation that reveals the exact psychodynamics of the genesis and reinforcement of her arthritic syndrome.

ROSSI: Oh, that person stopped giving you trouble when it . . . when it became . . .
CELESTE: Yeah.
ROSSI: . . . obvious that you were—what? Having a problem with your hands and stuff?
CELESTE: Yeah. And especially when it got diagnosed.
ROSSI: Especially when it got diagnosed. Well, *that was big of her,* wasn't it?
[The audience laughs at my humorous sarcasm in *"that was big of her."* Celeste has a big smile. I whistle a brief little tune.]

EXTENDING THE CREATIVE MOMENT WITH HUMOR

The Need to Assess Therapeutic Humor [48:26]

ROSSI: How about that other hand? I notice it is starting to lift and wants to get into the act. Is that right? Let's see, what's going on here now? Does this one feel like making a . . . that's it. How does this one feel now?

I am doing all I can to capitalize on Celeste's rapid series of insights and therapeutic movement by facilitating the nascent state of her creative edge via an exploration of the borderland between voluntary and involuntary behavior.

ROSSI: It seems to be recovering faster!
CELESTE: Yeah. This one was never as stiff. It's a little bit . . . it's

stiff, actually.

ROSSI: Yeah.

CELESTE: It is.

ROSSI: All right. My goodness. How long since you have really moved your hands this way?

CELESTE: I used to play the piano.

ROSSI: Really?

CELESTE: Yes.

ROSSI: How long has it been since you've played the piano?

CELESTE: Five years.

ROSSI: God! I suppose there isn't a piano around here!

[Observers of the videotape have remarked that I move my hands in interesting, humorous, and somewhat novel ways at this point. Everyone laughs as I suddenly turn and look to the side of the stage, as if looking for stage hands to bring in a piano.]

What is the staff doing!? [yelling in a mocking and humorous manner].

I am joking by suggesting that the staff supporting this workshop demonstration did not have the foresight to place a piano on stage that we could have used to test Celeste's therapeutic progress at this point. This again indicates a need for the developments of practical neuroscience tools to assess therapeutic progress in the creative moments of transformation in psychotherapy.

"ENERGY" AS A THERAPEUTIC METAPHOR

Play-Acting Boxing as an Activity-Dependent Therapeutic Exercise [49:28]

CELESTE: It feels good, that. Yeah. It feels like there is a lot of energy in both of them [referring to her arms and fists held up in a quasi-boxing position].

ROSSI: Yeah. A lot of energy in both of them. Go with the energy now. I mean, I really just want you to enjoy and have some delight in that energy. Yeah!

Celeste and I are now engage in pantomiming boxing movements (Figure 8.1). We are not pretending to shadow-box each other; rather, we both face more toward the audience than one another. We both grin broadly as we occasionally glance at each other, as if egging the other. Even when reviewing the tape, it is hard to determine who began this fun play of boxing movements. We are obviously having a good time together in this activity-dependent flow. Supporting fun and play seems to be the emerging attitude in psychology these days (Seligman, 2001; Seligman, & Csikszentmihalyi, 2000).

CELESTE: Mmm.

ROSSI: Go, team. go! [There is laughter all around.] Wow. Gosh. Boxing gloves! Staff!?

[Audience breaks out in laughter again as I humorously call for the staff of the Erickson Foundation to provide boxing gloves to further empower Celeste's therapeutically combative mood.]

CELESTE: I had boxing gloves. When I was a kid, my dad taught me how to box.

ROSSI: Really? Your dad taught you how to box! How was it? What was it like?

CELESTE: Well, it was one way to deal with my brothers.

ROSSI: Oh, my goodness. All right! All right. Whew. It would have been a hell of a good way to deal with her [referring to Celeste's former boss] too, wouldn't it?

CELESTE: Yeah.

ROSSI: Yes. Yes. Great. Whew.

[There is laughter all around as we continue our playful pantomiming of boxing movements, with great grins on our faces.]

ROSSI: Doesn't it feel good?

CELESTE: It does, yes.

ROSSI: Yeah. Mmm. Mmmm.

The Therapeutic Domino
Effect on the Symptom Path

Foolish Temptations of Demonstration Therapy [50:46]

CELESTE: Mmm. I don't feel that pain in my back anymore.
ROSSI: You don't feel that pain in your back anymore. Well . . .
CELESTE: Kind of surprising.
ROSSI: Not surprising at all. We are getting to the source of these hands, man.
CELESTE: Yeah. I guess so.

This surprising progress in the relief of back pain, which she did not even mention previously, is another indication of therapeutic progress on an implicit level that is typical of this kind of mindbody work. Does such symptom relief, without any direct suggestion for it by the therapist, imply that we have initiated a therapeutic process on the level of gene expression that is now continuing with its own momentum? Only research will tell for sure. This is what Erickson (Erickson & Rossi, 1989) called the "therapeutic domino effect" (pp. 237–240). Solving one problem or symptom can quickly lead to a more general creativity sweep that resolves many more problems at the same time.

ROSSI: My goodness. How long has it been since you've worked those fingers like that?
CELESTE: I don't think—I haven't.
ROSSI: What's that?
CELESTE: I haven't even tried.
ROSSI: Haven't even tried? Oh. All right.
[Pause as Celeste continues her vigorous finger workout.]
ROSSI: Oh. It hurts. Oh, I'm feeling afraid that it hurts. Ahh.

Suddenly, in a startling and humorous manner, I mock the "hurt" feeling in a high-pitched falsetto voice. I am treading a dangerously thin line here. It can only work well if the patient fully realizes the therapist is mocking the symptom, not the patient. This is not a therapeutic procedure that could be recommended for clinical practice. It's not something I ever do in my private practice. Perhaps it happened here because I am foolishly "showing off" in front of an audience—possibly this is a dangerous temptation in demonstration therapy.

> ROSSI: Does it hurt?
> CELESTE: In parts it hurts—in other parts it doesn't hurt at all.
> ROSSI: All right. Other parts, you feel the energy, the power, the strength?
> CELESTE: Lots of energy on it, it's incredible.
> ROSSI: Lots of energy. All right. That's it.
> CELESTE: Yeah.
> ROSSI: Rah! Rah!

I strongly support her replay of the other parts of herself that experience energy, power, and strength. I humorously cheer her on like a coach with this *"Rah! Rah!"* exhortation. Research is needed to determine if such "hurt" and "energy" are associated with therapeutic gene expression and neurogenesis.

NOT KNOWING AS A SIGN OF NOVEL BEHAVIOR

Implicate Dynamics of Healing on the Molecular-Genomic Level
[51:54]

> CELESTE: *I don't know how I ever did this.*
> [Everyone breaks into laughter.]

It seems like everyone is having fun, but is this serious psychotherapy? The in-group joke among therapists who use hypnosis is that all is going well when the healing appears to take place on an implicit, unconscious level so that the patient does not know how the "magic" of the hypnotic work took place. From a neuroscientific perspective, "not knowing" implies that the real healing is on the molecular-genomic level.

STAGE-FOUR VERIFICATION OF CHANGE AND HEALING

Facilitating Therapeutic Progress in the Future [52:00]

ROSSI: Tell me *now, who would you like to box first?*
CELESTE: In my life?
ROSSI: In your whole life.
CELESTE: Oh, I'll have to keep that quiet [laughter all around].
ROSSI: Okay. Good. That's all right.

By asking, *"now, who would you like to box first?"* I am shifting toward stage four of the creative cycle: that of verification. How will Celeste continue her therapeutic progress in the future? Of course, I eagerly support her privacy, so that she can go all the more deeply into the inner work of reframing her self-image and co-creating real-life possibilities.

ROSSI: I want you to go through the whole list privately, okay. Don't tell me at all. Don't tell anybody. Don't ever tell anybody. Just, pop! Pop!

I am humorously going through boxing motions as I restate that Celeste doesn't have to tell anybody—just "pop" (punch) them one in her imagination.

CELESTE: Yeah.
ROSSI: Oh, my gosh.

STAGE-FOUR:
REALITY-TESTING OF THERAPEUTIC PROGRESS

Time Distortion and Orienting to Conclude the Session [52:37]

CELESTE: Actually the only part that hurts right now . . .
ROSSI: Yeah?

CELESTE: . . . I have two tendons that ruptured in these two fingers.

ROSSI: Yes.

CELESTE: It means they ruptured—came off the joint one time.

ROSSI: The tendons came off the joint. Yeah.

CELESTE: So they have to be put on the joint. But the rest of them . . .

ROSSI: Are okay?

CELESTE: Don't hurt.

ROSSI: Okay, so we are going to have to go on working on those two little joints for a while.

CELESTE: Oh, gee [laughing].

ROSSI: Go! Anybody! Staff!

[I am, once again, play-acting as I turn and look to the side of the stage and mockingly call for help from anyone on the staff to help with boxing props or whatever. Celeste and the audience are in an uproar of laughter.]

CELESTE: Hmm.

[There is a long period of play as Celeste and I imitate each other in mock shadow-boxing movements.]

CELESTE: Feels like I should do this all night.

ROSSI: It feels like you should do this all night?

CELESTE: Yeah.

ROSSI: Yeah.

CELESTE: Maybe not up here? But . . .

ROSSI: All right.

CELESTE: Something like a Friday night.

ROSSI: [A quiet, barely audible whistle can be heard under my breath, as I continue to clench and unclench my hands in imitation of Celeste's movements.] Ahh [with satisfaction]. My goodness. I've never stretched my fingers like this. Ahh.

I genuinely enjoy the feeling in my hands as I stretch them up and outward as much as I can. I am modeling and verifying how good feelings can come from stretching hands and fingers.

ROSSI: Gosh. [Another quiet, barely audible whistle is detectable as I release my own body tension with the stretching movements.] Oh, Celeste, I'm looking at my clock. How long do you think we've been doing this?
[There is immediate audience laughter.]

Everyone in this professional audience knows that time distortion (time seems contracted when we are deeply absorbed in any activity) is a good indicator of a job well done, particularly in therapeutic hypnosis.

CELESTE: I'm not a good person to ask.

Validation of Altered State via Time Distortion

Closure of the Therapeutic Session [54:45]

ROSSI: [I humorously stretch out five fingers to acknowledge that I received the signal from the staff that there is only five minutes left for this session.] What are you doing with this [said to staff member]? That's what I was afraid of—she was going to cut us off [said to Celeste].
[I am having fun making a funny face in imitation of the frantic hand-waving signals I am getting from the staff off stage.]
CELESTE: How long have we been doing this?
ROSSI: How long does it seem, just subjectively? Just take a guess.
CELESTE: Mmm, 20? 25 minutes?
ROSSI: 20 or 25 minutes? Actually it's been, you know, almost an hour.
CELESTE: Oh, gee [laughing].
ROSSI: But the hell with time! We are not demonstrating any fake hypnotic phenomena. We are going for the real thing here. [There is an uproar of laughter from the audience].

The inside joke here is about the time distortion Celeste is experiencing when she estimates that she has been on stage for only 20 minutes.

Hypnotherapists interpret this as a sign that patients have been in an altered state of hypnosis, whether they know it or not.

CELESTE: Right. Right.

RAPID VERIFICATION OF INSIGHTS IN STAGE FOUR

Reframing Symptoms into Signals, and the Activity-Dependent Facilitation of Behavior Change [55:33]

ROSSI: So how will you continue this, Celeste?

CELESTE: Well, I'll keep moving my hands.

ROSSI: Keep moving your hands. Yeah! That's for starters, for sure.

CELESTE: Any other suggestions?

ROSSI: Yes.

CELESTE: Mmm.

ROSSI: Who are you going to bop? Bop! Bop! [Said humorously, with mock shadow boxing]

CELESTE: I can do that so easily.

ROSSI: I mean, do you see what this is all about? Do you really believe it? I mean, we haven't talked about psychodynamics, but I guess you are getting it, right?

CELESTE: Yeah. Yeah.

ROSSI: That it was a month after you came to grief with this woman, you couldn't say anything, so you let your poor hands begin speaking for you.

This is a psychodynamic reframing of the body language of arthritic symptoms into the language of behavior change. *I am attempting to reframe symptoms into signals for a change in behavior.*

CELESTE: Yeah.

ROSSI: And then a profound aspect of fate—*she eases up on you. It's like another positive reinforcement.*

This is the first real psychodynamic interpretation of her symptomatic behavior: that she developed the condition during a time when she was not able to speak up to her supervisor, and then the sympton is reinforced when Celeste is treated better by the supervisor. Notice that this interpretation is offered only now in stage four of the creative process, after the patient has experienced and expressed her own genuine insights in stage three. From the vantage point of maximizing creativity, the most typical therapist error is to offer such interpretations in stage two of the creative process, when the patient is still in an early phase of his or her inner work. Such premature interpretations tend to abort the patient's own creative process. This style of offering interpretations only at the end of the therapeutic process, almost as a casual after-thought, was highly characteristic of Erickson's approach to therapeutic hypnosis (Erickson & Rossi, 1979, 1981, 1989).

> CELESTE: Right.
>
> ROSSI: It's like your mindbody has a kind of a genius. It found just the kind of thing to get her off your back. Only, you know, at the price of crippling yourself.
>
> CELESTE: That's right.

INSIGHT INTO THE MULTIPLE DETERMINATION OF SYMPTOMS

Consolidating and Verifying Last-Minute Interpretations [56:33]

> ROSSI: Does that make sense in terms of what you know about yourself—that you tend to, you know, hurt yourself rather than . . . ?
>
> CELESTE: Yeah. There is one other piece to this, too. I was going with somebody at the time who made more of a commitment to me after the arthritis was diagnosed.
>
> ROSSI: Oh, my God—a double-header working here!
>
> CELESTE: I did not know it. I didn't know it that his mother had just died from rheumatic heart fever. Heart fever?
>
> ROSSI: Maybe?

CELESTE: Anyway, one of the symptoms of that is arthritis in your ankles.
ROSSI: I see.
CELESTE: And so he had tapped into me wobbling around on my ankles.
ROSSI: I see.
CELESTE: It just got out of . . .
ROSSI: I see. He couldn't stop loving his mother?
CELESTE: That's right. He finally had to.

When I say "a double-header working here," I am noting how her psychodynamics may be illustrating Freud's concept of the multiple determination of symptoms. Celeste's arthritic sympton was reinforced inadvertently when the person she "was going with" made more of a commitment to her when the arthritis began.

ROSSI: He had to. Okay. So we don't have any of those strings from the past hanging on. You are really free to continue unlocking yourself, literally, getting your energy out. Do you know how you are going to be getting your energy up in your life? How you are going to continue to expand your energy? Your power?
CELESTE: At this moment?
ROSSI: Yeah. Well, I guess it's kind of short [referring to the short time left for this demonstration]. But do you get the idea that this is what you need to do—that this is what you are all about? Does that make sense?
CELESTE: Sure.
ROSSI: So, I'm not telling you anything new. You need to get out there and find creative channels for your energy, find your power and stuff. Enough with being locked in, right? Now, that's tragic.

I'm offering the interpretation that Celeste needs to reframe and channel her energy and power into creative outlets rather than locking herself in with her arthritic symptoms. That is, I am attempting to prescribe new activity-dependent channels for her energy and power that can further fuel the novelty-numinosum-neurogenesis effect.

CELESTE: I knew it, but I haven't been told it quite so directly.

ROSSI: Yes.

CELESTE: But that's . . .

ROSSI: But you knew it. That's right. So, I'm just telling you what you already knew, but no one has ever told you this directly.

CELESTE: Yeah.

DRAMATIZING THE BREAKOUT HEURISTICS

The Hell with Being Nice and Sweet [58:12]

ROSSI: But the hell with being nice, sweet, "Help! Out! Celeste!" [humorous, falsetto voice again]. Right?

CELESTE: Yes.

ROSSI: Right.

CELESTE: Yes.

ROSSI: I mean, you have such a beautiful smile and face. Who wouldn't have compassion for you, right? It's easy to get compassion and love being in victim role. I mean . . .

CELESTE: Yeah. I don't like it much.

ROSSI: I know you don't like it much. It's really an affront to anyone's dignity.

CELESTE: Yeah.

My exclamation "Help! Out! Celeste!" in that curious falsetto voice is another effort to dramatize her experience of the breakout heuristic (Chapter 6).

FINAL BEHAVIORAL SELF-ASSESSMENT OF THERAPEUTIC PROGRESS [58:41]

ROSSI: You don't need that role anymore. Okay, so she [a staff member] is giving me the signal of one minute. Anything you would like to say in this last minute to solidify this. My goodness.

Figure 8.2 | Stage Four: Celeste experiences a satisfying self-empowerment with the support of a standing ovation from the therapeutic community. The therapist hopes this encounter with the experiential theater of demonstration therapy will be sufficiently numinous to activate zif-268 gene expression in her REM dreaming tonight to optimize the therapeutic reorganization and reconstruction of her mind and memories in keeping with the psychobiological dynamics of current neuroscience research.

[Celeste extended both hands forward and straight in an open gesture, as if to show the audience how well she was.] How does that look to you, the audience?

[There is loud, extended clapping from the audience, while Celeste smiles with delight and I clap my hands and gaze at her with appreciation. I then stand up, still clapping, inviting Celeste to stand with her hands upward in triumph (see Figure 8.2).]

ROSSI: Whoa! Whoa! Whoa!

Audience members stand, clapping and shouting with approval. Celeste, in a confident, calm manner, takes off her lapel microphone, and I do the same. So ends this exploratory experiential theater of demonstration therapy at 59:56—with four seconds to spare.

SUMMARY

In this chapter we have demonstrated some familiar and some new, *implicit processing heuristics* for facilitating the third and fourth stages of insight and verification in the creative cycle. As can be seen in Table 8.1, these heuristics are all adopted from the normal experiences of everyday life, wherein people attempt to facilitate or influence each other. The value of these heuristics is that they utilize and facilitate a person's inner creative processing on an implicit or unconscious level in ways that we usually do not understand. They tend to help us break out of the limitations of our previous worldview by facilitating Darwinian natural variation and selection in unpredictable ways. The unpredictability of these heuristics is both the source of their *strength in facilitating creativity* and the source of their *limitations as a method of controlling, programming, or manipulating human behavior.*

We can only speculate about the potential value, in this demonstration, of the novel, numinous, dramatic, experiential approaches for facilitating *activity-dependent gene expression, neurogenesis, and healing.* We came to the vivid realization that all forms of psychotherapy, as they are currently practiced, are handicapped by being limited to the phenotypic level of behavioral observation. In principle, however, we now can use brain imaging, DNA, and protein microarrays to assess the value of these novel mindbody approaches in facilitating the novelty-numinosum-neurogenesis effect. Meanwhile, psychotherapists of every school need to further develop and refine their use of implicit processing heuristics to help their patients *right now* as well as researchers *in the future*. We will turn to this task in the next two chapters.

Table 8.1 | Implicit Processing Heuristics that may facilitate gene expression to optimize neurogenesis and healing during the illumination and verification stages of creatively oriented psychotherapy and the healing arts.

- "And now?"

- Experience-inducing anecdotes, humor, questions, secrets, whispers, and ambiguity

- Courage to explore and stay with the process

- Creative edge between voluntary and involuntary, dissociated behavior

- Creative uncertainty of therapist and patient

- Incubation for facilitating natural variation and selection

- Metaphorical questions

- Metaphorical movements expressing the wisdom of the implicate body

- Not-knowing questions

- Novelty-numinosum-neurogenesis effect

- Playful games about not knowing

- Privacy and creative choice

- Questions evoking experience- and activity-dependent exploration

- Segmentalized trance to focus problem-solving and healing

- Swearing humorously *with* (not at!)

- Truism

- Symptom-scaling

- Ultradian mindbody sensitivity and body language

- Utilizing the patient's own words and worldview

- Weird, unusual, odd, strange, bizarre, scary, but interesting experiences

---

CHAPTER 9

IMPLICIT PROCESSING
HEURISTICS IN THE HEALING ARTS

The Language of Facilitating Creative Experience

"I am not yet concerned with application. I'm concerned with building on concrete, getting the basic science down. I don't want to make the same mistake which I think has been almost lethal in clinical, making application claims that far outstrip the science. . . . What we do in the clinical is identify and nurture the strengths of our clients, emphasizing and building of strengths, not only the repairing of weaknesses. . . . I am interested in supporting the science from which solid applications will come."

—Martin E. P. Seligman

"Psychological implication is a key that automatically turns the tumblers of a patient's associative processes into [creative patterns] without awareness of how it happened. It is important in formulating psychological implications to realize that the therapist only provides a stimulus; the hypnotic aspect of psychological implications is created on an unconscious level by the listener. The most effective aspect of any suggestion is that which stirs the listener's own associations and mental processes into automatic action. It is this autonomous activity of the listener's own associations and mental processes that creates hypnotic experience."

—M. Erickson, E. Rossi, & S. Rossi
Hypnotic Realities

The psychobiology of gene expression, neurogenesis, and healing is a highly individualized creative process in everyday life that is replayed in an infinite variety of experience and behavior. In this chapter we will explore the language of facilitating the psychodynamics of the four-stage creative cycle in psychotherapy and the healing arts—language that may generate gene expression, neurogenesis, and healing. The typical phenomenology of the client's experience and how the therapist can facilitate that experience is outlined at each stage of the creative process.

This is a creative art that has its sources in the Socratic method of the Western tradition, the Zen koan in the Eastern tradition, suggestion theory in hypnosis, and what we now call *implicit processing heuristics* from the current perspective of neuroscience. These heuristics engage and facilitate the natural psychodynamics of creative interaction between implicit (unconscious) and explicit (conscious) processes. This is a creative art that is easily corrupted, however. The basic intent of all these approaches to facilitating the creative process of replaying and resynthesizing human experience is *nurturance*. This intent is corrupted when it is superceded by a desire to manipulate and control social processes and a person's thinking and behavior. Because of this vulnerability, we need to focus carefully on how to utilize implicit processing heuristics in ways that facilitate each individual's unique personal style of learning how to optimize problem-solving and healing via gene expression and neurogenesis.

IMPLICIT PROCESSING HEURISTICS FACILITATING THE CREATVE PROCESS

We began our study of the relationship between the genomic level and human experience in Chapter 1 with the illustration of how a mother's touch can facilitate gene expression leading to the synthesis of proteins that generates physical development, neurogenesis, memory, learning, and behavior. Such neuroscience research is the kernel of a new view of the nature-nurture relationship: The *cooperative dynamics* of the nature-nurture equation are emphasized rather than the *competitive controversy* about whether nature or nurture is more important in human development. From the neuroscience perspective, all life forms are complex adaptive systems with *emergent characteristics of holism and cooperation* that supplement the *reductive-mechanistic view of competitive forces* pushing and pulling inert matter into shape.

The holistic perspective means that the "top," or whole, can modulate its parts; this *top-down perspective* on the guiding role of mind and human experience has traditionally characterized the humanistic, philosophical, ethical, and spiritual approaches to understanding life. The reductive-mechanistic perspective means that the parts modulate the whole; this is the *bottom-up perspective* of traditional Newtonian science. Both perspectives have spheres of validity for conceptualizing the patterns of life (Ben-Jacob & Levine, 2001), and both have constraints on their effectiveness in the healing arts. Modern molecular medicine, for example, engages the bottom-up perspective when it uses tranquilizers or other psychotropic drugs to facilitate healing when people seem unable to modulate their own psychobiology. When modern medicine takes steps in the direction of using advice, counseling, psychotherapy, therapeutic hypnosis, meditation, or the ultradian healing response, however, it is using the holistic top-down approach of the humanities and the spiritual arts.

From our currently developing neuroscience perspective, we seek to use implicit processing heuristics to integrate the best of the top-down and bottom-up perspectives (Rossi, 1996a). From a top-down perspective, implicit processing heuristics may be used to facilitate Darwinian natural variation and conscious selection on the highest levels of human creative choice. Using the bottom-up perspective, implicit processing heuristics may be used to facilitate the psychobiology of gene expression to generate neurogenesis, problem-solving, and healing. As a prelude to the art and science of learning how to use implicit processing heuristics in psychotherapy we will review how this concept evolved out of Erickson and Rossi's original formulation of the "implied directive" in therapeutic hypnosis (Erickson et al., 1976).

FROM INDIRECT HYPNOTIC
SUGGESTION TO IMPLICIT PROCESSING HEURISTICS

As reviewed in Chapter 5, the myth of classical hypnosis is that an authoritarian therapist induces patients to enter a special state of receptivity wherein they are relaxed, sleepy, or even in a somnambulistic-like state of sleepwalking. In this receptive state the patient is then programmed by the therapist's direct suggestions, which are believed to be the *raison d'etre* of hypnotherapy. There are varying degrees of authoritarian directness: Patients can be vigorously commanded, more gently persuaded, or gradually led through a series of inner experiences that enable them to give up their errant attitudes, bad habits, or symptoms.

There can be no denying that Erickson believed in and used many varieties of direct suggestion throughout his career. He carefully noted, however, that attempting to directly program people without understanding their individuality was "a very uninformed way of doing therapy" (Erickson & Rossi, 1979, p. 288). An important aspect of Erickson's therapeutic approach was to spell out the conditions for optimizing both direct and indirect suggestions by utilizing patients' belief systems and inner resources.

In our first collaborative effort we called Erickson's innovative approaches to accessing and utilizing patients' inner resources for problem-solving and healing "indirect suggestion" (Erickson et al., 1976). Of the many varieties we identified, the "implied directive" was found to be the most useful for conceptualizing the essence of therapeutic suggestion (Rossi, 1995a, 1995b). Erickson and I regarded *implication as the essence of the dynamics of suggestion because it is not what the therapist says that is important as much as what the patient does with what the therapist says* (Erickson et al., 1976):

> An understanding of how Erickson uses implication will provide us with the clearest model of his indirect approach to hypnotic suggestion. Since his use of "implication" may involve something more than the typical dictionary definition of the term, we will assume that he may be developing a special form of "psychological implication" in his work. For Erickson, *psychological implication is a key that automatically turns the tumblers of a patient's associative processes into predictable patterns without awareness of how it happened.* The implied thought or response seems to come up autonomously within patients, as if it were their own inner response rather than a suggestion initiated by the therapist. Psychological implication is thus a way of structuring and directing patients' associative processes when they cannot do it for themselves. The therapeutic use of this approach is obvious. If patients have problems because of the limitations of their ability to utilize their own resources, then implications are a way of bypassing these limitations. (pp. 59–60)
>
> It is important in formulating psychological implications to realize that the therapist only provides a stimulus; *the hypnotic aspect of psychological implication is created on an unconscious level by the listener.* The most effective aspect of any suggestion is that which stirs the listener's own associations and mental processes into automatic action. It is this autonomous activity of the listener's own associations and mental processes that creates hypnotic experience.

There are, to be sure, crude and mostly ineffective uses of implication in everyday life, where the speaker in a very obvious manner attempts to cast negative implications or aspersions on the listener. In such crude usage the implication is obviously created entirely by the speaker. In our use of psychological implication, however, we mean something quite different. In the psychological climate of the therapeutic encounter the patient is understood to be the center of focus. Every psychological truth, consciously or unconsciously, is received by the patient for its possible application to himself. Psychological implication thus becomes a valuable indirect approach for evoking and utilizing a patient's own associations to deal with his own problems. (p. 61)

Twenty years after we formulated this concept, the fundamental role of implication in human experience was recognized in neuroscience in the distinctions made between the implicate or nondeclarative (unconscious) and explicit or declarative (conscious) levels of functioning (reviewed in Chapter 3; Squire & Zola, 1996). Squire and Kandel (1999) summarize the central importance of these distinctions for an understanding of memory and cognition, as follows.

When one thinks about it, much of our knowledge about the world is in the form of categories. We operate comfortably with categories every day (birds, cars, buildings, music, and clouds), and appreciating the similarities among the things we see is every bit as important as appreciating the differences.

The question of interest is: What kind of memory underlies the ability to learn about categories? The surprising answer is that at least some kinds of category learning are nondeclarative. Category learning can be independent of and parallel to declarative memory, rather than simply derivative from it. People can acquire knowledge about categories *implicitly* even when their declarative memory for the instances that define the category is impaired. (p. 183, italics added)

The long history of learning theory's view of categorical learning and memory (Rossi, 1963, 1964; Rossi & Rossi, 1965) gradually evolved into behavioral science's general recognition of implicit memory and learning by the "cognitive unconscious" (Reber, 1993).

More recent research efforts to assess the value of Erickson's approaches to psychological implication in therapeutic suggestion remain controversial, however. A number of researchers failed to find any superiority of indirect suggestion over direct suggestion (Matthews, 2000), for example, but many clinicians dispute the validity of the research methods used (Peter & Revenstorf, 2000). In the perspective on implication we developed in Chapter 3, we noted that Erickson's indirect approaches to therapeutic suggestion involved the patient's "experience of re-associating and reorganizing his own experiential life that eventuates in a cure, *not the manifestation of responsive behavior, which can, at best, satisfy only the observer*" (Erickson, 1948/1980, p. 38, italics added). Most of the research that purports to demonstrate that indirect suggestion is not superior to direct suggestion used objective measures of *responsive behavior* on objective scales of hypnotic susceptibility. The use of this methodology may indeed show that indirect suggestion is not superior to direct suggestion in *manipulating responsive behavior in an objective manner.*

When investigators compared subjects' *subjective* perceptions of their own hypnotic experiences with the objective behavioral measures of direct and indirect suggestion, however, they came to these interesting conclusions (Matthews et al., 1985):

> While the behavioral data failed to show significant differences between the two methods, however, subjects did report feeling more deeply hypnotized during the indirect procedure than during the direct procedure. This result is in support of an Ericksonian view (Langton & Langton, 1983; Rossi, 1980) of indirect suggestion in that *a wider latitude for responding* is offered to subjects than is typically available with direct suggestion. *Thus, the subjects' unique responses to indirect suggestions can be interpreted by them to be an indication of trance depth.* If subjects fail to perform in accordance with a direct suggestion, however, there is some likelihood that they may perceive themselves as less deeply hypnotized. (pp. 222–223, italics added)

It is precisely this *"wider latitude for responding"* and "subjects' *unique responses to indirect suggestions"* that Erickson wished to facilitate as a creative "experience of re-associating and reorganizing [their] own experiential life [in a way] that eventuates in a cure." Whereas indirect suggestion can facilitate the reassociation and resynthesis of *subjective experience* characteristic of each individual's creativity, it is not necessarily useful in the manipulation and control of *objective behavior* as it is assessed with standardized scales of hypnotic susceptibility.

Even with research demonstrating the efficacy of indirect suggestion in facilitating personal creativity, the popular view of indirect hypnotic suggestion as a covert method of programming patients with the therapist's own goals remains pervasive. Because this confusion between programming by the therapist and creativity by the patient is an important issue in professional practice as well as public perception, we will explore how the concept of "hypnotic suggestions" can be complemented with the concept of "implicit processing heuristics" to facilitate the four-stage creative process within patients.

STAGE ONE: PREPARATION

Facilitating Personal Sources of Behavioral State-Related Gene Expression

The typical psychotherapeutic session ideally begins with patient and therapist cooperating in an effort to identify the problems and issues that the patient hopes to resolve. The therapist's role in this initial stage is to facilitate this search by using familiar, open-ended questions (such as those in Box 9.1) that serve as mini-rites of transition between the everyday world of congenial talk to the more focused creative work of the therapy session. From our current neuroscience perspective, these open-ended questions function as implicit processing heuristics that facilitate the replay of patients' personal history and the state-dependent sources and encoding of their problems. When emotional problems and highly numinous personal issues are involved, these motivated psychodynamics will naturally engage *immediate early genes, behavioral state-related genes, and activity-dependent gene expression*—all of which generate the possibility of natural variation and selection in the new cascades of protein synthesis, neurogenesis, problem-solving, and healing.

This series of basic accessing questions may seem conventional, trivial, and trite, but in the context of therapy they are actually important transition

Box 9.1 | Typical implicit processing heuristics to initiate transitions from congenial talk to focused creative work.

- How are you?

- What's happening?

- What is going on with you these days?

- What's most important to you today?

- What courageous questions come up?

- What issue is absorbing your attention today?

- What are you dreaming up these days?

- What would be the most fascinating possibilities of your life?

- What is most alive in you right now?

- What's new?

- What is your truth and beauty today?

- What feelings are coming up at this moment?

- What is most interesting to you right now?

- What are you experiencing within yourself?

- What would be the creative path for you?

- What would you like to do?

- What steps do you need to take to that goal?

phrases helping the patient move from everyday conversation to the expression of highly personal concerns. These open-ended questions empower people to present themselves from the worldview with which they are most comfortable. Focusing patients' attention on the most important issue of the moment is an effective approach for utilizing their mental preoccupations for creative work, problem-solving, and healing.

From our current neuroscience perspective, these open-ended questions function as implicit processing heuristics that facilitate the replay of the person's personal history and the state-dependent sources of their problems. The motivating psychodynamics of emotional problems and highly numinous personal issues naturally engage *immediate early genes, behavioral state-related genes, and activity-dependent gene expression* that generate natural variation and selection in new cascades of protein synthesis, neurogenesis, problem-solving, and healing.

Questions that touch upon what is most interesting may initiate enough motivation and emotional arousal to facilitate the novelty-numinosum-neurogenesis effect and engage the inner work of the 90–120-minute ultradian cycle. The therapist's task in this first stage is to facilitate a creative wave of ultradian dynamics, wherein patients' attention is focused on the source and solution of their problems with a sense of numinous fascination, self-discovery, and self-empowerment that has always been a hallmark of genuine psychotherapeutic states (Abraham & Gilgen, 1995; Rossi, 1989, 1996a, 2000a).

Experiential States Characterizing Stage One

Curiosity, confusion, uncertainty, hope, expectancy and an exploratory attitude are typical experiences at this initial stage of the creative process when the patient's major task is to identify which issues are most pressing in the here and now. Initially, many people express feelings of stress, anxiety, anger, hopelessness, frustration, and a variety of other negative attitudes about themselves and the therapeutic process in this initial stage.

While it is certainly necessary to establish a positive emotional rapport in the therapeutic encounter, it is *not* the task of the therapist at this early stage to try to reduce the person's anxiety and stress. Anxiety, uncertainty, tears, and stress can be cues of the patient's emotional arousal and readiness to embark on stage two of the creative process. The therapist's task in the beginning is to help people utilize their own anxiety and stress to motivate themselves toward problem-solving and healing. As people access the emotionally encoded state-dependent memories about the sources of their problems, they may engage behavioral state-related and acitivty-dependent gene expression associated with neurogenesis—which creates the possibility of problem-solving and healing at the molecular level.

STAGE TWO: INCUBATION

The Inner Journey: Arousal, Natural Variation, and Creative Replay

Is there any fundamental difference between shaman, priest, and the modern psychotherapist who initiates an inner process of problem-solving and healing with a therapeutic ritual? When we learn that thousands of people tune their TV sets to charismatic church services that offer faith healing, accompanied by music and singing, do we recognize any kinship between them and our psychotherapeutic science today? While the underlying worldview, belief systems, and styles of initiating the healing process may appear to be different, many researchers would agree that faith, hope, and expectancy of healing can evoke the therapeutic states of emotional arousal, creative replay, and resynthesis that are common to both ancient and modern approaches to healing.

As we have seen, one of Erickson's major contributions to the art of therapeutic hypnosis and psychotherapy, for example, was his focus on the naturalistic and utilization approaches (Erickson, 1958/1980, 1959/1980). Rather than initiating a ritual of hypnotic induction by focusing patients' attention outward on a candle flame, a swinging pendulum, or the sound and rhythm of a drum beat, Erickson focused their attention inward on their own moods, anxieties, emotions, stresses, thoughts, or whatever was uppermost in their ongoing experience of the moment. I have previously outlined dozens of relatively simple three- and four-step approaches to facilitating permissive replay that allows patients to maintain their own internal locus of control (Rossi, 1986/1993; Rossi & Cheek, 1988). Most of these are variations on the basic accessing questions and implicit processing heuristics that initiate the dynamics of self-reflection, which encourage people to pursue novel adventures along a therapeutic path.

The most important role of the therapist in stage two of the creative process is to encourage and support patients' inner journey, where the newly accessed state-dependent sources of their problems and symptoms may evoke negative memories and emotions. This is the stage where we witness the main difference between the typical failures of problem-solving in everyday life and the carefully guided focus on the creative replay of inner work in the psychotherapeutic situation.

In everyday life we often retreat from the labor of ultradian arousal that normally takes place whenever we are engaged in learning or problem-solving of any sort. Because problem-solving situations that involve states of arousal have been associated with stress, failure, and negative feelings in the past, we tend to avoid them in everyday life as well as in psychotherapy. To solve a problem, however, the mindbody needs a state of arousal to explore emergent natural variations in the replay of the inner experiencing. This is a process of inner search for new life possibilities. Just before we awaken in the morning, for example, we automatically experience the highest level of cortisol secretion (a hormonal messenger molecule) that signals the cells and tissues of the body to prepare for activity. Each of the subsequent ultradian peaks of psychobiological arousal, every few hours throughout the day, has a similar peak that optimizes the possibility of creative problem-solving (illustrated in Figure 6.1).

When this psychobiological arousal leads patients to access and experience past painful states, they may undergo varying degrees of *emotional catharsis*. A spontaneous catharsis is a signal that they probably are successfully accessing and replaying the important state-dependent memories and emotions that are encoding their problem. The therapist's role is to support this inner accessing and replay, so that patients do not break it off before the all-important process of problem-solving and symptom resolution has a chance to take place (stage three). Box 9.2 contains key implicit processing heuristics the therapist may use to facilitate a safe replay of arousal and motivated inner work during the person's transition from stage one to stage two.

Allowing the person to explore and replay privately seem counterintuitive to the mainstream of psychotherapy, but it is one of the most interesting innovations suggested by our current neuroscience perspective. Permitting private replay is an effective approach for facilitating a person's implicit processing. Privacy enhances the person's ability to access inner emotional issues without resistance because there is no concern about how to verbalize them for the therapist. This allows the personal material to be replayed on implicate levels in ways that cannot be easily verbalized. This approach empowers people to replay their own unique matrix of experiences, engaging the novelty-numinosum-neurogenesis effect and solving their own problems in their own way.

BOX 9.2 | KEY IMPLICIT PROCESSING HEURISTICS INITIATING INNER
SEARCH AND CREATIVE REPLAY.

- When you are ready to focus inward on that problem [issue, symptom], what
  will you actually experience as you review its sources and history?

- Can you let yourself continue to experience that for another moment or two
  in a *private* manner—only long enough to experience what it leads to next?

- Good, can you replay that again to learn what it is all about . . . ?

- Will it be okay to allow yourself to continue replaying that *privately* for a
  while, difficult though it may be, so you can learn what you need for healing [prob-
  lem-solving or whatever] . . . ?

- And will it be all right to replay that *secret* [problem, embarrassing moment,
  trauma, etc.] again in a way that you would really like to experience it . . . ?

Experiential States of Arousal
and Natural Variation in Creative Replay

Sympathetic system arousal, accompanied by sweating and increased heart
rate, pulse, and breathing, together with a feeling of heat ("as if burning up"),
are typical in the replaying of past problems in stage two. When ideodynamic hand approaches to replay are used (Rossi & Cheek, 1988; Rossi, 1995a,
1996a), as was demonstrated with Celeste in Chapters 7 and 8, patients may
ask, "Why am I experiencing this shaking [vibrating, pulsing]!" People are
surprised and sometimes a bit disturbed by these signs of arousal because in
the past, they were frequently associated with stress, discomfort, or failure.
Indeed, many patients have learned to block their own natural patterns of
psychobiological arousal that are important for problem-solving. This may be
why they cannot solve their problems—they have learned to shut off their
natural ultradian creative cycle to the point where they need a therapist to
help them relearn how to experience the normal tension of arousal and transition in everyday life. Therapists can facilitate fresh replay and creative expe-

---

BOX 9.3 | IMPLICIT PROCESSING HEURISTICS FACILITATING SAFE ACCESS-ING AND CREATIVE REPLAY OF DARWINIAN VARIATION AND CONSCIOUS SELECTION.

• Within the privacy of a person's mind, anything is possible.

• And knowing you can quietly review all that privately.

• Curious though it may be, you can privately review whatever is known only to yourself.

• The secret hopes of childhood . . .

• The whispers of things that have been and can become . . .

• Private promises and possibilities come in surprising ways . . .

• Secrets can be important when something new is coming up . .

• Sacred promises to yourself of what can be . . .

• Wisdom whispers of . . .

---

rience by using implicit processing heuristics that safely access state-dependent arousal, variation, and conscious selection such as those in Box 9.3.

If the patient demonstrates obvious behavioral cues of excessive arousal and anxiety, the therapist can reframe the arousal by using reassuring implicit processing heuristics that may facilitate natural variation, such as those in Box 9.4.

Reframing Negativity and Confusion
in the Natural Variations of Stage Two

Sometimes patients fall into negative response sets when experiencing the natural variations in their behavior. They may verbalize self-talk such as:

"This won't work with me."

"This is stupid!"

"I don't know how to do this."

"I'm blocked!"

Box 9.4 | Implicit processing heuristics facilitating natural variation with therapeutic dissociations.

- Can you actually enjoy your experience of energy [sweating, shaking, trembling, nervousness, confusion, uncertainty, or whatever] . . .

- What is coming up privately for a moment or two as a sign that you are on you way to dealing with whatever you need to?

- Have you ever let yourself have a good shakeup [or whatever]?

- Yes, changes in your breathing often mean your mind and body are getting ready to deal with important issues . . .

- Will you allow your changing experiences to continue for another moment or two until . . . ?

- Noticing the subtle shifts and changes taking place within yourself all by them selves . . .

"I can't do this—I can't think."

"I don't feel anything."

"I can't remember anything!"

Other patients complain of confusion, not knowing, doubts, and feeling dizzy, foggy, or misty. The therapist can recognize and reframe such experiences as important transition signs of the creative process by using implicit processing heuristics such as those in Box 9.5.

Sometimes patients experience anxiety, tension, stress, and fear to the point where therapists are concerned that patients will lose control and become overwhelmed by what they are experiencing. Expressions such as "I feel crazy-like! What is happening to me?" may raise the alarm in therapists unaccustomed to this activity-dependent experiential approach, where the locus of control remains within the patient. In my experience, this acute phase can be rapidly diffused by using *therapeutic dissociation* to facilitate patients' control over their own experiences. Therapeutic dissociation divides the patient's experience into parts: One part is obviously going through important emotions, while other parts watch safely from the sidelines and guide the

---

BOX 9.5. | IMPLICIT PROCESSING HEURISTICS REFRAMING NEGATIVITY AND CONFUSION AS CREATIVE TRANSITIONS.

• Are you aware of how your feelings of being stupid [dumb, inadequate, or whatever] are telling you that you're ready for something better?

• Have you ever experienced confusion before you learned something new?

• Do you often experience darkness before the light?

• The dark night of the soul can lead to light.

• Are your worries a clue to what needs to change?

• Yes, will your confusion be the first step toward wisdom?

• Have you experienced how risks can lead to rewards?

• Every cloud has a silver lining.

• They say every problem contains its own solution.

• Every difficulty opens up new possibilities, does it not?

---

emotional process with care, control, and wisdom, as illustrated by Davina's dreams in Chapter 4. From the perspective of nonlinear dynamics and chaos theory, therapeutic dissociations can be conceptualized as *creative bifurcations* of consciousness. Implicit processing heuristics that facilitate therapeutic dissociations to optimize creative replay and resynthesis within the patient are listed in Box 9.6.

There are also many wrong ways of attempting to facilitate the ongoing psychological work of stage two. Out of habit from previous models of training in the counseling process (where such questions may be entirely appropriate), the therapist's most common error at this early stage is to ask premature questions that tend to interrupt the ongoing flow: "What can I do to help you?" "Would you like a handkerchief?" "Do you need help?" "Would you like some advice?" Notice how these kinds of open-ended offers to help tend to shift patients' attention from their own experience to the therapist, which could interrupt the patient's ongoing personal process. Patients tend to imme-

---

Box 9.6 | Implicit processing heuristics facilitating therapeutic dissociations, creative replay, and reframing of the negative.

- Can one part of you feel that as fully as you need to, while another part of you carefully guides you safely?

- Will it be okay to continue experiencing that as much as necessary for another moment or so, while a healer within you directs you wisely?

- Can you continue to experience those tears (or whatever) as intensely as some part of you needs to, while another part of you observes yourself calmly and learns what it needs to help you?

- How does your inner child (competent adult, parent, wiser self, etc.) react to that?

- While one part of you feels (believes, experiences, identifies with, etc.) that, can you wonder what the other side of you feels?

- Suppose that is the negative (worst, depressed, devil, etc.) side of you at this moment. What is the better side that you could be?

- Given that as your past, what would you have your future be?

- Suppose you really did have your own private angel. How would she [he] help you recreate your past, present, and future?

- Um-mmm . . . and what would your better side do?

---

diately turn off their uncomfortable state of negative emotional arousal in hopes that the therapist will offer an easier solution. That is, such questions tend to shift the locus of control and healing to the therapist instead of allowing it to continue processing within the patient, where the creative dynamics of gene expression, neurogenesis, and healing need to take place. Furthermore, such questions may imply that patients are weak and need to fall back on the therapist's wisdom rather than continuing to explore their own inner resources for creative problem-solving.

Sometimes the beginning therapist is uncomfortable not knowing what is happening, particularly if the patient, with eyes closed, is frowning, crying, or caught in a negative affect. The art at this stage is in how to offer an implic-

---

BOX 9.7 | IMPLICIT PROCESSING HEURISTICS FACILITATING PERMISSIVE
AND CREATIVE INNER WORK DURING DISCOMFORT AND STRESS.

- Knowing you can continue receiving whatever comes all by itself and saying a
  few words about it whenever you need to.

- Will it be all right for you to share a sentence or two with me in a moment if
  you want to—only what I need to hear to help you?

- Will you have the courage to let that continue *for just a moment, or two,* so
  you can really experience everything necessary privately within yourself—sharing
  only a word or two if you need to?

---

it processing heuristic *without impeding the patient's inner process.* Implicit processing heuristics, such as those in Box 9.7, avoid placing interfering demands
upon patients, yet support them by allowing them to speak and request further help from the therapist if they feel it is necessary.

If the patient asks for help, the therapist can immediately provide it with
another appropriate implicit processing heuristic. The basic issue for the therapist now becomes how to come up with an appropriate implicit processing
heuristic. What can any therapist really know about what is going on in the
patient on so many levels simultaneously, from imagery, emotion, and cognition all the way to gene expression and neurogenesis? Who is the therapist
who feels wise enough to take control, direct and program the patient on all
these levels? The therapist can manifest some degree of wisdom, however, by
responding to patients' requests for help by using the safe, four-step, time-
binding implicit processing heuristics, as shown in Box 9.8.

Notice how much is being accomplished with these seemingly simple,
four-step sentences. First, the therapist is responding positively and support-
ively to the patient's request for help by immediately saying "yes." Second,
the therapist is supporting the patient's ongoing experience with a safe, time-
binding limitation, *"just for another moment or so."*

Third, there is a mild therapeutic dissociation implied in the words "allow
[or *let*] *yourself continue with those feelings.*" Using the wording *allow yourself*
implies that there is another stronger part the patient that is allowing his or

---

Box 9.8 | Four-step, time-binding implicit processing heuristics
to support the patient's inner creative replays and reframing.

---

1. Yes, I really want to help you. . . .

2. Can you *allow yourself* to continue with those feelings just *for another moment or so* . . .

3. . . . until you find yourself expressing a sentence or two . . .

4. . . . only what I need to hear to help you further?

Another four-step example of time-binding implicit processing heuristics:

1. Yes, continue *for just another moment* as you are . . .

2. . . . *letting yourself carefully* consider what kind of help you need . . .

3. . . . and then share it with me in a sentence or two . . .

4. . . . so I can help you in every way I can.

---

her dependent, hurt, and needy side to express itself. The wording *letting yourself* also implies that the patient is no longer lost in an emotional dependency over which he or she has no control. Patients are encouraged to maintain an internal locus of control by *allowing* the emotional experience to take place within their own self-directed therapeutic process.

Fourth, the therapist is asking the patient to do a major piece of therapeutic work with words such as, "until you *find yourself sharing a sentence or two*." The embedded implication is that patients are able to self-reflect on their inner situation and co-create an adequate report about it a possible solution. The therapist responds to whatever the patient says by feeding it back in a manner that facilitates accessing inner resources for problem-solving.

Suppose the patient says something like: "I feel blocked, just like I've felt all my life when I get hopeless about something. Can't you help me break through this block?" What therapist does not feel some uncertainty about how to help a patient break through a lifetime block within the limits of a

---

BOX 9.9 | FOUR-STEP IMPLICIT PROCESSING HEURISTIC UTILIZING THE AFFECT BRIDGE.

---

1. Yes, I want to help you break through that block. . . .

2. Can you allow yourself to experience another time in your life . . .

3. . . . when you found yourself with these kinds of feelings . . .

4. . . . and share just a sentence or two so I can facilitate the next stage of your creative work?

---

typical 50-minute therapeutic hour? Since emotions are usually high in stage two, the therapist could respond to such direct but daunting requests for help by creating four-step implicit processing heuristics (an example is shown in Box 9.9) that utilizes the affect bridge to access the state-dependent dynamics of the source of problems—which, in turn, may contain the seeds of their own resolution and resynthesis via creative replay (Rossi, 1986/1993; Watkins, 1978).

Whatever the patient's response, the therapist can usually facilitate a creative path along the state-dependent affect bridge. Replaying past problems and traumas within the safety of the therapeutic situation prepares the patient to move on to the third stage of the creative process: the moment of illumination is experienced as a new insight. This shift from stage two to stage three is often mediated by a particularly focused period of private concentration, wherein the patient seems to suddenly become quiet after a stressful period of arousal and conflict. This is usually a time for the therapist to remain silent, lest the patient's concentration is interrupted.

The Period of Private Creative Inner Work

An appropriate period of private inner work often serves as a bridge from the emotional stress and strain of replaying conflicting emotions in stage two to the quiet insights that signal stage three of the creative process. When this period of private inner work is going well, as indicated by a transition from

BOX 9.10 | IMPLICIT PROCESSING HEURISTICS FACILITATING PRIVATE CREATIVE INNER WORK.

• And knowing you can quietly review all that privately . . .

• Within the privacy of a person's mind, anything is possible . . .

• You can review what is known only to yourself about . . .

• The whispers of things that have been and can become . . .

• Private possibilities to come in surprising ways that . . .

• Secrets can be important when something new is coming up . . .

• Sacred promises to yourself of what can be . . .

• Wisdom whispers of . . .

the arousal and distress of stage two to a quieting and emotionally neutral or mildly positive attitude (often indicated by little smiles and head nodding), the therapist does not need to say anything. When the patient shifts uncertainly back and forth between the active distress of stage two and the positive transition to stage three, the therapist may offer a few implicit processing heuristics in a quiet tone to facilitate the private period of creative replay and co-creation. As can be seen in Box 9.10, many of these implicit processing heuristics are incomplete sentences that can facilitate a patient's private processing without having to check back with the therapist.

STAGE THREE: ILLUMINATION

Surprise and the Conscious Selection of Emergent Insights

During the transition period of creative inner work, the patient's overt behavior may replay many subtle shifts that are observable to the therapist. There may be sudden increases or decreases in the movements of the patient's closed eyes, for example. Sometimes there seems to be an inward pulling within the eyeball, and the patient's whole head may even pull backward slightly, as if in surprise.

While such shifts in eye behavior have been studied for centuries, there is no general agreement about their significance (Edmonston, 1986; Rossi, 1996a; Weitzenhoffer, 1971). When I notice periodic or momentary bursts of rapid eyelid vibration or a shifting of the eyeballs from side to side, as if the patient is following an inner moving scene, I usually remain silent, lest I disturb the private processing. On occasion, when I notice that all negativity and tension of stage two have really dissipated, I will support the delicate shift to stage three by using implicit processing heuristics such as those in Box 9.11.

There are many fascinating behavioral cues that reveal patients' changing states in this third stage of the creative cycle. Occasionally there may be a slight smile of surprise or a momentary grin. Slow and sideward body movements of various types during therapeutic hypnosis have been described as "strongly indicative of a shift in the balance of neural control from cortical

---

BOX 9.11 | IMPLICIT PROCESSING HEURISTICS FACILITATING SURPRISE, INSIGHT, AND CONSCIOUS SELECTION.

- Um-mm, surprising . . . ?

- Yes, the courage to receive the unexpected . . . ?

- Fun to experience personal secrets leading to private understanding . . . ?

- Courage to receive that too, can you not. . . . ?

- Yes, noticing interesting things . . . ?

- Um-hum, something important to recognize about that . . . ?

- Interesting options that come up . . . ?

- The best you can imagine about that . . . ?

- Possibilities you would like to explore . . . ?

to subcortical structures" (Weitzenhoffer, 1971, p. 120). I respond to such important moments of what Edelman and Tononi (2000, p. 84) might call "experiential selection [that] leads to changes in the connection strengths of synapses favoring some pathways over others" by softly offering implicit processing heuristics that may support the novelty-numinosum-neurogenesis effect during the patient's positive experiencing (Box 9.12).

Sometimes there is a moment of absolute stillness, as if the patient, with bated breath, were receiving something from within. This may be the moment of insight that has been described as the *"aha!"* experience in humanistic and scientific literature on the creative process (Rossi, 1972, 1985, 2000a; Sternberg & Davidson, 1996). There may be a slight smile and the head may slowly nod *yes* repetitively, with minimal movement. I softly support this positive therapeutic experiencing and behavior by using implicit processing heuristics such as those in Box 9.13.

It is not always easy to recognize the transition from the struggle of stage two to the moments of significant insight in stage three. It is fairly typical for patients to shift uncertainly for a while, back and forth, between stages two

---

Box 9.12 | Implicit processing heuristics facilitating the novelty-numinosum–neurogenesis effect via positive experiential selection.

• Sometimes a little light at the end of the tunnel . . . ?

• Um-hum . . . good allowing that to go on a bit longer . . . ?

• Pleasant to consider the possibilities . . . ?

• Yes, okay for that wonderful experiencing all by itself . . . ?

• Nice to experience that privately so you can really appreciate it . . . ?

• Choices your creative self can make . . . ?

• Pathways you would like to explore . . . ?

• Unexpected but fascinating possibilities . . . ?

• The wonder of how changes take place within you . . . ?

• The mysteries of . . . ?

and three, as if to compare and contrast the reality of their experiencing, before a settling into the positive possibilities of stage three and a movement toward stage four. Indications that a natural completion to this period of the inner work is being experienced often come from larger postural adjustments of the head, neck, arms, or legs, with movements suggesting that an opening, loosening or relaxation is taking place. Previous muscle constrictions and tensions evident in the jaws, hands, and arms seem to release, and some people may actually shake out their arms and legs. The person is moving from the postures of defensiveness, anger, frustration, confusion, sorrow, and depression so typical of stage two, to expressions of lifting, lightness, joy, happiness, and well-being in stages three and four.

Many people are not aware of the significance of the profound critical phase transitions they are experiencing between stages two and three; they often need help to recognize the value of the spontaneous creative breakthroughs and reframes that seem to take place all by themselves. They don't know that these creative experiences are the essence of their therapy; they are

---

BOX 9.13 | IMPLICIT PROCESSING HEURISTICS SUPPORTING THE "AHA!" INSIGHT OF STAGE THREE CREATIVE PROCESSING.

• That's right!

• A little light . . . ?

• All right . . . fully appreciating that surprise . . . ?

• Really receiving something you like . . . ?

• Yes, really worth receiving . . . !

• The best moments in life . . . ?

• Yes . . . and more . . . !

• Enjoying . . . !

• Accepting some of the best . . . ?

• Knowing what is really worthwhile . . . ?

still looking for *the* answer or waiting for some kind of magical healing to come from the therapist. The task of the therapist is to help such patients recognize the value of their own therapeutic transitions—and above all, to help patients recognize that the locus of creativity and therapy is within themselves during these moments of insight and spontaneous symptom relief. In short, *the therapist's major task in stage three is to help people learn how to recognize and facilitate their own personal experiences of choice and creativity!*

Many people suspect that their insights and numinous feelings during these therapeutic breakthrough moments of are only temporary fantasies, wishes, delusions, or a *mere* placebo response. They are all too ready to criticize the essence of their creative experience as unreal. Indeed, these creative experiences and insights are unreal in the sense that they are, as yet, untested in the real world. Many people can relate a personal history of how their creative moments were not recognized or supported during childhood by parents or teachers. It is valuable to encourage such patients to review the negative responses they received when they tried to express their creative experiences. Such a review helps them recognize the unfortunate realities of the past that are now changing. They never learned how to appreciate and select their own best insights and developing awareness in everyday life because their creative experience was so often accompanied by not knowing, confusion, rebellion, and emotional chaos (Rossi, 1968, 1972, 1985, 1998, 2000a).

The therapist's permissive and open-ended implicit processing heuristics in stage three provide a gentle, supportive environment that helps people recognize, select, and stabilize their still nascent creative state and emergent insights. Implicit processing heuristics that may help people self-reflect on the value of their new insights and select them for further development are listed in Box 9.14.

Positive Experiences, Humor, and Creative Resynthesis in Stage Three

The profoundly significant shift from the crisis and catharsis of stage two to the moments of insight and positive feeling in stage three is often accompanied by a sense of relief, surprise, and laughter. People experience a dramatic shift from what Maslow (1962, 1967) called "deprivation motivation" to "being motivation." They may happily whisper, *"It's wild, really strange, weird, odd!"* The usage of such words signifies that they are experiencing the nov-elty-numinosum-neurogenesis effect. They may even mention, *"Something really new, something I was never aware of before, suddenly popped into my mind."* In the classical descriptions of the creative process, a sense of light, illumina-tion, color, or fascinating and meaningful visual imagery is frequently men-tioned during these moments. *At this point we can only speculate that these creative moments are precisely the point at which activity-dependent gene expression is replayed for neurogenesis, problem-solving, and healing. The new proteins and synaptic connections synthesized at the cellular-genomic level in long-term potenti-ation make their effects known, on a conscious level, as an original psychological experience* (Rossi, 1972, 2000a). People recognize such creative moments in many ways—in silent wonder, bursts of joy, exclamations of awe, prayers of gratitude. Some people are full of questions about these moments of creative experiencing and need the therapist's ever-present assurance that they can make the creative choices that will really make a difference in their lives. The continuing mystery at this stage is how to select and safely try out what appears

---

BOX 9.14 | IMPLICIT PROCESSING HEURISTICS SELECTING AND STABILI-ZING STAGE THREE INSIGHTS.

• Interesting . . . ?

• Curious, isn't it . . . ?

• A little surprise . . . ?

• Learning to appreciate the unexpected . . . ?

• Yes, are you experiencing something a little different now . . . ?

• My goodness, is something really changing now . . . ?

• Experiencing the wonder of that . . . ?

• Mmm—really appreciating what continues all by itself . . . ?

• Okay to let yourself really appreciate that . . . ?

• Experiencing what it's really like to make your own choice.

to be of value in these deeply experienced moments of beauty and being motivation.

Personal Choice and the Selection
of Evolving States of Consciousness

There is a profound evolutionary significance in the questions people ask at these creative moments. The ability to ask such questions implies that a selection process is possible. At these critical choice points, people can select the emotional qualities as well as the contents of their present and future states of consciousness. To recognize, support, and utilize these precious moments of self-reflection and experiencing "the new" may be the most important function of consciousness—it may be the reason why consciousness evolved in the first place. At these delicate moments the therapist's task is to acknowledge and support the value of the person's experiencing and evolving self-awareness by using implicit processing heuristics such as those listed in Box 9.15.

Patients' need to evaluate and utilize the new brings them to stage four and the important work of verifying the value of their creative experiences and learning how to manifest them in real life.

BOX 9.15 | IMPLICIT PROCESSING HEURISTICS FACILITATING AND SELECTING CREATIVE MOMENTS IN TRANSITION BETWEEN STAGES THREE AND FOUR.

- Okay, simply continue receiving the best of that for a while . . . ?

- Yes, recognizing whatever is interesting about that . . . ?

- Wonderful, exploring this experience as much as you need to . . . ?

- Really appreciating the value of this kind of experience . . . !

- Selecting what is of greatest value for you!

- Really believing you can choose what you want?

- Yes, how will you take an active hand in co-creating yourself?

Stage Four: Verification

Co-Creation and Ratifying
the Reality of the Numinous and the New

After a period of novel experiences in stage three, the person may now make bigger postural adjustments that signal a shift into the evaluative and co-creative period of stage four. Sometimes people spontaneously stretch and open their eyes without being told to do so. The therapist never told them to close their eyes in the first place, so this spontaneous eye opening is an excellent validation that the locus of control is really within the patient—who is now ready to rejoin consensual reality and discuss whatever is necessary with the therapist. When patients do not open their eyes and awaken spontaneously, they may be under the impression that the therapist is supposed to tell them to wake up or give a signal to end the process. I usually satisfy this need by using a four-step implicit processing heuristic, such as the one contained in Box 9.16.

Mentioning *"appropriate times throughout the day"* and *"the right time to tune in"* are ultradian cues that may utilize the now-familiar basic rest–activity cycle that takes place every 90–120 minutes throughout the day. Such phrases are implicit processing heuristics that may help people access the state-dependent encoding of behavioral state-related gene expression, neurogene-

---

Box 9.16 | Stage four: Implicit processing heuristic to validate continuing creative work in everyday life.

---

1. When . . .

2. . . . a part of you knows it can continue this creative work entirely on its own, at *appropriate times throughout the day* . . .

3. . . . and when your conscious mind knows it can cooperate in helping you recognize when it is the *right time to tune in* . . .

4. . . . will that give you a feeling, a signal, that it's time for you to stretch, open your eyes, and come fully alert so you can discuss whatever is necessary for now?

sis, and healing that can take place most easily at "appropriate times." Stage four is a period of returning to normal consensual consciousness in which patients can now discuss strategies for trying out their new experiences. The therapist's task is to encourage the co-creation of their worldview, identity, and personality in their own way. Box 9.17 contains implicit processing heuristics that may facilitate the person's ability to co-create, that is, have creative dialogues wherein their explicit conscious experiences (words, feelings, motivations, and images) interact with their own more implicit, unconscious processing.

Verification of Positive Experiences in Stage Four

Patients invariably feel a sense of relief, happiness, and well-being in stage four. If symptom-scaling was introduced in stage one, it can be applied in stage four as a subjective check on the process of change. Stage four is an important time to ask patients to rescale their symptom, because the fact that their symptom is now usually less intense validates their therapeutic experience. If the symptom has disappeared completely, this is the time to discuss how the patient can continue to implement this type of creative inner healing at appropriate ultradian periods in everyday life. It is well to remind people that it is precisely when they are in their ultradian rest periods, when they are most likely to feel tired or discouraged, that they may have best access to

---

BOX 9.17 | IMPLICIT PROCESSING HEURISTICS FACILITATING CO-CREATION.

- Something you would like to share about that?

- How can this experience change your life?

- What is most significant about this for you?

- What does this lead you to now?

- How will you make changes in your life?

- What will you actually do in your life that is different this week?

such healing (Rossi & Nimmons, 1991). In this way, symptoms can be reframed into signals that the mindbody needs a quiet period of inner healing, and problems can be reframed into opportunities for accessing inner resources (Rossi, 1996a). It is likely that varying aspects of behavioral state-related and activity-dependent gene expression, neurogenesis, problem-solving, and healing take place in the ultradian rest phase of the basic rest–activity cycle. This is an area in which we urgently need the technologies of genomic neuroscience to validate exactly what is taking place and when in everyday life as well as in therapeutic work (Chin & Moldin, 2001).

If the symptom-scaling has dropped only a few points by stage four, it can be taken as an indication of partial success that needs to be developed further in future sessions. To prepare for further improvement, the patient is encouraged to explore the ultradian healing response in everyday life and keep a written record of experiences that can provide hints of the next step that is needed to facilitate further creative phase transitions in healing and problem-solving. I like to remind people of how many natural ultradian cycles of inner work and healing they can experience between sessions. If the therapy takes place once a week, for example, I may mention the somewhat startling fact that they will go through at least 84 creative cycles between now and when they return for the next session (7 days times approximately 12 two-hour ultradian rhythms a day yield 84 cycles of creative work each week). The only thing they have to do is tune into themselves, in a sensitive and sympathetic manner, about six to eight times a day and simply take note of anything new about their issues that presents itself since the last time they tuned in. As illustrated in Figure 2.11, nature is doing inner work and healing all the time. There are about four ultradian cycles during the typical sleep period of eight hours, when neurogenesis and healing can take place during the varying stages of sleep and dreaming. All people really need to do is to tune in now and then to notice how the novelty-numinosum-neurogenesis effect becomes manifest in their evolving states of consciousness, choice, and being.

People develop a positive outlook and a genuine sense of self-empowerment as they learn to recognize the sensory and emotional cues associated with their personal ultradian cycles of optimal healing and problem-solving. The essence of creatively oriented therapy is helping people learn to optimize their own skills in facilitating the evolution of self-reflection and co-creation. People sometimes acknowledge that their insights and healing experiences in the therapy session are not entirely novel—they've had similar experiences before but did not appreciate their value as seeds for future development. In stage four, however, they can develop a conviction about which options and choices are truly possible to facilitate their own development in a practical and realistic manner. People are usually able to give themselves their own behavioral prescriptions for the future.

SUMMARY

The infinite variety of human experience in the creatively oriented therapeutic arts will always defy simple summary. The major focus in this chapter has been on developing skills for recognizing the behavioral cues of each phase of the four-stage creative cycle and for utilizing implicit processing heuristics to facilitate each stage. This general summary integrates concepts from this and previous chapters for an overview of the psychosocial genomics of therapeutic hypnosis, psychotherapy, and the healing arts as illustrated in Figure 9.1.

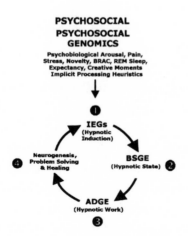

FIGURE 9.1 | The psychosocial dynamics of immediate-early genes (IEGs), behavioral state-related gene expression (BSGE), activity-dependent gene expression (ADGE), and the consolidation of neurogenesis, problem-solving, and healing in the four-stage creative process of therapeutic hypnosis, psychotherapy, and the healing arts.

Stage One:
Preparation: Immediate-Early Gene Expression in Hypnotic Induction

Immediate-early gene expression takes place with the initiation of any psychosocial state of arousal, which may occur in everyday life, as illustrated in Figure 9.1, as well as during therapeutic hypnosis and psychotherapy. Immediate-early gene expression begins with the arousal that usually takes place during the typical history-taking at the beginning of any psychotherapeutic process. More than mere words and talk therapy are involved. The patient's tears and distress in an initial interview indicate that he or she is already accessing and replaying state-dependent memory and emotional dynamics in the initial stages of a potentially healing adventure. *The therapist's task at this point is to recognize that the natural ultradian creative cycle of gene expression, neurogenesis, and healing has already begun, and to facilitate that cycle.* Implicit processing heuristics may be employed to optimize the entire psychogenomic process (sometimes without the therapist's knowing the nature of the patient's problem). The psychotherapeutic process may begin with a *symptom scaling* of the patient's currently experienced symptom, emotions, or attitudes. A subjective one-to-ten scale (ten representing the worst the symptom or problem was ever experienced; one, a completely satisfactory state) may be used to assess and validate the patient's experiential state before, during, and after the psychotherapeutic process to assess and validate it.

Stage Two: Incubation:
Behavioral State-Related Gene Expression in Hypnotic States

The second stage of the creative process is typically experienced as a period of incubation, meditation, or the valley of shadow and doubt ("the storm before the light") that is portrayed in poetry, song, and dance in the rituals of all cultures during important psychosocial transitions. These rituals—including those of therapeutic hypnosis, psychotherapy, and virtually all the healing arts—initiate behavioral state-related gene expression, neurogenesis, and the possibility of problem solving, and healing. The natural dynamics of Darwinian variation are usually evident as the patient reviews and explores the sources and history of his or her current issues. *The therapist's tasks during this second stage are (1) to offer open-ended therapeutic questions that may function as implicit*

processing heuristics designed to access the state-dependent memory and behavior-encod-
ing symptoms, and (2) to support the person through the sometimes painful arousal of
the natural ultradian cycle of creativity, problem solving, and healing in everyday life.
Less is often more at this stage of respectful listening rather than giving advice.

Stage Three: Illumination: Activity-Dependent Gene Expression and Conscious Selection of the Novel and Numinous

This is the famous "aha" or "eureka" experience celebrated in ancient and modern literature of the arts, culture, and sciences when the creative process is accessed. This is the essence of hypnotherapeutic work and psychotherapy. People are usually surprised when they experience a creative intuition. Those who automatically dismiss their originality as worthless never had their originality reinforced in their early life experience. *The therapist's task in stage three is to help the patient recognize and appreciate the value of the novel and the numinous that usually emerges spontaneously and unheralded after a period of inner struggle.* Often the patient may already have thought of the options for problem solving that come up at this stage but dismissed them because they were never validated.

Stage Four: Verification: Consolidating Target Gene Expression, Neurogenesis, Problem-Solving, and Healing

Most people are not aware of the need to facilitate and consolidate gene expression, neurogenesis, problem-solving, and healing in everyday life as well as in psychotherapy. This fourth stage of the creative process requires cooperation between the implicit, unconscious dynamics of target gene expression (the specific genes engaged in memory, learning, and healing) and the explicit, conscious dynamics of ratifying the reality of the new. Patients participate in this cooperative process but usually they do not know how to direct it. This is the greatest source of misunderstanding about the creative process by both patients and therapists: Patients and therapists are co-creators, not directors of the natural dynamics of gene expression, neurogenesis, and healing. *The patient's co-creative task is to recognize and value the new that arises in stage three and then plan in stage four how the new can be practiced in real life. The therapist's co-creative task in stage four is to (1) facilitate experiences to validate the value of the psychotherapeutic process and (2) help reframe and resynthesize symptoms into signals and psychological problems into inner resources.* Symptom scaling may be used to validate the therapeutic experience and what may need to be done in future sessions. In the next chapter we will utilize this four-stage outline of the creative cycle of replaying natural variation and conscious selection with a variety of innovative, activity-dependent approaches to facilitate novel experiences that could lead to gene expression, neurogenesis, and healing.

CHAPTER 10

NOVEL APPROACHES TO
ACTIVITY-DEPENDENT CREATIVE WORK

Exercises in Symmetry-Breaking, Self-Reflection, and Co-Creation

"*The shuttling to and fro of arguments and affects represents the transcendent functions of opposites. The confrontation of the two positions generates a tension charged with energy and creates a living, third thing—not a logical still birth in accordance with the principle tertium non datur but a movement out of the suspension between opposites, a living birth that leads to a new level of being, a new situation. The transcendent function manifests itself as a quality of conjoined opposites. So long as they are kept apart—naturally for the purpose of avoiding conflict—they do not function and remain inert.*

"*In whatever form the opposites appear in the individual, at bottom it is always a matter of a consciousness lost and obstinately stuck in one-sidedness, confronted with the image of instinctive wholeness and freedom. This presents a picture of the anthropoid and archaic man with, on the one hand, his supposedly uninhibited world of instinct and, on the other his often misunderstood world of spiritual ideas, who, compensating and correcting our one-sidedness, emerges from the darkness and shows us how and where we have deviated from the basic pattern and crippled ourselves psychically.*"

—Carl G. Jung
The Structure and Dynamics of the Psyche

"*Whether in dealing with the organization of systems or with the structure of languages, hardships with self-referential situations have the same root: the distinction between actor or operand, and that which is acted or operated upon, collapses. There seems to be an irreducible duality between the act of expression and the content to which this act addresses itself: self-referential occurrences blend these two immiscible components of our cognitive behavior and engender a dual nature which, apparently, succeeds in escaping this universal behavior and thus seems peculiar in our knowledge. Their peculiarity lies in being self-indicative in a given domain, in standing out of a background by their own means, in being autonomous as the strict meaning of the word enounces. . . .*

"By allowing an antimonic form (from the point of view of logic [a duality]) we have constructed a new larger domain akin to the complex plane, where new forms can be lodged, including those of the preceding primary domain found to be in conflict by the introduction of re-entering expressions. Again, rather than avoid the antimony, by confronting it, a new domain emerges.

"This intercrossing of domains at the point of self-referring, hence antimonic [duality], situations in a given domain, repeats itself. The most impressive instance being the appearance of living systems when a set of chemical productions closes onto itself to become a self-productive and self-constructive unity. Later on, when in a living system cognitive structures become capable of self-description, again a sig- nificant new domain emerges, that of self-consciousness. By uniting two constituents of a domain, producer and produced, description and describer, into a third state which blends the two preceding ones through circular closing, we see the appearance of a much more inclusive domain. It appears as if different, successively larger levels are connected and intercross at the point where the constituents of the now lower level refer to themselves, where antinomic forms appear, and time sets in. . . . We recognize this fact in ordinary speech. When trying to convey a description of a new domain we often construct apparent antimony to induce the listener's cognition in a way such as to compel his imagination towards the construction of a larger domain where the apparent opposites can exist in unity. A moral example: once you lose everything, you have everything; a philosophical one: a being is when it ceases to be."

—Francisco Varela
A Calculus for Self-Reference

"We have seen that neural systems display large-scale coherent patterns of activity in both space and time. In many cases these can be very well simulated by simple models displaying the same coherent, although chaotic, patterns of behavior. . . . This transition occurs by means of a symmetry-breaking instability. . . . symmetry- breaking involves the choice between two symmetric final states. Consider a ball perched precisely on the peak of a steep ridge: the tiniest chance fluctuation in the forces acting on it (such as a stray gust of wind) may send it on one of two widely divergent paths."

—Richard Solé and Brian Goodwin
Signs of Life: How Complexity Pervades Biology

Perhaps the greatest oversimplification about psychotherapy is to describe it as mere "talk therapy." It is rather quaint and interesting to learn that it was one of Freud's early patients, Anna O, who came up with this talk therapy idea. More than one hundred years later, however, it still leads us astray. Psychotherapy, the humanities, and the healing arts are deeply engaged in the psychobiology of gene expression, neurogenesis, and mind-molecular communication to a degree that previous generations, before the advent of the Human Genome Project, could not have conceived. Behind all the therapeutic talk and mutual support systems on the psychosocial level are the numinous psychobiological dynamics of self-reflection and co-creation on all levels from mind to gene. A meaningful life is facilitated by an awareness of how to remain in touch with the numinous "aha!" experiences that signal creativity in everyday life as well as in the arts and sciences.

This chapter begins to explore novel psychogenomic approaches to facilitating activity-dependent gene expression in the creative psychotherapeutic encounter. These activity-dependent approaches range from what Carl Jung called the "transcendent function" (which mediates between the polarities, dualities, and conflicts of the psyche) to the utilization approaches of Milton Erickson (which facilitate the reassociation, reorganization, and resynthesis of experience). From the perspective of psychosocial genomics, the activity-dependent approaches of this chapter optimize Darwinian variation and conscious selection by facilitating the natural symmetry-breaking dynamics characteristic of all emergent processes computing on the edge of deterministic chaos.

THE SYMMETRY-BREAKING HEURISTICS IN MATHEMATICS, LOGIC, PHYSICS, BIOLOGY, AND PSYCHOLOGY

The concept of symmetry-breaking was introduced in Chapter 1 as a unifying heuristic to integrate psychology with the foundations of logic, mathematics, physics, and biology. In this final chapter we utilize the related concepts of symmetry-breaking, duality, and resynthesis to facilitate novel, easily learned approaches to the four-stage process of activity-dependent creative work. Let us begin with a brief summary of the evolution of these concepts and their practical application to psychotherapy.

The story of Western philosophy and its relevance for psychotherapy began with Socrates' injunction, "Know thyself." How to know oneself as distinct in relationship to the outside world can be regarded as the essence of epistemology—the theory of the origin, nature, methods, and limits of knowledge. The beginning of logic in ancient Greece was a way of systematizing epistemology to find a way of proving that we have a path to true knowledge. George Boole is credited with the first systematic attempt to formalize logic as the laws of thought. One of the most important laws of thought was its essential *duality*—any statement supposedly could belong to one of two categories: true or false. This was also called the principle of the excluded middle—*tertium non datur: there was no middle way between true and false*. This was the idealistic intent of Boole's (1847) *The Mathematical Analysis of Logic*, and for a while it was the key to its utility. In the confusing Tower of Babel of human experience, the rational hope was that dual logic could discern truth from falsehood so that we could settle our conflicts within (psychology) and between (sociology) each other.

Alas, two millennia of scholarship and practical affairs have documented that this rationalist ideal does not lead to Utopia. Ironically, the source of the problem was recognized in the paradoxes of logic discovered by the early Greeks. The most famous of these is attributed to the Cretan philosopher Epimenides who remarked, "All Cretans are liars." This classical paradox is seditious to orthodox logic, however, "because it maintains an undesirable *autonomy* vis-à-vis any orthodox attempt to apprehend it: when apprehended as true, it turns out to be false; when apprehended as false, it turns out to be true" (Howe & Von Foerster, 1975, p. 1, italics added).

Such paradoxes, antinomies, or confusing conflicts of logic were not taken seriously for a long time, however. They were often regarded as playful but irrelevant pastimes of the scholarly, impractical—perhaps mere aberrations of thought—pathologies of language seized upon by querulous cranks with nothing better to do. This misunderstanding changed dramatically when Whitehead and Russell (1925) recognized that the contradictions inherent in paradox are central to all logical inquiry that involves "a certain kind of vicious cycle" (p. 37).

In all the above contradictions (which are merely selections from an infinite number) there is a common characteristic, which we may describe as *self-refer-*

ence or reflexiveness. The remark of Epimenides must include itself in its own scope. If *all* classes, provided that they are not members of themselves, are members of *w* this must also apply to *w*; and similarly for the analogous relational contradictions. (p. 61)

Whitehead & Russell then attempted to deal with the problems of paradox in logic by prohibiting them with their Theory of Logical Types that separated antinomies (the opposites) by putting them on different levels so they could not contradict each other. Howe & Von Foerster (1975) outline the profound historical implications of the prohibition imposed by the Theory of Logical Types for the evolution of the autonomous psychology of the individual in philosophy, ethics, and science as follows.

This prohibition not only eliminates the potential for formulating paradoxes of the above kind, it also eliminates the potential for contaminating utterances, statements, propositions, descriptions, etc. with properties of those who utter, state, propose, describe, etc. In other words, implicit in the Theory of Types is the proviso that is the ultimate protector of the Claim to Objectivity: "The properties of the observer shall not enter into the description of his observations . . . which exclude in principle the autonomy of paradox and the individual. In the scientific revolution that we now create and experience, however, we perceive a shift from causal unidirectional to mutualistic [and circular] systematic thinking, from a preoccupation with the properties of the observed to the study of the properties of the observer. The initiator of this shift was Kant, who placed the autonomy of the observer at the center of his philosophy, thus making this autonomy responsible for the properties of the observed. (pp. 1–2)

In placing the autonomy of the observer at the center of his philosophy, Kant's intention was not to effect a shift from objectivity to subjectivity but rather to initiate an ethic, for he clearly saw that without autonomy there could be no responsibility and hence no ethics. Ethics—and not subjectivity—is the complement of objectivity. (p. 3)

Bateson (1972) clearly recognized the experiential associations between paradox, contradiction, psychopathology, trance, and creativity as noted in his conversations with Brand (1974).

A paradox is a contradiction in which you take sides—both sides [a duality]. Each half of the paradox proposes the other . . . if you sweat out one of these paradoxes you embark . . . on a voyage, which may include hallucinations and trance. . . . But you come out knowing something you didn't know before, something about the nature of where you are in the universe. (pp. 9–36)

The implications of Bateson's view are profound: if one "sweats out"— that is, if one continues to do inner work replaying a paradox (contradiction, duality) one is actually engaged in a voyage that may fail (hallucination) to generate a creative trance wherein one may learn something new. Milton H. Erickson had previously discovered independently essentially the same thing in his use of a variety of double and triple binds to facilitate creative work via therapeutic hypnosis (Erickson, Haley, & Weakland, 1959). Erickson and Rossi (1975) then clarified the nature of a variety of double and triple therapeutic binds that could help patients transcend their learned limitations (psychopathology) by employing metalevels of experiencing. Multiple levels of cognition carried on simultaneously were recognized as the key to the therapeutic applications of the microdynamics of trance and suggestion (Bateson, personal communication, 1974–1975; Erickson & Rossi, 1976). This was the core of what Erickson & Rossi (1976, 1976/1980) initially called "the indirect forms of suggestion," which are more aptly now called "implicit processing heuristics" in this volume.

The mathematician Spencer-Brown (1972) resolved the conceptual foundations of the association between paradox and self-reference by breaking out of the limitations of Boolean dualistic logic. Spencer-Brown did this by vastly extending logic to include not just two but four classes of statement: true, false, meaningless, and imaginary. The use of the imaginary class is of particular interest to psychotherapists because of its analogy with the use of imagination as a fundamental aspect of the therapeutic process. Spencer-Brown established the utility of including the imaginary in the foundations of mathematics in analogy with the invention of complex values to resolve the paradoxes of simple self-referential equations (such as $x = -1/x$) by constructing the imaginary number i (where i is set equal to the square root of minus one). In his preface to his famous book, *The Laws of Form*, Spencer-Brown (1972) comments on the broad and deep implications of introducing the imaginary as a valid class of value in cognition as follows.

What is fascinating about the imaginary Boolean values, once we admit them, is the light they apparently shed on our concepts of matter and time. It is, I guess, in the nature of us all to wonder why the universe appears just the way it does. Why, for example, does it not appear more symmetrical? Well, if you will be kind enough, and patient enough, to bear with me through the argument as it develops itself in this text, you will I think see, even though we begin it as symmetrically as we know how, that it becomes, of its own accord, less and less so as we proceed. (pp. xv–xvi)

Spencer-Brown's ground-breaking work in the foundations of mathematics (called the "calculus of indications") supports our development of the symmetry-breaking approaches to novel activity-dependent creative work in psychotherapy introduced in this chapter and previous studies (Rossi, 1996a, 2000a). Spencer-Brown outlines the initial step in his calculus of indications as an act of "severance" (a duality or dissociation) in the creation of being.

The theme of this book is that a universe comes into being when a space is severed or taken apart. The skin of a living organism cuts off an outside from an inside [the beginning of duality]. So does the circumference of a circle in a plane. By tracing the way we represent such a severance, we can begin to reconstruct, with an accuracy and coverage that appear almost uncanny, the basic forms underlying linguistic, mathematical, physical, and biological science, and can begin to see how the familiar laws of our own experience follow inexorably from the original act of severance. The act is itself already remembered, even if unconsciously, as our first attempt to distinguish different things in a world where, in the first place, the boundaries can be drawn anywhere we please. At this stage the universe cannot be distinguished from how we act upon it, and the world may seem like shifting sand beneath our feet.

Although all forms, and thus all universes, are possible, and any particular form is mutable, it becomes evident that the laws relating such forms are the same in any universe. It is this sameness, the idea that we can find a reality which is independent of how the universe actually appears, that lends such fascination to the study of mathematics. That mathematics, in common with other art forms, can lead us beyond ordinary existence, and can show us something of the structure in which all creation hangs together, is no new idea. But mathematical texts generally begin the story somewhere in the middle, leaving the reader to pick up the thread as best he can. Here the story is traced from the beginning. (p. xxix)

Severance, indication, consciousness, and the appearance of the universe are described in this chapter as the initial breaking of a symmetry into dualities that are then reconstructed or resynthesized into new experiential realities via activity-dependent creative work generating gene expression, neurogenesis, and healing. A practical illustration of the act of indication was presented in Figure 7.2, wherein the psychotherapist simply points to the spontaneous appearance of vibration in the hand and finger movements of the patient as she enters stage two—the phase of psychobiological arousal, conflict, and inner-work so characteristic of the creative process.

Francisco Varela (1975) took the next step in the ongoing saga of the paradoxes of self-reference and the creative approaches to psychotherapy in a fundamental paper in which he introduces "a calculus for self-reference" to supplement Spencer-Brown's calculus of indications (Marks-Tarlow & Martininez, 2002).

Self-reference is awkward: one may find the axioms in the explanation, the brain writing its own theory, a cell computing its own computer, the observer in the observed, the snake eating its own tail [an ancient symbol] in a ceaseless generative process.

Stubbornly, these occurrences appear as outstanding in our experience. Particularly obvious is the case of living systems, where the self-producing nature of their entire dynamics is easy to observe, and it is this very fact that can be taken as a characterization for the organization of living systems. Similarly, the physiological and cognitive organization of a self-conscious system may be understood as arising from a circular and recursive neuronal network, containing its own description as a source of further descriptions. (p. 5)

It is generally assumed that self-reference leads inevitably to contradictions even in ordinary discourse, let alone in formal languages, and hence, as said, is carefully avoided. Yet, true as this may be, language *is* self-referential, and if we are not prepared to avail ourselves fully of self-referential notions, it is not possible to deal either with this aspect of discourse or with the many systems where self-reference is a central feature of their organization. . . .

Self-reference, in this calculus, can be identified with the notion of *re-entry* and in this way its basic form is recovered at this deep level, from which all its manifestations can be contemplated, whether in logic and formal systems, or in the organization of certain systems . . . (p. 6).

The new or third value introduced by re-entry as an imaginary state in the form, may be taken as a value in and extended arithmetic to arrive at a calculus capable of containing re-entering expressions . . . Let this state arise autonomously, that is, by self-indication. Call this third state appearing in a distinction, the autonomous state. (Varela, 1975, p. 7)

Of special interest for the practice of therapeutic hypnosis and psychotherapy is the way Varela identifies the third or imaginary state (that can resolve the antinomies, dualities, stress, and conflicts of paradox) as *autonomous*. Erickson et al. (1976; Erickson & Rossi, 1976) have identified the essential microdynamics of the therapeutic response as autonomous (taking place all by itself) on an implicit or unconscious level. Spencer-Brown (1972) can even account for the fundamental significance of time, oscillation, and rhythm that we have emphasized as so characteristic of the chronobiology of gene expression, neurogenesis, and healing outlined in Part I of this book.

We should not be surprised by the connection between infinity and time since the nature of a re-entering expression is precisely that of an infinite recursion in time of a closed system. Thus in a cell we deal with productions of production of productions and in self-consciousness with descriptions of descriptions of descriptions . . . we may take the states [whether indicated or not] as timeless constituents of a [self-reference] occurring as an oscillation in time. . . . We may note in this connection that by considering [self-reference] as an oscillation in time, we may also consider other re-entering of expressions as modulations of a basic frequency. (pp. 20–21)

It is interesting to note how logicians and mathematicians use the concepts of symmetry-breaking, duality, and reorganization as a re-entry process of recursive oscillations in time to describe the dynamics out of which something evolves out of nothing in the psychological as well as the physical world (Poundstone, 1985; Robertson, 1995, 1999). Physicists use symmetry-breaking and duality theory to describe the quantum dynamics of the big bang theory of the origin of the universe and its continuing thermodynamic evolution over time (Greene, 1999; Treiman, 1999; Wheeler, 1994; Zee, 1986). Smolin (2001) has recently described the relationship of the duality hypothesis and symmetry in association with Einstein's relativity theory and the field theories of the quantum view of the universe as follows.

> It is called the *hypothesis of duality*. I should emphasize that this hypothesis of duality is not the same as the wave-particle duality of quantum theory. But it is as important as that principle or the principle of relativity. Like the principles of relativity and quantum theory, the hypothesis of duality tells us that two seemingly different phenomena are just two opposite ways of describing the same thing. If true, it has profound implications for our understanding of physics. (p. 114)
>
> The idea of duality is still a major driving force behind research in elementary particle physics and string theory. Duality is the very simple view that there are two ways of looking at the same thing—either in terms of strings or in terms of fields. . . . It has been shown to be valid in very specialized theories, which depend on very specific simplifying assumptions. Either the dimensionality of space is reduced from three to one, or a great deal of additional symmetry is added, which leads to a theory that can be understood much more easily. (pp. 117-118)

Mathematicians and biologists use the symmetry-breaking concept to describe the emergence of self-organization in chaos theory and the complex adaptive systems view of life (Kauffman, 1995; Mainzer, 1994; Prigogine, 1980, 1997; Prigogine & Stengers, 1984). In Chapter 4 we used the Feigenbaum scenario as a mathematical model of symmetry-breaking to describe the evolution of creative choice in the dramas of dreams and imagination. We now extend the grounding metaphor (Lakoff & Núñez, 2000) *of symmetry-breaking, duality, and resynthesis in mathematics and physics to biology* as a heuristic for facilitating the psychosocial genomics of self-reflection and co-creation in psychology and the humanities. Before we do so, however, we need a brief excursion into the basic biology of gene expression in the symmetry-breaking dynamics of development to deepen our understanding.

Polarity and the Symmetry-Breaking Heuristic in Biology

How does the biological process of life begin? In the beginning there is the development of symmetry-breaking and polarity (or duality). The newly fertilized egg contains all the genes necessary for life and development. How does this genotype manifest itself in the many different outward phenotypic forms and functions we see in the adult? In normal development different sets of genes receive molecular signals (initially from the mother's body and later from the physical and psychosocial environment) to express themselves by generating proteins that give each cell, tissue, and organ its specific identity and function. *Symmetry-breaking of the original genome on the cellular level leads to varying patterns of gene expression, self-organization, and ultimately the cocreation of self-identity on the psychological level.*

The original fertilized egg divides (differentiates) to form a *symmetrical* sphere of cells (the blastula) in the first stage in the development of the organism. The blastula differentiates into three germ layers (ectoderm, mesoderm, and endoderm) out of which the organism will take shape. The symmetrical sphere of the blastula gradually develops two major *polarities or asymmetries* that are called the head–tail axis and the front–back axis. These polarities could be described as the initial major *symmetry-breaking process* that gives rise to the differentiation and development of the organism on biological levels that eventually generate the psychobiology of self-organization, self-reflection, and self-identity.

In the fetus the body plan (head, torso, limbs, etc.) is guided by gradients of hormones and tropic (growth) factors that function as molecular signals that originally come from the mother. These molecular gradients are gradually taken over by the developing fetus where they continue the process of signaling cells throughout the developing body. Genes that control this process of polarization and differentiation are called "homeotic" genes. Homeotic genes generate proteins that function as transcription factors that turn other genes on and off in the symphony of self-creation. These genes are also important on the psychological level (e.g., compulsive grooming behavior in mice, Greer & Capecchi, 2002).

The Opposites and the Symmetry-Breaking Heuristic in Psychology

Mesmer, Janet, Freud, Jung, Erickson, and other pioneers of psychotherapy and the healing arts knew nothing of this implicit, genotypic level of development that generates the phenotypic level of human experience and behavior. They could not have known that the observable phenotypic conflicts, polarities, dualities, and so-called pathological dissociations of human consciousness, experience, and behavior that they observed and attempted to heal were all manifestations of normal developmental symmetry-breaking processes at the levels of gene expression and neurogenesis. The early pioneers of psychotherapy recognized that the source of many human problems could be found in early childhood development, but they did not yet understand the extent to which gene expression, neurogenesis, problem-solving, and healing are continually generated at the molecular level in normal everyday adult life. They did not have the data documenting how the molecular genomic level is driven by the daily and hourly struggle of existence throughout life that we now call the "complex adaptive systems of mind-gene communication and healing" illustrated in Figures 2.11, 3.6, and 5.1.

The symmetry-breaking heuristic in math, physics, and biology was described in depth psychology as the antithetical meaning of primal words (Freud, 1910), the problem of the opposites (Jung, 1960), duality (Rank, 1941), as well as dissociation in therapeutic hypnosis, and conflict in the psychiatric context of psychopathology. The traditional techniques of psychotherapy ranging from suggestion, free association, active imagination, biofeedback, and imagery to the use of metaphor, meditation, and movement were all designed to help people work through the difficult early stages of symmetry-breaking, duality, dissociation, and conflict in the recursive dynamics and replays of identity formation and reformation. We now recognize all these traditional psychotherapeutic techniques as implicit processing heuristics to facilitate behavioral state-related and activity-dependent gene expression via the four-stage creative cycle. At the present time, neuroscience is approaching a threshold where the techniques of brain imaging and DNA microarrays will be used to assess the molecular dynamics of psychosocial genomics generated by these activity-dependent approaches (Lok, 2001; Phelps, 2001).

All the traditional techniques of psychotherapy can be used with a variety of the novel, sensory-motor, activity-dependent approaches to symmetry-breaking dynamics of development presented in this chapter. These novel approaches all involve symmetry-breaking and the nonlinear dynamics of chaos theory in psychology (Abraham & Gilgen,1995; Kauffman, 1995; Masterpasqua & Perna, 1997; Robertson & Combs, 1995; Vallacher & Nowak, 1994). They are infinite in their variety and application (Erickson et al., 1976; Erickson & Rossi, 1979, 1981, 1989). The main thing they add to traditional techniques is a general facilitation of behavioral state-related and activity-dependent experiencing that is channeled into permissive, self-directed processes of creative problem-solving and healing (Rossi, 1996a; Rossi & Cheek, 1988).

The natural symmetry-breaking dissociations, dualities, and polarities that are usually manifest as symptoms and psychological problems can be replayed, reframed, and resynthesized by helping people to experience all sides of their conflicts in the therapeutic encounter. *The therapist's initial task is to formulate an appropriate and safe setting to stage an experience of different sides of the contending forces of the person's personality so they may be engaged in a creative cycle of problem-solving and healing.* It is important to understand the intent of these activity-dependent exercises. They are all designed to facilitate the three factors that can optimize neurogenesis according to current neuroscience: novelty, environmental enrichment, and exercise, together with the numinous sense of wonder typically associated with creative experience. There is nothing stereotyped about these exercises. That is why so many are outlined. We need a large repertory because we hope to engage the subject's sense of novelty and the numinosum by never repeating these creative processes in exactly the same way with the same person. We always utilize the person's language, attitudes, and worldview in these approaches (Erickson & Rossi, 1979, 1981, 1985; Rossi, 1996a). I typically initiate people into the four-stage creative cycle to integrate dissociated parts of a problem by projecting them into their hands with an activity-dependent, symmetry-breaking approach somewhat as follows.

EXERCISE 1:
THE FOUR-STAGE CREATIVE CYCLE FACILITATED WITH ACTIVITY-DEPENDENT, HAND SYMMETRY-BREAKING

The purpose of this activity-dependent four-stage creative process, facilitated by the use of hand symmetry-breaking, is to access, replay, and resynthesize state-dependent memory, learning, and behavioral systems that encode significant life experiences. Typical hand positions of the four stages of the creative cycle are illustrated in Box 10.1, along with an initial implicit processing heuristic to facilitate each stage.

BOX 10.1 | THE FOUR-STAGE CREATIVE CYCLE FACILITATED BY ACTIVITY-DEPENDENT SYMMETRY-BREAKING USING THE PALMS-FACING-EACH-OTHER PROCESS.

1. Preparation: Sensitization and Ideodynamic Experiencing

"Lift up your hands with the palms facing each other in a symmetrical manner about six to eight inches apart [therapist demonstrates]. . . . With great sensitivity, notice what you begin to experience. . . . Is one hand warmer or cooler than the other? . . . Lighter or heavier? . . . More or less flexible? . . . Stronger or weaker?"

2. Incubation: Accessing State-Dependent Memory, Learning, and Behavior

"Will just one of those hands now begin to drift down slowly to signal that your inner nature will now explore some *private . . . even secret emotions and memories . . .*?"

3. Illumination: The Novelty-Numinosum-Neurogenesis Effect

"Will the other hand now drift down slowly as you explore possibilities of healing and problem-solving? . . . Will that hand go down slowly signaling when you are ready to begin *to experience something new? . . . Interesting? . . . Curious?*"

4. Verification: Reframing Symptoms into Signals and Problems into Resources

Review the entire session. Explore how symptoms can be reframed into signals and problems into creative inner resources to be developed further. Then close with:

"On a scale of 0–10, how confident are you of this new understanding?"

The four-stage creative cycle, as it is facilitated with activity-dependent hand symmetry-breaking, is infinite in the manner in which it can be experienced, even when an attempt is made to standardize the therapist's manner, verbalizations, and implicit processing heuristics. Recall that this approach does not involve an effort to program, prescribe, or condition people in a behavioristic manner. We are using heuristics to engage creative processes whose outcome cannot be predicted in advance. In the initial preparation stage, for example, it is of no concern whether the person actually experiences any of the sensations that are offered as basic accessing questions (lighter, cooler, etc.). The intent of the preparation stage is simply to focus the person inward with heightened self-sensitivity. There are implicit, complex adaptive systems operating within the person that continually evaluate, respond to, and modulate what the therapist is saying. Again, the therapist's words are *heuristics*—not suggestions, directives, covert demands, commands, or interpretations in the conventional sense. Heuristics, as defined by the Second College Edition of *Webster's New World Dictionary*, means "to invent or discover: see EUREKA; helping to discover or learn . . . using rules of thumb to find solutions or answers."

The therapist offers novel, enriching, and interesting implicit processing heuristics to help people self-reflect and co-create within the safe privacy of ideodynamic discovery within. *Ideodynamic*, a word that does not appear in most dictionaries, means that an *idea* can generate *dynamic* transformations of psychological experience (e.g., memories, sensations, images, emotions, cognitions, behaviors, and psychobiological states) that are surprising to consciousness. When things are going well, the person typically experiences these novel and surprising ideodynamic transformations as insights, intuitions, creativity, solutions, and healing.

A more detailed set of typical implicit processing heuristics to accompany each stage of the creative cycle follows. The ellipses usually mean that there is a pregnant pause to allow the person a few minutes to process the therapist's words. Recall that it requires at least a minute or two for immediate early genes to spring into action and generate the molecular dynamics of gene expression and its attendant creative cascades of neurogenesis and psychobiological work. Italicized words usually facilitate creative choices for implicit processing and conscious surprise.

1. Preparation:
Sensitization and Ideodynamic Experiencing

"Lift up your hands with the palms facing each other about six to eight inches apart [therapist demonstrates]. With great sensitivity, notice what you begin to experience. . . . Do the hands *feel the same or different? . . .* One hand may be *lighter or heavier? . . .* One hand may be *warmer or cooler? . . .* Something like a *magnetic force or energy pulling those hands together or pushing them apart? . . .* Or perhaps they seem to take on *a life of their own, moving in a special way? . . . Different feelings in one part of your body or other? . . .* Heat . . . sweating . . . vibration . . . shaking . . . or tingling? Whatever you are experiencing is a unique expression of your individuality. . . . It is nature's way of getting ready to do inner work."

2. Incubation:
Accessing State-Dependent Memory, Learning, and Behavior

"Will just one of those hands now begin to drift down slowly to signal that your inner nature will now explore some private . . . even *secret . . . emotions and memories that you will keep to yourself? . . .* The *courage* to simply *wonder and receive* everything that is necessary to help you? . . . Allowing yourself to continue experiencing that for another moment or two . . . only long enough to *learn what it is about so you can receive what comes next? . . .* Feeling as much or as little as you want *privately. . . .* Then letting it go . . . and welcoming what comes next. . . . One part of you experiencing that as fully as you need to . . . while *another part safely guides what is deeply meaningful to you . . . ?*"

3. Illumination:
The Novelty-Numinosum-Neurogenesis Effect

"Will the other hand now drift down slowly as you explore possibilities of healing and problem-solving? . . . Will that hand go down slowly, signaling when you are ready to begin *to experience something new? . . . Interesting? . . . Curious? . . .* Working well with yourself? . . . Something *pleasantly surprising you can look forward to? . . .* What *you really need for healing and problem-solving? . . .* Simply receiving and continuing to explore the *sources of strength for dealing successfully* with that problem. . . . Yes, experiencing healing and problem-solving as that hand finally comes to rest."

4. Verification:
Reframing Symptoms into Signals and Problems into Resources

First review the entire session and reframe symptoms into signals and problems into resources for healing and further development. Then use symptom-scaling.

"On a scale of 0–10 . . . where 10 is the worst you have ever felt about yourself and 0 *is a completely satisfactory state . . . how confident are you of this understanding . . . ?* What number comes up . . . ? When your inner nature knows you can continue these positive developments . . . and when you really know *you can take a break for about 20 minutes several times a day to transform symptoms into signals and problems into resources . . . will something within you review what you need to for a minute or so, to confirm your inner resolve?"*

Repeat if necessary: [If a number greater than 5 is reported, continue with the following.]

"If your inner nature knows it can do another unit of healing [problem-solving or whatever] right now so you can reach a completely satisfactory state, will you find yourself replaying that inner work again with new variations for a few moments so you can fully receive everything you need at this time?"

[If a number greater than 5 continues after two or three creative replays, then utilize the ultradian creative cycle to support the person's continuing work in everyday life.]

"You know that your mind and body go through *a natural cycle of ultradian healing and problem-solving every couple of hours throughout the day and even at night when you are dreaming.* Notice how *your progress continues all by itself,* and we will pick it up from there the next time we meet."

By using a three-dimensional projection that maps the relative amount of sensory-motor input to the cortex of the brain from the peripheral areas of the body, Figure 10.1 provides an interesting perspective on the neuroanatomical dynamics of activity-dependent approaches that utilizes the

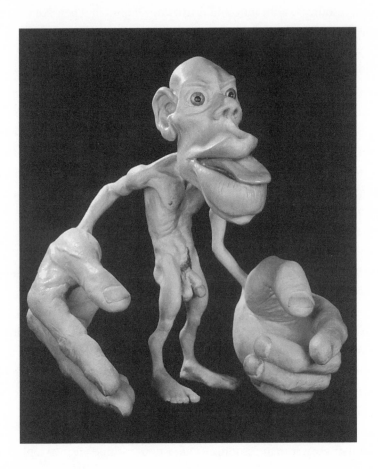

FIGURE 10.1 | Three-dimensional projection of mind-brain space of the sensory-motor homunculus. (Copyright 2001 The Natural History Museum, London)

ideodynamics of hand symmetry-breaking. Noteworthy in the mind-brain space of this so-called "sensory-motor homunculus" is the apparently grotesque oversize of certain body areas such as the hands, fingers, feet, face, and lips, indicating that these areas require particularly large areas of brain activity. Evolution has favored the significance of sensory-motor input from these body areas by giving them proportionally larger areas of the brain for more detailed mapping. This is one neuroanatomical rationale for utilizing the hands—one of the oversized sensory-motor areas of the body-brain maps—in the novel, activity-dependent approaches to creative experience outlined in this chapter. The hands have larger areas for receiving implicit associations from other areas of the brain—associations that engage them in the creative therapeutic process.

Figure 10.1 also illustrates something interesting about the "binding problem" in organizing sensory-perceptual experience. How does the brain integrate all the modalities (qualia) of phenomenological experience so that consciousness appears as a single picture or whole gestalt? The perceptual paradoxes and illusions of faces and vases, illustrated in elementary textbooks of psychology and neuroscience (Edelman & Tononi, 2000), indicate that consciousness is a continually updated and integrated process of analysis and synthesis that is in constant replay, iterating itself to better and better solutions (Rossi, 1989a, 1996a, 1997a, 1998). The implicit psychodynamics of focusing on one set of state-dependent neural networks to the partial exclusion of others in the natural flow of creative experience (Csikszentmihalyi, 1996; Sternberg & Davidson, 1996) involve microsecond fluxes in dissociation, replay, and the resynthesis of memory and learning that we finally register *as something new—an insight* in consciousness.

Conscious experience invariably involves natural implicate processes of "taking apart," "dissociation," or "polarization" that often leads to the experience of "emotional conflicts" in the initial stages of any creative process. People typically find themselves having psychological problems, stress, and symptoms of "psychopathology" when they experience themselves as being stuck, with no solution in sight, in the initial *preparation* and *incubation* stages of symmetry-breaking in the creative cycle. Putting things back together again—that is, replaying, reframing, and resynthesizing—via natural variation, selection, and conscious choice in the later stages of *insight* and *verification* of the creative cycle is experienced as successful problem-solving and healing.

These dynamics are examined in greater detail in the next exercise, which focuses on engaging duality, polarity, and conflict in the creative cycle.

EXERCISE 2:
ENGAGING DUALITY, POLARITY, AND
CONFLICT USING THE PALMS-UP PROCESS

The therapist begins by modeling the palms-up hand position, as illustrated in Box 10.2. *Notice that this position begins with hands and arms suspended in the air about six to eight inches above the lap.* Sometimes people begin by resting their hands on their lap because they do not notice that the therapist is holding hands and arms above the lap. Resting hands on the lap or on the arms of a chair or table could orient some people to passivity wherein they mistakenly believe this will be an exercise in relaxation or meditation and they can take their ease while the therapist somehow weaves some healing magic about them. Resting the hands in the lap tends to give an implicit cue for relaxation rather than creating a readiness for action. We seek to facilitate *immediate early, behavioral state-related,* and *activity-dependent gene expression* at this initial stage of the creative cycle, not rest. We encourage people to hold their hands up in a fairly symmetrical position so that their inner ideodynamic transformations can be projected into the outer sensory-motor psychodramas of symmetry-breaking hand movements, as they engage in the numinous therapeutic replay and resynthesis of significant life issues.

This particular position of the hands, with open palms facing up, will have varying symbolic meanings for people on an implicit level: that of supplication, receiving, openness, and so on. Whatever specific meaning they may attribute to the palms-up position, they are poised for a psychodramatic engagement in an activity-dependent process of creative self-facilitation.

People function on many psychological levels at the same time, even when they do not realize it. On one level they may be replaying and reexperiencing salient life events in a more or less involuntary manner. On another level, they are simultaneously observing themselves and responding to their self-generated inner motivations, emotional forces, and fantasies. On yet another level, they are apparently directing their own psychotherapeutic inner work and describing it to the therapist. These many levels of experiencing

BOX 10.2 | ACTIVITY-DEPENDENT CREATIVE FACILITATION: THE PALMS-UP SYMMETRY-BREAKING HAND PROCESS.

1. Preparation: Facilitating Self-Sensitivity

"When you are ready to do some important inner work on that problem will you lift your hands above your lap with your palms up . . . as when you are ready to *receive* something? [Therapist models.] As you *focus on those hands in a sensitive manner,* I wonder if you can begin by letting me know which hand seems to experience or express that fear (or whatever the negative side of the patient's conflict may be) more than the other . . . ?" [As soon as the person indicates that one hand is more expressive of the problem or symptom than the other, the therapist goes on to stage two.]

2. Incubation: Replaying Conflicts

"*Wonderful.* . . . Now I wonder what *you experience in your other hand, by contrast* . . . at the same time . . . ? What do you experience in that other hand that is the *opposite* of your problem [issue, symptom, etc.]? *Good, as you continue experiencing both sides of that conflict* [or whatever] . . . *at the same time* . . . will it be okay to let me know *what begins to happen next* . . . ? Reviewing and replaying that until . . . ?"

3. Insight: Intuition and Breakout

"Becoming more aware of . . . ? Interesting . . . ? Something changing . . . ? And is that going well . . . ? Is it really possible . . . ? Positive possibilities . . . ? Appreciating the value of . . . ? Something surprising . . . new?"

4. Verification: Reintegration and Reframing

"What does all this experience mean to you . . . ? How will you experience [behave, think, feel, or whatever] differently now . . . ? How will your life be different now . . . ? How will your behavior change now . . . ? What will you do that is different now . . . ? Will you be sharing some of this with other people in your life . . . ?"

and behaving imply that there are many opportunities for duality, conflict, polarization, and what Jung would call "the problem of the opposites."

The problem of getting stuck between the opposites or polarizing one's identity to one side or the other goes back to ancient times and has been addressed by many spiritual practices and meditative philosophies. Erich Neumann (1962), in his monumental two-volume work on the origin and history of consciousness, reviews the natural dualities of mental experiencing as they were projected into the earliest myths of how the world came to be and how human nature evolved—myths that typically began with the duality of the earth being separated from the sky or the waters, light from darkness, the gods from humankind, spirit from matter, etc. Myths expressed the earliest theories of how personal individuality evolved out of separating the natural dualities of mother from father, masculine from feminine, good from bad, the active versus the passive, etc. Individuality, personal consciousness, and personality are invariably portrayed as process of creation that resolves the inherent conflict of the active and passive aspects of human nature. Traces of this archetypal psychological experience of reconciling the opposites of the active and passive principles can be found in many ancient cultures—such as China (p'o versus hun) and Egypt (ba versus ka)—and religions—in Buddhism the duality is expressed as the kama-manas versus the buddhi-manas; in Zoroastrianism, the daena versus the urvan; in Judaism, the nephesh versus the ruach. Personal human destiny is intuited as the outcome of how the individual (in-divi-dual) resolves the natural clash of these apparent divisions in nature, mind, and spirit.

The search for the core of one's self, the one that transcends all dualities and conflicts, is often described as the *conflict between oneness and duality* on the path to enlightenment. Ken Wilber (1995), for example, summarizes a vast number of spiritual traditions that describes "the ultimate state of consciousness" as Brahman or the Absolute that is the "One without a second." Wilber (1995) explains:

> "All is Brahman" is not merely a logical proposition; it is more of an experiential revelation, and while the logic of the statement is admittedly quite flawed, the experience itself is not. And the experience of All-is-Brahman makes it quite clear that there is not one thing outside the Absolute, even though when trans-

lated into words, we are left with nonsense. But, as Wittgenstein would say, although It cannot be *said*, It can be *shown*. (p. 199)

It is precisely this insight by the philosopher Wittgenstein that provides a deeper rationale for our use of the activity-dependent, symmetry-breaking, and nonverbal somatic signaling approaches outlined in this chapter. In a sense, all these novel approaches engage state-dependent memory, learning, and behavior systems on a more implicit level than is usually available to explicit consciousness. They are all attempting to go beyond the typical cognitive processing emphasized by most schools of psychotherapy, where words are the primary medium of communication, problem-solving, and healing. These nonverbal, activity-dependent, and symmetry-breaking approaches allow people to experience the easier process of *showing* rather than the more elaborate and difficult cognitive work of *explaining what cannot be said.* Wilber (1995) notes:

> Now the insight that there is nothing outside of Brahman implies also that there is nothing *opposed* to it; that is to say, the Absolute is that which has no opposites. Thus, It is also called the Non-dual, the Not-two, the No-opposite. To quote the third Patriarch of Zen:
>
> "All forms of dualism
> Are ignorantly contrived by the mind itself.
> They are like visions and flowers in the air;
> Why bother to take hold of them?
> When dualism does no more obtain,
> Even Oneness itself remains not as such.
> The True Mind is not divided—
> When a direct identification is asked for,
> We can only say, 'Not-Two!' "(p. 199)

It is numinous, indeed, to realize how such ancient phenomenological descriptions of the conflict between oneness and duality on the path to enlightenment correspond so well to the experiences of the four-stage creative cycle. *From our psychobiological perspective, the conflict between oneness and duality is a normal developmental process that is typically experienced in stage two of this creative cycle.* Our activity-dependent, symmetry-breaking approaches facilitate creative psychodynamic work by utilizing the psychobiological arousal and energy generated by the tension between the polarities, duality, divisions, opposites, and conflicts that are described as error and delusion by ancient philosophies (Rossi & Jichaku, 1992).

Duality and division are error and delusion when they are prematurely accepted by consciousness as a basis for behavior—acting out conflicts in war, for example—rather than being understood as a transitional phase (stage two) of the creative cycle. This is why people sit in meditation rather than acting out every whim of the monkey mind engaged in the "shuttling to and fro of arguments and affects [that] represents the transcendent function of opposites" (Jung, 1960, p. 90). From our psychobiological perspective, the teaching "True Mind is not divided" refers to the reintegration that takes place in phase four of the creative cycle. These subtle dynamics are explored further in the next exercise.

EXERCISE 3:
REPLAYING AND REFRAMING A PROBLEM INTO A CREATIVE RESOURCE USING THE PALMS-DOWN PROCESS

The dynamics of hand polarity can be used with palms facing downward, as illustrated in Box 10.3. This variation of hand and arm polarity sometimes facilitates an evaluative experience as when a person needs to take a broad outlook to replay, reevaluate and reframe a significant identity or life transition issue.

1. Preparation: Engaging the Opposites

A spiritually oriented young woman complains of feeling weak and being a failure. I demonstrate an initial, symmetrical, palms-down hand position (as illustrated in Box 10.3) and ask,

"Can you tell me which hand feels weaker and more of a failure at this moment?"

Box 10.3 | Replaying and reframing a problem into a creative resource using the palms-down process.

1. Preparation: Engaging the Opposites
"Can you tell me which hand feels weaker and more of a failure at this moment . . . ? Wonderful. . . . Now I wonder what you experience in your other hand by contrast at the same time . . . ?"

2. Incubation: Sensory Experience and Replay
"My goodness, is that hand really trembling [or whatever] all by itself . . . ? I wonder if you have the courage to allow that to continue privately within yourself for another moment or two, so you can experience what takes place next?"

3. Illumination: Insight, Intuition, and Breakout
"Quite a surprise . . . ? And is that now going well . . . ? Continuing to receive the best of that . . . ? New ideas of what you can do . . . ? Sense something . . . ? Intuitions of . . . ?"

4. Verification: Reframing Problems into Resources
Absentminded "*hand play*" as she looked up . . . and said, "*Such stuff as dreams are made of . . .*"

After a moment of self-reflection, she hesitantly acknowledges that her left hand feels a bit weaker. I then ask,

"And what do you experience in your other hand by contrast?"

She takes a slow deep breath and admits that her other side, her right hand, seems lost in daydreams of heroic adventures. *The opposites are thus successfully activated. She is now ready to engage in creative inner work.* I then proceed with,

> **"Wonderful, how you can experience both sides of yourself at the same time. . . . Now I wonder if you can let yourself continue to experience both sides at the same time and explore what begins to happen between them?"**

2. Incubation: Sensory Experience and Replay

After a few moments, her "weaker" left hand begins to tremble, and after a few more moments I solicitously ask,

"My goodness, is that hand really trembling all by itself?"

Blushing somewhat, she admits that the left hand is shaking all by itself because it is so nervous about a fantasy that *it* is experiencing. *I deeply respect her privacy by not asking her to tell me what the fantasy is.* Instead, I quietly support her obvious struggle by asking,

> **"I wonder if you have the courage to allow that to continue *privately* within yourself for another moment or two, so you can *experience* what takes place next?"**

Her eyes close at this point; many private emotions cross her face as her hands slowly approach each other and then retreat just before they touch. Her hands continue oscillating from side to side and up and down for a few moments. Finally they do come together and touch in a seemingly accidental manner. As they do so, her entire body goes through a slight startle response.

3. Illumination: Insight, Intuition, and Breakout

Noticing the slight startle response, I quietly question,

"Quite a surprise . . . ?"

She merely nods her head yes slowly a few times, as her hands now begin to touch each other lightly, again and again in a tentative, exploratory manner. Finally, the still trembling left "fantasy" hand covers the right hand. After a few moments the trembling stops and the anxious expression on her face is replaced by calm and the slightest of smiles. After a few more moments, when her apparently calm and satisfactory state seems stabilized, I quietly ask,

"And is that now going well . . . ? Continuing to receive the best of that . . . ?"

After about 30 seconds of inner preoccupation, she silently nods her head yes and murmurs quietly to herself, "I wonder if that is really possible?" She does not communicate anything about what she is wondering, however; she clearly accepted my earlier suggestion about *private experiencing.*

After a few more minutes of silent contemplation, she finally opens her eyes, stares at the floor, and slowly stretches and touches herself on the arms, head, sides, and legs in a manner that is very characteristic of people coming out of deep inner states. To my silent look of inquiry, she describes how she had read recently that humans are essentially "herd animals" who take comfort in each other's presence and touch. She now comments that during this inner work, she explored the question of how it might be possible to touch a young man she had met recently "without acting like a creep and scaring him off." She explains how she finally imagined going on a hike with him, and as they moved through more and more difficult terrain, their hands might reach out to each other and then finally touch to support each other.

4. Verification: Reframing Problems into Resources

The final task of this therapeutic session was to engage the young woman in a discussion about the possibility that she could, in fact, ask her male friend out for just such a hike. She acknowledged that she could and would. At the same time, her right hand reached over and absentmindedly caressed her formerly trembling but now calm left hand. After a few moments of such absentminded *hand replay*, she looked up at me sheepishly and said, *"Such stuff as dreams are made of . . ."* We both smiled at the allusion to Shakespeare, stretched, looked at the clock, and realized it was time for the session to end.

We spent the next few sessions discussing how her occasional feelings of being weak and lost in fantasy were actually important cues, a kind of mind-body signal that she was going into a natural period of ultradian rest. The ultradian rest could mean she was going into a creative state of introversion, wherein she could experience and recognize how she needed to take new steps to move forward in her life. Her previous "symptom" of being a weak failure lost in fantasy was thereby reframed into a therapeutic mindbody signal that she needed to explore the realistic possibilities underlying her fantasies. It was entirely natural for her to take an ultradian healing break and get lost in a pleasing fantasy for a few minutes every few hours or so throughout the day. That is what introverts are like. She could be proud that this was her way of self-reflecting and co-creating herself. In this way, her former problem of being weak and lost in fantasy was reframed into a *creative resource for problem-solving.*

What are the actual dynamics of exploring and resolving problems by externalizing internal hopes and conflicts into observable sensory-motor behavior in this manner? In what way are these polarity or bifurcation dynamics similar to those in Sigmund Freud's *free association*, Carl Jung's *active imagination*, Fritz Perl's *Gestalt dialogue*, Carl Roger's *reflecting* what the client says, and Milton Erickson's *naturalistic and utilization approaches* to problem-solving and healing?

Careful study indicates that the main line in the evolution of psychotherapeutic technique from Freud to Erickson over the past 100 years has been away from direct suggestion and analysis by the therapist. The current trend is toward the facilitation of implicit processing and creative mindbody communication within the patient. Many of the therapeutic innovations of this century were designed to avoid outside suggestion and influence as much as possible so that patients could have an opportunity to explore the creative transitions of their own evolving personality and consciousness. The symmetry-breaking approaches take another step in this direction. The ideal is to facilitate creative, private processing on deeply implicit levels so that people can learn how to optimize their own problem-solving and healing in their own way. Neuroscience research is now needed to determine the degree to which these symmetry-breaking approaches can set the stage for: (1) *immediate early gene and behavioral state-related gene entrainment* of ultradian arousal and relaxation for problem-solving (see Chapters 1 and 2), and (2) *activity-dependent gene expression* that may generate neurogenesis and encode new processes of memory, learning, and behavior as well as healing (see Chapters 3, 4, and 5).

Therapist's Role

Here is a summary of the therapist's role in facilitating each of the four stages of the creative cycle.

1. The therapist recognizes and utilizes the creative moment when the person is approaching a psychobiological state of ultradian arousal or apparent rest, wherein an opportunity for problem-solving or symptom resolution may be at hand.
2. The therapist offers implicit processing heuristics in the form of Socratic questions that utilize the person's arousal and complaint by projecting them into one hand and, simultaneously, projecting another aspect of themselves (usually a more constructive or healing aspect of the personality) into the other hand.
3. The therapist facilitates peoples' suspension of their usual everyday reality orientation in order to experience the novelty-numinosum-neurogenesis effect as they search for solutions. The therapist recognizes and supports the potential values of the novel solutions that seem to come up spontaneously in varying mixtures of pure fantasy and possible realities.
4. The therapist validates the novel solutions with symptom-scaling and discussions of how to test and practice them in everyday life.

Variations on a Theme

Here are a few more variations of implicit processing heuristics, usually verbalized by the therapist slowly and quietly but with a certain sense of drama and determination, while demonstrating the initial palms-up or palms-down hand processes illustrated in Boxes 10.2 and 10.3.

Variation for Emotional Problems

"To what extent can you actually experience your anger [or whatever] in one hand more than the other? [Pause until the patient acknowledges one hand or the other.] Wonderful! And now, at the same time, what do you experience in the other hand, by contrast?"

Variation for Psychosomatic Symptoms

"Can you tell me which hand seems to sense or experience [whatever psychosomatic symptom or complaint] the most? [pause] And now, what does the other hand experience that is different?"

Variation for Stress

"As you let yourself continue experiencing that stress [posttraumatic symptoms, etc.] for another moment or two, I wonder if you can tell me which hand or part of your body feels it most strongly? [pause] Good! Now tell me what the other hand experiences, by way of contrast?"

Variation for Addictions and Habits

"Tell me which hand is the one that reaches for a cigarette [alcohol, drugs, etc.], and which represents the side of you that wants to change to help yourself?"

As soon as the person acknowledges experiencing two or more sides of the symptom, problem, or personality in the hands or other parts of the body, the therapist can assume that behavioral state-related and activity-dependent creative experiencing are being engaged. The person's inner dynamics have been primed for a numinous, self-therapeutic engagement that is now ready to take place. The therapist then merely wonders aloud what will happen next between the different forces or parts that are being experienced. The therapist encourages a sense of wonder about whatever ideodynamic processes become evident.

Typically after about 10 to 20 minutes of inner self-encounter, many people spontaneously remark, with either a sense of satisfaction or disappointment, that the inner process is now over. They are satisfied when they feel they have received an important insight; they are wistfully disappointed when they feel they have received something of value but intuit that there is much more that needs to come. I tend to warmly support the reality of such evaluations when they appear to be constructively oriented, even though incomplete. Some patients really do solve a significant problem completely in a single session. Others may make some progress, but their inner world needs more time to synthesize the new psychobiological realities that are evolving within themselves. *The most important learning for people is how to recognize the stages and the psychological reality of their creative experience rather than any particular problem they may have solved. It is often helpful to talk about the four stages of the creative process so that they learn how to appreciate the ongoing evolution that is taking place on an implicit level within themselves at all times.*

People occasionally report that their creative inner process, which was as engrossing as a dream for about 10 or 20 minutes, seems to have vanished like a puff of smoke or a bubble popping. They are often puzzled about why it ended just when it did. It is precisely this sense of the *experiential reality and semi-autonomous nature* underlying inner processes that come and go in an unexpected manner, that leads us to believe that they engage a psychobiologically distinct state of consciousness similar to the REM state (Hobson, 1997, 1999; Rossi, 2000a). The current neuroscience challenge is to identify measurable psychobiological correlates of gene expression and neurogenesis during such creative encounters of self-reflection and co-creation.

EXERCISE 4:
TRANSFORMING A SYMPTOM
INTO A SIGNAL USING THE ARM LEVER PROCESS

A process analogous to reframing a problem into a resource is that of transforming a symptom into a signal illustrated in Box 10.4. What are the dynamics of this process of transforming a symptom into a signal? Milton Erickson

BOX 10.4 | TRANSFORMING A SYMPTOM INTO A SIGNAL USING THE ARM-LEVER PROCESS AND SYMPTOM-SCALING.

1. Preparation: Symptom-Scaling

"Can you hold your arm out, as if it is a lever that can tell us how strongly you are experiencing those feelings ...? [Pause] And will you let me know how much you're experiencing those symptoms on a scale of 0–10, where ten is the worst ...?"

2. Incubation: Arousal and Creative Replay

"As you continue watching your arm ... can you let yourself be so sensitive that your arm goes up if the feelings get more intense, and your arm goes down when they fade out? That's right, the *courage* to allow that to continue all by itself for another moment or two until ...? What number are you experiencing ...?"

3. Illumination: Transformation, Insight, and Reframe

"Until you experience a little surprise ...? Um-mm ... Changing in an interesting way ...? Surprising ...? Experiencing something different ...? Learning everything you need ...? And what number is it at now ...?"

4. Verification: Reintegration and Evaluation

"What about that is most valuable to you ...? How much sense does it make to you ...? Will you be able to do this for yourself ...? How can this help you from now on ...? What changes will you now make ...? What will you do differently this week ...? What number do you experience now?"

frequently explored polarities and state-dependent divisions in his patients' experiences, although he never used such words to describe them. Erickson described the process of transforming a symptom into a signal as a "yo-yoing of consciousness" or a "yo-yoing of symptoms" so that they would be experienced as alternately getting worse and better.

The paradox, Erickson believed, is that patients do not realize that as they allow symptoms or pains to get worse for a moment and then better, they are actually *gaining control over them*. I have speculated that this is not a paradox but rather the best way for people to engage their own state-dependent memory, learning, and behavior systems by replaying and reframing alternating states of ultradian arousal and relaxation. These alternations are associated with the release of hormones and messenger molecules that initiate the gene expression and neurogenesis to mediate psychotherapeutic and healing processes (Rossi, 1986/1993, 1996a).

It is now well known that all addicting drugs achieve their effects by mimicking the molecular structure and functions of the mindbody's natural hormones and messenger molecules (neurotransmitters and neuromodulators). Many of these are the same messenger molecules that encode stress and traumatic experiences in a state-dependent manner and are responsible for the amnesias, dissociated states and general symptoms of the addictions (Rossi, 1987, 1986/1993, 1996a; A. Rossi, 2001). Replaying one's own state-dependent systems of memory, learning, and behavior allows Darwinian natural variation and selection to take place, creating new reframes of mind and behavior. *Every recall is a potentially therapeutic reframe*. This is the direct implication of the research by Nader et al. (2000a, 2000b) reviewed in Chapter 3.

What is the difference between creative replay—which generates natural variation and conscious selection leading to therapeutic responses—versus the obsessive-compulsive syndromes wherein there is obviously too much repetition of the same pattern, with little or no variation and selection of something new? Brain imaging methods may help us determine differences in functional brain anatomy; assessment with DNA (gene) and proteomic (protein) microarray methods may help us determine to what degree there are differences in gene expression and neurogenesis during creative replay versus obsessive repetition. When these objective methods of therapeutic assessment become available, we can expect that therapists will become more effective in helping people focus on the implicit processing heuristics that will be most useful in helping them break out of their obsessive preoccupations and move into creative replay. Meanwhile, the therapist proceeds empirically on a psychological level, trying one implicit processing heuristic after another—in search of the heuristic that turns the tumblers of the patient's emotional complexes and opens them to therapeutic vistas of activity-dependent creative replay and resynthesis.

EXERCISE 5:
IMAGERY FACILITATING
CREATIVE REPLAYS OF ULTRADIAN DYNAMICS

A substantial body of data indicates that there are significant relationships between visual imagery, its neural correlates (Kosslyn et al., 2001), and psychophysiological responses (Achterberg et al.,1994; Battino, 2000; Kreiman et al., 2000; Rossi, 1996a, 2000a). We need more scientific data, however, on the question of whether visual imagery can be combined with the ultradian periodicity of the 90–120-minute basic rest-activity cycle to optimize performance and/or healing (Rossi, 1996a; Wallace, 1993; Wallace & Kokoszka, 1995). Box 10.5 contains approaches I have explored to utilize the replay of the creative cycle of alternating ultradian arousal and rest in clinical practice.

It is important to realize that it is the therapist's intent to recognize, facilitate, and utilize whatever ultradian state is emerging naturally in the person during the psychotherapeutic session. It is not the therapist's intent to try to manipulate the person to shift from one psychobiological state to another. *Attempting to manipulate or program another person's behavior is a source of stress in both the manipulator and the person manipulated. The major source of stress in the psychotherapist comes from the well-intentioned but often misguided effort to manipulate people without full understanding of what they really need and when they need it.* The implicit processing heuristics of Box 10.5 are fail-safe in the sense that they facilitate a neutral attitude in the therapist so that the patient's natural ultradian state can signal what needs to be experienced.

It will be important to gather normative data on the types of mindbody signals (sensations, emotions, cognitions, imagery, etc.) children, adults, and seniors experience at different times of the day in response to these implicit processing heuristics that facilitate natural ultradian states (as described in Box 10.5; Pratt, 1996; Rossi & Nimmons, 1991). We would predict that there are

BOX 10.5 | IMAGERY FACILITATING THE ULTRADIAN CREATIVE CYCLE.

1. Preparation: Implicit Processing Heuristics

Begin with a series of implicit processing heuristics that may evoke imagery and inner exploratory sets. People are usually not expected to respond to these questions verbally. These implicit processing heuristics set the stage for the creative work that is about to take place.

"Would you like to explore what you need to experience right now ...? How can we allow your inner nature to guide you ...? Wonder how to improve your performance [problem-solving, healing, etc.] at this time ...?"

2. Incubation: Imagery

Offer imagery to explore whether the patient's mind-body is moving in the direction of either psychobiological arousal or relaxation to optimize some aspect of performance.

"Will you begin with an inner journey taking you on a path somewhere in nature. ... Suppose you found yourself in a favorite place.... Will you experience yourself *moving upward ... or downward ... or along an even path ...?* Knowing you can share just as much or as little of what you are experiencing as you wish to ... sharing only what I need to hear to help you further ...? *Can you see yourself* moving upward with *energy and exhilaration* on a mountain trail ... or downward toward a *pleasant and restful* place or ...? Wonder whether your inner nature takes you *upward and away to a high-energy adventure ... or relaxing in a comfortable* and beautiful setting ...?"

3. Illumination: Self-Selection of Arousal or Rest

Support the person's self-selected psychobiological state toward optimizing ultradian arousal and performance or facilitating ultradian rest and healing facilitation.

"Yes, recognizing how that continues all by itself for a while ...? Um-mm, allowing yourself to experience what feels right ...? Appreciating and really going along with that ... as it continues ...? Sensitive to what's natural now ...?"

4. Verification: Validation and Reintegration

Support the value of the person's ultradian arousal, relaxation, or both. Discuss how the alternating states of optimum arousal and performance and rest and healing can be recognized and facilitated in everyday life. Review personal examples of how stress is the result of attempting to go against our natural mindbody signals for rest or activity.

"What was of most value for you in this experience ...? What are your typical mind-body signals that you are entering ultradian arousal [alertness, new insights, high motivation, restlessness, anxiety, etc.] ...? What are your typical mind-body signals that you are entering ultradian rest [fatigue, sleepiness, memory, performance failure, etc.] ...? How will you continue learning to recognize and utilize your mind-body signals ...? When and where will it be most valuable for you to explore these natural states again ...?"

natural ultradian shifts in mood and mentation every two hours or so in keeping with the basic rest–activity cycle discussed in Chapter 2 (Lloyd & Rossi, 1992, 1993; Wallace, 1993; Wallace et al., 1992, 1995).

EXERCISE 6:
VISIBLE AND HIDDEN ACTIVITY-DEPENDENT PSYCHO-DYNAMICS USING ARMS IN FRONT AND BACK PROCESS

We are all familiar with the adage, "The right hand does not know what the left hand is doing." To integrate consciousness with what is known only on implicit or unconscious levels is the essence of many schools of psychotherapy. An interesting activity-dependent approach that facilitates the integration of Darwinian unconscious variation with conscious selection of what is currently evolving within the person is illustrated in Box 10.6. The person is asked to stand up for this novel approach, so that there will be a greater freedom of movement. As is the case with all the activity-dependent approaches to healing, the therapist models each stage of the creative cycle.

The first stage involves an initial exploration of which side of the body is more expressive of sensations or feelings and which side is less expressive or more hidden; the less expressive or "secret side" of the body is dramatized by putting an arm behind the body in a hidden position. As usual in stage two of the creative cycle, the person is encouraged to receive and explore in a private manner whatever comes up all by itself. The therapist is ever alert to notice and support anything that is evidently surprising to the patient in stage three, and then encourages discussion of how this experience can help formulate a prescription for real life changes in stage four.

Many people experience a burst of energy and freedom when engaged in this activity-dependent accessing and replaying of their issues. They are often delighted with the conscious clarity of their dramatic replay of what is usually processed on more implicit, hidden, or unconscious levels. The full-body replays while standing are effective in transforming unconscious psychodynamics into conscious understanding. A good rule to follow is to initiate people into this activity-dependent process whenever they say, *"I don't know."* Not knowing or confusion is often a pointer to the creative edge of a person's consciousness. Self-reflection and co-creative dynamics are facilitated in

BOX 10.6 | VISIBLE AND HIDDEN PSYCHODYNAMICS IN THE CREATIVE CYCLE USING ARMS-IN-FRONT-AND-BACK PROCESS.

1. Preparation: Sensitivity and Awareness

"As you stand there, notice which side of your body has more sensations [pause]. By *contrast*, notice which side has less sensation and feeling . . . ? Which side of your body you are more aware of . . . and which side seems more hidden [pause]. Allow the arm of the more aware side to remain in front, like this [therapist demonstrates], while the more hidden or secret side of you is expressed by placing your other arm behind your back like this [therapist demonstrates]. That's right . . ."

2. Incubation: Psychodramatic Replays

"And now, allowing each side to have a life of its own . . . the front side is what you know about yourself, and what you wonder about is still hidden in back for a moment . . . ? The freedom of the arms, hands, head, and fingers to express themselves in a psychodrama . . . in any way they please . . . as if each has a life of its own [therapist demonstrates slow body movements]. The courage to allow yourself to experience whatever in a safe way . . . wonderful . . . ! Really allowing that to continue more or less all by itself . . . ?"

3: Illumination: Novelty and Reframing

"Surprising . . . ? Continuing until you become aware of something interesting . . . ? The unexpected . . . Some meaning in all this . . . ? What this is all leading you to . . . ? How something changes for you . . . ? Yes, allowing that to continue until it comes to a natural ending . . . until you become aware of the meaning of all this for your life . . . ?"

4. Verification: Supporting the New

"How this experience can become a guide for changing your expectations . . . ? What behavior do you need to change . . . ? What will you do now that is different . . . ? What is the next step for you in real life . . . ? In relation to yourself . . . ? Toward others . . . ?"

these sensitive replays and creative reframes of past, present, and possible futures. Research is now required to determine to what extent that which is hidden on implicit, unconscious levels of functioning is more associated with *behavioral state-related gene expression* in contrast to the more *activity-dependent processes of gene expression and neurogenesis* that may become engaged with this novel approach.

EXERCISE 7:
THE BUDDHA BENEFICENCE AND
FEAR-NOT PSYCHODYNAMIC PROCESS

This is an excellent psychodynamic process for people who are obviously fearful and needy, on the one hand, and spiritually oriented on the other. It is a good introductory approach to learning how to experience ideodynamic movements. It is one of the fastest ways I know of helping people experience the creative replay and therapeutic reframes of the natural polarities, dualities, and emotional conflicts within themselves. It serves as an easily crossed bridge between implicit and explicit experiences, wherein many people can access the new and the numinous in the evolution of their own self-understanding.

I originally thought I had invented this symmetry-breaking process about 15 years ago as a simpler and more easily motivated variation of David Cheek's ideomotor finger signaling and 20 questions approach to resolving problems (Cheek, 1994; Rossi & Cheek, 1988). I was wrong to think I could be so original, however. I finally saw the light when I visited the Po Lin Monastery, situated in the Ngong Ping Plateau on the island of Lantau, near Hong Kong, a few years ago. You can imagine my surprise when I first gazed upon a statue of the Buddha several stories high, meditating in what I had been calling the "fear-not" process. I quickly back-peddled from my hubris born of ignorance. I now humbly acknowledge the Buddha's 5,000-year priority in all this. That's why I now call it the "Buddha beneficence and fear-not process." There can be many variations in the words and attitudes used to initiate and facilitate this approach to activity-dependent problems-solving and healing by utilizing the person's own language and worldview. A general approach is illustrated in Box 10.7

BOX 10.7 | THE BUDDHA BENEFICENCE AND FEAR-NOT PSYCHODYNAMIC PROCESS.

1. Preparation: Wanting and Not Wanting

"Everyone wants something very much *on the one hand* . . . [therapist models by holding out one hand with the palm facing upward as if receiving]. At the same time, most people have some idea of what they do not want, *on the other hand* . . . [therapist models the other hand with a palm facing outward position, as in a "stop" gesture]. As you tune into yourself, you can begin to wonder which hand feels like the side of you that wants something, and which hand expresses what you do not want. . . . That's right, test one hand and the other to experience which expresses two opposite sides of your nature." [Therapist models by making a few tentative, alternating gestures of the *receiving* and *stop* gestures with one hand and the other, conveying an exploratory attitude, to find which side feels right expressing which part.]

2. Incubation: Accessing Motivations and Creative Replay

"That's right, as if each hand and arm has a mind of its own . . . each going their own way . . . ? [Therapist pantomimes a dramatic confrontation between the two hands with slow exploratory movements.] Sometimes together . . . sometimes apart . . . ? Allowing that to continue until . . . ?"

3. Illumination: Discovery and Surprise

[Therapist continues to model a psychodrama between the hands to encourage the person with supportive implicit processing heuristics.] "That's right . . . how does it come to its own natural resolution . . . ? Letting that inner drama play itself out in its own way. Yes, honestly letting it guide you in ways that may seem surprising . . . ? All right, what was most surprising [curious, unexpected, meaningful, etc.] to you about all that . . . ?"

4. Verification: Assessment and Change

[The person usually experiences an ending of the inner journey in anywhere from a few moments to 15 or 20 minutes, and usually feels ready to say something about it.] "Having had this experience . . . how will it change your life? . . . What will you now do differently . . . ?"

My experience in rediscovering at least a part an ancient spiritual path in this psychodynamic process suggests that there is an incredible richness in the diversity of our human heritage illustrated in the many manuals and images of yoga, meditation, and healing practices of many cultures. To what extent these diverse practices have universal or archetypal aspects are questions to be answered by the new field of neuroscience that is being called "neuro-theology" (d'Aquili & Newberg, 1999; Newberg & d'Aquili, 2001).

EXERCISE 8:
IDEODYNAMIC HEAD, HAND, FOOT, OR FINGER SIGNALING

Just how easy is it to create a new school of psychotherapy? Could we, right here and now, develop a new modality of creative activity-dependent psychotherapy as a standardized approach to be used in neuroscience research on gene expression, neurogenesis, and healing? Suppose we decide to create a new therapeutic modality to be called "Finger Family Therapy?"

This proposal for a new activity-dependent approach to facilitating family dynamics is no joke. Indeed, it actually happened! I do admit, however, that it began by accident as one of the most amazing examples of the creative madness of crowds I have ever experienced. In this case, it was a crowd of psychotherapists in a professional workshop on the healing arts, who perhaps had become a bit overly enthusiastic about the therapeutic possibilities of the ideodynamic approaches.

After an introductory talk about permissive approaches to creative inner work, I asked for a volunteer to demonstrate how much therapy a person could experience in privacy in front of a group. A woman in her 60s volunteered and sat beside me on stage. I silently modeled by holding up one of my hands as illustrated in Box 10.8. I noticed that as she held her hand up, she maintained a fixed eye focus on my hand rather than her own. I was about to "correct" her by asking her to focus on her own hand instead of mine. Before I could say anything, however, I noticed one or two of her fingers beginning to quiver slightly but rapidly in that apparently involuntary manner so characteristic of people in a state of intense inner focus that we call "hypnotic." I had enough quick wit to hold my tongue and simply watch her

BOX 10.8 | A TIME-BINDING IMPICIT PROCESSING HEURISTIC IN ACTIVITY-DEPENDENT IDEODYNAMIC SIGNALING.

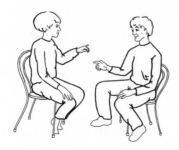

1. Preparation: A Time-Binding Introduction

"As soon as . . . [or *when*] [pause]

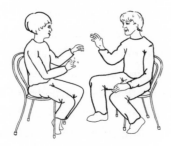

2. Incubation: Private Implicit Processing

" . . . some part of you has reached the sources of that issue [emotion, problem, symptom, etc.] . . . [pause]

3. Illumination: Ideodynamic Signaling and Insight

" . . . your finger [hand, arm, or foot] can lift [head nod] or move a bit, more or less, by itself . . . as you make internal reviews [replays] until . . . [pause] you understand something about yourself in a new way . . . [pause]

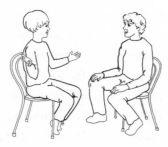

4. Verification: Self-Prescription and Co-Creation

". . . so that you can do what you need to help your life change for the better."

wide-eyed and growing sense of wonder, as she slowly shifted her gaze to her now gently dancing fingers.

After a few minutes I said,

"And knowing you can share as much or as little of that as you wish . . . only what I would need to hear to help you further."

She then began a quiet commentary about what was happening between her fingers. Indeed, she said, the fingers were moving all by themselves. Her fore-finger was herself, her thumb was her husband, the little finger of one hand was one of her children, and the little finger on her other hand was another of her children. In a bewildering array, and more quickly then I could fol-low, she continued exploring her family dynamics by describing how each of her fingers represented other family members: aunts, grandmothers, brothers, and so on. She expressed many shifting emotions and vocal patterns as she dramatized what they were saying to each other and the issues they were deal-ing with in their ever-changing relationships (Gray, 1994).

After another few minutes of these alternating periods wherein she ver-balized her ongoing family dynamics and engaged in silent creative play between the fingers of both of her hands, I realized she certainly needed no help from me. I gazed at the audience to see if they were appreciating the value of what they were seeing. To my surprise I noticed that about a third of the audience members were deeply absorbed in their own creative ideo-dynamic finger play. I tried not to show my surprise, but by now the die was cast and there was nothing I could do about it. For the remainder of the workshop, volunteer after volunteer silently went into what we began to call "Finger Family Therapy." Everyone somehow assumed that this novel form of family therapy was a new psychotherapeutic modality I had invented.

At the present time, however, we do not have a well-developed theory of ideodynamic signaling in family therapy. Such a theory would integrate data from a wide range of disciplines: from sensory-motor development and language acquisition to the psychosocial dynamics of the family. Jean Piaget (1954), for example, studied children's early learning of conceptual behavior and cognitive development in their "construction of reality"; he proposed that language develops out of sensory-motor actions of learned behavior.

Chomsky (1957), in contrast, emphasized the deep genetic source of language. The consensus that is currently emerging among many linguists and cognitive psychologists is that both are right. There is a deep genetic structure that underlies language, but it emerges first in sensory-motor actions and gradually develops into spoken language.

Current cognitive science research has developed these ideas further and documented how human ideas are grounded in sensory-motor experience and metaphor. This view is expressed by Lakoff and Núñez (2000):

> These metaphor studies mesh with studies showing that the conceptual system is embodied—that is, shaped by—the structure of our brains, our bodies, and everyday interactions in the world. In particular they show that abstract concepts are grounded, via metaphorical mappings, in the sensory-motor system and that abstract inferences, for the most part, are metaphorical projections of sensory-motor inferences. (p. 101)

This primary role of sensory-motor systems and metaphor in language and cognitive development suggests that ideodynamic signaling may access areas of state-dependent memory and learning encoded in early life that are not easily available to the spoken words of the adult (Cheek, 1994; Rossi & Cheek, 1988). The predominantly nonverbal, ideodynamic, and symmetry-breaking approaches to the metaphorical dramas of self-reflection and co-creation are effective because their activity-dependent behavior accesses the sensory-motor sources of the embodied mind (Lakoff & Johnson, 1999). Gene expression and neurogenesis contain many levels of implicit processing of the embodied mind that are not usually available to the explicit levels of verbal consciousness that are relied upon in traditional "talk therapy."

Current neuroscience research has come to interesting conclusions after comparing the dynamics of verbal language with the nonverbal, sensory-motor communication of American Sign Language (Collins, 2001). Sign language has phonological, morphological, and syntactic levels of language organization that are homologous to those of verbal language and conveys the full range of grammatical expressiveness and semantic and conversational rules found in spoken languages. Petitto et al. (2000), using PET brain imaging with profoundly deaf subjects, found that their signing, which engages hands and upper body positions, used specific sites of their left frontal cortex for higher-order linguistic processing that did not depend on the presence of sound. In addition, brain activity was noted bilaterally in an area of the superior temporal gyrus, the planum temporale. These brain imaging studies are models of what now needs to be done to investigate the activity-dependent ideodynamic approaches to mindbody signaling developed here. "Sixty-four Research Projects in Search of a Graduate Student" have been proposed to advance our understanding of ideodynamic signaling in psychotherapy and therapeutic hypnosis (Rossi & Cheek, 1988).

A general paradigm for utilizing ideodynamic signaling that permissively allows people to explore the natural variations in their mindbody talents and the selection of their own modalities of self-expression is presented in Box 10.8. This general paradigm utilizes the implied directive (Erickson et al., 1976) as an implicit processing heuristic to facilitate the four-stage creative cycle in a series of therapeutic replays.

Notice how the entire four-stage implicit processing heuristic of Box 10.8 can be offered as a single sentence, said slowly with careful emphasis and pauses of 30 seconds, or so, where indicated. The theory and practice of using implied directives in activity-dependent ideodynamic signaling with a variety of clinical issues, such as psychosomatic symptoms, pain, sexual dysfunctions, dreams, motivational problems, stress, and posttraumatic syndromes have been described in elaborate detail (Cheek, 1994; Erickson et al., 1976; Rossi, 1985/1993, 1995a, 1996a; Rossi & Cheek, 1988) but still await controlled neuroscience research to determine the areas of brain activation (via fMRI, PET) and the parameters of gene expression, neurogenesis, and healing that are facilitated with this therapeutic modality.

EXERCISE 9:
THE EVOLUTION OF THERAPEUTIC
TOUCH IN THE HEALING ARTS

The rich history of therapeutic touch spans the rituals of many cultures in antiquity to current research on how maternal touch can facilitate gene expression within a couple of minutes (see Chapter 1). Touch has been used in therapeutic hypnosis in a variety of ways, from those of Mesmer and Esdaile (Edmonston, 1986) to current standardized hypnotic susceptibility scales (Spiegle & Spiegel, 1978).

An interesting source of the use of therapeutic touch can be found in the beginning of Freud's pioneering efforts at healing, which eventually evolved into psychoanalysis. In one of his first publications (Breuer & Freud, 1895/1957), he described his procedure of using therapeutic or hypnotic touch in his early experiments:

> This astonishing and instructive experiment served as my model. I decided to start from the assumption that my patients knew everything that was of any pathogenic significance and that it was only a question of obliging them to communicate it.
>
> Thus when I reached a point at which, after asking a patient some question such as: "How long have you had this symptom?" or: "What was its origin?" *I was met with the answer: "I really don't know."* I proceeded as follows.
>
> I placed my hand on the patient's forehead or took her head between my hands and said: "You will think of it under the pressure of my hand. At the moment at which I relax my pressure, you will see something in front of you or something will come into your head. Catch hold of it. It will be what we are looking for.—Well, what have you seen or what has occurred to you?" (p. 110, italics added)

Notice that Freud used this approach when his patients reached the edge of their self-understanding by coming to the point of not knowing ("I really don't know.") Not knowing points to a possible growing edge of consciousness and self-identity. It is the ideal time to facilitate patients' initiation into a deeper experience of their own creative edge. Patients' responses to Freud's query sometimes came in the form of early memories of psychological traumas and crises, and these then became the content of classical psychoanalysis. The replay of these memories in successive sessions, leavened with the analyst's interpretations, gradually reframed the patient's attitudes and self-experience toward problem-solving and healing. The only problems seemed to be the so-called "transference" and "countertransference" reactions between patient and analyst (what current-day adaptive complexity theory would call "mutually adaptive players," Axelrod & Cohen, 1999) and patients' "resistance" (what we call stage two of the creative cycle). By "working through" these challenging areas, however, patients would eventually move on to develop skills in self-efficacy and in facilitating their own creative cycle. The classical Freudian psychoanalyst no longer uses therapeutic touch, but innovative therapists, such as Milton Erickson, invented new variations of touch to facilitate the inner accessing of the sources of problems and their therapeutic resolutions, as quoted and illustrated in Box 10.9.

Photographs of Erickson teaching how to use minimal touch in inducing therapeutic trance and his descriptions of the process are an art best learned in professional teaching workshops for therapists, but an outline can be provided here (Erickson & Rossi, 1981):

> You take hold of the wrist very, very gently. What is your purpose? Your purpose is to let the patient feel your hand touching his wrist. That is all. The patient has muscles that will enable him to lift his arm, so why should you do it for him? *The body has learned how to follow minimal cues. You utilize that learning. You give your patient minimal cues. When he starts responding to those minimal cues, he gives more and more attention to any further cues you offer him. As he gives more and more attention to the suggestions you offer, he goes deeper into trance. The art of deepening the trance is not necessarily yelling at him to go deeper and deeper; it is giving minimal suggestion gently, so the patient pays more and more attention to the process within himself and thus goes deeper and deeper.*
>
> I think all of you have seen me take hold of a patient's arm and lift it up and move it in various fashions. I induce a trance in that way. I have tried to teach a number of you how to take hold of the wrist, how to take hold of the hand. You do not grip with all the strength in your hand and squeeze down on the patient's wrist. What you do is take hold of it so as to very, very gently suggest a grip on his wrist, but you don't actually grip it; you just encircle the wrist with your thumb and index finger with light touches. You suggest a movement of the wrist with only the slightest pressure. You suggest a movement of the hand upward. And how do you suggest it upward? You press with your thumb just lightly, while at the same time you move your index finger this way to give a balance. You move your fingers laterally, and while the patient gives attention to that, you have your thumb actually lifting the hand. This is essentially a distraction technique: while the thumb very lightly and consistently directs the hand upward, your other fingers make touches and distracting movements in a variety of other directions that tend to cancel out each other.
>
> Another approach to guiding the hand upward is to attract the patient's conscious attention with a firm pressure by your fingers on top of this hand and only a gentle guiding pressure by your thumb on the underside of his hand. The

BOX 10.9 | MILTON ERICKSON'S UTILIZATION OF MINIMAL TOUCH CUES (EXCERPTED FROM ERICKSON AND ROSSI, 1981, PP. 43–46).

1. Preparation: Gentle, Minimal Touch
You take hold of the wrist very, very gently. . . . You suggest a movement of the wrist with only the slightest pressure.

2. Incubation: Tactile Cues Implying Movement and Creative Replay
You suggest a movement of the hand upward. And how do you suggest it upward? You press with your thumb just lightly, while at the same time you move your index finger this way to give a balance.

3. Illumination: Ideodynamic Movements and/or Catalepsy
When she starts responding to those minimal cues, you give more and more attention to any further cues you offer. As she gives more and more attention to the suggestions you offer, she goes deeper into trance.

4. Verification: Catalepsy and Responsiveness
If you have balanced muscle tonicity throughout the body, catalepsy throughout the body, you have reduced the sensations that exist within the body to those sensations that go into maintaining that catalepsy. A patient then becomes decidedly responsive to a wealth of other ideas.

only way the firm touch can remain firm is for the patient to keep moving his hand against your fingers. At the same time the lower touch of your thumb is kept gentle by the patient by constantly moving upward away from it. The therapist needs to practice these movements over and over because they are one of the quickest and easiest ways of distracting the conscious mind and securing the fixation of the unconscious mind.

You lift the hand in that fashion, letting your fingers linger here and there so that the patient unconsciously gets a sense of the lingering of your hand. You want the patient to have that nice comfortable feeling of the lingering of your hand because you want his attention there in his hand and you want the development of that state of balanced muscle tonicity which is catalepsy. *Once that state of balanced muscle tonicity is established to achieve catalepsy, you have enlisted the aid of the unconscious mind throughout the patient's body.* Because you can get catalepsy in one hand, there is a good possibility there will be catalepsy in the other hand. If you get catalepsy in the other hand, then you probably have catalepsy in the right foot, in the left foot, and throughout the body, face and neck. As soon as you get that balanced tonicity of the muscles then you have a physical state that allows the patient to become unaware of fatigue, unaware of any disturbing sensations. It is normally hard to maintain that balanced muscle tonicity and pay attention to pain. *You want your patient giving all of his attention to that balanced muscle tonicity because that distracts him from pain and other proprioceptive cues so that numbness, analgesia, and anesthesia are frequently experienced in association with catalepsy. If you have balanced muscle tonicity throughout the body, catalepsy throughout the body, you have reduced the sensations that exist within the body to those sensations that go into maintaining that catalepsy. A patient then becomes decidedly responsive to a wealth of other ideas.* (pp. 43–46, italics added)

Catalepsy is a curious word with antithetical meanings. *Webster's* Second College Edition Dictionary defines catalepsy from Greek sources meaning to "seize, grasp or take down." In psychology, catalepsy is described as a condition in which consciousness and feeling are suddenly and temporarily lost. In psychiatry, catalepsy is regarded as a condition in which muscles of the body become stiff and rigid, as occurs in epilepsy and in schizophrenia with catatonia. The condition of catalepsy is often cyclic, with phases of stupor alternating with excitement. This cyclic activity, together with the apparently opposite manifestations of stupor and excitement, remind us of the alternating psychobiological phases of the creative cycle, wherein private inner work (stage two, incubation) alternates with the numinous and exciting experience of new ideas (stage three, illumination). This alternation of consciousness and activity level appears, at least in part, similar to bipolar mood disorders. The depressive phase occurs when the person remains overlong in the incubation of stage two, while the manic phase is an overextended experience of illumination of stage three, utilizing catalepsy.

Into this flux of meanings, Erickson introduced a new constructive view: catalepsy as a condition of "balanced muscle tonicity" in which *"a patient then becomes decidedly responsive to a wealth of other ideas."* This, of course, is exactly the sort of activity we want to facilitate in psychotherapy and the healing arts. When a person becomes "responsive to a wealth of other ideas," we are witnessing a Darwinian increase in natural variation on implicit levels and selection on more conscious levels. Research is now needed to assess how our activity-dependent processes facilitate gene expression and neurogenesis during problem-solving and healing utilizing catalepsy.

It is precisely here, in the possibility of facilitating the creative cycle of natural variation and selection on all levels, from gene expression to problem-solving, that we find the deep psychobiological rationale for many of Erickson's innovative approaches to what neuroscientists call "activity-dependent processes." All of our *novel* and apparently *strange* illustrations of setting up *peculiar* hand and body processes, accompanied by permissive but *mildly provocative implicit processing heuristics*, are ways of optimizing gene expression and neurogenesis. Our goal is to help people find their own creative edge and make it safe for them to experience the natural variation and selection that will enable them to outgrow their learned limitations. From the perspective of art, this is a domesticated kind of wild Fauvism. From the perspective of nonlinear dynamics, chaos, and catastrophe theory, these are the dynamics of creative living on the edge of catastrophe (Rossi, 1996a; Solé & Goodwin, 2000).

EXERCISE 10:
FULL-BODY CREATIVE REPLAY
IN INDIVIDUAL AND GROUP ACTIVITY

As we move from conventional psychosocial constraints to increasing creative freedom in experiencing the psychobiological dynamics of natural variation and selection, we seek ever more flux throughout the mindbody. After experiencing the safety of the gentle and subtle finger, head, hand, and arm ideo-dynamic movements, outlined above, many people are inspired to explore broader body processes. For didactic purposes, we will again use the four-stage creative cycle to explore some of the principle parameters and bound-ary conditions of what are essentially infinite processes of natural variation and selection on all levels, from mind to gene, in therapeutic work. Box 10.10 illustrates typical body positions and movements that can be used (with both individuals and groups) in the creative replay of life issues. The therapist begins by standing and modeling how to go through the four-stage creative process using these full-body movements.

The therapist and participants are mutually adaptive players engaged in a dance of co-evolving replay and resynthesis, which can be facilitated with verbal implicit processing heuristics, somewhat as follows.

1. Preparation:
Accessing the State-Dependent Encoding of Issues

"Let's begin in a somewhat crouched position, like this, with one arm outstretched to one side as you get ready to deal with some important issues. . . . Consider what is an appropriate starting point for you . . . How did that problem begin . . . ?"

The therapist models this initial position, as illustrated in Box 10.10, and engages in creative movements in which the entire body flows upward from stages one to two.

"Memories . . . images . . . feelings . . . positions . . . or move-ments that express your experience . . . how you were moving. . . . Yes, privately replaying what was most important about all that . . . ?"

2. Incubation:
Replaying via Activity-Dependent Experiences with New Variations

The therapist continues modeling body movements that gradually shift upward and out of the initial, slightly crouched body posture to standing movements in stage two, while verbalizing these implicit processing heuristics.

BOX 10.10 | FULL–BODY FOUR–STAGE CREATIVE REPLAY IN INDIVIDUAL AND GROUP ACTIVITY.

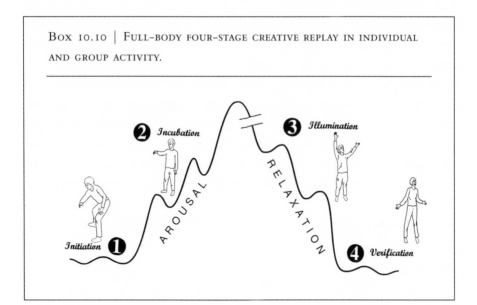

"I don't know if this takes you back to the beginning of ... ?
Childhood ... adolescence ... or your first ... ? I don't know
what private memories ... feelings ... come up for you ...
thoughts ... sensations ... movements that seem to want to take
place all by themselves ... as you experience the truth of your
own experience ... welcoming ... receiving ... uncertainty ... ?
An adventure ... of your own ... life ... meanings ...? Yes ...
and is there a private early life experience ... or a more current
situation ... that something within you ... finds itself coming up
over and over again ... when you least expect it ... exploring ...?
Getting ready to replay something ... in a way that could become
more ... satisfying ... healing for you ... ?"

To some degree or other, therapist and participants are using their personal body language to signal their changing psychobiological states as they
move through the four-stage creative process. They are "mutually adaptive players" (Axelrod & Cohen, 2000, p. xii) involved in a co-evolving replay and resynthesis in privacy and communion, on all levels from mind to gene. They are
engaging private, deeply intuitive, ideodynamic processes that tap into state-
dependent memories and life experiences on implicit levels. The first few

replays of traumatic early life experiences or adult posttraumatic problems may activate negative experiences of tears, despair, anger, struggle, or other psychobiological indications of conflict. This is the dark night of the soul of stage two experienced before the dawn of new light in stage three.

Through successive replays, the difficult or negative aspects of stage two are gradually replaced with more positive feelings as the therapist encourages people to reexperience their original trauma with new variations. People learn to welcome the process of creatively replaying their problematic experience with new variations within themselves. They are learning to receive and explore natural Darwinian variations in their emergent experience as they replay the past until they break through to surprising developments of numinous interest in stage three.

3. Illumination:
The Novelty-Numinosum-Neurogenesis Effect

The quixotic quest of stage two comes to its crisis or high point of private creative work in the transition to stage three illumination with a wide range of positive emotions: quiet appreciation, reverence, happiness for whatever reason, joy, and exhilaration. People learn to receive and appreciate the positive in whatever unexpected way it comes. As they do so, they may be integrating the behavioral state-related and activity-dependent gene expression that facilitates neurogenesis in those parts of the brain associated with positive affect—*which is the essence of the novelty-numinosum-neurogenesis effect*. The therapist's main task here is to help people recognize and support the shift from the negative experiences of stage two to the delicate, nascent, and still tentative stage-three discoveries and positive states of being, as illustrated in Davina's new world dream in Chapter 4.

As illustrated in Chapter 8 and previous boxes in this chapter, the therapist may utilize verbal implicit processing heuristics to facilitate this stage, somewhat as follows.

> **"That's right . . . the courage to break out of the box of your usual experience. . . . Um-mm, allowing yourself to experience what feels right on that path . . . ? Appreciating and really going along with that . . . as it continues . . . ? Welcoming the best of your insights . . . ? Yes, a little surprise at the unexpected . . . ?**

Exploring your best feeling as you become aware of what you
most enjoy about . . . ? A little light breaking through the
clouds . . . ?"

There comes a moment toward the end of stage three when it feels good
to stretch upward, as illustrated in Box 10.10. Obviously we all experience
stage-three illumination in our own way as an important facet of our evolving and unique individuality.

4. Verification:
Supporting New World of Experience

The therapist is more active during the first few replays of this recreation
process for an individual or a group. As people become more experienced
and learn to enjoy facilitating their own creative dynamics, they need less
modeling by the therapist. Just as no two artists worthy of the name would
copy each other—people seeking an aesthetic pleasure in their own movement and creative ideodynamic processing do not copy each other. Ultimately,
these highly idiosyncratic movements are completely satisfying to no one but
ourselves. It is a marvel to witness this creative process in groups of people
who are all learning to enjoy doing their own thing in their own way—
together.

In therapist training workshops, this entire four-stage process may be
replayed in rapid succession anywhere from three to six times depending on
the mood of the group. Each replay takes about 5–10 minutes. Some participants will replay the same early life trauma over and over again, experiencing new variations in stage two and selecting new possibilities for reframing
and resynthesizing the experience in a more satisfying way in stage three.
After half a dozen, or so, replays, even the most traumatic life events usually
can be reframed and experienced in a therapeutic manner. Other participants
may find themselves experiencing and reframing different life issues each time
the four-stage creative cycle is replayed.

At the end of each replay I typically ask participants to

"shake yourself out . . . with any-which-way movements of your
whole body. . . . For a moment . . . let go of that replay . . . and
get ready to create a better one next time!"

After another moment or two, I ask them to

"let yourself experience a private moment of slowing down . . . letting your movements come to a comfortable rest for a few moments . . . until you feel ready to replay that whole issue in a new way again."

As they move their body back to the initial position for another replay (see the first image in stage one of Figure 10.10), I ask them first to explore a few original micromovements very slowly. I typically model these original movements as I verbalize implicit processing heuristics, such as:

"I can enjoy careful attention to how it really feels to move my arm and fingers very slowly in a way I have never experienced before. . . . Does this movement of my leg really allow me to explore new sensations? . . . Yes . . . does this slow rolling of my head feel interesting in ways that I am aware of for the first time?"

This combination of nonverbal movement and verbal implicit processing heuristics encourages people to seek and explore more natural variation and selection in the next creative replay of the life issues with which they are dealing. After half a dozen replays, or so, most group participants are ready to take a break. This is a very important ultradian time period, indeed—time for what neuroscientists would call "environmental enrichment," wherein people listen to each other and learn about other possibilities for expanding their experiential boundaries. After a few moments of quiet rest, some individuals may feel inclined to share parts of their experiences—particularly the surprising moments of creative transition from stage two to stage three. Group sharing is an important process of psychosocial and cultural enrichment that can plant the seeds for future implicit processing in participants, so that they will be surprised at how they experience new variations on an explicit, conscious level during the next replay.

This type of experiential enrichment in therapeutic groups epitomizes our rediscovery and exploration of the cultural rituals that involve healing and spiritual practices of many peoples in many periods throughout human history (Greenfield, 2000; Keeney, 1999–2000; Neumann, 1962; Wilber, 1993, 1995). The current creative edge in the neuroscience technologies of assessing gene expression, neurogenesis, and healing in such numinous group encounters and phenomenological play will be our contribution to the ever-expanding evolution of the human condition.

SUMMARY

Psychotherapy, the humanities, and the healing arts are all deeply engaged in facilitating the psychobiology of gene expression, neurogenesis, and mind-molecular communication to a degree that previous generations could not have dreamed. The symmetry-breaking approaches to problem-solving and healing developed in this chapter have deep sources in the evolution of physical and biological realms as well as the phenomenology of human consciousness. From a cultural perspective our symmetry-breaking approaches are a rediscovery and exploration of the holistic rituals of healing in the spiritual practices of many peoples throughout human history. The psychobiology of gene expression contributes an understanding of the natural creative replay and resynthesis of human nature by integrating neuroscience, neurogenesis, and the numinosum in the humanities as well as therapeutic hypnosis, psychotherapy, and the holistic healing arts.

EPILOGUE

The Psychosocial Genomics of Creative Experience

Consciousness, choice, culture, humanities, art, science, gene expression, neurogenesis, psychotherapy, and healing: How do they all come together? We explored a variety of answers to this question that points toward a new worldview of how we can continually co-create ourselves in cooperation with nature. The oversimplistic idea of genomic determinism in human experience is incomplete: It needs to be amended to include the complementary concept that human experience can modulate gene expression. Current research documents how our environment, behavior, consciousness, and lifestyle choices can modulate gene expression to optimize creative adaptation to novel circumstances. The science of psychosocial genomics explores how creative human experience modulates gene expression, as well as vice versa. Here are a few key concepts of psychosocial genomics that bring together the art, culture, ethics, philosophy, and science of facilitating the human condition.

- *Immediate Early Gene Expression.* This special class of genes can respond to psychosocial cues and significant life events in an adaptive manner within minutes. Immediate early genes have been described as the newly discovered mediators between nature and nurture: They receive signals from the environment to activate the genes that code for the formation of proteins, which then carry out the adaptive functions of the cell in health and illness. Immediate early genes integrate mind and body; they are key players in psychosomatic medicine, mind-body healing, and the therapeutic arts.
- *Behavioral State-Related Gene Expression.* Different states of behavior and consciousness—waking, sleeping, dreaming, emotions, motivation, and stress—are all associated with different patterns of behavioral state-related

gene expression. This level of gene expression provides a fundamental link between psychology and biology and is of essence in exploring the psychobiology of consciousness. Behavioral state-related gene expression is a genetic source of behavior that can be modulated by psychosocial cues and cultural rituals to facilitate health, performance, and healing.

• *Activity-Dependent Gene Expression.* Learning to do something new initiates cascades of molecular-genomic processes that are termed *activity-dependent gene expression.* This special class of genes generates the proteins and growth factors that signal *stem cells* in the brain to differentiate into newly functioning neurons with new connections between them. Likewise, stem cells in tissues throughout the body receive psychogenomic signals that enable them to replace injured cells with healthy ones: This process is proposed to be a basic dynamic of the *healing placebo response.*

• *The Novelty-Numinosum-Neurogenesis Effect.* Novelty, enriching life experiences, and exercise associated with a positive sense of curiosity and wonder can turn on activity-dependent gene expression to construct and reconstruct the brain and the way it works throughout our entire lifetime. This is the psychobiological essence of the relationship between the creative psychological experience, gene expression, and neurogenesis: It is a psychosynthetic process of updating and recreating ourselves in everyday life as well as in the arts, humanities, and sciences. The novelty-numinosum-neurogenesis effect documents how highly motivated states of consciousness can turn on and focus gene expression, protein synthesis, neurotransmitters, and neurogenesis in our daily creative work of building a better brain.

• *Creative Replay and Resynthesis as the Essence of Psychotherapy.* Traumatic and fear memories, when reactivated by recall, return to a labile state wherein gene expression and new proteins are synthesized for the potential healing reorganization and reconsolidation of memory as it grows more complex with the addition of new information. *Replaying the four-stage creative process in this resynthesis of experience is the fundamental dynamic of psychotherapy and the healing arts.* In our emerging model of creativity, optimal performance, stress, and healing, Darwinian variation and conscious selection are engaged on all levels, from mind to gene, in the natural ultradian flow of human experiencing. Immediate early, behavioral state-related, and activity-dependent gene expression are bridges between body, brain, and mind that can be accessed to facilitate the creative replays of therapeutic hypnosis, psychotherapy, and the healing arts.

• *Individual Response-Ability and Ethical Self-Realization.* Although it is generally recognized that we are all 99.9% alike in our genetic legacy, it is still not generally understood that there are at least three million small variations in our genes—called *single nucleotide polymorphisms*—that are expressed in our individuality. This implies that we all have a profoundly unique psychogenomic entitlement to the personal perceptions, potentials, and problems that we alone can recognize and realize as our ultimate response-ability. Of necessity, each person is responsible for the facilitation of his or her unique psychogenomic endowment. Parents and teachers may help us find paths, but ultimately only we alone can know when we are really okay in our quests for ethical self-realization—the mindful integration of our personal psychogenomic potentials with those of society and culture.

• *Social and Cultural Response-Ability.* The ultimate gifts of art, music, dance, and the humanities are their evocative effects on gene expression, neurogenesis, and healing in the co-evolution of consciousness and culture. Play, imagination, fantasy, and dreams are all natural exploratory efforts in the creative replay and resynthesis of life experiences on all levels. Gaia, gene, individual, and society co-create each other in the self-reflective replays of the ever-emergent dynamics of becoming. The ultimate response-ability of leadership on all levels is to facilitate this natural evolution of the goals, philosophies, and ethics of life.

• *Positive Psychology and the New Decade of Behavior: 2000–2009.* Our current *Decade of Behavior*, as announced by the American Psychological Association, is a wonderful complement to the previous *Decade of the Brain (1900–1999)* for integrating psyche and soma with the new spirit of *positive psychology.* Experiences of creativity, happiness, humor, uplifting surprise, awe, and that special spiritual sense of the *numinousum—fascination, mysteriousness, and the tremendous*—are all associated with immediate early, behavioral state-related, and activity-dependent gene expression, neurogenesis, healing, and self-realization. The ethical challenge is to discover new research methods for the deepening exploration and the practical implementation of these insights for the rediscovery and recreation of human nature. Let's all go for it!

REFERENCES

Abraham F., & Gilgen, A. (Eds.). (1995). *Chaos theory in psychology*. Westport, CT: Greenwood.

Achterberg, J. (1985). *Imagery and healing: Shamanism and modern medicine*. Boston: Shambhala.

Achterberg, J., Dossey, B., & Kolkmeier, L. (1994). *Rituals of healing: Using imagery for health and wellness*: New York: Bantam.

Aczel, A. (1997). *Fermat's last theorem*. New York: Dell.

Ader, R. (1997). The role of conditioning in pharmacology. In A. Harrington (Ed.), *The placebo effect: An interdisciplinary exploration* (pp. 138–165). Cambridge, MA: Harvard University Press.

Ader, R. (2000). The placebo effect: If it's all in your head, does that mean you only think you feel better. *Advances in Mind-Body Medicine, 16*, 7–46.

Akil, H., & Morano, M. (2000). Stress. *American College of Neuropsychopharmacology: The fourth generation of progress* [On-line]. Available: http://www.acnp.org/G4/GN401000073/Nav.htm

Aldrich, K., & Bernstein, D. (1987). The effect of time of day on hypnotizability. *International Journal of Clinical & Experimental Hypnosis, 35*(3), 141–145.

Alexander, F. (1950). *Psychosomatic medicine*. New York: W.W. Norton.

Alman, B., & Lambrou, P. (1991). *Self-hypnosis: The complete manual for health and self-change*. New York: Brunner/Mazel.

Altman, J. (1962). Are new neurons formed in the brains of adult mammals? *Science, 135*, 1127–1128.

Amabile, T. (2001). Beyond talent: John Irving and the passionate craft of creativity. *American Psychologist, 56*, 333–335.

Amigo, S. (1994). Self-regulation therapy and the voluntary reproduction of stimulant effects of epinephrine: Possible therapeutic applications. *Contemporary Hypnosis, 11*(3), 108–120.

Andersen, J., Schjerling, P., & Saltin, B. (2000). Muscle, genes, and athletic perform-ance. *Scientific American, 283*, 48–55.

Antell, S., & Keating, D. (1983). Perception of numerical invariance in neonates. *Child Development, 54*, 695–701.

Aserinsky, E., & Kleitman, N. (1953). Regularly occurring periods of eye motility and concomitant phenomena during sleep. *Science, 118*, 273–274.

Autelitano, D. J. (1998). Stress-induced stimulation of pituitary POMC gene expres-sion is associated with activation of transcription factor AP-1 in hypothalamus and pituitary. *Brain Research Bulletin, 45*(1), 75–82.

Axelrod, R., & Cohen, M. (1999). *Harnessing complexity: Organizational implications of a scientific frontier.* New York: Free.

Azar, B. (1999). New pieces filling in addiction puzzle. *APA Monitor, 30*, 1–15.

Bailey C., Bartsch, D., & Kandel E. (1996). Toward a molecular definition of long-term memory storage. *Proceedings of the National Academy of Sciences, USA, 93*, 13445–13452.

Bailey, C., & Kandel, E. (1995). Molecular and structural mechanisms underlying long-term memory. In M. Gazzaniga (Ed.), *The cognitive sciences* (pp. 19–36). Cambridge, MA: MIT Press.

Balsalobre, A., Brown, S., Marcacci, L., Tronche, F., Kellendonk, C., Reichardt, H., Schütz, G., & Schibler, U. (2000). Resetting of circadian time in peripheral tis-sues by glucocorticoid signaling. *Science, 289*, 2344–2347.

Balter, M. (2000). Celebrating the synapse. *Science, 290*, 424.

Balthazard, C., & Woody, E. (1992). The spectral analysis of hypnotic performance with respect to "absorption." *International Journal of Clinical and Experimental Hypnosis, 40*(1), 21–43.

Barabasz, A., & Barabasz, M. (1996). Neurotherapy and alert hypnosis in the treat-ment of attention deficit hyperactivity disorder. In S. Lynn, J. Kirsch, & J. Rhue (Eds.), *Clinical hypnosis handbook* (pp. 271–291). Washington, DC: American Psychological Association.

Barabasz, A., & Lonsdale, C. (1983). Effects of hypnosis on P300 olfactory-evoked potential amplitudes. *Journal of Abnormal Psychology, 92*, 520–523.

Barber, J. (1990). Miracle cures? Therapeutic consequences of clinical demonstrations. In J. Zeig & S. Gilligan (Eds.), *Brief therapy: Myths, methods, and metaphors* (pp. 437–422). New York: Brunner/Mazel,.

Barber, J. (2000). Where Ericksonian legend meets scientific method. *The International Journal of Clinical and Experimental Hypnosis, 48*, 427–432.

Barber, T. (1969). *Hypnosis: A scientific approach.* New York: Van Nostrand Reinhold.

Barber, T. (1984). Changing unchangeable bodily processes by (hypnotic) suggestions: A new look at hypnosis cognitions, imagining, and the mindbody problem. *Advances, 1,* 7–40

Barber, T. (2000). A deeper understanding of hypnosis: Its secrets, its nature, its essence. *American Journal of Clinical Hypnosis, 42,* 208–273.

Bardin, A., Visintin, R., & Amon, A. (2000). A mechanism for coupling exit from mitosis to partitioning of the nucleus. *Cell, 102,* 21–31.

Barinaga, M. (2002). How the brain's clock gets daily enlightenment. *Science, 295,* 955–957

Barrio, R., Zhang, L., & Maini, P. (1997). Hierarchically coupled ultradian oscillators generating robust circadian rhythms. *Bulletin of Mathematical Biology, 59,* 517–532.

Barrow, J. (2000). Mathematical jujitsu: Some informal thoughts about Gödel and physics. The limits of mathematical systems. *Complexity, 5*(5), 28–34.

Bartolome, J., Wang, S., Schanberg, S., & Bartolome, M. (1999). Involvement of c-myc and max in CNS beta-endorphin modulation of hepatic ornithine decarboxylase responsiveness to insulin in rat pups. *Life Sciences, 64,* 87–91.

Bassett, D., Eisen, M., & Bogudki, M. (1999). Gene expression informatics-it's all in your mine. *Nature Genetics Supplement, 21,* 51–55.

Bateson, G. (1972). *Steps to an ecology of mind.* New York: Ballantine.

Battino, R. (2000). *Guided imagery and other approaches to healing.* Carmarthen, UK: Crown House.

Baudry, M., Davis, J., & Thompson, R. (2000). *Advances in synaptic plasticity.* Cambridge, MA: MIT Press.

Baudry, M., & Thompson, R. (2000). Synaptic plasticity: From molecules to behavior. In M. Baudry, J. Davis, & R. Thompson (Eds.), *Advances in synaptic plasticity* (pp. 291–319). Cambridge, MA: MIT Press.

Beckmann, A., Matsumoto, I., & Wilce, P. (1997). AP-1 and Egr DNA-binding activities are increased in rat brain during ethanol withdrawal. *Journal of Neurochemistry, 69,* 306.

Beggs, J., Brown, T., Byrne, J., Crow, T., LeDoux, J., LeBar, K., & Thompson, R. (1999). Learning and memory: Basic mechanisms. In M. Zigmond, F. Bloom, S. Landis, J. Roberts, & L. Squire (Eds.), *Fundamental neuroscience* (pp. 1411–1454). New York: Academic Press.

Begley, S. (2000, March 27). The nature of nurturing: A new study finds how parents treat their children can shape which of his genes are turned on. *Newsweek,* 64–66.

Begun, B. (2000, September 25). Mind games. *Newsweek*, 60–61.

Bélair, J., Glass, L., an der Heiden, U., & Milton, J. (Eds.). (1995). *Dynamical disease: Mathematical analysis of human disease*. New York: American Institute of Physics.

Ben-Jacob, E., & Levine, H. (2001). Pattern: The artistry of nature. *Nature, 409*, 985–986.

Benloucif, S., Masana, M., & Dubocovich, M. (1997). Light-induced phase shifts of circadian activity rhythms and immediate early gene expression in the suprachiasmatic nucleus are attenuated in old C3H/HeN mice. *Brain Research, 747*(1), 34–42.

Benson, H. (1983). The relaxation response: Its subjective and objective historical precedents and physiology. *Trends in Neuroscience, 281–284.*

Bentivoglio, M., & Grassi-Zucconi, G. (1999). Immediate early gene expression in sleep and wakefulness. In R. Lydic & H. Baghdoyan (Eds.), *Handbook of behavioral state control: Cellular and molecular mechanisms* (pp. 235–253). New York: CRC.

Benzer, S. (1967). Behavioral mutants of Drosophila isolated by countercurrent distribution. *Proceedings of the National Academy of Sciences, USA, 58*, 112–119.

Benzer, S. (1971). From genes to behavior. *Journal of the American Medical Association, 218*, 1015–1022.

Benzer, S. (1973). Genetic dissection of behavior. *Scientific American, 229*(6), 24–37.

Berdanier, C. (Ed.). (1996). *Nutrients and gene expression*. New York: CRC.

Berman, D., & Dudai, Y. (2001). Memory extinction, learning anew, and learning the new: Dissociations in the molecular machinery of learning in the cortex. *Science, 291*, 2417–2419.

Bernheim, H. (1886/1957). *Suggestive therapeutics: A treatise on the nature and uses of hypnotism*. Westport: Associated Booksellers.

Bijeljac-Babic, R., Bertonici, J., & Mehler, J. (1991). How do four-day-old infants categorize multisyllabic utterances. *Developmental Psychology, 29*, 711–721.

Bittman, B., Berk, L., Felten, D., Westengard, J., Simonton, C., Pappas, J., & Ninehouser, M. (2001). Composite effects of group drumming music therapy on modulation of neuroendocrine-immune parameters in normal subjects. *Alternative Therapies, 7*, 38–47.

Bjick, S. (2001). Accessing the power in the patient with hypnosis and EMDR. *American Journal of Clinical Hypnosis, 43*(3), 203–216.

Björklund, A., & Lindvall, O. (2000). Self-repair in the brain. *Nature, 405*, 892–895.

Black, I. (1991). *Information in the brain: A molecular perspective*. Cambridge, MA: MIT Press.

Black, I. (2001). *The dying of Enoch Wallace: Life, death and the changing brain.* New York: McGraw-Hill.

Bliss, T., & Lomo, T. (1973). Long-lasting potentiation of synaptic transmission in the dentate area of anesthetized rabbit following stimulation of the preforant path. *Journal of Physiology, 232,* 331–356.

Boivin, D., Duffy, J., Kronauer, R., & Czeisler, C. (1996). Dose-response relationships for resetting the human circadian click by light. *Nature, 379,* 540–542.

Bongartz, E. (Ed.). (1992). *Hypnosis: 175 years after Mesmer: Recent developments in theory and application.* Konstanz, Germany: Universitatsvergag.

Boring, E. (1950). *A history of experimental psychology.* New York: Appleton-Century-Crofts.

Boole, G. (1847). *The mathematical analysis of logic.* London: Walton & Maberly.

Boone, T., & Cooper, R. (1995). The effect of massage on oxygen consumption at rest. *American Journal of Chinese Medicine, 23*(1), 37–41.

Born, J., Hansen, K., Marshall, L., Molle, M., & Fehm, H. (1999). Timing the end of nocturnal sleep. *Nature, 397,* 29-30.

Bourdieu, P. (1977). *Outline of a theory of practice.* Chicago: University of Chicago Press.

Bourguinon, E. (1968). *A cross-cultural study of dissociational states: Final report.* Columbus OH: Research Foundation.

Bourguinon, E. (1973). *Religion, altered states of consciousness and social change.* Columbus, OH: Ohio State University Press.

Bourguinon, E., & Evascu, T. (1977). Altered states of consciousness within a general evolutionary perspective: A holocultural analysis. *Behavior Science Research, 12,* 199–216.

Bower, B. (1997). The power of limited thinking: Small scale minds may pay non-random dividends. *Science News, 152,* 334–335.

Bower, B. (2001). Learning in waves. *Science News, 159,* 172–174.

Braid, J. (1855/1970). The physiology of fascination and the critics criticized. In M. Tinterow (Ed.), *Foundations of hypnosis* (pp. 369–372). Springfield, IL: C. C. Thomas.

Brand, S. (1974). *Both sides of the necessary paradox. Conversations with Gregory Bateson. Cybernetic frontiers.* New York: Random House.

Brandenberger, G. (1992). Endocrine ultradian rhythms during sleep and wakefulness. In D. Lloyd & E. Rossi (Eds.), *Ultradian rhythms in life processes: A fundamental inquiry into chronobiology and psychobiology* (pp. 123–138). New York: Springer Verlag.

Brennan, M., & Weitz, J. (1992). Lymphedema: 30 years after radical mastectomy. *American Journal of Physical Medicine & Rehabilitation, 71*(1), 12–14.

Breuer, J., & Freud, S. (1895/1955). *Studies on hysteria*. In J. Strachey (Ed. & Trans.), *The standard edition of the complete psychological works of Sigmund Freud, Vol. II*. New York: W. W. Norton.

Brey, D. (1995). Protein molecules as computational elements in living cells. *Nature, 376*, 307–312.

Brivanlou, A., & Darnell, J. (2002). Signal transduction and the control of gene expression. *Science, 295*, 813–818.

Brodsky, V. (1992). Rhythms of protein synthesis and other circahoralian oscillations: The possible involvement of fractals. In D. Lloyd & E. Rossi (Eds.), *Ultradian rhythms in life processes: A fundamental inquiry into chronobiology and psychobiology* (pp. 23–40). New York: Springer-Verlag.

Broughton, R. (1975). Biorhythmic variations in consciousness and psychological functions. *Canadian Psychological Review: Psychologie Canadienne, 16*, 217–239.

Brown, J., & Duguig, R. (2000). *Information technology: The social life of information*. Boston: Harvard Business School Press.

Brown, J., Ye, H., Bronson, R. T., Dikkes, P., & Greenberg, M. E. (1996). A defect in nurturing in mice lacking the immediate early gene fosB. *Cell, 86*(2), 297–309.

Brown, P. (1991a). Ultradian rhythms of cerebral function and hypnosis. *Contemporary Hypnosis, 8*(1), 17–24.

Brown, P. (1991b). *The hypnotic brain: Hypnotherapy and social communication*. New Haven, CT: Yale University Press.

Brown, T. (1999). *Genomes*. New York: Wiley-Liss.

Brush, F., & Levine, S. (1989). *Psychoendocrinology*. San Diego: Academic Press.

Búcan, M., & Abel, T. (2002). The mouse: Genetics meets behavior. *Nature Reviews Genetics, 3*, 114–123.

Bucke, M. (1901/1967). *Cosmic consciousness*. New York: Dutton.

Buijs, R., & Kalsbeek, A. (2001). Hypothalmic integration of central and peripheral clocks. *Nature Reviews: Neuroscience, 2*, 521–526.

Buss, D. (2000). The evolution of happiness. *American Psychologist, 55*, 15–23.

Cafarelli, E., & Flint, F. (1999). The role of massage in preparation for and recovery from exercise: An overview. *Sports Medicine, 14*(1), 1–9.

Cahill, L., Prins, B., Weber, M., & McGaugh, J, (1994) B-Adrenergic activation and memory for emotional events. *Nature, 371*(20), 702–704.

Cairns, J., & Foster, P. (1991). Adaptive reversion of a frameshift mutation in escherichia coli. *Genetics, 128*, 695–701.

Cairns-Smith, A. (1996). *Evolving the mind: On the nature of matter and the origin of consciousness*. Cambridge, UK: Cambridge University Press.

Callard, R., George, A., & Stark, J. (1999). Cytokines, chaos, and complexity. *Immunity, 11*, 507–513

Campbell, J. (1956). *The hero with a thousand faces*. Cleveland, OH: World.

Campbell, J. (1959). *The masks of god: Primitive mythology*. New York: Viking.

Campbell, J. (Ed.). (1960). *Spiritual disciplines: Papers from the Eranos Yearbooks*, No. 4. New York: Pantheon.

Campbell, J. (1962). *The masks of god: Oriental mythology*. New York: Viking.

Campbell, J. (1964). *The masks of god: Occidental mythology*. New York: Viking.

Campbell, J. (1968). *The masks of god: Creative mythology*. New York: Viking.

Campeau, S., Falls, W., Cullinan, W., Helmreich, D., Davis, M., & Watson, S. (1997). Elicitation and reduction of fear: Behavioral and neuroendocrine indices and brain induction of the immediate-early gene c-fos. *Neuroscience, 78*(4), 1087–1104.

Cantor, G. (2000). Fighting the wrong battle. *Nature, 403*, 831.

Carpenter, S. (2001). When at last you don't succeed . . . *Monitor on Psychology, 32*, 70–71.

Carroll, S. (2000). Endless forms: The evolution of gene regulation and morphological diversity. *Cell, 101*, 577–580.

Castes, M., Hagel, I., Palenque, M., Canelones, P., Corano, A., & Lynch, N. (1999). Immunological changes associated with clinical improvement of asthmatic children subjected to psychosocial intervention. *Brain & Behavioral Immunology, 13*(1), 1–13.

Chaitin, G. (2000). A century of controversy over the foundations of mathematics. *Complexity, 5*(5), 12–21.

Cheek, D. (1994). *Hypnosis: The application of ideomotor techniques*. Boston: Allyn & Bacon.

Cheverton, R. (2001, January 28). An artist at the abyss. *Los Angeles Times Magazine*, 20–22.

Chicurel, M. (2001a). Can organisms speed their own evolution? *Science, 292*, 1824–1827.

Chicurel, M. (2001b). Windows on the brain. *Nature, 412*, 266–268.

Chin, H., & Moldin, S. *Methods in genomic neuroscience*. Boca Raton, FL: CRC.

Cho, K. (2001). Chronic "jet lag" produces temporal lobe atrophy and spatial cognitive deficits. *Nature Neuroscience, 4*, 567–568.

Chomsky, N. (1957). *Syntactic Structures*. The Hague: Mouton Publishers.

Christensen, D. (2001). Fatty findings: Genes that regulate fat may affect a variety of diseases. *Science News, 159,* 238–239.

Christopher, H., Stanley D., Pamela, R., Masafumi, O., Brian, E., Phyllis D., Stevenson, D. K., & Benaron, D. (1997). Visualizing gene expression in living mammals using a bioluminescent reporter. *Photochemistry and Photobiology, 66,* 523–531.

Ciccone, C. (1995). Basic pharmacokinetics and the potential effect of physical therapy interventions on pharmacokinetic variables. *Physical Therapy, 75*(5), 343–351.

Cirelli, C. Pompeiano, M., & Tononi, G. (1995). Sleep deprivation and c-fos expression in the rat brain. *Journal of Sleep Research, 4,* 92–106.

Cirelli, C., & Pompeiano, M., & Tononi, G. (1996). Neuronal gene expression in the waking state: A role for the locus coeruleus. *Science, 272,* 1211–1215.

Cirelli, C, Pompeiano, M., & Tononi, G. (1998). Immediate early genes as a tool to understand the regulation of the sleep-waking cycle. In R. Lydic (Ed.), *Molecular regulation of arousal states.* New York: CRC.

Clark, R., & Squire, L. (1998). Classical conditioning and brain systems: The role of awareness. *Science, 280,* 77–81.

Clayton, D. (2000). The genomic action potential. *Neurobiology of memory and learning, 74,* 185–216.

Collins, F. (2000). In Siri, C., Human genome project director says psychologists will play a critical role in the initiatives success. *Monitor on Psychology, 31,* 14–15.

Collins, P. (2001). The hands that hold the keys. *Nature Reviews: Neuroscience, 2,* 76.

Conklin, B. (1999, November 17). *Making geneChip probe array data relevant to biologists and pharmacologists from disease models to biochemical pathways.* Affymetrix Technology Symposium, Santa Clara, California.

Cooper, L., & Erickson, M. (1959). *Time distortion in hypnosis: An experimental and clinical investigation.* Baltimore, MD: Williams & Wilkins.

Corsini, R. (Ed.). (2001). *Handbook of innovative therapy* (2nd ed.). New York: Wiley.

Coué, E. (1922). *Self mastery through conscious autosuggestion* (A. Van Orden, Trans.). New York: Malkan.

Coué, E. (1923). *My method, including American impressions.* Garden City, NY: Doubleday.

Crabtree, G. (1989). Contingent genetic regulatory vents in T lymphocyte activation. *Science, 243*(20), 355–361.

Crasilneck, H. (1997). *Pain: The case of Janelle.* Advanced workshop presented at the 14th International Congress of Hypnosis, San Diego, California.

Csikszentmihalyi, M. (1996). *Creativity*. New York: HarperCollins.

Cvitanović, P. (1989). *Universality in chaos*. Philadelphia, PA: Institute of Physics Publishing.

Damasio, A. (1999). *The Feeling of what happens: Body and emotion in the making of consciousness*. New York: Harcourt.

Damasio, R., Grabowski, T., Bechara, A., Damasio, H., Ponto, L., Parvizi, J., & Hichwa, R. (2000). Subcortical and cortical brain activity during the feeling of self-generated emotions. *Nature Neuroscience, 3*, 1049-1056.

Darwin, C. (1859/1999). *On the origin of species by means of natural selection: Or the preservation of favored races in the struggle for life*. New York: Bantam.

David, D., King, B., & Borkardt, J. (2001). Is a capacity for negative priming correlated with hypnotizability? *The International Journal of Clinical and Experimental Hypnosis, 49*(1), 30–37.

Davson, H., & Segal, M., (1996). *Physiology of the CFS and blood-brain barriers*. Boca Raton: CRC.

Dawkins, R. (1976). *The selfish gene*. Oxford: Oxford University Press.

d'Aquili, E., & Newberg, A. (1999). *The mystical mind: Probing the biology of religious experience*. Minneapolis, MN: Fortress Press.

Day, J., Mason, R., & Chesrown, S. (1987). Effect of massage on serum level of beta-endorphin and beta-lipotropin in healthy adults. *Physical Therapy, 67*(6), 926–930.

Dearden, P., & Akam, M. (2000). Segmentation in silico. *Nature, 406*, 131–132.

DeBenedittis, G., Cigada, M., Bianchi, A., Signorini, M., & Cerutti, S. (1994). Autonomic changes during hypnosis: A heart rate variability power spectrum analysis as a marker of sympatho-vegal balance. *International Journal of Clinical and Experimental Hypnosis, 42*(2), 140–152.

DeGrandpre, R. (2000). A science of meaning: Can behaviorism bring meaning to psychological science? *American Psychologist, 55*, 721–739.

Deikman, A. (1980a). *Deautomatization and the mystic experience. Understanding mysticism*. Garden City, NJ: Image Books.

Deikman, A. (1980b). *Bimodal consciousness and the mystic experience: Understanding mysticism*. Garden City, NJ: Image Books.

Deikman, A. (1982). *The observing self: Mysticism and psychotherapy*. Boston: Beacon.

de la Fuente-Fernández, R., Ruth, T., Sossi, V., Schultzer, M., Calne, T., & Stoessl, A. (2001). Expectation and dopamine release: Mechanism of the placebo effect in Parkinson's disease. *Science, 293*, 1164–1166.

Delander, G. E., Schott, E., Brodin, E., & Fredholm, B. B. (1997). Spinal expression of mRNA for immediate early genes in a model of chronic pain. *Acta Physiologia Scandanavia, 161*(4), 517–525.

Dement, W. (1972). *Some must watch while some must sleep.* Stanford, CA: Stanford Alumni Association, 8-12.

Dement, W. (1992). *The sleepwatchers.* Stanford, CA: Stanford Alumni Association.

DeSouza, E., & Grigoriadis, D. (2000). Corticotropin-releasing factor: Physciology, phramacology, and role in central nervous system and immune disorders. In *The American College of Neuropsychopharmacology: Psychopharmacology, The Fourth Generation of Progress* [On-line]. Available: www. Acnp.org/G4

Desvergne, B., & Wahli. W. (1999). Peroxisome proliferator-activated receptors: Nuclear control of metabolism. *Endocrine Reviews, 20,* 649–688.

Devlin, K. (1999). *The language of mathematics: Making the invisible visible.* New York: W. H. Freeman.

Devlin, K. (2000). *The math gene: How mathematical thinking evolved and why numbers are like gossip.* New York: Basic.

Dhabhar, F., & McEwen, B. (1999). Enhanced versus suppressive effects of stress hormones on skin immune function. *Proceedings of the National Academy of Sciences, 96,* 1059–1064.

Dhabhar, F., Satoskar, A., Bluethmann, H., David, J., & McEwen, B. (2000). Stress-induced enhancement of skin immune function: A role for gamma interferon. *Proceedings of the National Academy of Sciences, 97,* 2846–2851.

DiAntonio, A. (2000). Translating activity into plasticity. *Nature, 405,* 1011–1012.

Diener, E. (2000). Subjective well-being. *American Psychologist, 55,* 34–43.

Dinges, D., & Broughton, R. (1989). *Sleep and alertness.* New York: Raven.

Dobrunz, L., & Garner, C. (2002). Priming plasticity. *Nature, 415,* 277–278.

Dolan, Y. (1985). *A path with heart: Ericksonian utilization with resistant and chronic patients.* New York: Brunner/Mazel.

Dossey, L. (1993). *Healing words: The power of prayer and the practice of medicine.* San Francisco, CA: HarperSanFrancisco.

Dragunow, M. (1995). Differential expression of immediate-early genes during synaptic plasticity, seizures and brain injury suggests specific functions for these molecules in brain neurons. In T. R. Tölle, J. Schadrack, W. Zieglgansberger (Eds.), *Immediate early genes in the CNS* (pp. 35–50). New York: Springer-Verlag.

Dudai, Y. (2000). The shaky trace. *Nature, 406,* 686–687.

Duman, R., & Neslter, E. (2000). Role of gene expression in stress and drug-induced neural plasticity. In *TEN: The economy of neuroscience* [On-line]. Available: www.tenmag.com/TEN400_Duman.html

Dunlap, J. (1999). The molecular basis for circadian clocks. *Cell, 96,* 271–290.

Dyson, F. (1999). *Origins of life* (2nd ed.). New York: Cambridge University Press.

Edelman, G. (1987). *Neural Darwinism: The theory of neuronal group selection.* New York: Basic.

Edelman, G., & Tononi, G. (2000). *A universe of consciousness: How matter becomes imagination.* New York: Basic.

Edmonston, W. (1986). *The induction of hypnosis.* New York: Wiley.

Ehrlich, P. (2000). *Genes, cultures, and the human prospect.* New York: Island Press.

Eigen, M. (1971). Self-Organization of matter and the evolution of biological macro-molecules. *Naturwissenschaften 58,* 465–523.

Eigen, M., & Schuster, P., (1977). The hypercycle: A principle of natural self-organization. Part A: The emergence of the hypercycle. *Naturwissenschaften 64,* 541–565.

Eigen, M., Gardiner, W., Schuster, P., & Winkler-Oswatitsch, R. (1981). *Scientific American, 244*(4), 88–118.

Eigen, M., & Winkler, R. (1981). *Laws of the game: How the principles of nature govern chance.* New York: Harper & Row.

Eigen, M., & Winkler-Oswatitsch, R. (1992). *Steps towards life: A perspective on evolution.* New York: Oxford University Press.

Eisenberg, L. (2000). The role of social experience in transforming geneotype into phenotype. In *TEN: The Economy of Neuroscience* [On-line]. Available: www.tenmag.com/TEN400_Duman.html

Elena, S., Cooper, V., & Lenski, R. (1996). Punctuated evolution caused by selection of rare beneficial mutations. *Science, 272,* 1802–1804.

Eliska, O., & Eliskova, M. (1995). Are peripheral lymphatics damaged by high pressure manual massage? *Lymphology, 28*(1), 21–30.

Ellenberger, E. (1970). *The discovery of the unconscious.* New York: Basic.

Engel, G., & Reichsman, F. (1956). Spontaneous and induced depression in an infant with gastric fistula. *Journal of the American Psychoanalytic Association, 4,* 428–452.

Ernst, E. (1999). Massage therapy for low back pain: A systematic review. *Journal of Pain Symptom Management, 17,* 65–69.

Erickson, M. (1932/1980). Possible detrimental effects of experimental hypnosis. In E. Rossi (Ed.), *The collected papers of Milton H. Erickson on hypnosis: I. The nature of hypnosis and suggestion* (pp. 493–497). New York: Irvington.

Erickson, M. (1938/1980). A study of clinical and experimental findings on hypnotic deafness: I. Clinical experimentation and findings. In E. Rossi (Ed.), *The collect-ed papers of Milton E. Erickson on hypnosis: II. Hypnotic alteration of sensory, percep-tual, and psychophysical processes* (pp. 81–99). New York: Irvington.

Erickson, M. (1948/1980). Hypnotic psychotherapy. In E. Rossi (Ed.), *The collected papers of Milton H. Erickson on hypnosi: Vol. 4. Innovative hypnotherapy* (pp. 35–48). New York: Irvington.

Erickson, M. (1952/1980). Deep hypnosis and its induction. In E. Rossi (Ed.), *The collected papers of Milton H. Erickson on hypnosis: Vol. 1. The nature of hypnosis and suggestion* (pp. 139–167). New York: Irvington.

Erickson, M. (1954/1980). The development of an acute limited obsessional hyster-ical state in a normal hypnotic subject. In E. Rossi (Ed.), *The collected papers of Milton E. Erickson on hypnosis: II. Hypnotic alteration of sensory, perceptual, and psy-chophysical processes* (pp. 51–80). New York: Irvington.

Erickson, M. (1958/1980). Naturalistic techniques of hypnosis. In E. Rossi (Ed.), *The Collected papers of Milton H. Erickson on hypnosis: Vol. I. The nature of hypnosis and suggestion* (pp. 168–176). New York: Irvington.

Erickson, M. (1959/1980). Further clinical techniques of hypnosis: Utilization tech-niques. In E. Rossi (Ed.), *The collected papers of Milton H. Erickson on hypnosis: Vol. I. The nature of hypnosis and suggestion* (pp. 177–205). New York: Irvington.

Erickson, M. (1963/1980). An application of implications of Lashley's researches in a circumscribed arteriosclerotic brain condition. In E. Rossi (Ed.), *The collected papers of Milton H. Erickson on hypnosis: IV. The nature of hypnosis and suggestion* (pp. 315–316). New York: Irvington.

Erickson, M. (1964a/1980). Pantomime techniques in hypnosis and the implications. In E. Rossi (Ed.), *The collected papers of Milton H. Erickson on hypnosis: I. The nature of hypnosis and suggestion* (pp. 331–339). New York: Irvington.

Erickson, M. (1964b/1980). The "surprise" and "my-friend-john" techniques of hyp-nosis: Minimal cues and natural field experimentation. In E. Rossi (Ed.), *The col-lected papers of Milton H. Erickson on hypnosis: I. The nature of hypnosis and suggestion* (pp. 340–359). New York: Irvington.

Erickson, M. (1964c/1980). An hypnotic technique for resistant patients: the patient, the technique, and its rationale and field experiments. In E. Rossi (Eds), *The col-lected papers of Milton H. Erickson on hypnosis: I. The nature of hypnosis and sugges-tion* (pp. 229–330). New York: Irvington.

Erickson, M. (1964d/1980). The burden of responsibility in effective psychotherapy. In E. Rossi (Ed.), *The collected papers of Milton H. Erickson on hypnosis: IV. The nature of hypnosis and suggestion* (pp. 207–211). New York: Irvington.

Erickson, M. (1965/1980a). A special inquiry with Aldous Huxley into the nature and character of the various states of consciousness. In E. Rossi (Ed.), *The collected papers of Milton H. Erickson on hypnosis: I. The nature of hypnosis and suggestion* (pp. 83–107). New York: Irvington.

Erickson, M. (1965/1980b). Provocation as a means of motivating recovery from a cerebrovascular accident. In E. Rossi (Ed.), *The collected papers of Milton H. Erickson on hypnosis: Vol. IV. The nature of hypnosis and suggestion* (pp. 321–327). New York: Irvington.

Erickson, M. (1966/1980). The interspersal hypnotic technique for symptom correction and pain control. In E. Rossi (Ed.), *The collected papers of Milton H. Erickson on hypnosis: I. The nature of hypnosis and suggestion* (pp. 262–278). New York: Irvington.

Erickson, M. (1976/1980). Two-level communication and the microdynamics of trance and suggestion. In E. Rossi (Ed.), *The collected papers of Milton H. Erickson on hypnosis: I. The nature of hypnosis and suggestion* (pp. 430–451). New York: Irvington.

Erickson, M. (1980a). Facilitating new identity. In E. Rossi (Ed.), *The collected papers of Milton H. Erickson on hypnosis: IV. Innovative hypnotherapy section nine* (pp. 446–548). New York: Irvington.

Erickson, M. (1980b). An introduction to the study and application of hypnosis for pain control. In E. Rossi (Ed.), *1*

Erickson, M. (1954/1980c). The development of an acute limited obsessional hysterical state in a normal hypnotic subject. In E. Rossi (Ed.), *The collected papers of Milton E. Erickson on hypnosis: II. Hypnotic alteration of sensory, perceptual, and psychophysical processes* (pp. 51–80). New York: Irvington.

Erickson, M., & Erickson, E. (1941/1980). Concerning the nature and character of post-hypnotic behavior. In E. Rossi (Ed.), *The collected papers of Milton H. Erickson on hypnosis: I. The nature of hypnosis and suggestion* (pp. 360–365). New York: Irvington.

Erickson, M., & Erickson, E. (1958/1980). Further considerations of time distortion: Subjective time condensation as distinct from time expansion. In E. Rossi (Ed.), *The collected papers of Milton H. Erikson on hypnosis: II. Hypnotic alteration of sensory, perceptual and psychophysical processes* (pp. 291–298). New York: Irvington.

Erickson, M., Haley, J., & Weakland, J. (1959). A transcript of a trance induction with commentary. *American Journal of Clinical Hypnosis, 2,* 49–84.

Erickson, M., & Rossi, E. (1975). Varieties of double bind. *The American Journal of Clinical Hypnosis, 17*(3), 143–157.

Erickson, M., & Rossi, E. (1976). Two-level communication and the microdynamics of trance and suggestion. *The American Journal of Clinical Hypnosis, 18,* 153–171.

Erickson, M., & Rossi, E. (1977/1980). Autohypnotic experiences of Milton H. Erickson. In E. Rossi (Ed.), *The collected papers of Milton H. Erickson on hypnosis: I. The nature of hypnosis and suggestion* (pp. 108–132). New York: Irvington.

Erickson, M., & Rossi, E. (1979). *Hypnotherapy: An exploratory casebook.* New York: Irvington.

Erickson, M., & Rossi, E. (1980). Indirect forms of suggestion in hand levitation. In E. Rossi (Ed.), *The collected papers of Milton H. Erickson on hypnosis: I. The nature of hypnosis and suggestion* (pp. 478–490). New York: Irvington.

Erickson, M., & Rossi, E. (1981). *Experiencing hypnosis: Therapeutic approaches to altered states.* New York: Irvington.

Erickson, M., & Rossi, E. (1989). *The February man: Evolving consciousness and identity in hypnotherapy.* New York: Brunner/Mazel.

Erickson, M., Rossi, E., & Rossi, S. (1976). *Hypnotic realities.* New York: Irvington.

Erickson, M. (1983). *Healing in hypnosis: I. The seminars, workshops and lectures of Milton H. Erickson* (E. Rossi, M. Ryan, & F. Sharp, Eds.). New York: Irvington.

Erickson, M. (1985). *Life reframing in hypnosis: II. The seminars, workshops and lectures of Milton H. Erickson* (E. Rossi & M. Ryan, Eds.). New York: Irvington.

Erickson, M. (1986). *Mind-body communication in hypnosis: III. The seminars, workshops and lectures of Milton H. Erickson* (E. Rossi & M. Ryan, Eds.). New York: Irvington.

Erickson, M. (1992). *Creative choice in hypnosis: IV. The seminars, workshops and lectures of Milton H. Erickson* (E. Rossi & M. Ryan, Eds.). New York: Irvington.

Eriksson, P., Perfilieva, E., Björk-Eriksson, T., Alborn, A-M., Nordborg, C., Peterson, D., & Gage, F. (1998). Neurogenesis in the adult human hippocampus. *Nature Medicine, 4,* 1313–1317.

Escera, C., Cilveti, R., & Grau, C. (1992). Ultradian rhythms in cognitive operations: Evidence from the P300 component of the event-related potentials. *Medical-Science-Research, 20*(4), 137–138.

Ewin, D. (1986). Emergency room hypnosis for the burned patient. *American Journal of Clinical Hypnosis, 29,* 7–12.

Eysenck, H. (1991). Is suggestibility? In J. Schumaker (Ed.), *Human suggestibility: Advances in theory, research and application* (pp. 76–90). New York: Routledge.

Fahy, G. (1999). Aging revealed! In *Life Extension* [On-line], *5,* 52–60. Available: www.lef.org

Feher, S., Berger, L., Johnson, J., & Wilde, J. (1989). Increasing breast milk production for premature infants with a relaxation/imagery audiotape. *Pediatrics, 83*, 57–60.

Feigenbaum, M. (1980). Universal behavior in nonlinear systems. *Los Alamos Science, 1,* 4-27. (Reprinted in Cvitanović, P. [1989]. *Universality in Chaos*, pp. 49–84, Philadelphia, PA: Institute of Physics Publishing.)

Felker, B., & Hubbard, J. (1998). Influence of mental stress on the endocrine system. In J. Hubbard & E. Workman (Eds.), *Handbook of stress medicine: An organ system approach* (pp. 69-86). New York: CRC.

Fernandez, J., & Hoeffler, J. (1999). *Gene expression systems.* New York.: Academic.

Ferrell-Torry, A., & Glick, O. (1993). The use of therapeutic massage as a nursing intervention to modify anxiety and the perception of cancer pain. *Cancer Nursing, 16*(2), 93–101.

Fraser, J., & Kerr, J, (1993). Psychophysiological effects of back massage on elderly institutionalized patients. *Journal of Advanced Nursing, 18*(2), 238–245.

Freud, S. (1910). The antithetical sense of primal words. In Riviere (Trans.), *The collected papers of Sigmund Freud* (pp. 184–191). London: Hogarth.

Field, T. (Ed.). (1995). *Touch in early development.* New York: Lawrence Erlbaum.

Folkman, J. (2001, February 19). In Kalb, C., Folkman looks ahead. *Newsweek*, 44–45.

Fox, K. (2000). A moving experience. *Nature, 404*, 825-827.

Fox, K., Henley, J., & Isaac, J. (1999). Experience-dependent development of NMDA receptor transmission. *Nature Neuroscience, 2*(4), 297–299.

Frank, M., Issa, N., & Stryker, M. (2001). Sleep enhances plasticity in the developing visual cortex. *Neuron, 30*, 275–287.

Frankland, P., O'Brien, C., Ohno, M., Kiekwood, A., & Silva, A. (2001). Alpha-CaMKll-dependent plasticity in the cortex is required for permanent memory. *Nature, 411*, 309–313

Fredrickson, B. (2001). The role of positive emotions in positive psychology. *American Psychologist, 56*, 218–226.

Freeman, W. (1995). *Societies of brains.* New York: Lawrence Erlbaum.

Freeman, W. (2000). *How brains make up their minds.* New York: Science News Books.

Freidman, S. (1972). On the presence of a varient form of instinctual regression. *Psychoanalytic Quarterly, 41*, 364–383.

Friedman, S. (1978). A psychophysiological model for the chemotherapy of psychosomatic illness. *The Journal of Nervous & Mental Diseases, 166*, 110–116.

Friedman, S., & Fischer, C. (1967). On the presence of a rhythmic diurnal, oral instinctual drive cycle in man: A preliminary report. *Journal of the American Psychoanalytic Association, 15*, 317–343.

Fromm, E. (1992). An ego-psychological theory of hypnosis. In E. Fromm & M. Nash (Eds.), *Contemporary hypnosis research* (pp. 131–148). New York: Guilford.

Fuchs, E., & Segre, J. (2000). Stem cells: A new lease on life. *Cell, 100*, 143–155.

Gage, F. (2000a). Mammalian neural stem cells. *Science, 287,* 1433–1438.

Gage, F. (2000b). Growing new brain cells, *Life Extension, 6*, 49–54.

Galli, R., Borello, U., Gritti, A., Minasi, M., Bjornson, C., Coletta, M., Mora, M., De Angelis, M., Fiocco, R., Cossu, G., & Vescovi, A. (2000). Skeletal myogenic potential of human and mouse neural stem cells. *Nature Neuroscience, 3,* 986–991.

Gallo, V., & Chittajallu, R. (2001). Unwrapping glial cells from the synapse: What lies inside? *Science, 292*, 872–873.

Gardner, L. (1972). Deprivation dwarfism. *Scientific American, 227*, 76–82.

Gershon, M. (1998). *The second brain.* New York: Harper Perennial.

Gillett, L. (2001). *Sleep your way to success: Working smarter in the information age.* Hamilton Central Qld, Australia: Time Out Seminar Company.

Glaser, R., Kennedy, S., Lafuse, W., Bonneau, R., Speicher, C., Hillhouse, J., & Kiecolt-Glaser, J. (1990), Psychological stress-induced modulation of interleukin 2 receptor gene expression and interleukin 2 production in peripheral blood leukocytes. *Archives of General Psychiatry, 47*, 707–712.

Glaser, R., Lafuse, W., Bonneau, R., Atkinson, C., & Kiecolt-Glaser, J. (1993). Stress-associated modulation of proto-oncogene expression in human peripheral blood leukocytes. *Behavioral Neuroscience, 107*, 525–529.

Glass, L., & Mackey, M. (1988). *From clocks to chaos: The rhythms of life.* Princeton, NJ: Princeton University Press.

Glenn, S., Ellis, J., & Greenspoon, J. (1992). On the revolutionary nature of the operant as a unit of behavioral selection. *American Psychologist, 47*, 1329–1336.

Glik, D. (1993). Beliefs, practices, and experiences of spiritual healing adherents in an American industrial city. In W. Andritsky (Ed.), *Yearbook of cross-cultural medicine and psychotherapy, 1992* (pp. 199–223). Berlin: Verlag für Wissenschaft und Bildung.

Goldbeter, A. (1996). *Biochemical oscillations and cellular rhythms: The molecular bases of periodic and chaotic behavior.* Cambridge, UK: Cambridge University Press.

Goodenough, U. (1998). *The sacred depths of nature.* New York: Oxford University Press.

Goodwin, B. (1965). Oscillatory behavior in enzymatic control processes. In G. Weber (Ed.), *Advances in enzyme regulation:* Vol. 3 (pp. 425–438). Oxford, UK: Pergamon.

Gorton, B. (1949). The physiology of hypnosis: A review of the literature. *The Psychiatric Quarterly, 23*, 317–343, 457–485.

Gorton, B. (1957). The physiology of hypnosis: I. *Journal of the Society of Psychosomatic Dentistry, 4*(3), 86–103.

Gorton, B. (1958). The physiology of hypnosis: Vasomotor activity in hypnosis. *Journal of the American Society of Psychosomatic Dentistry, 5*(1), 20–28.

Gottheil, E. (Ed.). (1987). *Stress and addiction.* New York: Brunner/Mazel.

Gould E., Tanapat, P., McEwen, B., Flügge, G., & Fuchs, E. (1998). Proliferation of granule cell precursors in the dentate gyrus of adult monkeys is diminished by stress. *Proceedings of the National Academy of Sciences, 95*, 3168–3171.

Gould E., Tanapat, P., Reeves, A., & Shors, T. (1999a). Learning enhances adult neurogenesis in the hippocampal formation. *Nature Neuroscience, 2*(3), 260–265.

Gould E., Alison, J., Reeves, A., Michael, S., Graziano, S., & Gross, C. (1999b). Neurogenesis in the neocortex of adult primates. *Science, 286,* 548–552.

Gould, S., & Eldredge, N. (1977). Punctuated equilibria: the tempo and mode of evolution reconsidered. *Paleobiology, 3*, 115–151.

Granott, N., & Parziale, J. (Eds.). (in press). *Microdevelopment: Transition processes in development and learning.* New York: Cambridge University Press.

Grant, B. (1993). Evolution of Darwin's finches caused by a rare climatic event. *Proceedings of the Royal Society of London (B), 251*, 111–117.

Gray, J. (1994). *What you feel you can heal: A guide to enriching relationships.* Berkley, CA: Publisher Group One.

Green, R., & Green, M. (1987) Relaxation increases salivary immunoglobulin A. *Psychological Reports, 61*, 623–629.

Greene, B. (1999). *The elegant universe: Superstrings, hidden dimensions, and the quest for the ultimate theory.* New York: Norton.

Greenfield, S. (1994). A model explaining Brazilian spiritist surgeries and other unusual religious-based healing. *Subtle Energies, 5*, 109–141.

Greenfield, S. (2000, September). *Religious altered states and cultural-biological transduction in healing.* Paper presented at the Congresso Brasilerio de Etnopsiciatria, Fortaleza CE, Brazil.

Greenspan, R., Hall, J., & Tully, T. (1995). *Genes and behavior.* Princeton, NJ: Princeton University Press.

Greer, J., & Capecchi, M. (2002). Hoxb8 is required for normal grooming behavior in mice. *Neuron, 33,* 23–34.

Gross, C. (2000). Neurogenesis in the adult brain: Death of a dogma. *Nature Reviews Neuroscience, 1*, 67–73.

Gruber, H., & Wallace, D. (2001). Creative work: The case of Charles Darwin. *American Psychology, 56*, 346–349.

Guastello, S. (1995). *Chaos, catastrophy, and human affairs.* Hillsdale, NJ: Lawrence Erlbaum.

Gura, T. (2000). Tracing leptin's partners in regulating body weight. *Science, 287*, 1738–1741.

Hadamard, J. (1954). *The Psychology of invention in the mathematical field.* New York: Dover.

Hall, J. et al. (2001). Cellular imaging of zif-268 expression in the hippocampus and amygdala during contextual and cued fear memory retrieval: Selective of hippocampal CA1 neurons during the recall of contextual memories. *Journal of Neuroscience, 21*, 2186–2193.

Haley, J. (1963). *Strategies of psychotherapy.* New York: Grune & Stratton.

Haley, J. (1985). *Conversations with Milton H. Erickson, M.D.* New York: Norton.

Haley, J. (1993). Erickson hypnotic demonstration: 1964. In *Haley on Milton H. Erickson* (pp. 136–175). New York: Brunner/Mazel.

Hamadeh, H., & Afshari, C. (2000). Gene chips and functional genomics. *American Scientist, 88*, 508–515.

Hameroff, S., Kaszniak, A., & Scott, A. (1996). *Toward a science of consciousness: The first Tucson discussions and debates.* Cambridge, MA: MIT Press.

Hammond, D. (2001). Treatment of chronic fatigue with neurofeedback and self-hypnosis. *Neurorehabilitation, 16*(4), 295–300

Hammond, D. (2002). Integrating hypnosis and neurofeedback. *American Society of Clinical Hypnosis Newsletter, 43*(1), 3.

Hammond, K., & Diamond, J., (1997). Maximal sustained energy budgets in humans and animals. *Nature, 386*, 457–462.

Hardin, P., & Sehgal, A. (1999). Molecular components of a model circadian clock: Lessons from Drosophila. In R. Lydic & H. Baghdoyan (Eds.), *Handbook of behavioral state control: Cellular and molecular mechanisms* (pp. 61–74). Boca Raton: CRC.

Hargrove, J., & Berdanier, C. (1993). *Nutrition and gene expression.* Boca Raton: CRC.

Harrington, A. (Ed.). (1997). *The placebo effect.* Cambridge, MA: Harvard University Press.

Harris, G. (1948) Neural control of the pituitary gland. *Physiological Review, 28*, 139–179.

Harris, R., Porges, S., Clemenson Carpenter, M., & Vincenz, L. (1993). Hypnotic susceptibility, mood state, and cardiovascular reactivity. *American Journal of Clinical Hypnosis, 36*, 15–25.

Hautkappe, H., & Bongartz, W. (1992). Heart-rate variability as in indicator for posthypnotic amnesia in real and stimulating subjects. In E. Bongartz (Ed.), *Hypnosis: 175 years after Mesmer: Recent developments in theory and application.* Konstanz, Germany: Universitatsvergag.

Hebb, D. (1949). *The organization of behavior: A neuropsychological theory.* New York: Wiley.

Henig, R. (2000). *The monk in the garden: The lost and found genius of Gregor Mendel, the father of genetics.* New York: Houghton Mifflin.

Hetter, G., & Herhahn, F. (1983). Experience with "lipolysis": The Illouz technique of blunt suction lipectomy in North America. *Aesthetic Plastic Surgery, 7*(2), 69–76.

Hiatt, J., & Kripke, D. (1975). Ultradian rhythms in waking gastric activity. *Psychosomatic Medicine, 37*, 320–325.

Hilgard, E. (1965). *Hypnotic susceptibility.* New York: Harcourt, Brace.

Hilgard, E. (1973). The domain of hypnosis: With some comments on alternative paradigms. *American Psychologist, 28*, 972–982.

Hilgard, E. (1977). *Divided consciousness: Multiple controls in human thought and action.* New York: Wiley.

Hilgard, E. (1981). Hypnotic susceptibility scales under attack: An examination of Weitzenhoffer's criticisms. *The International Journal of Clinical and Experimental Hypnosis, 24*, 24–41.

Hilgard, E. (1982). Hypnotic susceptibility and implications for measurement. *International Journal of Clinical & Experimental Hypnosis, 30*, 4, 394–403.

Hilgard, E. (1991). Suggestibility and suggestions as related to hypnosis. In J. Schumaker (Ed.), *Human suggestibility: Advances in theory, research and application* (pp. 37–58). New York: Routledge.

Hilgard, E., & Hilgard, J. (1983). *Hypnosis in the relief of pain.* Los Altos, CA: William Kaufmann.

Hill, C. (1993). Is massage beneficial to critically ill patients in intensive care units? A critical review. *Intensive & Critical Care Nursing, 9*(2), 116–121.

Hobson, J. (1997). Consciousness as a state-dependent phenomenon. In J. Cohen & J. Schooler (Eds.), *Scientific approaches to the question of consciousness* (pp. 379–396). Mahwah, NJ: Lawrence Erlbaum.

Hobson, J. (1999). *Consciousness.* New York: Scientific American Library.

Hofstadter, D. (1995). *Fluid concepts and creative analogies.* New York: Basic.

Holden, C. (2002). Drugs and placebos look alike in the brain. *Science, 292,* 947–949.

Hollander, H., & Bender, S. (2001). ECEM (eye closure eye movements): Integrating aspects of EMDR with hypnosis for treatment of trauma. *American Journal of Clinical Hypnosis, 43*(3)/*43*(4), 187–202.

Hood, D. (2001). Plasticity in skeletal, cardiac, and smooth muscle. Invited review: Contractile activity-induced mitochondrial biogenesis in skeletal muscle. *Journal of Applied Physiology, 90,* 1137–1157.

Hoppenbrouwers, T. (1992). Ontogenesis of human ultradian rhythms. In D. Lloyd, & E. Rossi (Eds.), *Ultradian rhythms in life processes: A fundamental inquiry into chronobiology and psychobiology* (pp. 173–197). New York: Springer-Verlag.

Horgan, J. (1997). Profile: Ronald L. Graham, juggling act. *Scientific American, 276*(3), 28–30.

Hovind, H., & Nielsen S. (1974). Effect of massage on blood flow in skeletal muscle. *Scandinavian Journal of Rehabilitation Medicine, 6,* 74–77.

Howe, R., & Von Foerster, H. (1975). Introductory comments to Francisco Varela's calculus for self-references. *International Journal of General Systems, 2,* 1–3.

Hubbard, J., & Workman, E. (Eds.). (1998). *Handbook of stress medicine: An organ system approach.* New York: CRC Press.

Hull, C. (1933/1968). *Hypnosis and suggestibility.* New York: Appleton-Century-Crofts.

Hunziker, M., Saldana, R., & Neuringer, A. (1996). Behavioral variability in SHR and WKY rats as a function of rearing environment and reinforcement contingency. *Journal of the Experimental Analysis of Behavior, 65,* 129–144.

Huttunen, P., Hyypia, T., Vihinen, P., Nissinen, L., & Heino, J. (1998). Echovirus 1 infection induces both stress and growth-activated mitogen-activated protein kinase pathways and regulates the transcription of cellular immediate-early genes. *Virology, 250*(1), 85–93.

Hwang, J. (2001). Nutrient control of Insulin-stimulated glucose transport in 3T3-L1 adipocytes. In N. Moustaïd-Moussa & C. Berdanier (Eds.), *Nutrient-gene interactions in health and disease: The effect of macronutrients and micronutrients on gene expression* (pp. 163–175). Boca Raton, FL: CRC.

Huxley, A. (1970). *The perennial philosophy.* New York: Harper & Row.

Hyden, H., & Egyhazi, E. (1963). Glial RNA changes during a learning experiment in rats. *Proceedings of the National Academy of Sciences, U.S., 49,* 618–624.

Iino, M., Goto, K., Kakegawa, W., Okado, H., Sudo, M., Ishiuchi, S., Miwa, A., Takayasu, Y., Saito, I., Tsuzuki, K., & Ozawa, S. (2001). Glia-synapse interaction through Ca^{2+} permeable AMPA receptors in bergmann glia. *Science, 292,* 926–929.

Incyte Corporation. (1999). UniGEMTMV experiment analysis: Incyte's time course expression experiment [On-line]. Available: http://gem.incyte.com/gem/data/unigemv/UniGEMVexperiment.pdf

Iranmanesh, A., Lizarralde, G., Johnson, M., & Veldhuis, J. (1989). Circadian, ultradian, and episodic release of B-endorphin in men, and its temporal coupling with cortisol. *Journal of Clinical Endocrinology & Metabolism, 68*, 1019–1026.

Iranmanesh, A., Veldhuis, J., Johnson, M., & Lizarralde, G. (1989). 24-hour pulsatile and circadian patterns of cortisol secretion in alcoholic men. *Journal of Andrology, 10*, 54–63.

Ironson, G., Field, T., Scafidi, F., Hashimoto, M., Kumar, M., Kumar, A., Price, A., Goncalves, A., Burman, I., Tetenman, C., Patarca, R., & Fletcher, M. (1996). Massage therapy is associated with enhancement of the immune system's cytotoxic capacity. *International Journal of Neuroscience, 84*(1–4), 205–217.

Izquierdo, I., Netto, C., Dalmaz, D., Chaves, M., Pereira, M., & Siegfried, B. (1988). Construction and reconstruction of memories. *Brazilian Journal of Medical and Biological Research, 21*, 9–25.

Jackson, E. (2000). The unbounded vistas of science: Evolutionary limitations. *Complexity, 5*(5), 35–44.

James, W. (1890). *The principles of psychology.* New York: Dover

Janet, P. (1925). *Psychological healing: A historical and clinical study.* (E. Paul & C. Paul, Trans.). New York: Macmillan.

Jenkins, D. (1989). Nibbling versus gorging: Metabolic advantages of increased meal frequency. *The New England Journal of Medicine, 321*, 929–935.

John, R., & Surani, M. (2000). Genomic imprinting, mammalian evolution, and the mystery of egg-laying mammals. *Cell, 101*, 585–588.

Johnson, J. (1999, November, 17). *The effects of chronic, high-glucose exposure on total gene expression in human pancreatic islets as measured by Affymetrix GeneChip Probe Array.* Affymetrix Technology Symposium, Santa Clara, CA.

Johnston, D., & Maio-sin Wu, S. (1995). *Foundations of cellular neurophysiology.* Cambridge, MA: MIT Press.

Jones, M. et al., (2001). A requirement for the immediate early gene Zif268 in the expression of late LTP and long term memories. *Nature Neuroscience, 4*, 289–296.

Jordan, K. (2001). Fats for life. *Life Extension, 7*, 20–30.

Jung, C. (1916). The transcendent function. In R. F. C. Hull (Trans.), *The collected works of C. G. Jung: Vol. 8. The structure and dynamics of the psyche* (pp. 67–91). Princeton, NJ: Princeton University Press.

Jung, C. (1923). *Psychological types or the psychology of individuation.* New York: Pantheon.

Jung. C. (1943/1996). The synthetic or constructive method. In R. C. F. Hull (Trans.), *The collected works of C. G. Jung: Vol. 7. Two essays on analytical psychology* (pp. 80-89). Princeton, NJ: Princeton University Press.

Jung, C. (1950). *The collected works of C. G. Jung: Vol. 18. The symbolic life.* (R. F. C. Hull, Trans.). Princeton, NJ: Princeton University Press.

Jung, C. (1953). *The collected works of C. G. Jung: Vol. 12. Psychology and alchemy.* (R. F. C. Hull, Trans.). Princeton, NJ: Princeton University Press.

Jung, C. (1960). *The collected works of C. G. Jung: Vol. 8. The structure and dynamics of the psyche.* (R. F. C. Hull, Trans.). Princeton, NJ: Princeton University Press.

Jung, C. (1966). *The collected works of C. G. Jung: Vol. 7. Two essays on analytical psychology.* (R. F. C. Hull, Trans.). Princeton, NJ: Princeton University Press.

Kalt, H. (2000). Psychoneuroimmunology: An interpretation of experimental case study evidence towards a paradigm for predictable results. *American Journal of Clinical Hypnosis, 43*, 1, 41–52.

Kandel, E. (1983). From metapsychology to molecular biology: Explorations into the nature of anxiety. *American Journal of Psychiatry, 140*(10), 1277–1293.

Kandel, E. (1989). Genes, nerve cells, and the remembrance of things past. *Journal of Neuropsychiatry, 1*(2), 103–125.

Kandel, E. (1998). A new intellectual framework for psychiatry? *American Journal of Psychiatry, 155,* 457–469.

Kandel, E. (2000). Cellular mechanisms of learning and the biological basis of individuality. In E. Kandel, J. Schwartz, & T. Jessell, (Eds.), *Principles of neural science* (4th ed, pp. 1247–1279). New York: Elsevier.

Kandel, E. (2001). The molecular biology of memory storage: A dialogue between genes and synapses. *Science, 294,* 1030–1038.

Katz, L., Nathan, L., Kuhn, C., & Schanberg, S. (1996). Inhibition of GH in maternal separation may be mediated through altered serotonergic activity at 5-HT2a and 5-HT2c receptors. *Psychoneuroendocrinology, 21*, 219–235.

Kaufer, D., Friedman, A., Seidman, S., & Soreq, H. (1998). Acute stress facilitates long-lasting changes in cholinergic gene expression. *Nature, 393,* 373–377.

Kauffman, S. (1993). *The origins of order: Self-organization and selection in evolution.* Oxford: Oxford University Press.

Kauffman, S. (1995). *At home in the universe: The search for the laws of self-organization and complexity.* Oxford: Oxford University Press.

Keeney, B. (1999–2000). *Profiles in healing.* (Vols. 1–4). Philadelphia, PA: Ringing Rock.

Kelner, K., & Bloom, F. (2000). *The best of science: Neuroscience.* Washington, DC: American Association for the Advancement of Science.

Kempermann, G., & Gage, F. (1999). New nerve cells for the adult brain. *Scientific American, 280*, 48–53.

Kempermann, G., Kuhn, G., & Gage, F. (1997). More hippocampal neurons in adult mice living in an enriched environment. *Nature, 386*, 493–495.

Kennedy, B., Ziegler, M., & Shannahoff-Khalsa, D. (1986). Alternating lateralization of plasma catecholamines and nasal patency in humans. *Life Sciences, 38*, 1203–1214.

Kersten, S., Desvergne, B., & Wahli, W. (2000). Roles of PPARs in health and disease. *Nature, 405*, 421–424.

Kiger, A., Jones, D., Schulz, C., Rogers, M., & Fuller, M. (2001). Stem cell self-renewal specified by JAK-STAT activation in response to a support cell cue. *Science, 294*, 2542–2545.

Kihlstrom, J. (1980). Posthypnotic amnesia for recently learned material: Interactions with "episodic" and "semantic" memory. *Cognitive Psychology, 12*, 227–251.

Kihlstrom, J. (2000). *Hypnosis and pain: Time for a new look* [On-line]. Available: http://www.institute-shot.com/hypnosis_pain_newlook.htm

Kimura, M. (1983). *The neutral theory of molecular evolution.* New York: Cambridge University Press.

Kirsch, I. (2000). The response set theory of hypnosis. *American Journal of Clinical Hypnosis, 42*(3)/*42*(4), 274–292.

Kirsch, I., & Lynn, S. (1995). The altered state of hypnosis: Changes in the theoretical landscape. *American Psychologist, 50*(10), 846–858.

Kirsch, I., & Lynn, S. (1999). Automaticity in clinical psychology. *American Psychologist, 54*, 504–515.

Kirsch, I., & Saperstein, G. (1999). Listening to Prozac but hearing placebo: A meta-analysis of antidepressant medication. In I. Kirsch (Ed.), *How expectancies shape behavior* (pp. 303–320). Washington, DC: American Psychological Association.

Kirsch, J. (1966). *Shakespeare's royal self.* New York: C. G. Jung Foundation for Analytical Psychology.

Klein, M. (1980). *Mathematics, the loss of certainty.* New York: Oxford University Press.

Klein, R., & Armitage, R. (1979). Rhythms in human performance: 1 1/2 hour oscillations in cognitive style. *Science, 204*, 1326–1328.

Klein, R., Pilon, D., Prosser, S., & Shannahoff-Khalsa, D. (1986). Nasal airflow asymmeteries and human performance. *Biological Psychology, 23*, 127–137.

Kleitman, N. (1963). *Sleep and wakefulness as alternating phases in the cycle of existence.* Chicago, IL: University of Chicago Press.

Kleitman, N. (1969). Basic rest-activity cycle in relation to sleep and wakefulness. In

A. Kales (Ed.), *Sleep: Physiology and pathology* (pp. 33–38). Philadelphia, PA: Lippincott.

Kleitman, N. (1970). Implications of the rest-activity cycle: Implications for organizing activity. In E. Hartmann (Ed.), *Sleep and dreaming.* Boston: Little, Brown.

Kleitman, N., & Rossi, E. (1992). The basic rest-activity cycle—32 years later. An interview with Nethaniel Kleitman at 96. In D. Lloyd & E. Rossi (Eds.), *Ultradian rhythms in life processes: A fundamental inquiry into chronobiology and psychobiology* (pp. 303–306). New York: Springer-Verlag.

Klevecz, R. (1988) Time and life incorporated. *Cell, 53,* 499–501.

Klevecz, R., & Braly, P. (1987) Circadian and ultradian rhythms of proliferation in human ovarian cancers. *Chronobiology International, 4,* 513–523.

Koch, C., & Crick, F. (2001). The zombie within. *Nature, 411,* 893.

Kohara, K., Kitamura, A., Morishima, M., & Tsumoto, T. (March, 2001). Activity-dependent transfer of brain-derived neurotrophic factor to postsynaptic neurons. *Science, 291,* 2419–2423.

Koonin, E., Aravind, L., & Kondrashov, A. (2000). The impact of comparative genomics on our understanding of evolution. *Cell, 101,* 573–576.

Krech, D., Rosenzweig, M., & Bennett, E. (1964). Chemical and anatomical plasticity of the brain. *Science, 146,* 610–619.

Kreiman, G., Koch, C., & Itzhak, F. (2000). Imagery neurons in the human brain. *Nature, 408,* 357–361.

Kripke, D., Garfinkel, L., Wingard, D., Klauber, M., & Marler, M. (2002). Mortality associated with sleep duration and insomnia. *Archives of General Psychiatry, 59,* 131–136.

Kronauer, R. (1984). Modeling principles for human circadian rhythms. In M. Moore-Ede, & C. Czeisler (Eds.), *Mathematical models of the circadian sleep-wake cycle* (pp. 105–128). New York: Raven.

Kucherlapati, R., & DePinho, R. (2001). Telomerase meets its mismatch. *Nature, 411,* 647–648.

Kuhl, D. (2000). Learning about activity-dependent genes. In M. Baudry, J. Davis, & R. Thompson (Eds.), *Advances in synaptic plasticity* (pp. 1–31). Cambridge, MA: MIT Press.

Kuhn, C., & Schanberg, S. (1998). Responses to maternal separation: mechanisms and mediators. *International Journal of Developmental Neuroscience, 16,* 261–270.

Kuhn, C., Schanberg S., Field T., Symanski, R., Zimmermann E., Scafidi, F., & Roberts, J. (1991). Tactile-kinetic stimulation effects on symapthetic and adrenocortical function in preterm infants. *Journal of Pediatrics, 119,* 434–40.

Kupfer, D., Monk, T., & Barchas, J. (1988). *Biological rhythms and mental disorders*. New York: Guilford.

Kupfermann, I., Kandel, E., & Iversen, S. (2000). Motivational and addictive states. In Kandel, E., Schwartz, J., & Jessell, T. (Eds.), *Principles of neural science* (4th ed., pp. 999–1013). New York: Elsevier.

LaBar, K., & Disterhoft, J. (1998). Conditioning, awareness, and the hippocampus. *Hippocampus, 8*, 620–626.

LaBerge, S. (1985). *Lucid dreaming: The power of being awake & aware in your dreams*. Los Angeles: Tarcher.

LaBerge, S. (1990). Lucid dreaming: Psychophysiological studies of consciousness during REM sleep. In J. Bootsen, J. Kihlstrom, & D. Schacter (Eds.), *Sleep and cognition* (pp. 109–126). Washington, DC: American Psychological Association.

LaBerge, S., & Rheingold, H. (1990). *Exploring the world of lucid dreaming*. New York: Ballantine.

Lakoff, G., & Johnson, M. (1999). *Philosophy in the flesh*. New York: Basic.

Lakoff, G., & Núñez, R. (2000). *Where mathematics came from: How embodied mind brings mathematics into being*. New York: Basic.

Lander, E., & Weinberg, R., (2000). Genomics: Journey to the center of biology. *Science, 287*, 1777–1782.

Langton, S., & Langton, C. (1983). The answer within: A clinical framework of *Ericksonian hypnotherapy*. New York: Brunner/Mazel.

Lavie, P., & Kripke, D. (1977). Ultradian rhythms in urine flow in waking humans. *Nature, 269*, 142–144.

Lazarus, A., & Mayne, T. (1991). Relaxation: Some limitations, side effects, and proposed solutions. *Psychotherapy, 27*, 261–266.

Lee, C., Klopp, R., Weindruch, R., & Prolla, T. (1999). Gene expression profile of aging and its retardation by caloric restriction. *Science, 285*, 1390–1393.

Lei, X. (2001). Differential regulation and function of glutatione peroxidases and other selenoproteins. In N. Moustaïd-Moussa & C. Berdanier (Eds.), *Nutrient-gene interactions in health and disease: The effect of macronutrients and micronutrients on gene expression* (pp. 425-448). Boca Raton, FL: CRC.

Lendval, B., Stern, E., Chen, B., & Svoboda, K. (2000). Experience-dependent plasticity of dendritic spines in the developing rat barrel cortex in vivo. *Nature, 404*, 876–880.

Lenski, R., & Travisano, M. (1994). Dynamics of adaptation and diversification: A 10,000 generation experiment with bacterial populations. *Proceedings of the National Academy of Sciences, USA, 91*, 6808–6814.

Leslie, M. (2000). When cells listen to the wrong hormones, prostate cancer turns deadly [On-line]. Available: http://www.stanford.edu/dept/news/report/news/june28/prostate-628.html

Levine, J., Kurtz, R., & Lauter, J. (1984). Hypnosis and its effects on left and right hemisphere activity. *Biological Psychiatry, 19*, 1461–1475.

Lewontin, R. (2000). *The triple helix: Genes, organism, and environment.* Cambridge, MA: Harvard University Press.

Libet, B. (1993). *Neurophysiology of consciousness: Selected papers and new essays.* Boston: Birkhäuser.

Lisman, J., & Morris, G. (2001). Why is the cortex a slow learner? *Nature, 411*, 248–249.

Lippincott, B. (1992). Owls and larks in hypnosis: Individual differences in hypnotic susceptibility relating to biological rhythms. *American Journal of Clinical Hypnosis, 34*, 185–192.

Lippincott, B. (1993). The temperature rhythm and hypnotizability: A brief report. *Contemporary Hypnosis, 10*, 155–158.

Lloyd, D., & Rossi, E. (Eds.). *Ultradian rhythms in life processes: A fundamental inquiry into chronobiology and psychobiology.* New York: Springer-Verlag.

Lloyd, D., & Rossi, E. (1993). Biological rhythms as organization and information. *Biological Reviews, 68*, 563–577.

Lockhart, D., & Winzeler, E. (2000). Genomics, gene expression and DNA arrays. *Nature, 405*, 827–836.

Loewenstein, W. (1999). *The touchstone of life: Molecular information, cell communication, and the foundations of life.* New York: Oxford University Press.

Lok, C. (2001). Picture perfect. *Nature, 412*, 372–374.

Lovelock, J. (1988). *The ages of Gaia: A biography of our living earth.* New York: Norton.

Lüscher, C., Nicoll, R., Malenka, C., & Muller, D. (2000). Synaptic plasticity and dynamic modulation of the postsynaptic membrane. *Nature Neuroscience, 3*, 545–567.

Lydic, R. (Ed.). (1998). *Molecular regulation of arousal states.* Boca Raton: CRC.

Lydic, R., & Baghdoyan, H. (Eds.). *Handbook of behavioral state control: Cellular and molecular mechanisms.* New York: CRC.

Lynch, J. (2000). *A cry unheard: New insights into the medical consequences of loneliness.* Baltimore, MD: Bancroft.

Lynn, S., & Sherman, S. (2000). The clinical importance of sociocognitive models of hypnosis. *American Journal of Clinical Hypnosis, 42(3)/42(4)*, 294–315.

Lyubomirsky, S. (2001). Why are some people are happier than others? *American Psychologist, 56*, 239–249.

Macilwain, C. (2000). AAAS members fret over links with theological foundation. *Nature, 403*, 819.

Macklis, J. (2001). New memories from new neurons. *Nature, 410,* 314–315.

Magavi, S., Leavitt, B., & Macklis, J. (2000). Induction of neurogenesis in the neocortex of adult mice. *Nature, 405,* 951–955.

Mainzer, K. (1994). *Thinking in complexity: The complex dynamics of matter, mind, and mankind.* New York: Springer-Verlag.

Malarkey, W., Glaser, R., Kiecolt-Glaser, J., & Marucha, P. (2001). Behavior: The endocrine-immune interface and health outcomes. *Advances in Psychosomatic Medicine, 22,* 104–115.

Maldonado, R., Blendy, J. A., Tzavara, E., Gass, P., Roques, B. P., Hanoune, J., & Schutz, G. (1996). Reduction of morphine abstinence in mice with a mutation in the gene encoding CREB. *Science, 273*(5272), 657–659.

Malenka, R., & Nicoll, R., (1999). Long-term potentiation—A decade of progress? *Science, 285*(5435), 1870–1874.

Mangun, G., Jha, A., Hopfinger, J., & Handy, T. (2000). The temporal dynamics and functional architecture of attentional processes in human extrastriate cortex. In Gazzaniga, M. (Ed.), *The new cognitive neurosciences* (2nd ed., pp. 121–137) Cambridge, MA: MIT Press.

Mann, B., & Sanders, S. (1995). The effects of light, temperature, trance length and time of day on hypnotic depth. *American Journal of Clinical Hypnosis, 37*(3), 43–53.

Maquet, P., Laureys, S., Peigneux, P., Fuchs, S., Petiau, C., Phillips, C., Aerts, A., Del Fiore, C., Degueldre, G., Meulemans, T., Luxen, A., Granck, G., Van Der Linden, M., Smith, C., & Cleeremans, A. (2000). Experience-dependent changes in cerebral activation during human REM sleep. *Nature Neuroscience, 3*, 831–836.

Marijuan, P. (1996). The cell as a problem-solving "engine." In R. Cuthbertson, M. Holcombe, & R. Paton (Eds.), *Computation in cellular and molecular biological systems*, 183-194. London: World Scientific.

Margulis, L. (1998). *Symbiotic planet: A new look at evolution.* New York: Basic.

Margulis, L., & Sagan, D. (1986). *Microcosmos: Four billion years of evolution from our microbial ancestors.* New York: Simon & Schuster.

Marks-Tarlow, T., & Martininez, M. (2002) Commemorative: Francisco Varela (1946–2001). *Society of Chaos Theory in Psychology and Life Sciences Newsletter, 9,* 3–5.

Marmont, M., Davey Smith, G., Stansfield, S., Patel, C., North, F., & Head, J. (1991). Health inequalities among British civil servants: The Whitehall II study. *Lancet, 337*, 1387–1393.

Martin, K., Bartsch, D., Bailey, C., & Kandel, E. (2000). Molecular mechanisms underlying learning-related long-lasting synaptic plasticity. In M. Gazzaniga (Ed.), *The new cognitive neurosciences* (2nd ed., pp. 121–137). Cambridge, MA: MIT Press.

Martindale, C. (2001). Oscillations and analogies: Thomas Young, MD, FRS, genius. *American Psychology, 56,* 342–345.

Maslow, A. (1962). *Toward a psychology of being.* New York: Van Nostrand.

Maslow, A. (1967). A theory of meta motivation: The biological rooting of the value-life. *Journal of Humanistic Psychology, 7,* 93–127.

Masterpasqua, F., & Perna, P. (1997). *The psychological meaning of chaos: Translating theory into practice.* Washington, DC: American Psychological Association.

Matthews, W. (1985). Indirect versus direct hypnotic suggestions—an initial investigation: A brief communication. *The International Journal of Clinical and Experimental Hypnosis, 34,* 219–223.

Matthews, W. (2000). Ericksonian approaches to hypnosis and therapy: Where are we now? *The International Journal of Clinical and Experimental Hypnosis, 48,* 418–426.

Matus, A. (2000). Actin-based plasticity in dendritic spines. *Science, 290,* 754–758.

Mayford, M., Bach, M., Huang, Y., Wang, L., Hawkins, R., & Kandel, E. (1996). Control of memory formation through regulated expression of a CaMKll transgene. *Science, 274,* 1678–1683.

Mayr, E. (1988). *Toward a new philosophy of biology: Observations of an evolutionist.* Cambridge, MA: Harvard University Press.

McEwin, B. (2000). Stress, sex, and the structural and functional plasticity of the hippocampus. In M. Gazzaniga (Ed.), *The new cognitive neurosciences* (2nd ed., pp. 171–197). Cambridge, MA: MIT Press.

McGuffin, P., Riley, B., & Plomin, R. (2001). Toward a behavioral genetics. *Science, 291,* 1232–1249.

McGaugh, J. (1989). Involvement of hormonal and neuromodulatory systems in the regulation of memory storage. *Annual Reviews Neuroscience, 12,* 255–287.

McGaugh, J. (2000). Memory—A century of consolidation. *Science, 287,* 248–251.

McKenzie, G., Peter L., Lombardo, M., Hastings, P., & Rosenberg, S. (2001). SOS mutator DNA polymerase IV functions in adaptive mutation and not adaptive amplification. *Molecular Cell, 7,* 571–579.

McLaren, A. (2000). Cloning: Pathways to a pluripotent future. *Science, 288,* 1775–1780.

McKechnie, A., Wilson, F., Watson, N., & Scott, D., 1983. Anxiety states: a preliminary report on the value of connective tissue massage. *Journal of Psychosomatic Research, 27*(2), 125-129.

McKernan, M., & Shinnick-Gallagher, P. (1997). Fear conditioning induces a lasting potentiation of synaptic currents in vitro. *Nature, 390*, 607-611.

Meek, S. (1993). Effects of slow stroke back massage on relaxation in hospice clients. *Journal of Nursing Scholarship, 25*(1), 17–21.

Meier-Koll, A., (1992). Ultradian behavior cycles in humans: Developmental and social aspects. In D. Lloyd & E. Rossi (Eds.), *Ultradian rhythms in life processes: A fundamental inquiry into chronobiology and psychobiology* (pp. 243–282). New York: Springer-Verlag.

Mejean, L., Bicakova-Rocher, A., Kolopp, M., Villaume, C., Levi, F., Debry, G., Reinberg, A., & Drouin, P. (1988). Circadian and ultradian rhythms in blood glucose and plasma insulin of healthy adults. *Chronobiology International, 5*, 227–236.

Mendel, G. (1865/1965). *Experiments in plant hybridization.* Cambridge, MA: Harvard University Press.

Merchant, K. (1996). *Pharmacological regulation of gene expression in the CNS.* Boca Raton, FL: CRC.

Merchant-Nancy, H., Vazquez, J., Aguilar-Roblero, R., & Drucker-Colin, R. (1992). C-fos proto-oncogene changes in relation to REM sleep duration. *Brain Research, 579*, 342–346.

Merchant-Nancy, H., Vazquez, J., Garcia, F., & Drucker-Colin, R. (1995). Brain distribution of c-fos expression as a result of prolonged rapid eye movement (REM) sleep period duration. *Brain Research, 681*, 15–22.

Metzgar, D., & Wills, C. (2000). Evidence for the adaptive evolution of mutation rates. *Cell, 101*, 581–584.

Miller, G. (1956). The magic number seven, plus or minus two: Some limits on our capacity for processing information. *Psychological Review, 63*, 81–97.

Miller, N., & Neuringer, A. (2000). Reinforcing variability in adolescents with autism. *Journal of Applied Behavior Analysis, 33*, 151–165.

Milner, B., Squire, L., & Kandel, E. (1998). Cognitive neuroscience and the study of memory. *Neuron, 20*, 445–468.

Moerman, D., & Jonas, W. (2000). Toward a research agenda on placebo. *Advances in mind-body medicine, 16*, 7–46. [Special edition]

Moller, H., & Volz, H. (1996). Drug treatment of depression in the 1990s. *Drugs, 52*, 625–638.

Montgomery, G., DuHamel, K., & Redd, W. (2000). A meta-analysis of hypnotically induced analgesia: How effective is hypnosis? *The International Journal of Clinical and Experimental Hypnosis, 48*, 138–153.

Moore-Ede, M., Sulzman, F., & Fuller, C. (1982). *The clocks that time us.* Cambridge, MA: Harvard University Press.

Morimoto, R. (2001). Stress induced gene expression [On-line]. Available: www.grc.uri.edu/programs/2001/stress.htm

Morimoto, R., & Jacob, S. (1998). *Stress-inducible gene expression conference: Book of abstracts.* New York: Cognizant Communication Corp.

Morgan, T. (1910). Sex limited inheritance in Drosophilia. *Science, 32*, 120–122.

Morgan, T. (1911). The origin of five mutations in eye color in Drosophilia and their modes of inheritance. *Science, 33*, 534–537.

Morimoto, R., Tissières, A., & Georgopoulos, C. (1990). *Stress proteins in biology and medicine.* Woodbury, New York: Cold Spring Harbor Press.

Morowitz, H., & Singer, J. (1995). The mind, the brain, and complex adaptive systems. *Proceedings of the Santa Fe Institute Studies in the Sciences of Complexity.* New York: Addison-Wesley.

Morris, K., Wang, Y., Kim, S., & Moustaïd-Moussa, N. (2001). Dietary and hormonal regulation of the mammalian fatty acid synthase gene. In N. Moustaïd-Moussa, & C. Berdanier (Eds.). *Nutrient-gene interactions in health and disease: The effect of macronutrients and micronutrients on gene expression* (pp. 1–23). Boca Raton, FL: CRC.

Morrison, R., & Farmer, S. (2001). Nutrition and adipocyte gene expression. In N. Moustaïd-Moussa, & C. Berdanier (Eds.), *Nutrient-gene interactions in health and disease: The effect of macronutrients and micronutrients on gene expression* (pp. 25–48). Boca Raton, FL: CRC.

Moustaïd-Moussa, N., & C. Berdanier (Eds.). (2001). *Nutrient-gene interactions in health and disease: The effect of macronutrients and micronutrients on gene expression.* Boca Raton, FL: CRC.

Murray, A., Solomon, M., & Kirschner, M. (1989). The role of cyclin synthesis and degradation in the control of maturation promoting factor activity. *Nature, 339*, 280–286.

Nader, K., Schafe, G., & Le Doux, J. (2000a). Fear memories require protein synthesis in the amygdala for reconsolidation after retrieval. *Nature, 406*, 722–726.

Nader, K., Schafe, G., & Le Doux, J. (2000b). The labile nature of consolidation theory. *Nature Reviews: Neuroscience, 1*, 216–219.

Naish, P. (Ed.). (1986). *What is hypnosis? Current theories and research*. Philadelphia, PA: Open University Press, Milton Keynes.

Nakamura, J., & Csikszentmihalyi, M. (2001). Catalytic creativity: The case of Linus Pauling. *American Psychology, 56*, 337–341.

Nash, M. (2001). The truth and the hype of hypnosis. *Scientific American, 285*, 47–55.

Neisser, U. (1967). *Cognitive psychology*. New York: Appleton-Century-Crofts.

Nester, E. (2001). Molecular basis of long-term plasticity underlying addiction. *Nature Reviews: Neuroscience, 2*, 119–128.

Neumann, E. (1959). *Art and the creative unconscious*. Princeton: Princeton University Press.

Neumann, E. (1962). *The origins and history of consciousness* (Vols. I & 2). New York: Harper Torchbook.

Neumann, E. (1968). Mystical man. In J. Campbell (Ed.), *The mystic vision. Papers from the Eranos yearbooks, No. 6* (pp. 375-415). New York: Pantheon.

Neuringer, A., Kornell, N., & Olufs, M. (2001). Stability and variability in extinction. *Journal of Experimental Psychology-Animal Behavior Processes, 27*, 79–94.

Newberg, A., & d'Aquili, E. (2001). *Why God won't go away*. New York: Ballantine.

Newman, J. (2001). "I have seen cancers disappear" (Steven Rosenberg as interviewed by Newman). *Discover, 22*, 44–51.

Newtson, D. (1994). The perception and coupling of behavior waves. In R. Vallacher & J. Nowak (Eds.), *Dynamical systems in social psychology*. New York: Academic.

Nguyen, H., Le, L., Tran, M., Nguyen, T., & Nguyen, N. (1995). Chomassi: A therapy advice system based on chrono-massage and acupression using the method of ZiWuLiuZhu. *Medinfo, 8*, Pt 2, 998.

Nørretranders, T. (1998). *The user illusion*. New York: Viking.

Numan, M., Numan, M. J., Marzella, S. R., & Palumbo, A. (1998). Expression of c-fos, fos B, and egr-1 in the medial preoptic area and bed nucleus of the stria terminalis during maternal behavior in rats. *Brain Research, 792*(2), 348–352.

Núñez, R., & Freeman, W. (2000). *Reclaiming cognition: The primacy of action, intention, and emotion*. New York: Imprint Academic.

Nusse, H., & York, J. (1998). *Dynamics: Numerical explorations*. New York: Springer-Verlag.

Odum, H. (1988). Self-organization, transformity, and information. *Science, 242*, 1132–1139.

Offer, D. (2000). Memories are made of this and that. *Science, 288*, 1961.

Okura, M., Honda, K., Riehl, J., Mignot, E., & Nishnio, S. (1999). Roles of diencephalic dopaminergic cell groups in regulation of cataplexy in canine narcolepsy. *Sleep, 22,* 1.

Oliet, S., Piet, R., & Poulain, D. (2001). Control of glutamate clearence and synaptic efficiency by glial coverage of neurons. *Science, 292,* 923–926.

Osowiec, D. (1992). *Ultradian rhythms in self-actualization, anxiety, and stress-related somatic symptoms.* Unpublished doctoral dissertation, California Institute of Integral Studies, San Francisco, CA.

O'Neill, M., Hicks, C., Shaw, G., Parameswaran, T., Cardwell, G., & O'Neill, M. (1998). Effects of 5-hydroxytryptamine2 receptor antagonism on the behavioral activation and immediate early gene expression induced by dizocilpine. *Journal of Pharmacological Experimental Therapy, 287*(3), 839–846.

Oswald, S., Wallace, C., & Worley, P. (2000). Synaptic regulation of messenger RNA Trafficking within neurons. In M. Baudry, J. Davis, & R. Thompson (Eds.), *Advances in synaptic plasticity* (pp. 33–52). Cambridge, MA: MIT Press.

Othmer, S., Othmer, S., & Kaiser, D. (In Press). EEG biofeedback: A generalized approach to neuroregulation [On-line]. Available: www.minderlabs.com/kall.htm

Otto, R. (1923/1950). *The idea of the holy.* New York: Oxford University Press.

Ottoson, D., Ekblom, A., & Hansson, P. (1981). Vibratory stimulation for the relief of pain of dental origin. *Pain, 10*(1), 37–45.

Palmeirin, I., Henrique, D., Ish-Horowicz, D., & Pourquié, O. (1997). Avian hairy gene expression identifies a molecular clock linked to vertebrate segmentation and somitogenesis. *Cell, 91,* 639–648.

Palmer, D., & Donahoe, J. (1992). Essentialism and selectionism in cognitive science and behavior analysis. *American Psychologist, 47,* 1344–1358.

Papez, J, (1937). A proposed mechanism of emotion. *Archives of Neurology & Psychiatry, 38,* 725–744.

Pardue, M., Feramisco, J., & Lindquist, S. (1989). *Stress induced proteins.* New York: Alan Liss.

Pavlov, I. (1927). *Conditioned reflexes: An investigation of the physiological activity of the cerebral cortex.* London: Oxford University Press.

Peitgen, H., Jürgens, H., & Saupe D., (1992). *Chaos and fractals: New frontiers of science.* New York: Springer-Verlag.

Pennebaker, J. (1997). *Opening up: The healing power of expressing emotions* (Rev. ed.). New York: Guilford.

Pert, C., Ruff, M., Weber, R., & Herkenham, M. (1985). Neuropeptides and their receptors: A psychosomatic network. *The Journal of Immunology, 135*(2), 820s–826s.

Pert, C., & Ruff, M., Spencer, D., & Rossi, E. (1989). Self-reflective molecular Psychology. *Psychological Perspectives, 20*(1), 213–221.

Peter, B., & Revenstorf, D. (2000). Commentary on Matthews "Ericksonian approaches to hypnosis and therapy: Where do we go now? *The International Journal of Clinical and Experimental Hypnosis, 48*, 433–437.

Petitto, L., & Zatorre, R. (2000). Speech-like cerebral activity in profoundly deaf people processing signed languages: Implications for the neural basis of human language. *Proceedings of the National Academy of Sciences, USA, 97*, 13961–13096.

Petrovic, P., Kalso, E., Petersson, K., & Ingvar, M. (2002). Placebo and opioid analgesia—imaging a shared neuronal network. *Science, 295*, 1737–1740.

Pfays, J. G., & Heeb, M. M. (1997). Implications of immediate-early gene induction in the brain following sexual stimulation of female and male rodents. *Brain Research Bulletin, 44*(4), 397–407.

Phelps, M. (2001). Positron emission tomography provides molecular imaging of biological processes. *Proceedings of the National Academy of Sciences, USA, 97*, 9226–9233.

Piaget, J. (1954). *The construction of reality in the child.* New York: Basic.

Pisella, L., Gréa, H., Tilikete, C., Vighetto, A., Desmurget, M., Rode, G., Boisson, D., & Rossetti, Y. (2000). An "automatic pilot" for the hand in human posterior parietal cortex: Toward reinterpreting optic ataxia. *Nature Neuroscience, 3*, 729–736.

Plomin, R., DeFries, J., McClearn, G., McGuffin, P. (2001). *Behavioral genetics.* New York: Freeman.

Polston, E. K., & Erskine, M. S. (1995). Patterns of induction of the immediate-early genes c-fos and egr-1 in the female rat brain following differential amounts of mating stimulation. *Neuroendocrinology, 62*(4), 370–384.

Pompeiano, M., Cirelli, C., & Tononi, G. (1994). Immediate-early genes in spontaneous wakefulness and sleep: Expression of c-fos and NGFI-A mRNA and protein. *Journal of Sleep Research, 3*, 80–96.

Pompeiano, M., Cirelli, C., & Tononi, G. (1998). Reverse transcription mRNA differential display: A systematic molecular approach to identify changes in gene expression across the sleep-waking cycle. In R. Lydic (Ed.), *Molecular regulation of arousal states* (pp. 157-166). Boca Raton: CRC.

Poon, C., & Merrill, C. (1997). Decrease of cardiac chaos in congestive heart failure. *Nature, 389*, 492–495.

Porkka-Heiskanen, T., Toppila, J., & Stenberg, D. (1998). In situ hybridization of messenger RNA in sleep research. In R. Lydic (Ed.), *Molecular regulation of arousal states* (pp. 167–179). New York: CRC.

Poundstone, W. (1985). *The recursive universe: Cosmic complexity and the limits of scientific knowledge.* Chicago: Contemporary Books.

Powell, G., Brasel, J., & Hansen, J. (1967a). Emotional deprivation and growth retardation simulating idiopathic hypopituitiarism: I. Clinical evaluation of the syndrome. *New England Journal of Medicine, 276*, 1271–1278.

Powell, G., Hopwood, N., & Baratt, E. (1973). Growth hormone studies before and during catch-up growth in a child with emotional deprivation and short stature. *Journal of Clinical Endocrinology and Metabolism, 37*, 674–679.

Pratt, M. (1993). *The use of ideodynamic concepts with children for personal adjustment problem-solving in the counseling process.* Unpublished doctoral dissertation, University of Michigan, Ann Arbor.

Pratt, M. (1996). Adapting the three-step approach (Rossi, 1986/1993) for use with children. *American Journal of Clinical Hypnosis, 39*, 48–61.

Price, D. (1998). Mechanisms of pain reduction by hypnotic and placebo suggestions. *The Science and Practice of Mindbody Interactions.* Sedona Conference, Sedona, AZ.

Prigogine, I. (1980). *From being to becoming: Time and complexity in the physical sciences.* San Francisco: Freeman.

Prigogine, I. (1997). *The end of certainty: Time, chaos, and the new laws of nature.* New York: Free.

Prigogine, I., & Stengers, I. (1984). *Order out of chaos.* New York: Bantam.

Puntschart, A., Wey, E., Jostarndt, M., Vogt, M., Wittwer, H., Widmer, H., Hoppeler, H., & Billeter, R. (1998). Expression of fos and jun genes in human skeletal muscle after exercise. *American Journal of Physiology: Cell Physiology, 274*, C129–C137.

Quitkin, F., McGrath, P., & Stewart, J. (1996). Chronological milestones to guide drug change: When should clinicians switch antidepressants? *Archives of General Psychiatry, 53*, 785–792.

Rama, S., Ballentine, R., & Ajaya, S. (1976). *Yoga and psychotherapy.* Honesdale, PA: Himalayan International Institute.

Rainville, P., Ducan, G., Price, D., Carrier, B., & Bushnell, C. (1997). Pain affect encoded in human anterior cingulate but not somatosensory cortex. *Science, 277*, 968–871.

Rainville, P., Hofbauer, R., Paus, T., Duncan, G., Bushnell, M., & Price, D. (1999). Cerebral mechanisms of hypnotic induction and suggestion. *Journal of Cognitive Neuroscience, 11*(1), 110–125.

Rank, O. (1941). *Beyond psychology*. New York: Dover.

Rao, S., & Potdar, A. (1970). Nasal airflow with the body in various positions. *Journal of Applied Psychology, 28,* 162–165.

Rayl, A. (2001). Imaging gene expression in humans: The race is on. *The Scientist, 15,* 15–16.

Rayner, P., & Rudd, B. (1973). Emotional deprivation in three siblings associated with functional pituitary growth hormone deficiency. *Australian Pediatric Journal, 9,* 79–84.

Reber, A. (1993) *Implicit learning and tacit knowledge: An essay on the collective unconscious*. New York: Oxford University Press.

Reick, M., Garcia, J., Dudley, C., & McKnight. (2001). NPAS2: An analog of clock operative in the mammalian forebrain. *Science, 293,* 506–509.

Reilly, M., Fehr, C., & Buck, K. (2001). Alcohol and gene expression in the central nervous system. In N. Moustaïd-Moussa & C. Berdanier (Eds.), *Nutrient-gene interactions in health and disease: The effect of macronutrients and micronutrients on gene expression* (pp. 131–162). Boca Raton, FL: CRC.

Reiss, D., Neiderhiser, J., Hetherington, E., & Plomin, R. (2000). *The relationship code: deciphering genetic and social influences on adolescent development*. Cambridge, MA: Harvard University Press.

Reuters News (2002, March 4). *Vechta: German town finds 20 minute naps can boost workers' efficiency* [On-line]. Available: http://printerfriendly.abcnews.com

Ribeiro, S., Goyal, V., Mello, C., & Pavlides, C. (1999). Brain gene expression during REM sleep depends on prior waking experience. *Learning & Memory, 6,* 500–508.

Richards, K. (1998). Effect of a back massage and relaxation intervention on sleep in critically ill patients. *American Journal of Critical Care, 7,* 288–299.

Richardson, K. (2000). *The making of intelligence*. New York: Columbia University Press.

Ridley, M. (1996). *The origins of virtue: Human instincts and the evolution of cooperation*. New York: Penguin.

Ridley, M. (1999). *Genome: The autobiography of a species in 23 chapters*. New York: HarperCollins.

Ridley, M. (2001). *The cooperative gene: How Mendel's deamon explains the evolution of complex beings*. New York: Free.

Rietman, E. (1994). *Genesis redux: Experiments creating artificial life.* New York: McGraw-Hill.

Rioult-Pediotti, M., Friedman, D., & Donoghue, J. (2000). Learning-induced LTP in neocortex. *Science, 290,* 533–536.

Robertson, D. (2000). Gödel's theorem, the future of everything, and the future of science and mathematics. *Complexity, 5*(5), 22–27.

Robertson, R. (1995). *Jungian archetypes: Jung, Gödel, and the history of archetypes.* York Beach, ME: Nicholas-Hays.

Robertson, R. (1999). Some-thing from no-thing: G. Spencer-Brown's laws of form. *Cybernetics & Human Knowing, 6*(4), 43–55.

Robertson, R., & Combs, A. (Eds.). (1995). *Chaos theory in psychology and the life sciences.* Hillside, NJ: Lawrence Erlbaum.

Rosa, L., Rosa, E., Sarner, L., & Barrett, S. (1998). A close look at therapeutic touch. *Journal of the American Medical Association, 279,* 1005–1010.

Rosales, C., & Juliano, R. (1996). Integrin signaling to NF-kappa B in monocytic leukemia cells is blocked by activated oncogenes. *Cancer Research, 56*(10), 2302–2305.

Rosen, S. (1982). *My voice will go with you: The teaching tales of Milton H. Erickson.* New York: Norton.

Rosen, J., Fanselow, M., Young, S., Sitcoske, M., & Maren, S. (1998). Immediate-early gene expression in the amygdala following foot shock stress and contextual fear conditioning. *Brain Research, 796*(1–2), 132–142.

Rosenberg, S., & Barry, J, (1992). *The transformed cell: Unlocking the mysteries of cancer.* New York: Putnam/Chapmans.

Rossi, A. (2002). *Taming the flame: Transformation from Dionysian archetypal possession to individuation.* Unpublished doctoral dissertation, Pacifica Graduate Institute; Carpenteria.

Rossi, E. (1964). The development of classificatory behavior. *Child Development, 35,* 137–142.

Rossi, E. (1967). Game and growth: Two dimensions of our psychotherapeutic zeitgeist. *Journal of Humanistic Psychology, 7,* 136–154.

Rossi, E. (1968). The breakout heuristic: A phenomenology of growth therapy with college students. *Journal of Humanistic Psychology, 8,* 16–28.

Rossi, E. (1972). *Dreams and the growth of personality: Expanding awareness in psychotherapy.* New York: Pergamon.

Rossi, E. (1973a). Psychological shocks and creative moments in psychotherapy. The *American Journal of Clinical Hypnosis, 16,* 9–22.

Rossi, E. (1973b). The dream-protein hypothesis. *American Journal of Psychiatry, 130,* 1094–1097.

Rossi, E. (Ed.). (1980). *The collected papers of Milton H. Erickson on hypnosis: I. The nature of hypnosis and suggestion.* New York: Irvington.

Rossi, E. (1981). Hypnotist describes natural rhythm of trance readiness. *Brain Mind Bulletin, 6,* 1.

Rossi, E. (1982). Hypnosis and ultradian cycles: A new state(s) theory of hypnosis? *The American Journal of Clinical Hypnosis, 25*(1), 21–32.

Rossi, E. (1985). *Dreams and the growth of personality: Expanding awareness in psychotherapy* (2nd ed.). New York: Brunner/Mazel.

Rossi, E. (1986a). Altered states of consciousness in everyday life: The ultradian rhythms. In B. Wolman & M. Ullman (Eds.), *Handbook of altered states of consciousness* (pp. 97–132). New York: Van Nostrand.

Rossi, E. (1986b). The indirect trance assessment scale (ITAS). In M. Yapko (Ed.), *Hypnotic and strategic interventions: Principles and practice.* New York: Irvington.

Rossi, E. (1986/1993). *The psychobiology of mindbody healing* (Rev. ed.). New York: Norton.

Rossi, E. (1987). From mind to molecule: A state-dependent memory, learning, and behavior theory of mindbody healing. *Advances, 4*(2), 46–60.

Rossi, E. (1988). The psychobiology of mindbody healing: The vision and state of the art. In J. Zeig & S. Lankton (Eds.), *Developing Ericksonian therapy: State of the art* (pp. 127–148). New York: Brunner/Mazel.

Rossi, E. (1989a). Archetypes as strange attractors. *Psychological Perspectives, 20,* 4–14.

Rossi, E. (1989b). Mindbody healing, not suggestion, is the essence of hypnosis. *American Journal of Clinical Hypnosis, 32,* 14–15.

Rossi, E. (1990a). From mind to molecule: More than a metaphor. In J. Zeig & S. Gilligan (Eds.), *Brief therapy: Myths, methods and metaphors* (pp. 445–472). New York: Brunner/Mazel.

Rossi, E. (1990b). Mind-molecular communication: Can we really talk to our genes? *Hypnos, 17*(1), 3–14.

Rossi, E. (1990c). The new yoga of the west: Natural rhythms of mindbody healing. *Psychological Perspectives, 22,* 146–161.

Rossi, E. (1990d). The eternal quest: Hidden rhythms of stress and healing in everyday life. *Psychological Perspectives, 22,* 6–23.

Rossi, E. (1991). The wave nature of consciousness. *Psychological Perspectives, 24,* 1–10.

Rossi, E. (1992a). Periodicity in self-hypnosis and the ultradian healing response: A pilot study. *Hypnos, 19,* 4–13.

Rossi, E. (1992b). The wave nature of consciousness: A new direction for the evolution of psychotherapy. In J. Zeig (Ed.), *The evolution of psychotherapy: The second conference* (pp. 216–235). New York: Brunner-Mazel.

Rossi, E. (1993). *The psychobiology of mindbody healing* (2nd ed.). New York: Norton.

Rossi, E. (1994a). Hypnose und die neue Homoostase: Auf der suche nach einem mathematischen modell fur Erickson's naturalistischen ansatz. *Hypnose und Kognition*, Band 11, Heft 1 und 2, 167–189.

Rossi, E. (1994b). Ericksonian psychotherapy-then and now. In J. Zeig (Ed.), *Ericksonian methods: The essence of the story* (pp. 46–76). New York: Brunner/Mazel.

Rossi, E. (1994c). The emergence of mind-gene communication. *European Journal of Clinical Hypnosis, 3*, 4–17.

Rossi, E. (1995a). The essence of hypnotherapeutic suggestion: Part one. The basic accessing question and ultradian dynamics in single session psychotherapy. *European Journal of Clinical Hypnosis, 2*, 6–17.

Rossi, E. (1995b). The essence of hypnotherapeutic suggestion: Part two. Ultradian dynamics of the creative process in psychotherapy. *The European Journal of Clinical Hypnosis, 2*(4), 4–16.

Rossi, E. (1995c). The chronobiological theory of therapeutic suggestion: Towards a mathematical model of Erickson's naturalistic approach. In M. Kleinhauz, B. Peter, S. Livnay, V. Delano, K. Fuchs, & A. Lost-Peter (Eds.), *Jerusalem lectures on hypnosis and hypnotherapy.* Munich: Hypnosis International Monographs, Vol. 1.

Rossi, E. (1995d). Accessing the creative process. Ernest Rossi interviewed by Barry Winbolt. *The Therapist, 3*(1), 24–29.

Rossi, E. (1996a). *The symptom path to enlightenment: The new dynamics of self-organization in hypnotherapy.* New York: Zeig, Tucker, Theisen.

Rossi, E. (1996b). Research on process oriented psychobiological therapy: Mind-gene communication in psychobiological work and healing. *Advances, 12*(2), 29–31.

Rossi, E. (1996c). The psychobiology of mindbody communication: The complex, self-organizing field of information transduction. *BioSystems, 38*, 199–206.

Rossi, E. (1996d). The creative process in naturalistic ultradian hypnotherapy. *Hypnosis International Monographs, 2*, 1–15.

Rossi, E. (1996e). The essence of hypnotherapeutic suggestion: Part three. Polarity and the creative dynamics of change. *The European Journal of Clinical Hypnosis, 2*(3), 2–17.

Rossi, E. (1997a). The Feigenbaum scenario in a unified science of life and mind. *World Futures, 50*, 633–645.

Rossi, E. (1997b). The symptom path to enlightenment: the psychobiology of Jung's constructive method. *Psychological Perspectives, 36*, 68–84.

Rossi, E. (1998a). Mindbody healing in hypnosis: Immediate-early genes and the deep psychobiology of psychotherapy. *Japanese Journal of Hypnosis, 43*, 1–10.

Rossi, E. (1998b). The Feigenbaum scenario as a model of the limits of conscious information processing. *Biosystems, 40*, 1–10.

Rossi, E. (1999a). Sleep, dream, hypnosis and healing: Behavioral state-related gene expression and psychotherapy. *Sleep and Hypnosis: An International Journal of Sleep, Dream, and Hypnosis, 1*(3), 141-157.

Rossi, E. (1999b). The co-creative dynamics of dreams, consciousness and choice. *Psychological Perspectives, 38*, 116–127.

Rossi, E. (1999c). Identity creation from mind to gene. *Psychological Perspectives, 39*, 116–124.

Rossi, E. (1999d). The Feigenbaum scenario in a unified science of life and mind. In W. Hofkirchner (Ed.), *The quest for a unified theory of information* (pp. 411–423). Amsterdam: Gordon & Breach.

Rossi, E. (1999e). Exploring gene expression in sleep, dreams, and hypnosis with the new DNA microarray technology: A call for clinical-experimental research. *Sleep and Hypnosis: An International Journal of Sleep, Dream, and Hypnosis, 2*(1), 40–46.

Rossi, E. (2000a). *Dreams, consciousness, spirit: The quantum experience of self-reflection and co-creation.* New York: Zeig, Tucker, Theisen.

Rossi, E. (2000b). Exploring gene expression in sleep, dreams and hypnosis with the new DNA microarray technology: A call for clinical-experimental research. *Sleep and Hypnosis: An International Journal of Sleep, Dream, and Hypnosis, 2*(1), 40–46.

Rossi, E. (2000c). In search of a deep psychobiology of hypnosis: Visionary hypotheses for a new millennium. *American Journal of Clinical Hypnosis, 42*(3)/*42*(4), 178–207.

Rossi, E. (2000d). The numinosum and the brain: The weaving thread of consciousness. *Psychological Perspectives, 40*, 94–103.

Rossi, E. (2001a). The deep psychobiology of psychotherapy. In R. Corsini (Ed.), *Handbook of innovative therapy* (2nd ed., pp. 155–165). New York: Wiley.

Rossi, E. (2001b). Updating Milton Erickson's neuro-psycho-physiological dynamics of therapeutic hypnosis and psychotherapy. *The Milton H. Erickson Foundation Newsletter*, 10–13.

Rossi, E., & Cheek, D. (1988). *Mindbody therapy: Ideodynamic healing in hypnosis.* New York: Norton.

Rossi, E., & Jichaku, P., (1992). Creative choice in therapeutic and transpersonal double binds. In E. Rossi & M. Ryan (Eds.), *Creative choice in hypnosis* (pp. 225–253). New York: Irvington.

Rossi, E., & Lippincott, B. (1992). The wave nature of being: Ultradian rhythms and mindbody communication. In D. Lloyd & E. Rossi (Eds.), *Ultradian rhythms in life processes: A fundamental inquiry into chronobiology and psychobiology* (pp. 371–402). New York: Springer-Verlag.

Rossi, E., & Lippincott, B. (1993). A clinical-experimental exploration of Erickson's naturalistic approach: Ultradian time and trance phenomena. *Hypnos, 20*, 10–20.

Rossi, E., Lippincott, B., & Bessette, A. (1994). Ultradian dynamics in hypnotherapy: Part one. *The European Journal of Clinical Hypnosis, 2*(1), 10–20.

Rossi, E., Lippincott, B., & Bessette, A. (1995). Ultradian dynamics in hypnotherapy: Part two. *The European Journal of Clinical Hypnosis, 2*(2), 6–14.

Rossi, E., & Nimmons, D. (1991). *The twenty minute break: The ultradian healing response.* New York: Zeig, Tucker, Theisen.

Rossi, E., & Ryan, M. (Eds.). (1986). *Mindbody communication in hypnosis: Vol. 3. The seminars, workshops, and lectures of Milton H. Erickson.* New York: Irvington.

Rossi, E., & Ryan, M. (Eds.). (1992). *Creative choice in hypnosis: Vol. 4. The seminars, workshops, and lectures of Milton H. Erickson.* New York: Irvington.

Rossi, E., & Smith, M. (1990). The eternal quest: Hidden rhythms of mindbody healing in everyday life. *Psychological Perspectives, 22*, 146–161.

Rössler, O. (1992a). The future of chaos. In J. Kim & J. Stringer (Eds.), *Applied chaos* (pp. 457–465). New York: Wiley.

Rössler, O. (1992b). Interactional bifurcations in human interaction: A formal approach. In W. Tschhacher, G. Schiepek, & E. Brunner (Eds.), *Self-organization and clinical psychology: Empirical approaches to synergetics in psychology* (pp. 229–236). New York: Springer-Verlag.

Rössler, O. (1994a). Micro constructivism. *Psysica D, 75*, 438–448.

Routtenberg, A., & Meberg, P. (1998). A novel signaling system from the synapse to the nucleus. *Trends in Neurosciences, 21*, 106.

Ruoff, P., & Rensing, L. (1996). The temperature compensated Goodwin model stimulates many circadian clock properties. *Journal of Theoretical Biology, 179*, 275–285.

Russell, D. (1985). Ornithine decarboxylase: A key regulatory enzyme in normal and neoplastic growth. *Drug Metabolism Reviews, 16*, 1–88.

Russell, R. (1966). Biochemical substrates of behavior. In R. Russell (Ed.), *Frontiers in physiological psychology* (pp. 185–246). New York: Academic Press.

Rutter, J., Reick, M., Wu, L., & McKnight, S. (2001). Regulation of clock and NPAS2 DNA binding by the redox state of NDA cofactors. *Science, 293*, 510–514.

Ryabinin, A., Melia, K., Cole, M., Bloom, F., & Wilson, M. (1995). Alcohol selectively attenuates stress-induced expression of c-fos expression in rat hippocampus. *Journal of Neuroscience, 15,* 721–730.

Rypma, B., & D'Esposito, M. (2000). Isolating the neural mechanisms of age-related changes in human working memory. *Nature Neuroscience, 3*(5), 509-515.

Saito, T., & Kano, T. (1992). The diurnal fluctuation of hypnotic susceptibility. The *Japanese Journal of Hypnosis, 37*(1), 6–12.

Salant, J. (2001, June 18). FAA worried about tired pilots: Fliers say they're pushed to work on little sleep. *The Tribune* (San Louis Obispo County), section A, p. 1.

Samad, T., Moore, K., Sapirstein, A., Billet, S., Allchorne, A., Poole, S., Bonventre, J., & Woolf, C. (2001). Interleukin-1-mediated induction of Cox-2 in the CNS contributes to inflammatory pain hypersensitivity. *Nature, 410*, 471–475.

Sanders, S. (1991). Unpublished personal communication on survey of members of the American Society of Clinical Hypnosis.

Sanders, S., & Mann, B. (1995). The effects of light, temperature, trance length and time of day on hypnotic depth. *American Journal of Clinical Hypnosis, 37*(3) 43–53.

Sapolsky, R. (January, 1990). Stress in the wild. *Scientific American, 262*(1), 116–123.

Sapolsky, R. (1992). *Stress, the aging brain, and the mechanisms of neuronal death.* Cambridge, MA: MIT Press.

Sapolsky, R. (1996). Why stress is bad for your brain. *Science, 273,* 749–750

Sarbin, T., & Coe, W. (1972). *Hypnosis: A social psychological analysis of influence communication.* New York: Holt, Reinhart & Winston.

Schacter, D. (1996). *Searching for memory: The brain, the mind and the past.* New York: Basic.

Schäfer, C., Rosenblum, M., & Kurths, J. (1998). Heartbeat synchronized with ventilation. *Nature, 392*, 239-240.

Schanberg, S. (1995). The genetic basis for touch effects. In T. Field (Ed.), *Touch in early development* (pp. 67–79). New York: Lawrence Erlbaum.

Scharrer, E., & Scharrer, B., (1940). Secretory cells within the hypothalamus. *Research publications of the association of nervous and mental diseases.* New York: Hafner.

Schibler, U., Ripperger, J., & Brown, S. (2001). Chronobiology—Reducing time. *Science, 293*, 437–438.

Schlingensiepen, K., Kunst, M., Gerdes, W., & Brysch, W. (1995). Complementary expression patterns of c-jun and jun B in rat brain and analysis of their function with antisense oligionucleotides. In T. R. Tölle, J. Schadrack, & W. Zieglgansberger (Eds.), *Immediate early genes in the CNS* (pp. 132–145). New York: Springer-Verlag.

Schmitt, F. (1984). Molecular regulators of brain function: A new view. *Neuroscience, 13,* 991–1001.

Schneider, P. (1959). Miró. *Horizon: A Magazine of the Arts, 1*(4), 70–81.

Schneider, S. (2001). In search of realistic optimism. *American Psychologist, 56,* 250–263.

Schore, A. (1994). *Affect regulation and the origin of the self: The neurobiology of emotional development.* Hillside, NJ: Lawrence Erlbaum.

Schrödinger, E. (1944). *What is life? & Mind and matter.* New York: Cambridge University Press.

Schwartz, M., Woods, S., Porte, D., Seeley, R., & Baskin, D. (2000). Central nervous system control of food intake. *Nature, 404,* 661–671.

Seligman, M. (2001). As quoted in Kogan, M., Where happiness lies. *Monitor on Psychology, 32,* 74–76.

Seligman, M., & Csikszentmihalyi, M. (2000). Positive psychology: An introduction. *American Psychologist, 55,* 5–14.

Selye, H. (1956). *The stress of life.* New York: Longmans, Green.

Selye, H. (1974). *Stress without distress.* New York: Signet.

Senba, E., & Ueyama, T. (1997). Stress-induced expression of immediate early genes in the brain and peripheral organs of the rat. *Neuroscience Research, 29*(3), 183–207.

Seppa, N. (2001). Does lack of sleep lead to diabetes? *Science News, 160,* 31.

Shannahoff-Khalsa, D. (1991). Lateralized rhythms of the central and autonomic nervous systems. *International Journal of Physiophysiology, 11,* 225–251.

Shapiro, F. (1999). Eye movement desensitization and reprocessing (EMDR) and the anxiety disorders: Clinical and research implications of an integrated psychotherapy treatment. *Journal of Anxiety Disorders, 13,* 35–67.

Shapiro, F., & Forrest, M. (1997). *EMDR: Eye movement desensitization and reprocessing.* New York.: HarperCollins.

Sheldon, K., & King, L. (2001). Why is positive psychology necessary? *American Psychology, 56,* 216–217.

Sheline, Y., Wang, P., Gado, M., Csernansky, J., & Vannier, M. (1996). Hippocampal atrophy in recurrent major depression. *Proceedings of the National Academy of Sciences, 93,* 3908–3913.

Shimizu, E, Tang, Y., Rampon, C., & Tsien, J. (2000). NMDA receptor-dependent synaptic reinforcement as a crucial process for memory consolidation. *Science, 290,* 1170–1174.

Shin-Ichiro, I., Armstrong, C., Kaeberlein, M., & Guarente, L. (2000). Transcriptional silencing and longevity protein Sir2 is an NAD-dependent histone deacetylase. *Nature, 403*, 795–800.

Shors, T., Beylin, A., Wood, G., & Gould, E. (2000). The modulation of Pavlovian memory. *Behavior and Brain Research, 110*, 39–52.

Shors, T., Miesegaes, G., Beylin, A., Zhao, M., Rydel, T., & Gould, E. (2001). Neurogenesis in the adult is involved in the formation of trace memories. *Nature, 410*, 372–376.

Siegel, B. (1986). *Love, medicine and miracles.* New York: Harper & Row.

Siegel, B. (1989). *Peace, love and healing.* New York: Harper & Row.

Siegel, B. (1993). *How to live between office visits.* New York: HarperCollins.

Siegler, R. S. (1998). *Children's thinking* (3rd ed.). Englewood Cliffs, NJ: Prentice Hall.

Siegfried, J., Bourdeau, H., Davis, A., Luketich, J., & Shriver, S. (2000). Expression of gastrin-releasing peptide receptor, but not neuromedin B receptor, is related to sex, smoking history, and risk for lung cancer. *Proceedings of the American Association of Cancer Research, 41*, 147.

Sigrist, S., Thiel, P., Reiff, D., Lachance, P., Lasco, P., & Schuster, C. (2000). Postsynaptic translation affects the efficacy and morphology of neuromuscular junctions. *Nature, 405*, 1062–1065.

Simpkins, C., & Simpkins, A. (2000). *Effective self-hypnosis: Pathways to the unconscious.* San Diego: Radiant Dolphin.

Singer, B., & Ryff, C. (Eds.). (2001). *New horizons in health: An integrative approach.* Washington, DC: National Academy Press.

Skinner, B. (1981). Selection by consequences. *Science, 281*, 501–504.

Smaglik, P. (2000). Placebos could improve link between medical outlooks. *Nature, 408*, 349.

Smith, S. (1995). Getting into and out of mental ruts: A theory of fixation, incubation, and insight. In R. Sternberg & J. Davidson (Eds.), *The nature of insight* (pp. 229–251). Cambridge, MA: MIT Press.

Smith, A., & Tart, C. (1998). Cosmic consciousness experience and psychedelic experiences: A first person comparison. *Journal of Consciousness Studies, 5*, 97–107.

Smolin, L. (2001). *Three roads to quantum gravity.* New York: Basic.

Snow, C. (1993). *The two cultures.* New York: Cambridge University Press.

Snowdon, D. (2001). *Aging with grace.* New York: Bantam.

Solé, R., & Goodwin, B. (2000). *Signs of life: How complexity pervades biology.* New York: Basic.

Sommer, C. (1993). Ultradian rhythms and the common everyday trance. *Hypnos, 20,* 135–140.

Soreq, H., & Friedman, A. (1997, May). Images of stress in the brain. *Discovery,* May, 19–20.

Spencer-Brown, G. (1972). *Laws of form.* New York: Dutton.

Spencer-Brown, G. (1979). *Laws of form* (Rev. ed.). New York: Dutton.

Sperling, G. (1960). The information available in brief visual presentations. *Psychological Monographs, 74.* Washington, DC: American Psychological Association.

Spiegel, D. (1999). A 43-year-old woman coping with cancer. *Journal of the American Medical Association, 282,* 371–378.

Spiegel, D., & Barabasz, A. (1988). Effects of hypnotic instructions on P300 event-related-potential amplitudes: research and clinical applications. *American Journal of Clinical Hypnosis, 31,* 11–16.

Spiegel, H., & Spiegel, D. (1978). *Trance and treatment: Clinical uses of hypnosis.* New York: Basic.

Squire, L., & Kandel, E. 1999. *Memory: From mind to molecules.* New York: Scientific American Press.

Stampi, C. (1992). *Why we nap.* Boston: Birkhäuser.

Starkey, P., & Cooper, R. (1980). Perception of numbers by human infants. *Science, 210,* 1033–1035.

Starkey, P., Spelky, E., & Gelman, R. (1983). Detection of intermodal numerical correspondences by human infants. *Science, 222,* 179–181.

Stefano, G., Fricchione, G., Slingsby, B., & Benson, H. (2001). The placebo effect and relaxation response: Neural processes and their coupling to constitutive nitric oxide. *Brain Research Reviews, 35,* 1–9.

Sternberg, E. (2000). *The balance within: The science of connecting health and emotions.* New York: Freeman.

Sternberg, R. (2001). What is the common thread of creativity? *American Psychologist, 56,* 360–362.

Sternberg, R., & Davidson, J. (1995). *The nature of insight.* Cambridge, MA: MIT Press.

Stewart, I. (1999). Designer differential equations for animal locomotion. *Complexity, 5*(2), 12–22.

Steward, O., Wallace, C., & Worley, P. (2000). Synaptic regulation of messenger RNA trafficking within neurons. In M. Baudry, J. Davis, & R. Thompson (Eds.), *Advances in synaptic plasticity* (pp. 33–52). Cambridge, MA: MIT Press.

Stokes, P. (2001). Variability, constraints, and creativity. *American Psychologist, 56,* 335–359.

Stokkan, K., Yamazaki, S., Tei, H., Sakaki, Y., & Menaker, M. (2001). Entrainment of the circadian clock by feeding. *Science, 291*, 490–493.

Strickgold, R., Malia, A., Maguire, D., Roddenberry, D., & O'Connor, M. (2000a). Replaying the game: Hypnogogic images in normals and amnesiacs. *Science, 290*, 350–353.

Strickgold, R., Whidbee, D., Schirmer, B., Patel, V., & Hobson, J. (2000b). Visual discrimination task improvement: A multi-step process occurring during sleep. *Journal of Cognitive Neuroscience, 12*(2), 246–254.

Strogatz, S. (1986). Lecture notes in biomathematics. *The Mathematical Structure of the Human Sleep-Wake Cycle.* New York: Springer-Verlag.

Stupfel, M. (1992). Metabolic and behavioral long period ultradian rhythms. In D. Lloyd, & E. Rossi, E. (Eds.), *Ultradian rhythms in life processes: A fundamental inquiry into chronobiology and psychobiology.* New York: Springer-Verlag.

Sturgis, L., & Coe, W. (1990). Psychological responsiveness during hypnosis. *International Journal of Clinical and Experimental Hypnosis, 38*(3), 196–207.

Sturis, J., Polonsky, K., Shapiro, E., Blackman, J., O'Meara, N., & Van Cauter, E. (1992). Abnormalities in the ultradian oscillations of insulin secretion and glucose levels in Type 2 (non-insulin-dependent) diabetic patients. *Diabetologia, 35*, 681–689.

Sturtevant, A. (1913). The linear arrangement of six sex-linked factors in Drosophilia, as shown by their mode of association. *Journal of Experimental Zoology, 14*, 43–59.

Sturtevant, A. (1965). *A history of genetics.* New York: Harper & Row.

Suinn, R. (1990). *Anxiety management training: A behavior therapy.* New York: Plenum.

Suinn, R. (2001). The terrible twos—anger and anxiety: Hazardous to your health. *American Psychologist, 56*, 27–36.

Suinn, R., & Richardson, F. (1971). Anxiety management training: A non-specific behavior therapy program for anxiety control. *Behavior Therapy, 2*, 498–512.

Sullivan, P., Neale, M., & Kendler, K. (2000). Genetic epidemiology of major depression: review and meta-analysis. *American Journal of Psychiatry, 157*, 1552–1562.

Sullivan, S., Williams, L., Seaborne, D., & Morelli, M. (1991). Effects of massage on alpha motorneuron excitability. *Physical Therapy, 71*(8), 555–560.

Suomi, S. (1995). Touch and the immune system in Rhesus monkeys. In T. Field (Ed.), *Touch in early development* (pp. 89–103). New York: Lawrence Erlbaum.

Szabo, S. (1998). Hans Selye and the development of the stress concept: Special reference to gastroduodenal ulcerogenesis. *Annals of the New York Academy of Science, 851*, 19–27.

Szasz, T. (1997). The healing word: Its past, present, and future. In J. Zeig (Ed.), *The evolution of psychotherapy: the third conference* (pp. 299–306). New York: Brunner/ Mazel.

Tache, J., Morley, M., & Brown, M. (1989). *Neuropeptides and stress.* New York: Springer-Verlag.

Taddei, F., Matic, I., & Radman, M. (1995). cAMP-dependent SOS induction and mutagenesis in resting bacterial populations. *Proceedings of the National Academy of Sciences, 92,* 11736–11740.

Takahashi, J., & Hoffman, M. (1995). Molecular biological clocks. *American Scientist, 83,* 158–165.

Tay, C., Glasier, A., & McNeilly, A. (1996). Twenty-four hour patterns of prolactin secretion during lactation and the relation to sucking and the resumption of fertility in breastfeeding women. *Human Reproduction, 11,* 950–955.

Temple, S. (2001). Stem cell plasticity—building the brain of our dreams. *Nature Reviews: Neuroscience, 2,* 513–520.

Tinterow, M. (1970). *Foundations of hypnosis.* Springfield, IL: Thomas.

Todorov, I. (1990). How cells maintain stability. *Scientific American, 263,* 66–75.

Tölle T. R., Schadrack J., & Zieglgansberger, W. (1995). *Immediate early genes in the CNS.* New York: Springer-Verlag.

Toma, D., Bloch, G., Moore, D., & Robinson, G. (2000). Changes in period mRNA levels in the brain and division of labor in honey bee colonies. *Proceedings of the National Academy of Sciences, USA, 97,* 6914–6919.

Tononi, G., Cirelli, C., & Pompeiano, M. (1995). Changes in gene expression during the sleep-wake cycle: A new view of activating systems. *Archives Italiennes de Biology, 134,* 21–37.

Toppila J., Stenberg, D., Alanko L., Asikainen, M., Urban, J., Turek, F., & Porkka-Heiskanen, T. (1995). REM sleep deprivation induces galanin gene expression in the rat brain. *Neuroscience Letters, 183,* 171–174.

Tortonese, D., Brooks, J., Ingleton, P., & McNeilly, A. (1998). Detection of prolactin receptor gene expression in the sheep pituitary gland and visualization of the specific transition of the signal in gonadotrophs. *Endocrinology, 139,* 5212–5223.

Trieman, S. (1999). *The odd quantum.* Princeton, NJ: Princeton University Press.

Tsien, J. (2000a). Enhancing the link between Hebb's coincident detection and memory formation. *Current Opinion in Neurobiology, 10,* 2.

Tsien, J. (2000b). Building a brainier mouse. *Scientific American, 282,* 62–68.

Tsogoev, A., Liu, Q., & Wu, J. (2000). Acupuncture effect on gene expression of c-fos mRNA, ppENK mRNA, iNO mRNA, and iNOS activity in the mouse spinal cord. *World Journal of Acupuncture-Moxibustion.*

Tsuji, Y., & Kobayshi, T. (1988). Short and long ultradian EEG components in day-time arousal. *Electroencephalography and Clinical Neurophysiology, 70,* 110–117.

Tully, T. (1996). Discovery of genes involved with learning and memory: An experimental synthesis of Hirschian and Benzerian perspectives. *Proceedings of the National Academy of Sciences, USA, 93,* 13460–13467.

Underhill, E. (1963). *Mysticism.* New York: World.

Unterweger, E., Lamas, J., & Bongartz, W. (1992). Heart-rate variability of high and low susceptible subjects during administration of the stanford scale, form C. In E. Bongartz (Ed.), *Hypnosis: 175 years after Mesmer: Recent developments in theory and application* (pp. 85–90). Konstanz: Universitatsvergag.

Vallacher, R., & Nowak, J. (1994). *Dynamical systems in social psychology.* New York: Academic.

Van Cauter, E., Desir, D., Decoster, C., Fery, F., & Balasse, E. (1989) Nocturnal decrease in glucose tolerance during constant glucose infusion. *Journal of Clinical Endocrinology and Metabolism, 69,* 604–611.

Van Dellen, A., Blakemore, C., Decon, R., York, D., & Hanna, A. (2000). Delaying the onset of Huntington's in mice. *Science, 404,* 721–722.

Van Praag, H., Kempermann, G., & Gage, F. (1999). Running increases cell proliferation and neurogenesis in the adult mouse dentate gyrus. *Nature Neuroscience, 2,* 266–270.

Van Praag, H., Kempermann, G., & Gage, F. (2000). Neural consequences of environmental enrichment. *Nature Reviews: Neuroscience, 1,* 191–198.

Van Praag, H., Schinder, A., Christie, B., Toni, N., Palmer, T., & Gage, F. (2002). Functional neurogenesis in the adult hippocampus. *Nature, 415,* 1030–1034.

Varela, F. (1975). A calculus for self-reliance. *International Journal of General Systems, 2,* 5–24.

Veldhuis, J. (1992). A parsimonious model of amplitude and frequency modulation of episodic hormone secretory bursts as a mechanism for ultradian signaling by endocrine glands. In D. Lloyd & E. Rossi (Eds.), *Ultradian rhythms in life processes: A fundamental inquiry into chronobiology and psychobiology* (pp. 139–172). New York: Springer-Verlag, p. 139–172.

Veldhuis, J., & Johnson, M. (1988). Operating characteristics of the hypothalamopituitary-gonadal axis in men: Circadian, ultradian, and pulsatile release of prolactin and its temporal coupling with luteinizing hormone. *Journal of Clinical Endocrinology and Metabolism, 67*, 116–123.

Venter, J. et al. (2001). The sequence of the human genome. *Science, 291*, 1304–1351.

Verhoef, M., & Page, S. (1998). Physicians' perspectives on massage therapy. *Canadian Family Physician, 44*, 1018–40.

Vogel, G. (2000a). New brain cells prompt new theory of depression. *Science, 290*, 258–259.

Vogel, G. (2000b). Stem cells: New excitement, persistent questions. *Science, 290*, 1672–1674.

Von Dassow, G., Meir, E., & Odell, G. (2000). The segment polarity network is a robust developmental module. *Nature, 406*, 188–192.

Vukmirovic, O., & Tilghman, S. (2000). Exploring genome space. *Nature, 405*, 820–822.

Vygotsky, L. (1962). *Thought and language.* Cambridge, MA: MIT Press.

Waelti, P., Dickinsen, A., & Schultz, W. (2001). Dopamine responses comply with basic assumptions of formal learning theory. *Nature, 412*, 43–48.

Wagstaff, G. (1986). Hypnosis as compliance and belief: a socio-cognitive view. In P. Naish (Ed.), *What is hypnosis? Current theories and research* (pp. 57–84). Philadelphia, PA: Open University Press, Milton Keynes.

Wallace, B., (1993). Day persons, night persons, and variability in hypnotic susceptibility. *Journal of Personality and Social Psychology, 64*, 827–833.

Wallace, B., & Kokoszka, A., (1995). Fluctuations in hypnotic susceptibility and imaging ability over a 16-hour period. *The International Journal of Clinical and Experimental Hypnosis, 16*(1), 7–19.

Wallace, B., Turosky, D., & Koloszka, A., (1992). Variability in the assessment of vividness. *Journal of Mental Imagery, 16*, 221–230.

Wallace, C., Lyford, G., Worley, P., & Steward, O. (1998). Differential intracellular sorting of immediate early gene mRNAs depends on signals in the mRNA sequence. *Journal of Neuroscience, 18*, 26–35.

Wallas, G. (1926). *The art of thought.* New York: Harcourt.

Wallenstein, G., Eichenbaum, H., & Hasselmo, M. (1998). The hippocampus as an associator of discontiguous events. *Trends in Neuroscience, 21*, 317–323.

Wang S., Bartolome, J., & Schanberg, S. (1996). Neonatal deprivation of maternal touch may suppress ornithine decarboxylase via downregulation of the protooncogenes C-myc and max. *Journal of Neuroscience, 16*(2):836–842.

Ward, T. (2001). Creative cognition: Conceptual combination, and the creative writing of Stephen R. Donaldson. *American Psychology, 56*, 350–354.

Wasserman, S., & DiNardo, S. (2001). Staying a boy forever. *Science, 245*, 2495–2497.

Watkins, J. (1978). *The therapeutic self.* New York: Human Sciences Press.

Watson, J., & Crick, F. (1953a). A structure for deoxyribose nucleic acid. *Nature, 171*, 737.

Watson, J., & Crick, F. (1953b). Genetical implications of the structure of deoxyribonucleic acid. *Nature, 171*, 964.

Weiner, J. (1994). *The beak of the finch: A story of evolution in our time.* New York: Knopf.

Weiner, J. (1999). *Time, love, memory: A great biologist and his quest for the origins of behavior.* New York: Vintage.

Weinstein, E., & Au, P. (1991). Use of hypnosis before and during angioplasty. *American Journal of Clinical Hypnosis, 34*, 29–37.

Weinstein, S. (1988). *Secrets to profiting in bull and bear markets.* New York: McGraw-Hill

Wen, X., Fuhrman, S., Michaels, G., Carr, D., Smith, S., Barker. J., & Somogyi, R. (1998). Large-scale temporal gene expression mapping of central nervous system development. *Proceedings of the National Academy of Sciences, 95*, 334–339.

Weitzenhoffer, A., & Hilgard, E. (1967). *Revised stanford profile scales of hypnotic susceptibility.* Palo Alto, CA: Consulting Psychologists Press.

Weitzenhoffer, A. (1971). Ocular changes associated with passive hypnotic behavior. *The American Journal of Clinical Hypnosis, 14*, 102–121.

Weitzenhoffer, A. (1980). Hypnotic susceptibility revisited. *The American Journal of Clinical Hypnosis, 22*, 130–146.

Weitzenhoffer, A. (2000). *The practice of hypnotism* (2nd ed). New York: Wiley.

Werntz, D. (1981). *Cerebral hemispheric activity and autonomic nervous function.* Unpublished doctoral dissertation, University of California, San Diego.

Werntz, D., Bickford, R., Bloom, F., & Shannahoff-Khalsa, D. (1983). Alternating cerebral hemispheric activity and lateralization of autonomic nervous function. *Human Neurobiology, 2*, 39–43.

Werntz, D., Bickford, R., Bloom, F., & Shannahoff-Khalsa, D. (1987). Selective hemispheric stimulation by unilateral forced nostril breathing. *Human Neurobiology, 6*, 165–171.

Wesson, R. (1991). *Beyond natural selection.* Cambridge, MA: MIT Press.

Wheeler, J. (1994). *At home in the universe.* Woodbury, NY: American Institute of Physics.

White, J. (1972). *The highest state of consciousness.* New York: Doubleday.

White, J. (1995). *What is enlightenment: Exploring the goal of the spiritual path.* New York: Paragon.

Whitehead, A., & Russell, B. (1925). *Principia mathematica* (2nd ed.). Cambridge, UK: Cambridge University Press.

Whitmore, D., Foulkes, N., & Sassone-Corsi, P. (2000). Light acts directly on organs and cells in culture to set the vertebrate circadian clock. *Nature, 404,* 87–91.

Wilber, K. (1993). *The spectrum of consciousness* (2nd ed.). Wheaton, IL: Quest.

Wilber, K. (1995). The ultimate state of consciousness. In J. White (Ed.), *What is enlightenment: Exploring the goal of the spiritual path* (pp. 196–208). New York: Paragon.

Williams, J., Vincent, S., & Reiner, P. (1998). Measurment of nitric oxide in the brain using the hemoglobin trapping technique coupled with in vivo microanalysis. In R. Lydic (Ed.), *Molecular regulation of arousal states.* New York: CRC.

Winstead, E. (2001). Beyond insomnia: strategies of circadian genomics [On-line]. Available: http://www.celera.com/genomics/news/articles/05_01/Clock.cfm

Wolfenbarger, L., & Phifer, P. (2000). The ecological risks and benefits of genetically engineered plants. *Science, 290,* 2088–2093.

Wolpe, J. (1969). *The practice of behavior therapy.* New York: Pergamon.

Wong, M., & Licinio, J. (2001). Research and treatment approaches to depression. *Nature Reviews: Neuroscience, 2,* 343–351.

Wood, R., Mitchell, M., Sgouros, J., & Lindahl, T. (2001). Human DNA repair genes. *Science, 291,* 1284–1289.

Woody, E., & Farvolden, P. (1998). Dissociation in hypnosis and frontal executive function. *American Journal of Clinical Hypnosis, 40,* 206–216.

Wright, C. (1995). Massage by nurses in the United States and the People's Republic of China: A comparison. *Journal of Transcultural Nursing, 7*(1), 24–27.

Xu, L., Anwyl, R., & Rowan, M. (1997). Behavioral stress facilitates the induction of long-term depression in the hippocampus. *Nature, 387,* 497–505.

Yaghoubi, S., Barrio, J., Dahlbom, M., Iyer, M., Namavari, M., Satyamurthy, N., Goldman, R., Herschman, H., Phelps, M., & Gambhir, S. (2001). Human pharmacokinetic and dosimetry studies of [18F]FHBG: A reporter probe for imaging herpes simplex virus type-1 thymidine kinase reporter gene expression. *Journal of Nuclear Medicine, 42,* 1225–1234.

Yamuy, J., Mancillas, J., Morales, F., & Chase, M. (1993). C-fos expression in the pons and medulla of the cat during carbacol-induced active sleep. *Journal of Neuroscience, 13,* 2703–2718.

Yeats, W. (1932/1962). Vacillation. In M. Rosenthal (Ed.), *The selected poems and two plays of William Butler Yeats* (pp. 134–137). New York: Macmillan.

Yerkes, R., & Dodson, J. (1908). The relationship of strength of stimulus to rapidity of habit formation. *Journal of Comparative Neurology and Psychology, 18*, 459–482.

Young, M. (1998). The molecular control of circadian behavioral rhythms and their entrainment in Drosophila. *Annual Review of Biochemistry, 67*, 135–152.

Young, M. (2000). The tick-tock of the biological clock. *Scientific American, 282*, 64–71.

Young, R. (2000). Biomedical discovery with DNA arrays. *Cell, 102*, 9–15.

Zee, A. (1986). *Fearful symmetry: The search for beauty in modern physics.* New York: Macmillan.

Zeig, J. (1985). *Experiencing Erickson: An introduction to the man and his work.* New York: Brunner/Mazel.

Zeig, J. (1990). *A teaching seminar with Milton H. Erickson, M.D.* New York: Brunner/ Mazel.

Zeig, J. (Ed.). (1994). *Ericksonian methods: The essence of the story.* New York: Brunner/ Mazel.

Zeig, J. (1997). Experiential approaches to clinician development. In J. Zeig (Ed.), *The evolution of psychotherapy: The third conference* (pp. 161–177). New York: Brunner/Mazel.

Zeig, J. (1999). The virtues of our faults: A key concept of Erickson therapy. *Sleep and Hypnosis: An International Journal of Sleep, Dream, and Hypnosis, 1*(3), 129–138.

Zeig, J., & Geary, B. (Eds.). (2000). *The letters of Milton H. Erickson.* Phoenix, AZ: Zeig, Tucker & Theisen.

Zeig, J., & Gilligan, S. (1990). *Brief therapy: Myths, methods and metaphors.* New York: Brunner Mazel.

Zeki, S. (1999). *Inner vision: An exploration of art and the brain.* Oxford, UK: Oxford University Press.

Zeki, S. (2001). Artistic creativity and the brain. *Science, 293*, 51–52.

Zhao, X., Malloy, P., Krishnan, A., Swami, S., Navone, N., Peehl, D., & Feldman, D. (2000). Glucocorticoids can promote androgen-independent growth of prostate cancer cells through a mutated androgen receptor. *Nature Medicine, 6*, 703–706.

Zilboorg, G., & Henry, G. (1941). *A history of medical psychology.* New York: Norton.

NAME INDEX

Abel, T., 35
Abraham, F., 400, 436
Achterberg, J., 18, 243, 458
Aczel, A., 277
Ader, R., 246
Afshari, C., 20
Akil, H., 46
Aldrich, K., 211
Alexander, F., 16
Alman, B., 194
Altman, J., 109
Amabile, T., 263, 267
Amigo, S., 196
Antell, S., 136
Armitage, R., 82
Aserinsky, E., 61
Au, P., 196
Auerbach, J., 101–2
Autelitano, D. J., 236
Axelrod, R., 20, 469, 475

Bailey, C., 61, 110, 239
Balsalobre, A., 64
Balter, M., 114
Balthazard, C., 198
Barabasz, A., 208, 211, 318
Barabasz, M., 208
Barber, J., 229, 243
Barber, T., 191, 334
Barinaga, M., 60
Barry, J., 215
Bartolome, J., 14
Bassett, D., 36
Bateson, G., 429–30
Battino, R., 458
Beckmann, A., 83
Beggs, J., 116
Bélair, J., 212
Bender, S., 126

Ben-Jacob, E., 394
Bennett, E., 109–10
Benson, H., 244–45
Bentivoglio, M., 42, 48, 49–50, 154, 201, 239
Benzer, S., 51–52
Berdanier, C., 91
Berman, D., 127–28, 235, 246
Bernheim, H., 193
Bernstein, D., 211
Bessette, A., 71
Bijeljac-Babic, R., 136
Bittman, B., 212, 243–44
Bjick, S., 126
Black, I., 188, 356
Bliss, T., 115
Bloom, F., 48, 53
Bongartz, W., 206
Boole, G., 428
Born, J., 202
Bower, B., 184, 185, 279
Braid, J., 188, 189, 190–91, 206
Braly, P., 75
Brand, S., 429
Brandenberger, G., 55, 56, 61
Breuer, J., 100–101, 469
Brey, D., 47
Brivanlou, A., 199, 236
Brodsky, V., 47, 60, 61
Broughton, R., 98, 293–94
Brown, P., 212
Brown, T., 36, 217
Brush, F., 46
Búcan, M., 35
Bucke, M., 267, 272, 277
Buijs, R., 209
Buss, D., 265

Cahill, L., 235
Cairns, J., 265–66

Campbell, J., 267
Capecchi, M., 435
Carpenter, S., 273, 274
Castes, M., 64, 216
Chaitin, G., 128
Charcot, J., 193
Cheek, D., 67, 75, 82, 84, 101, 102, 104, 146, 192, 210, 234, 280, 308, 332–33, 348, 401, 403, 437, 462, 466, 468
Cheverton, R., 274
Chicurel, M., 155, 265, 266
Chin, H., 38, 370, 420
Chittajallu, R., 117
Cho, K., 214
Chomsky, N., 466
Christensen, D., 94
Christopher, H., 71
Cirelli, C., 48–49, 134, 201, 239
Clark, R., 130
Clayton, D., 115
Coe, W., 196, 203–4
Cohen, M., 20, 469, 475
Collin, P., 466
Collins, F., 6
Combs, A., 436
Conklin, B., 36
Cooper, L., 145
Corsini, R., 133, 234, 239
Coué, É., 193–94
Crabtree, G., 60, 61
Crasilneck, H., 216
Crick, F., 29, 129
Csikszentmihalyi, M., 263, 265, 379, 443
Cvitanovic, P., 182

Damasio, A., 40, 43, 44
d'Aquili, E., 245, 464
Darnell, J., 199, 236
Davidson, J., 265, 267, 413, 443
Davson, H., 66
Dawkins, R., 10
DeBenedittis, G., 196, 238
DeGrandpre, R., 25
Deikman, A., 43, 60, 293
De la Fuente-Fernández, R., 247–48, 333
Dement, W., 75, 78
DePinho, R., 40
De Souza, E., 46
Devlin, K., 56, 203
Dhabar, F., 217
Diamond, J., 184
Dickinson, A., 107
DiNardo, S., 229
Dinges, D., 98, 293–94
Disterhoft, J., 130

Dodson, J., 203
Dolan, Y., 146
Donahoe, J., 25
Dossey, L., 243, 282
Doucet, J., 297
Dragunow, M., 47
Dudai, Y., 123, 127–28, 235, 246, 308, 375
Duman, R., 18, 37, 65
Dunlap, J., 52, 53
Dyson, F., 29–30

Edelman, G., 26, 44, 134, 276, 443
Edmonston, W., 18, 67, 126, 189, 195, 209, 412, 468
Egyhazi, E., 110
Eisenberg, L., 65
Eldridge, N., 269
Elena, S., 268, 275
Ellenbeger, E., 100–101
Ellenberger, E., 126, 193
Engel, G. 56, 16
Erickson, E., 81, 95, 145
Erickson, M., 18, 67, 73, 80, 81, 85, 90, 92, 95, 98, 101, 103, 104, 105, 110, 141–50, 192, 195, 208, 229, 235, 240–42, 244, 265, 280, 283, 286–88, 300, 303, 304, 308, 315–16, 322, 323, 328, 331, 333, 334, 335, 339, 360, 369, 386, 392, 394, 395, 397–98, 401, 430, 433, 436, 452, 455–57, 468, 473
Eriksson, P., 60, 65
Escera, C., 211
Ewin, D., 61, 220–22
Eysenck, H., 197–98

Farmer, S., 94
Farvolden, P., 235
Feher, S., 90
Feigenbaum, M., 173, 180–86
Feldman, D., 199–201
Felker, B., 46, 69–70
Field, T., 19
Fischbach, G., 246–47
Fischer, C., 91, 95
Folkman, J., 6
Forrest, M., 126
Foster, P., 265–66
Frank, M., 158
Frankland, P., 132
Fredrickson, B., 111, 265, 282
Freed, G., 293
Freeman, W., 66, 91
Freidman, S., 91, 95
Freud, S., 100–101, 101, 126, 369, 375–76, 436, 452, 469
Friedman, A., 233
Fuchs, E., 229

Gage, F., 65, 108, 150, 229
Gallo, V., 117
Gardner, L., 16
Geary, B., 145, 192
Gilgen, A., 400, 436
Gillett, L., 194, 282
Glaser, R., 64, 215, 216–17
Glass, L., 212
Glenn, S., 25, 274
Glik, D., 245
Goldbeter, A., 59, 60
Goodwin, B., 17, 20, 60–61, 63, 64, 203, 206, 212, 426, 473
Gorton, B., 195–96
Gottheil, E., 79
Gould, E., 110, 220, 244
Gould, S., 269
Graham, R., 185
Granott, N., 280
Grassi-Zucconi, G., 42, 48, 49–50, 154, 201, 239
Gray, J., 465
Green, M., 60, 212
Green, R., 60, 212
Greene, 433
Greenfield, S., 18, 126, 243, 244, 478–79
Greenspan, R., 53
Greer, J., 435
Grigoriadis, D., 46
Gross, C., 109, 121–23
Gruber, H., 274
Guastello, S., 186, 203

Hadamard, J., 178
Haley, J., 145, 146, 430
Hall, J., 155
Hamadeh, H., 20
Hameroff, S., 129
Hammond, D., 243
Hammond, K., 184
Hargrove, J., 91
Harrington, A., 246
Harris, G., 44
Harris, R., 196
Hautkappe, H., 206
Hebb, D., 109
Henry, G., 67, 100–101, 126, 189
Hiatt, J., 91, 95
Hilgard, E., 185, 191, 193, 197, 204, 249
Hilgard, J., 204, 249
Hobson, J., 284
Hoffman, M., 103
Holden, C., 249
Hollander, H., 126
Hood, D., 229

Hoppenbrouwers, T., 61
Horgan, J., 185
Howe, R., 428, 429
Hubbard, J., 46, 69–70
Hull, C., 195
Hunziker, M., 273
Huxley, A., 277, 283
Hwang, J., 91, 93
Hyden, H., 110
Hyman, S., 247

Iranmanesh, A., 69–70, 210
Ironson, G., 19
Izquierdo, I., 123

Jacob, S., 215, 236
James, W., 111
Janet, P., 126, 133
Jenkins, D., 93
Jichaku, P., 315
Johnson, J., 36
Johnson, M., 90, 466
Johnston, D., 60, 116
Jonas, W., 246
Jones, M., 155
Jordan, K., 94
Jung, C., 124, 133, 139, 148, 168, 178, 179, 186, 257, 267, 272, 285, 346, 376, 425, 436, 452

Kalsbeek, A., 209
Kandel, E., 5, 61, 110, 111, 113, 114, 117, 129, 133, 396
Kano, T., 212
Kaufer, D., 101, 121, 214
Kauffman, S., 434, 436
Keating, D., 136
Keeney, B., 243, 244, 267, 478–79
Kelner, K., 48, 53
Kempermann, G., 108, 110, 150, 244
Kennedy, B., 82
Kersten, S., 94
Kiger, A., 229
Kihlstrom, J., 184, 222
King, L., 257–58
Kirsch, J., 144, 145, 198
Klein, M., 128
Klein, R., 82
Kleitman, N., 59, 61, 75, 196, 209–10, 286
Klevecz, R., 75
Kobayashi, T., 103
Kobayshi, T., 295
Koch, C., 129
Kohara, K., 246
Kokoszka, A., 212, 458
Kosslyn, S., 458

Krech, D., 109–10
Kripke, D., 91, 95, 294
Kronauer, R., 205–6
Kucherlapati, R., 40
Kuhn, C., 14, 16–17
Kupfer, D., 75, 103
Kupfermann, I., 53
Kurtz, R., 82

LaBar, K., 130
LaBerge, S., 174, 281
Lakoff, G., 183, 434, 466
Lambrou, P., 194
Langton, C., 145, 397
Langton, S., 145, 397
Lauter, J., 82
Lavie, P., 91
Lazarus, A., 196, 244–45
Lee, C., 222, 223
Lei, X., 96
Lenski, R., 268, 275
Leslie, M., 200
Levine, H., 394
Levine, J., 82, 85
Levine, S., 46
Libet, B., 129
Licinio, J., 40
Lindquist, S., 266
Lippincott, B., 71, 212
Lisman, J., 131, 132
Lloyd, D., 30, 52, 55, 59, 60, 61, 69, 158, 209, 210, 220, 294, 460
Lockhart, D., 50
Loewenstein, W., 21
Lok, C., 211, 217, 220, 230, 236, 436
Lomo, T., 115
Lovelock, J., 22
Lowe, J., 274
Lüscher, C., 60, 117
Lydic, R., 47, 60
Lynch, J., 16, 62
Lynn, S., 144, 145, 198, 208
Lyubomirsky, S., 111

McEwen, B., 120, 217
McGaugh, J., 131
McGuffin, P., 36
McKenzie, G., 266
McKernan, M., 239
Mackey, M., 212
McLaren, A., 226
Mainzer, 434
Maio-Sin Wu, S., 116
Malarkey, W., 215
Malenka, R., 209

Mann, B., 212
Maquet, P., 155–56
Margulis, L., 22, 32, 33, 41
Marijuan, P., 47
Marks-Tarlow, T., 432
Marmont, M., 260
Martinez, M., 432
Maslow, A., 265, 276, 415
Masterpasqua, F., 436
Matthews, W., 334, 397
Matus, A., 117–19
Mayford, M., 114
Mayne, T., 196, 244–45
Mayr, E., 25
Meberg, P., 233
Meek, S., 19
Meier-Koll, A., 58, 61
Mejean, L., 91
Mendel, G., 9, 26–27, 35–36
Merchant, K., 47, 235–36
Merchant-Nancy, H., 49
Merrill, C., 212
Miller, G., 184
Miller, N., 274
Milner, B., 123
Moerman, D., 246
Moldin, S., 38, 370, 420
Moller, H., 246
Montgomery, G., 222
Moore-Ede, M., 78
Morano, M., 46
Morimoto, R., 64, 215, 236, 275
Morowitz, H., 212, 336
Morris, G., 131, 132
Morris, K., 94
Morrison, R., 94
Moustaïd-Moussa, N., 91, 96
Murray, A., 60

Nader, K., 125, 127, 155, 235, 308, 314, 335, 375, 457
Naish, P., 197
Nakamura, J., 263
Nakao, M., 272
Nash, M., 192
Neisser, U., 184
Neslter, E., 18, 37, 65
Nester, E., 79
Neumann, E., 267, 446, 478–79
Neumann, J. von, 29
Neuringer, A., 273, 274
Newberg, A., 245, 464
Newman, J., 215
Newtson, D., 59
Nicoll, R., 209

Nimmons, D., 60, 70, 74, 78, 79, 80, 97, 99, 103, 282, 284, 419–20, 458
Nørretranders, T., 129
Nowak, J., 59, 186, 436
Nuñez, R., 183, 434, 466

O'Dell, W., 216
Odum, H., 283
Okura, M., 72
Oliet, S., 117
Osowiec, D., 212
Othmer, S., 243
Otto, R., 139, 243, xvii–xviii

Palmeirin, I., 70
Palmer, D., 25
Papez, J., 44
Pardue, M., 220
Parziale, J., 280
Pavlov, I., 195
Pearls, F., 299
Peitgen, H., 182, 183–84, 186
Pennebaker, J., 60, 293
Perl, F., 452
Perna, P., 436
Pert, C., 31, 46, 199
Peter, B., 334, 397
Petitto, L., 466–68
Petrovic, P., 249, 324, 353
Phelps, M., 211, 220, 436
Piaget, J., 136, 465
Plomin, R., 36
Poincaré, H., 178, 184
Pompeiano, M., 48–49
Poon, C., 212
Porkka-Heiskanen, T., 199
Potdar, A., 82
Poundstone, 433
Pratt, M., 458
Price, D., 235
Prigogine, I., 20, 434
Puntschart, A., 236

Quitkin, F., 246

Rainville, P., 235, 249, 308
Rama, S., 82
Rank, O., 436
Rao, S., 82
Rayl, A., 236
Rayner, P., 16
Reber, A., 396
Reichsman, F., 16
Reick, M., 213
Reilly, M., 83

Rensing, L., 63
Revenstorf, D., 334, 397
Rheingold, H., 174, 281
Ribeiro, S., 154, 156, 235, 285
Richardson, K., 47, 50, 124, 144
Richarson, F., 126
Ridley, M., 10, 257, 289
Rietman, E., 26
Rioult-Pediotti, M., 115
Robertson, R., 128, 433, 436
Rogers, C., 300, 452
Rosa, L., 18
Rosen, J., 145
Rosenberg, S., 215
Rosenzweig, M., 110
Rossi, A., 267, 457
Rossi, E., 18, 20, 30, 31–32, 36, 43, 44, 46, 47, 50, 51, 52, 55, 58, 59, 60, 61, 62, 63, 67, 69, 70, 71, 73, 74, 75, 78, 79, 80, 81, 82, 84, 90, 92, 93, 95, 97, 98, 99, 100, 101, 102, 103, 104, 105, 109, 115, 121, 123, 124, 125, 129, 130, 135, 139, 143, 144, 145, 147, 152, 153, 155, 157, 158, 162, 166, 173, 178, 182, 185–86, 192, 195, 196, 201, 203, 204, 206, 208, 209–10, 211, 212, 214, 216, 218, 220, 234, 235, 236, 237, 238, 265, 266, 267, 276, 279, 280, 282, 284, 294, 300, 301, 303, 309, 312, 315, 322, 323, 324, 332–33, 341, 346, 348, 363, 367, 369, 376, 386, 392, 394, 395, 397, 400, 401, 403, 410, 412, 413, 415, 416, 419–20, 430, 433, 436, 437, 443, 454, 457, 458, 462, 466, 468
Rossi, S., 80, 90, 98, 303, 322, 323, 392, 396
Rössler, O., 186
Routtenberg, A., 233
Rudd, B., 16
Ruofff, P., 63
Russell, B., 428–29
Russell, D., 14
Russell, R., 110, 153
Rutter, J., 213
Ryabinin, A., 83
Ryan, M., 104, 105, 234, 235
Ryff, C., 9, 78, 90, 97, xvii

Sagan, D., 22, 33
Saito, T., 212
Salant, J., 214
Samad, T., 216, 220–22
Sanders, S., 60, 212
Sapolsky, R., 120, 260–63
Sarbin, T., 203–4
Schäfer, C., 212
Schanberg, S., 14, 16–17, 65, 230
Scharrer, B., 44
Scharrer, E., 44
Schibler, U., 213

Schlingensiepen, K., 216
Schmitt, F., 233
Schneider, P., 297
Schneider, S., 111
Schore, A., 40
Schrödinger, E., 28, 107, 128, 132, 156, 169
Schultz, W., 107
Segal, M., 66
Segre, J., 229
Seligman, M., 265, 379
Seligman, M. E. P., 392
Selye, H., 259–64
Seppa, N., 76
Shannahoff-Khalsa, D., 61, 82
Shapiro, F., 126, 185
Sheldon, K., 257–58
Sheline, Y., 120
Sherman, S., 144, 145, 208
Shimizu, E., 121–23, 123–24, 235, 308, 375
Shinnick-Gallatgher, P., 239
Shors, T., 111, 131
Siegfried, J., 79
Siegler, R. S., 279
Simpkins, A., 194
Simpkins, C., 194
Singer, B., 9, 78, 90, 97, xvii
Singer, J., 212, 336
Skinner, B., 25
Smaglik, P., 246
Smith, A., 60, 293
Smith, S., 260, 270
Smolin, L., 433–34
Snow, C., 140
Snowdon, D., 111
Solé, R., 17, 20, 61–62, 63, 64, 203, 206, 212, 426, 473
Soreq, H., 233
Spencer-Brown, 430–32, 433
Sperling, G., 184
Spiegel, D., 79, 86, 208, 211, 235, 318, 468
Spiegel, H., 79, 86, 208, 235, 468
Squire, L., 111, 113, 129, 130, 133, 396
Stampi, C., 60
Starkey, P., 135
Stefano, G., 246
Stengers, I., 20, 434
Sternberg, E., 16, 220
Sternberg, R., 265, 267, 413, 443
Steward, O., 121
Stokes, P., 274
Stokkan, K., 51
Strauss, S., 246
Strickgold, R., 156, 157–58, 235, 285
Strogatz, S., 205
Stupfel, M., 60

Sturgis, L., 196
Sturis, J., 57
Suinn, R., 126
Sullivan, P., 40
Szaasz, T., 144

Taché, J., 64, 93, 275
Taddei, F., 266
Takahashi, J., 103
Tart, C., 60, 293
Tay, C., 89
Temple, S., 226, 229
Tilghman, S., 51
Tinterow, M., 189, 190–91, 363
Todorov, I., 90
Tölle, T. R., 47, 235–36
Toma, D., 65
Tononi, G., 44, 48–49, 134, 276, 443
Toppila, J., 199
Tortonese, D., 89
Treiman, S., 433
Tsien, J., 209
Tsogoev, A., 230
Tsuji, Y., 103, 295
Tully, T., 239

Underhill, E., 245
Unterweger, E., 206

Vallacher, R., 59, 186, 436
Van Cauter, E., 91
Van Praag, H., 110, 244
Varela, F., 425–26, 432–33
Velduis, J., 61, 90
Venter, J., 35
Vogel, G., 120, 226
Volz, H., 246
Von Foerster, H., 428, 429
Vukmirovic, O., 50–51
Vygotsky, L., 135

Waelti, P., 107, 113, 136–37, 248, 333
Wagstaff, G., 196–97
Wallace, B., 212, 458, 460
Wallace, C., 61, 121
Wallace, D., 274
Wallas, G., 178, 265
Wallenstein, G., 131
Wang, S., 14
Ward, T., 274
Wasserman, S., 229
Watkins, J., 410
Watson, J., 29
Weakland, J., 430
Weiner, J., 8, 25, 51

Weinstein, E., 196
Weinstein, S., 265
Weitzenhoffer, A., 193, 413
Wen, X., 37, 38, 39, 70
Werntz, D., 61, 82
Wesson, R., 276
Wheeler, 433
Wheeler, J., 185
White, J., 267, 272, 277, 279
Whitehead, A., 428–29
Wilber, K., 43, 267, 276, 277, 279, 446, 447, 478–79
Wiles, A., 277
Williams, J., 199
Winstead, E., 76
Winzeler, E., 50
Wolpe, J., 126
Wong, M., 40

Wood, R., 33
Woody, E., 198, 235
Wynne, P., 262

Yaghoubi, S., 230
Yamuy J., 49
Yeats, W., 60
Yerkes, R., 203
Young, M., 53

Zee, A., 433
Zeig, J., 145, 192, 300
Zeki, S., 9, 135, 138
Zhao, X., 199
Zilboorg, G., 67, 100–101, 126, 189, 193
Zola, S., 396

SUBJECT INDEX

Page numbers in *italics* are illustrations.

abandonment, 167
acetylcholine, 155–56
acknowledgment, of symptoms, 454–55
active imagination, 133
activity-dependent exercises
    arm lever process, 455–58
    arms front and back, 460–62
    Buddha, 462–64
    full body creative replay, 473–79
    hand symmetry-breaking, 437–44
    ideodynamic, 464–68
    imagery facilitating, 458–60
    palms-down process (reframing), 448–55
    palms-up process (duality), 444–48
    touch, 468–73
activity-dependent gene expression
    and behavior-state-related gene expression, 164, 231
    and consciousness, 13
    and demonstration therapy, *364*
    and depression, 119–21
    and dreaming, 165, 167, 169–78
    and emotional problems, 398–400
    future studies, 281
    and hypnotic suggestion, 191
    and hypnosis, 231
and mitochondria, 229
    and neurogenesis, 12–13, 110–11, 229
    psychosocial dynamics, *421*
    and psychotherapy, 121–28, 149
    role of, 482
    and shock, 368
    time frame, 12
acupuncture, 212, 230
adaptability, 62–64, 70, 335–36
addictions
    implicit processing heuristic, 454
    mechanism, 457
    and reeducation, 142

    and ultradian rhythm, 79
    withdrawal syndromes, 83
    see also case studies
adenosine triphosphate (ATP), 10–11
adrenocorticotropic hormone (ACTH)
    and alarm clock effect, 202
    description, 46
    and psychotherapy, 101
    and ultradian rhythm, 56, 210
aging, 222–23, 224
    see also elderly persons; stem cells
airline personnel, 214
alarm clock effect, 202
alcohol, 31
alcoholism, 82–85, 142
alertness, 213
alpha-CaMKII, 132
Alzheimer's disease, 110–11
ambiguity, 332–33, 341–42
amnesia, 156–57, 234
AMPA receptors, 116, 118, 119
amphetamines, 31, 236
amplitude, 205
amygdala, 125
analgesia, 249
    see also pain
androgens, 199–201
anesthesia, 141–43
anger, 126, 365, 476
anterior cingulated cortex, 249
anxiety, 126, 400
arginine vasopressin (AVP), 46
arm lever process, 455–58
arms front and back, 460–62
arousal
    chronic stress, 260–62
    and creative cycle, 320
    and cultural transitions, 301
    in demonstration therapy, 363

and gene expression, 12
and healing, 243–45
in hypnotherapy, 143–44, 287
and IEGs, 237
and immune system, 243–44
level of, 39
and memory, 235
peak of, 363
and performance, 203–6
and problem-solving, 402
versus relaxation, 301, 320
sexual, 66
and swearing, 367–69
see also long-term potentiation
arthritis, 305–7, 374–75
see also demonstration therapy
artists, 274–75, 297
asthma, 216
astrocytes, 213
attention deficit/hyperactivity disorder (ADHD), 208, 273–74
attitudes, 305–7
autism, 274
avoidance, 402
awakening, 239, 285–86, 289
awareness
bifurcation of, 166
and creative cycle, 273, 277
of mind-body communication, 103–4
of novelty-numinosum-neurogenesis, 372
of ultradian rhythm, 418–20, 442
baboons, 260, 262
barbiturates, 31
basic rest-activity cycle
and creativity, 284–97
in demonstration therapy, 310
and hypnosis, 196, 203–6, 209–14
and implicit processing heuristics, 418–19
and learning, 158
mathematical model, 203–6
and neurogenesis, 140
and novelty, 140
and numinosum, 140
quasi-periodicity, 59
and sleep, 285–86
and stress, 214–15, 261
and ultradian rhythm, 69
behavior
activity-dependent change, 385–86
destructive, 178–79
and gene expression, 35, 257
inherited aspect, 273
patterns of, 269
and subjective experience, 398
voluntary, 355–56, 377–78

behavior, state-dependent
accessing, 440
and CNS neural networks, 38
and creative cycle, 270, 281
in demonstration therapy, 305
and messenger molecules, 67, 232–34
as mind analogue, 30–31
behavioral state-related gene expression
and activity-dependent gene expression, 164, 231
and arousal, 12
and consciousness, 13
defined, 42
in demonstration therapy, 323–24, *326*
and dreaming, 153, 163–65, 169–74
and early mornings, 289
and emotional problems, 398–400
facilitating, 398–400
future studies, 281
and hypnosis, 145, 231
and implicit processing heuristics, 398–400, 422–23
role of, 482
therapeutic use, 71, 148–50 (*see also* case studies)
behavior therapy, 126
beta-adrenergic response, 235
beta-endorphin, 46, 101, 210
Big Bang, 22–23
biofeedback, 212, 229
bioinformatics, 27
biosynthesis, 224–26 (table), 227–28 (table)
blood-brain barrier, 233
body language, *see* nonverbal cues
body work, 229
Boolean logic, 430–31
brain
atrophy of, 18
C-fos in, 49
development, 37–40, 117–18
emotion mapping, 43
and gene expression, 37–40
interchanges within, 134
mind space, 442–43
neurogenesis, 108–12, 229
and placebo effect, 249
reward systems, 137
and sensory-motor input, 442–43, 466–68
and stress, 229
brain-breath link, 85–88
brain-derived neurotropic factor (BDNF), 18, 37–38
breakout heuristic, 266–68, 388
see also creative cycle
Buddha, 462–64
burns, 220–22, 223

caffeine, 31
calcium, 114, 224–26 (table), 227–28 (table)
calculus of indications, 430–32
caloric restriction, 222–23, 227–28 (table)
cancer, 199–201, 215–16
case studies
    alcoholism, 82–85
    breast feeding, 89–90
    diabetes, 9193
    dreaming, 158–81
    electric shock memory, 100–101
    encopresis, 95–98, 105
    Erickson transcript, 146–50
    headaches and nausea, 85–88
    insomnia, 76–79
    narcolepsy, 72–75
    novelty-numinosum-neurogenesis, 240–42
    obesity, 93–95
    ovulation, 98–99
    shock and trauma, 102
    sinus congestion, 99
    smoking, 79–82
    stress, 79–81, 85–90, 91–95
    stroke victim, 240–42
    throat constriction, 100–101
    see also demonstration therapy
catalepsy
    and creative cycle, 72–75, 472–73
    and hypnosis, 353–55, 362–63
catharsis, 229–30, 402
cell membrane, 114
cells
    and clock genes, 52–53
    extracellular signals, 47
    mitosis, 54
    and ultradian rhythm, 54
cerebellum, 116
cerebral hemispheres, 82
cerebrovascular accident, 240–42
c-fos
    and alcohol, 83
    in brain, 49
    and exercise, 236
    and stress, 48
    and waking, 239
change, 279–80, 289
    see also creative cycle; transitions
chaos, deterministic, 182–86, 204, 335–36, 346
chaos theory, 61–62, 166, 183–84, 436–37
childhood
    and consciousness, 135–36
    learning in, 279–80, 290–91
    transition from, 161–63
choice, see conscious selection; Feigenbaum scenario

cholesterol, 284, 289
circadian rhythm
    and clock genes, 11, 53
    and creativity, 284–97
    and gene expression, 12
    hormonal peaks, 69
    and hypnosis, 210
    see also day, analysis of
C-jun, 236
clock genes, 11, 51–56
c-myb, 216
c-myc, 14, 17, 216
cocaine, 31, 85–86, 236
co-creation
    implicit processing heuristics, 419 (table)
    of intelligence, 135–36
    and psychodynamics, 460–62
    of rapport, 342–43
    of self-identity, 435
    in verification stage, 420–21
    see also inner work; privacy; social influences
cognitive-behavioral approach, 125–27, 229–30
coincidence regenerator, 124
combination theory, 27
communication
    mind-body (\I\see\R\ mind-body communi-cation)
    mind-gene, 324
    nonverbal, 303–4, 310, 311, 361–63 (see also nonverbal cues)
    patient with therapist, 342–43, 361–63, 411–13
    and swearing, 369–71, 375–76
    therapist with patient, 302–7, 310, 311, 317–18, 320–22, 342–43 (see also implicit processing heuristics; questions)
    and dissociation, 408–9
complex adaptive systems
    and arousal level, 39
    and chaos theory, 61–62
    consciousness as, 43–44
    and demonstration therapy, 335–36
    and gene expression, 17, 20–23
    and hypnosis, 212–13
    see also mind-body communication
conditioned response, 123, 127–28
conditioned taste aversion, 127
conflict, 178–79, 476
conflict resolution, 322–27
confusion, 460
consciousness
    and behavioral-state related expression, 201–3
    bifurcation of, 87
    and child development, 135–36
    as complex adaptive system, 43–44
    and confusion, 460

creative replay, 134
and creative replay, 134
versus Darwin, 26
defined, 13
and dreaming, 168–69
duality *versus* oneness, 446–48
evolution of, 185–86
and explicit memory, 129
and Feigenbaum scenario, 183–86
and gene expression, 13, 128–29, 130–33
hippocampal role, 130
implicit *versus* explicit, 141–43, 150
Jungian view, 425
main function, 417
molecular dynamics, 136–38
and natural selection, 26–28, 132, 134
and natural variation, 8–9
new patterns, 174–78
novelty-seeking, 134
reentry, 134
and sensory unification, 124
twentieth century views, 128–29
yo-yoing of, 323–24
conscious selection
and creative replay, 403–4
versus creative replay, 457–58
implicit processing heuristics, 413 (table), 417
(table)
and natural variation, 403–4
new situations, 417
and surprise, 411–15
control, 457
cortex
anterior cingulated, 249
and hippocampus, 131–32
insular, 127
orbitofrontal, 239
see also neocortex
corticotrophin-releasing hormone (CRH), 46
cortisol
DHEA ratio, 244
and early morning, 289
and problem-solving, 402
and prostate cancer, 200
and stress, 214, 244
and ultradian rhythm, *56*, 69–70, 210, 402
cosmos, creation of, 23
countertransference, 343, 469
couturier, 297
C-peptide, 57
creation, of cosmos, 22–23
creative cycle
and arousal, 320
in arts and sciences, 267
breakout heuristic, 266–68, 388

and catalepsy, 472–73
conflict indications, 476
and experimental psychology, 265
facilitation of (*see* activity-dependent exercises)
in groups, 473–79
illumination stage, 275–80, 476 (*see also* illumination)
incubation stage, 269–75, 401–11, 422–23
and natural selection, 276
and natural variation, 274, 276, 476
and novelty, 268, 275–80
and positive psychology, 264–66
preparation stage, 268–69, 303, 398–400, 422
self-organizational phase, 278
symmetry-breaking, 437–44
and transitions, 268, 273, 277
and ultradian rhythm, 284–97
verification stage, 280–82, 382–90, 418–21
see also day, analysis of; therapists
creative edge, 352–55, 377–78, 460
creative experience(s), 415
creative play, 356–58, *364*, 378–82, 383
creative questioning, 302–4
creative replay
and consciousness, 134
versus conscious selection, 457–58
and cultural healing, 301
and deepening insight, 358–61
versus dissociation, 314–15
and dreams, 155–58, 285
full body, 473–79
of gene-protein cycle, 28–32
and healing, 324
of hypnotic suggestions, 191–92, 338
implicit processing heuristics, 338, 359, 401–11
and natural selection, 457
and natural variation, 403–4
in pain control, 314–15
and physiological shock, 367–69
and problem-solving, 401–2, 448–55
role of, 482
and self-empowerment, 366–69
sensory-motor, 351–52
and state-dependent experience, 324
of trauma, 475–76
and ultradian rhythm, 458–60
creativity
basic rest-activity cycle, 284–97
neurogenesis, 416
novelty, 138–40
numinosum, 138–40, 268, 275–80
protein synthesis, 416
and sleep, 284–86
and stress, 264–66, 270–75
CREB genes, 140, *364*

Crick, F., 29
critical phase transitions, 163, 165, 169–78
cues, 47, 418–21
    see also nonverbal cues; psychosocial cues
cultural activities, 13, 483
cultural rituals, 301
cyclic AMP, 114
cyclooxygenase-2 (cox-2), 220–22
CYP17, 289, 310
Darwin, Charles, 8–9, 23–26
    see also natural selection; natural variation
day, analysis of
    afternoon, 293–94
    breaking point, 294–95
    deep sleep, 284–85
    early morning, 289–90
    evening, 295–96
    late morning, 291–93
    lunch, 293
    mid-morning, 290–91
    naps, 293–94
    physical exercise, 294
    restless sleep, 285–87
    very early morning, 286–89
dehydroepiandrosterone (DHEA), 244
demonstration therapy
    audience role, 368, 390
    break from, 361–63
    catalepsy, 353–55, 362–63
    closure, 384–90
    conflict resolution, 322–27
    creative edge, 353–55
    creative play, 356–58, 364, 378–82, 383
    creative replay, 301, 314–15, 324, 358–61,
        366–69
    dissociation, 314–15, 320–23, 327–30, 340–41
    double-bind, 333–34, 341–42
    emerging insight, 347–49
    empowerment, 349–50
    enriching experiences, 319
    expectancy, 305
    first response, 315–16
    and gene expression, 323–24, 390
    humor (see humor)
    hypnosis induction, 307–9
    implicit processing heuristics, 333–36
    and natural selection, 313, 345–47
    and natural variation, 372–73
    and neurons, 356–58
    nonverbal cues, 303–4
    novelty-numinosum- neurogenesis in, 327–30,
        331, 332–33, 347–50, 390
    patient role, 316–18, 339, 372
    preparation, 304–7
    privacy, 344–45, 348, 372

psychodynamics, 374–77
    questioning, 302–3, 304–7
    self-empowerment, 363–67, 374–75, 389
    and shock, 367–69
    and state-dependent memory, 305, 347–49
    symptom origins, 385–88
    symptom path, 307–13, 314–15, 345–47,
        360–61
    symptom scaling, 307–9, 314, 323–24, 355–56
    theory, 300–302
    time distortion, 382–85
    uncertainty, 342–44, 347, 352, 371, 381
    verification stage, 382–90
    warmth in, 324–25, 326
dendrites, 117–19
deoxyribonucleic acid (DNA), 30
    see also DNA microarrays
depression
    case study, 72–75
    and creative cycle, 271–73
    long-term, 115, 116
    and sleep, 294
    theory of, 119–21
Descartes, René, 289–90
despair, 476
development, blocks to, 269
diabetes, 91–93
dialogue, 133
diet, 293
dissociation, therapeutic
    in case study (headache), 87
    in demonstration therapy, 314–15, 320–23,
        327–30, 340–41
    and hypnotic suggestions, 191–92
    and implicit processing heuristics, 322–23,
        405–6, 407 (table), 408–9
    and natural variation, 346
    and symmetry-breaking, 436, 443
DNA microarrays
    and aging, 222–23
    description, 34–35
    and gene expression, 37, 217–20, 222–23
    and hypnosis, 203, 211
    ultradian rhythm, 217–20
dopamine, 137, 248, 332–33
dopamine hypothesis, 238
double-bind, 333–34, 341–42
dreaming
    and activity-dependent gene expression, 153,
        164
    and behavior-state-related gene expression, 164
    and creative cycle, 277, 279, 282, 288–89
    as creative replay, 155–58, 285
    Davina case, 158– 181
    Feigenbaum scenario, 180–86

and hippocampus, 156–57
and IEGs, 42, 49–50
and natural selection, 178
and neurogenesis, 174–78
and novelty, 154, 156, 159–60
and personality, 159–65
and phase transitions, 161–65
preparation for, 296
and protein synthesis, 157, 173
recall of, 124
and transitions, 161–69
(un)conscious in, 168–69
drugs, 31, 79
duality, 425–26, 428, 430–37, 444–48
dysrhythmia, 294
eating, 93–95
egr-1, 154
elderly persons, 78, 294
    see also aging
electric shock, memory of, 100–101
electroconvulsive (ECT) therapy, 120–21
emergency rooms, 220–22
emotional problems, 454
empowerment, 349–50
    see also self-empowerment
encopresis, 95–98, 105
energy, 366–68, 378–79, 460
entrainment coefficient, 205–6
environment
    and brain cell atrophy, 18
    enrichment of, 133, 135–36, 437–44, 478–79
    and genes, xvi–xvii, 47
epinephrine, 233
estradiol, 289
eustress, 265
evenings, 296
evolution
    and creative cycle, 269
    as genetic information, 29–30
    neo-Darwinian, 41
    and novelty, 138
    and stress, 265–66
    see also natural selection; natural variation
exercise, *see* activity-dependent exercises; physical
    exercise
expectancy, 305
experience(s)
    and behavior, 398
    creative, 415
    enriching, 319
    and memory, 157
    and neurogenesis, xv–xvi, 12–13
    once in a lifetime, 301
    previous patterns, 178
    and protein synthesis, 152, 153, 157

sensorimotor, 466–68
sensory-perceptual, 442–43
state-dependent, 234–35
time frames, 60–61
verification of, 419–21
    see also creative replay; life events; novelty
experience-dependent expression, *see* activity-
    dependent expression
experiential reality, 455
eye movement, 411–12, 418
    EMDR, 126
failure-to-thrive, 16
Fas, 94
fatigue, 291
fatty acids, 94
fear
    of abandonment, 166–67
    memory of, *112*, 125
    and NPAS2, 213
    and *zif-268*, 155
Feigenbaum scenario, 180–86, 229, 434
Feigenbaum's constant, 183
fetus, 434–35
fight or flight, 235
finger signaling, 464–68
flashbacks, 272, 360–61
flexibility, 62–64, 70
fluoxetine, 120
food, 293
foot signaling, 464–68
fos-B, 83
Fos protein, 47–48, 236
free association, 126
free will, 345–47
    see also consciousness
functional genomics
    defined, 7
    and human experience, 7–8
    versus Mendelian genetics, 35–36
    see also gene expression
Gaia-gene-body-mind system, 21–23, 32–34, 265
galanin, 199
gastrointestinal disorders, 90–91
gene(s)
    BDNF, 18
    cancer-related, 216
    clock, 11, 51–56, 53
    CREB, 140
    early activated, 11
    and environment, xvi–xvii, 47
    homeotic, 435
    immediate early (*see* immediate early genes)
    late response, 114
    and mind, 44–46
    ODC, 14–18, 30, 37

gene(s) *(continued)*
  selfish, 10, 41
  target, 35
gene expression
  activity-dependent, 145, 153, 164, 231, 304, 373
  and aging, 222–23, 224
  and analgesia, 249
  and brain, 37–40
  and complex adaptive systems, 17, 20–23
  and consciousness, 13, 128–29, 130–33
  cycles of response, 10
  and depression, 119–21
  and development, 10, 14–17, 41
  and diabetes, 91–93
  and ECT, 121
  and environment, xvii
  and exercise, 236
  and experience, xv, 234–35
  and fear memory, 125
  and hypnosis, 198–203, 223, 234–35, 238
  and implicit processing heuristics, 391 (table)
  and interleukin-2, 216
  and long-term memory, 114–15
  and memory, xv
  and mothering, 7, 14–17, 30
  and natural selection, 132, 134
  and novelty, 159–60
  of NPAS2, 213–14
  and numinosum, 13
  and nutrition, 93–95
  and pain, 220–22, 369–71
  protein synthesis cycle, 10–11
  and psychotherapy, 5
  and rat brain, 37–40
  and stem cells, 223–32
  and stress, 64–65, 215–17, 222–23, 224–26 (table), 227–28 (table)
  and symmetry-breaking, 435
  temporal waves, 37
  time frames, 8
  and ultradian rhythm, 70
  and Venn diagrams, 230–32
  and waking, 48–49
  waves of, 37–40
  see also activity-dependent expression; behavioral state-related expression; functional genomics
general adaptation syndrome, 259–63, 271
GeneSpring, 18
genomic action potential, 115
genotype, 369–70
gestures, 280
gill-withdrawal reflex, 113–14
glucose, 57, 91–93, 94
glycolysis, 213

group therapy, 473–79
growth, blocks to, 269
growth hormone, 69, 284
guilt, 99
hand signaling, 464–68
hand symmetry-breaking, 437–44
  see also palms-down process; palms-up process
headaches, 85–88
head signaling, 464–68
healing
  and gene expression, 230–32
  holistic, 212, 229–30, 393–94
  and hypnosis, 192–98
  and immediate early genes, 235–38
  and novelty-numinosum, 141–50, 301
  patient role, 315–18, 339, 372
  and prostate cancer, 199–201
  and religion, 229–30, 245
  rituals, 243–45
  and state-dependence, 31–32
  and stem cells, 223–32
  and stroke victim, 240–42
  by touch, 14–20
  and ultradian rhythm, 104–5, 205, 291, 292 (*see also* case studies)
  see also creative replay; immune system
health, 62
heart rate, 206
Heat Shock protein, 223, 224
Hebbian synapses, 109, 132, 209
heredity, 9, 24, 26–27
heuristics, 439
  see also implicit processing heuristics
hippocampus
  in amnesial subjects, 156–57
  and consciousness, 130
  and depression, 120
  and learning, 115, 209
  and memory, 111, 123–24, 130–31
  and presynaptic neurons, 116
  and REM sleep, 156–57
  and stress, 120, 214
holistic healing, 212, 229–30, 393–94
homeostasis, 263
homeotic genes, 435
homunculus, sensory-motor, 442–43
hormones
  and intentionality, 202
  neurotropic, 356–58
  primary messenger, 44–46, 56
  sexual, 66, 199–201 (*see also* testosterone)
  and stress, 66, 229, 260–62
Human Genome Project, 7
humanities, 138–40
humor

and body language, 373
conspiratorial, 341–42
and creative edge, 377–78
devil's advocacy, 345
hyperbole, 327
non-knowing, 352
and self-empowerment, 374–75
swearing, 367–69, 375–76
hybridization, 35
hyperbole, 327
hypnagogic imagery, 156–58
hypnosis
and ADHD, 208
and analgesia, 248–50
and arousal, 143–44, 287
background, 189–96
in case studies, 79–81, 86, 93
and catalepsy, 353–55
and complex adaptive systems, 212–13
and creative replay, 307–9
and DNA microarrays, 35
in emergency room, 220–22
Erickson approach, 141–43, 145–46, 335, 339, 401
falsifiability testing, 210–11
four-stage therapy, 204
Freudian view, 101
and gene expression, 198–203, 223, 234–35, 238
and healing, 192–96
and heart rate, 206
high- and low-phase, 203–7, 210, 356–58
and immediate early genes, 236–38
induction, 307–9
Janet system, 126
mathematical model, 205–6
and mind-body communication, 232–34
for narcolepsy, 73
and neurogenesis, 191–92, 238–43
and novelty, 238–43
and numinosum, 238–43
and passivity, 192–96, 239
phase locking, 206–8
single session, 300–301
of smoker, 79–81
and state-dependence, 70
of stroke victim, 240–42
susceptibility, 206–9, 212
and theater, 300
and time distortion, 384–85
and touch, 18
and ultradian rhythm, 212
see also demonstration therapy
hypothalamus, 44
identity, 161–66, 169–72, 176–81
ideodynamics, 439, 464–68, 475

illumination
in demonstration therapy, 341–44, 347–81, 364
implicit processing heuristics, 402, 411–17
moment of, 275–80, 476
imagery, 212, 216, 229, 458–60
imaginary state, 430–31, 433
imagination, 133
immediate early genes (IEGs)
and arousal, 237
and emotional problems, 398–400
and environment, 47
and Erickson approach, 150
and healing, 235–38
and hypnosis, 236–38, 422
and long-term memory, 114
and maternal touch, 14, 17
and mind-body communication, 47–51
proto-oncogenes, 216
psychosocial dynamics, 421
role of, 13–14, 17, 481
in sleeping and waking, 42, 48–49, 154
and withdrawal syndromes, 83
see also c-fos; c-jun
immune system
and arousal, 243–44
and relaxation techniques, 216
and stress, 11, 64–65, 215–17
implication, Ericksonian, 395–96
implicit processing heuristics
for addictions, 454
ambiguity as, 332–33, 341–42
for conscious selection, 413 (table), 417 (table)
and creative replay, 338, 359, 401–11
and dissociation, 322–23, 405–6, 407 (table), 408–9
double-binding questions, 333–34, 341–42
for emotional problems, 454
evening activities as, 296
and experiential selection, 413 (table)
indirect suggestions, 141–43
for inner work, 411 (table)
and inward focus, 403 (table)
and natural variation, 394, 404 (table), 405 (table)
and negativity, 405–6, 406 (table), 407 (table)
nonverbal cues as, 311
origins, 393–98
permissive, 316–18, 325, 330–31, 348
in preparation stage, 398–400
for psychosomatic symptoms, 454
replay as, 306–7
for stage three insights, 416 (table)
for stress, 454
versus suggestion, 333–36, 338, 397–98, 430
symptom scaling as, 355–56

implicit processing heuristics *(continued)*
   tables, 337, 391
      time-binding, 313–14, 324, 344–45
      value of, 390
implied directive, 394
inadequacy, sense of, 269
individuation, 161, 167–68, 178
infants, 14–17
inflammation, 220–22
inner work, 403 (table), 410–11, 430, 454–55
insomnia, 76–79
insulin, 57, 94
intelligence, co-creation of, 135–36
interleukin-1b, 220–22
interleukin-2, 11, 215–17
interleukin-10, 219
intracellular redox potential, 213
jet lag, 214–15
Jun protein, 236
krox-24, 154
lactate, 213
lactation, 66, 89–90
language, 465–68
late response genes, 114
laughter, 244, 415–17
learning
   in children, 279–80, 290–91
   and dreaming, 155–56
   and Feigenbaum constant, 184
   and gestures, 280
   microgenetic approach, 279
   and placebo response, 248
   and sleep, 157–58
   in young *versus* old mammals, 208–9
   and *zif-268*, 155
learning, activity-dependent
   classical view, 136
   and dendritic growth, 119
   long-term potentiation, 115
   molecular model, 113–14
   new learning *versus* re-learning, 127–28
   prediction error, 136–37
learning, state-dependent
   accessing, 440
   and CNS neural networks, 38
   and creative cycle, 270, 281
   in demonstration therapy, 305
   and messenger molecules, 67, 232–34
   and mind analogue, 30–31
leptin, 94
life, origin of, 30
life events
   and dreaming, 165
   and memory, 67, 127–28
   and molecular-genetic level, 117

and phase transition, 173
   and protein synthesis, 152, 153
   traumatic, 127–28
lifestyle, creative, 283
light, 53
limbic-hypothalamic-pituitary-adrenal system, 44–46, 48
lithium, 120
locus coeruleus, 134
logical types, theory of, 429
long-term depression, 115, 116
long-term potentiation (LTP), 115–17, 132, 239
low amplitude dysrhythmia, 294
lymphokines, 243–44
mathematical models
   Feigenbaum scenario, 180–86
   and symmetry breaking, 427–34
   for therapeutic hypnosis, 205–6
   uses, 63
   Venn diagrams, 230–31
   Yerkes-Dodson function, 203
max, 14, 17
meaning, 66–67
meditation, 212, 230, 245, 288–89
melatonin, 284
memory(ies)
   and arousal, 235
   and creative cycle, 272
   and dendritic growth, 119
   of electric shock, 100–101
   explicit *versus* implicit, 129
   extinction of, 125, 127–28
   of fear, *112*, 125
   and gene expression, xv, 114–15, 125
   and hippocampus, 111, 123–24, 130–31
   integration with present, 157
   lapses of, 375–76
   limitations of, 184–85
   long- *versus* short-term, 111–17, 123–24
   molecular, 66–68, 117
   and neurogenesis, xv, 130–31
   and protein synthesis, 114–15
   recall of, 123, 127–28
   and sleep, 155–56, 158
   and synapses, 109, 112, 113–16, 123–24
   and ultradian rhythm, 112–13
   variability, 123
   and *zif-268*, 155
memory, state-dependent
   accessing, 440
   and CNS neural networks, 38
   and creative cycle, 270, 272, 281
   in demonstration therapy, 305, 347–49
   and messenger molecules, 67, 232–34

as mind analogue, 30–32
messenger molecules, 66–69, 114, 233
messenger RNA (mRNA)
    after adventure, 12
    cox-2, 222
    description, 10–11
    role, 30
metabolism, 224–26 (table), 227–28 (table)
metaphor, 466–68
microarrays, *see* DNA microarrays
microgenetic approach, 279
mind-body communication
    clock gene role, 51–56
    four stages, 199
    and healing, 56–59
    and hypnosis, 232–34
    ideodynamic view, 464–68
    and immediate early genes, 47–51
    and interleukin-2, 215
    and mathematics, 59
    mind-gene, 44–46
    molecular memory, 66–68, 117
    non-linear dynamics, 204
    patient awareness, 103–4
    secondary messengers, 47–51
    in sleep, 284
    and stress, 262–63
    time frames, 59–64
mitochondria, 33, 229
mitosis, 54
mothering, 7, 14–17, 30
motivation, 415–15
music therapy, 244
mutations, 265–66
myths, 178, 186, 278–79, 446
naps, 293–94
narcolepsy, 72–75
natural contingencies, 25–26
natural killer cells, 216, 243–44
natural selection
    and consciousness, 26–28, 132, 134
    and creative cycle, 276
    and creative replay, 457
    and demonstration therapy, 313, 345–47
    description, 23–26
    and gene expression, 132, 134
    and life experiences, 8–9
    see also conscious selection; dreaming; evolution
natural variation
    and catalepsy, 473
    and consciousness, 8–9
    and creative cycle, 274, 276, 476
    and creative replay, 403–4
    in demonstration therapy, 372–73
    and dissociation, 346

and implicit processing heuristics, 394, 404 (table), 405 (table)
nature
    and behavioral state-related expression, 12, 201
    cooperation with, 32–34
    versus nurture, 9, 50, 153–54, 393–94
nausea, 85–88
negativity, 404–11, 415
neocortex, 116
nerve growth factor (NGF), 356–58
neurogenesis
    and activity-dependent expression, 12–13, 110–11, 229
    in brain, 108–12
    and creativity, 416
    and demanding tasks, 131
    and dendrites, 117–19
    and depression, 119–21
    and dissociation, 340–41
    and dreaming, 174–78
    and experiences, xv–xvi, 12–13
    genes associated with, 38
    and hypnosis, 191–92, 238–43
    and implicit processing heuristics, 337 (table)
    and long-term potentiation, 115–17
    and memory, xv, 130–31
    and novelty, 12–13, 248–43
    reversibility, 118, 119
    and shock, 368
    and stress, 214
    and suggestion, 191–92
    see also demonstration therapy; novelty; numinosum
neuronal factors, 224–26 (table), 227–28 (table)
neuronal PAS domain protein (NPAS2), 213
neurons, 109, 131, 356–58
neuropeptide hypothesis, 101
neurotransmitters
    and addictions, 457
    and depression, 120
    and REM sleep, 155–56
    and short-term memory, 114
neurotropic factors, 356–58
nicotine, 79–81
n-methyl-d-asparate (NMDA), 118, 119, 208–9
nonlinear dynamics, 204
    see also Feigenbaum scenario
nonverbal cues
    body language, 373, 464–68
    and creative cycle, 411–13, 475
    hand position, 444
    from patient, 361–63, 400
    from therapist, 303–4, *310*, 311, 320, 439
noradrenaline, 233, 238
not-knowing, 341–42, 381, 460

novelty
    and basic rest-activity cycle, 140
    and child development, 135–36
    and creative cycle, 268, 275–80
    and creativity, 138–40
    and cultural healing, 301
    and dissociation, 340–41
    and dreaming, 154, 156, 159–60
    Erickson approach, 146–50
    hand symmetry-breaking, 437–44
    and hypnosis, 238–43
    and neurogenesis, 12–13, 238–43
    and placebo effect, 246–50
    role of, 482
    see also demonstration therapy; numinosum
novelty-numinosum-neurogenesis effect, 138, 238–40,
    242–3, 246
numinosum
    and basic rest-activity cycle, 140
    and creativity, 138–40, 268, 275–80
    defined, xvii–xviii
    Erickson approach, 146–50
    and gene expression, 13
    and hand-symmetry breaking, 437–44
    and healing arts, 141–50
    and hypnosis, 238–43
    Jungian view, 139
    Otto view, 139
    patient attitude toward, 415
    and placebo effect, 246–50
    role of, 482
    see also demonstration therapy; neurogenesis;
    novelty
nurture
    and activity-dependent gene expression, 12–13,
    201
    versus nature, 9, 50, 153–54, 393–94
nurturing, 7, 14–17, 30
nutrition, 93–95
obesity, 93–95
oncogenes, 216
opiates, 236
orbitofrontal cortex, 239
ornithine decarboxylase (ODC), 14–18, 30, 37
ovulation, 98–99
oxytocin, 66
pain
    anesthesia recall, 141–43
    and creative replay, 314
    Erickson approach, 286–88
    and gene expression, 220–22, 369–71
    and hypnosis, 220–22
    and placebos, 249
    see also demonstration therapy
palms-down process, 448–55

palms-up process, 444–48
pantomime, 280
paradox, 355–56, 428–30
parents, 272–73
    see also mothering
patterns, 269
Pavlovian conditioning, 123, 129, 136–37
period, 205
periodicity, 59–61, 102–3
peripheral blood mononuclear cells, 218–20
perixosome, 94
permissive approach, 316–18, 325, 330–31, 348
personality, 159–65
phase, 205
phase locking, 206–8
phenotype, xvii, 369–71
phosphate, 114
physical exercise
    and cultural healing, 301
    and elderly, 294
    and gene expression, 229, 236
    and neurogenesis, 133
    requirement, 294
placebo response, 246–50
play-acting, 378–79, 383
polarity, see duality
polymorphisms, 40, 117, 483
pontinegeniculo-occipital (PGO) spikes, 49
positive psychology
    and creative cycle, 264–66
    defined, 257–58, 263
    goal, 255
positron emission tomography (PET), 155–56, 211,
    249
post traumatic stress syndrome (PTSD), 121, 122,
    272
prayer, 229–30, 245
prediction error, 136–37
Presley, Elvis, 285–86
primitive psychodynamics, 358–59
privacy
    in demonstration therapy, 344–45, 348, 372
    and problem-solving, 402, 410–12
problem-solving
    and indirect suggestions, 141–43
    and memory, 402
    reframing, 448–55
    and ultradian rhythm, 102–5, 401–2
projection, three-dimensional, 442–43
proopiomelanocortin (POMC), 46, 101
propranolol, 235
prostate cancer, 199–201
protein kinases, 114
protein metabolism, 224–26 (table), 227–28 (table)
protein synthesis

and adventure, 12
and body evolution, 28–30
and clock genes, 52
and creativity, 416
and dreaming, 157, 173
and fear memories, 125
and life events, 152, 153, 173
and long-term memory, 114–15
and shock, 368
and stem cells, 223–32
and stress, 13–14, 64
time frame, 10–11
and ultradian rhythm, 70
psychoanalysis, 126
psychodynamics, 358–59, 374–77, 460–62
psychoimmunology, 215–16
psychological implication, 395–96
psychological levels, 444–46
psychosocial cues
and asthma, 216
and cancer, 216
and hypnosis, 210–11
and suggestion, 70
psychosocial dwarfism, 16
psychosocial dynamics, *421*
psychosocial genomics
and Darwin, 22–26
and DNA microarrays, 34–35, 37
and evolution of body, 29–30
and gene expression waves, 37–40
and gene-protein cycle, 28–30
goals, 13
and healing, 31–32
and Mendel, 26–27, 35–36
and nature, cooperation with, 32–33
and symmetry-breaking, 434
see also mind-body communication
psychosomatic medicine, 16
psychosomatic network, 232–34
psychosomatic symptoms, 105, 230, 454
psychotherapy
and activity-dependent expression, 121–28
and creative uncertainty, 343
and genes, 5, 369–70
limitations, 369–71
patient role, 144–46, 147
psychobiological, 68
response to, 105, 141
symmetry-breaking, 427–37
and synapses, 117–19
top-down *versus* bottom-up, 393–94
trends, 452–53, 483
and ultradian rhythm, 68, 71, 102–6
see also demonstration therapy
quasi-periodicity, 59–61

quest, sense of, 146
questions
basic accessing, 302–7
double-binding, 314, 333–34, 341–42
not-knowing, 341–42
open-ended, 314, 398–400
permissive, 330–31
see also implicit processing heuristics
reality, experiential, 455
reality-testing, 280–82
rebelliousness, 272
receptivity, 278–79
reframing
of negativity, 404–10
and problem-solving, 448–55
and replay, 457
and self-realization, 460–62
smoking and stress, 79–81
rehabilitation, 238
rehearsal, 133
reinforcement, 25–26
relaxation, 244–45, 301
religion, 229–30, 245
resistance, 105, 469
responsibility, 483
rheumatoid arthritis, 305–7, 374–75
see also demonstration therapy
ribonucleic acid (RNA), 30
rituals, 244–45, 301
schizophrenia, 238
self, sense of, 40
self-actualization, 161
self-consciousness, 273
self-empowerment, 363–67, 366–69, 374–75, *389*,
420–21
self-esteem, 216
self-image, 382
self-organization, 167–68, 169–72, 278
self-realization, 420–21, 460–62, 483
self-reference, 432–33
self-reflection, 161–65, 305, 308, 346
sensory-motor humunculus, 442–43
serotonin, 120
seven, 184–86
sexual arousal, 66
shamans, 18, 243, 301
shock, 367–69
signals
finger, 464–68
from symptoms, 455–58
see also cues
sign language, 466–68
single nucleotide polymorphisms (SNPs), 40, 117,
483
sinus congestion, 99

sleep
   awakening from, 202, 285–86, 289
   and *c-fos*, 239
   and creativity, 284–86
   and elderly, 294
   and immediate early genes, 42, 48–49, 154
   insomnia, 76–79
   and learning, 157–58
   naps, 293–94
   and novel experience, 154
   preparation for, 296
   REM, 49–50, 69, 154–57, 284–85
   slow-wave, 154
   time frames, 48–49
   see also dreaming
sleep apnea, 62
sleep disorders, 72–75, 76–79
snail, 113–14
social deprivation, 16
social influences, 257, 483
social patterns, 269
soul, 146
starvation, 265–66
state dependence, *see* behavior, state-dependent;
   experience; learning, state-dependent; memory,
   state-dependent
stem cells, 223–32, 368
stories, 341–42
stress
   adaptive response, 13–14
   and aging, 222–32
   and basic rest-activity cycle, 214–15
   and blood-brain barrier, 233
   case studies, 79–81, 85–90, 91–95
   and *C-fos*, 48
   chronic, 259–62, 270
   creative potential, 264–66, 270–75
   and gene expression, xvii–xviii, 64–65, 215–17,
      222–23, 224–26 (table), 227–28 (table)
   and hippocampus, 120, 214
   and hormones, 66, 229, 260–62
   and immune system, 11, 64–65, 215–17
   implicit processing heuristic, 454
   late afternoon, 294–95
   long-term potentiation, 239
   and metabolism, 224–26 (table), 227–28 (table)
   and mind-body communication, 262–63
   and neurogenesis, 214
   and protein synthesis, 13–14, 64
   psychosocial, 260
   and psychosomatic symptoms, 105
   Selye view, 259–63
   and state-dependence, 31–32
   and ultradian rhythm, 104, 291–93
   utilization of, 400

stroke victim, 240–42
suggestion
   and burn patients, 220–22, 224
   direct *versus* indirect, 334
   Ericksonian, 143–44, 315–16, 334
   and gene expression, 191
   heightening, 130–31
   versus implicit processing heuristics, 333–36,
      338, 397–98, 430
   with implicit processing heuristics, 398–424
   (post)hypnotic, 191–92, 202–3, 338
   and prostate cancer, 199–201
   and psychosocial cues, 70
surprise, 411–15
swearing, 367–69, 375–76
symmetry-breaking
   hand exercise, 437–44
   heuristics, 256
   mathematical theory, 427–34
   therapist role, 452–53
symptom(s)
   acknowledgment of, 454–55
   and anger, 365
   control of, 457
   identification of, 103–4
   interpretation of, 385–88
   natural flux, 360–61
   psychosomatic, 105, 230, 454
   as signals, 455–58
   transforming, 455–58
   worsening, 104, 355–56, 457
symptom-scaling
   in arm-lever process, *456*
   in demonstration therapy, 307–9, 314, 323–24,
      355–56
   purpose, 104
   in verification stage, 419–20, 441–43
synapses
   Hebbian, 109, 132
   and memory, 109, 112, 113–16, 123–24
   neurogenesis, 116
   re-formation, 117–19
synaptic re-entry reinforcement, 124
taste aversion, 127
T-cell growth factor, 215
tears, 347–49, 476
temperature, 53, 143
temporal waves, 37–40
testosterone, 69, 260, 262, 289
Tetris experiment, 156–58
theater, 299
therapeutic change, 141
therapeutic touch
   in encopresis case, 97–98, 105
   evolution of, 468–73

and gene expression, 14–20
therapeutic period, 212
therapists
 in creative cycle, 453, 476–77
 in illumination stage, 411–15
 in incubation stage, 402–3, 405–8
 in verification stage, 422–23, 477
 initial task, 417
 in symmetry-breaking, 452–53
 see also under communication; demonstration
 therapy
three-dimensional projection, 442–43
throat constriction, 100–101
time-binding, 313–14, 324, 344–45, *467*
time distortion, 382–85
time frames
 to activity-dependent expression, 12
 basic rest-activity cycle, 59, 69
 of clock gene regulation, 53
 of gene expression, 8, 11–12
 of human experiences, 60–61
 and mind-body communication, 59–64
 for molecular-genetic memory, 117
 of ODC expression, 30
 of RNA evolution, 30
 ultradian rhythm, 54, 55
 waking and sleeping, 48–49
 \I\see also\R\ day, analysis of
tiredness, 213, 291
\I\tis-8,\R\ 154
tongue slip, 376–77
touch, \I\see\R\ therapeutic touch
transcription factors, 236, 435
transference, 343, 469
transfer RNA (tRNA), 30
transitions
 to adulthood, 161–65
 and arousal, 301
 and chaos theory, 184
 and creative cycle, 268, 273, 277
 critical, 163, 165, 169–78
 and cultural rituals, 301
 and dreams, 161–69
 Yeats poem, 291
trauma, 475–76
tree of life, 32
twins anecdote, 14, 19
ultradian healing response, 291, 292, 293–94
ultradian rhythm

awareness of, 418–20, 442
behavioral-sociocultural level, 58
and cerebral dominance, 82
and clock genes, 11, 54, 55, 56
cognitive-behavioral level, 58
and cortisol, 70
and creative cycle, 284–97
creative replay, 458–60
cues from, 418–21
cycle description, 69, 70
defined, 12
and demonstration therapy, 361–63
DNA microarray study, 217–20
endocrine-energy-metabolic level, 57
and gastrointestinal function, 90–91
and healing, 104–5, 205, 244–45
and hypnosis, 212
and immediate early genes, 48–49
and long term memory, 112–13
nonlinear dynamics, 204
and problem solving, 102–5, 401–2
and protein synthesis, 70
in psychotherapy, 68, 71, 102–6 (*see also* case
 studies)
and PTSD, 122
and senior citizens, 76–79
working with, 291–93
see also day, analysis of
ultradian stress syndrome, 104, 291, 292
uncertainty
 creative, 342–43, 347, 371
 not-knowing, 352, 381
unconscious
 and dreaming, 168–69
 and Feigenbaum point, 183–84
"un-humm," 317–18
unreality, 277
urination, 91
variability, 336, 346
 see also natural variation
Venn diagrams, 230–32
wakefulness, 48–49
waking, 239, 285–86, 289
warfare, 178–79
wholeness, 425–26
withdrawal syndromes, 83
Yeats, William Butler, 291
Yerkes-Dodson function, 203–6
zif-268, 83, 112, 154–56, *389*